# Lecture Notes in Earth System Sciences

More information about this series at http://www.springer.com/series/10529

Olga V. Frank-Kamenetskaya
Elena G. Panova · Dmitry Yu. Vlasov
Editors

# Biogenic—Abiogenic Interactions in Natural and Anthropogenic Systems

Springer

*Editors*
Olga V. Frank-Kamenetskaya
Department of Crystallography
Saint Petersburg State University
Saint Petersburg
Russia

Elena G. Panova
Department of Geochemistry
Saint Petersburg State University
Saint Petersburg
Russia

Dmitry Yu. Vlasov
Department of Botany
Saint Petersburg State University
Saint Petersburg
Russia

ISSN 2193-8571 ISSN 2193-858X (electronic)
Lecture Notes in Earth System Sciences
ISBN 978-3-319-79714-4 ISBN 978-3-319-24987-2 (eBook)
DOI 10.1007/978-3-319-24987-2

Springer Cham Heidelberg New York Dordrecht London

Softcover re-print of the Hardcover 1st edition 2016


Printed on acid-free paper

Springer International Publishing AG Switzerland is part of Springer Science+Business Media (www.springer.com)

# Preface

The book is a collection of papers presented at the Fifth International Symposium on Biogenic–Abiogenic Interactions in Natural and Anthropogenic Systems, which was held on 20–22 October 2014 in Saint Petersburg, Russia. Organizers of the symposium were Saint Petersburg State University (the largest educational-scientific center in Russia) and Saint Petersburg Society of Naturalists (the oldest Russian scientific society). The previous symposiums on this topic took place in Saint Petersburg in 2002, 2004, 2007, and 2011.

Saint Petersburg State University, which was founded by Peter the Great in 1724, is the oldest University in Russia. The Russian government has granted national heritage status to Saint Petersburg State University. The university is not only the largest educational center of Russia, but it also is a great scientific center. It is connected with the Saint Petersburg Society of Naturalists—one of the oldest natural science societies of Russia, being founded in 1868 by the Emperor Alexander II. From its foundation through the present day, the society has been associated with Saint Petersburg State University. The first president of the society, Karl Kessler, was the rector of Saint Petersburg State University. Presidents of the Saint Petersburg Society of Naturalists have included famous scientists such as Professor Beketov, Academician Vernadsky, Academician Ukhtomsky, and others. The priorities of the society currently include complex interdisciplinary study in the different fields of earth science and life science, support of scientific communication among scientists, and support for young researchers of nature. The society is a member of the World Union for Conservation.

The papers in this book cover a wide range of topics, covering the interactions between biogenic and abiogenic components in the lithosphere, biosphere, and technosphere. The authors are experts in different fields of science: earth sciences, biology, soil science, materials science, chemistry, environmental protection, and the preservation of cultural heritage.

This book has an interdisciplinary character and contains six parts:

1. Biomineralization in Geosystems
2. Geochemistry of Biogenic–Abiogenic Systems

3. Biomineral Interactions in Soil
4. Bioweathering and Destruction of Cultural Heritage
5. Minerals in Living Organisms and Biomimetic Materials
6. History of Science

This book should appeal to a wide range of readers because it is devoted to the interaction of living and nonliving matter throughout the development of the earth.

# Contents

## Part III Biomineral Interactions in Soil

# Contributors

**Evgeniy V. Abakumov** Saint Petersburg State University, Saint Petersburg, Russia

**Konstantin N. Abrosimov** V.V. Dokuchaev Soil Science Institute, Russian Academy of Sciences, Moscow, Russia

**Alexey V. Alekseenko** Department of Geoecology, National Mineral Resources University (Mining University), Saint Petersburg, Russia

**Vladimir A. Alekseenko** Southern Federal University, Rostov-on-Don, Russia; Kuban State Agrarian University, Krasnodar, Russia

**Andery O. Alekseev** Institute of Physicochemical and Biological Problems in Soil Science RAS, Pushchino, Russia

**Tatiana V. Alekseeva** Institute of Physicochemical and Biological Problems in Soil Science RAS, Pushchino, Russia

**Olga N. Alexanderova** Ural Federal University, Ekaterinburg, Russia

**Levon A. Araratyan** The Center for Ecological-Noosphere Studies, NAS, Yerevan, Republic of Armenia

**Marieta H. Avetisyan** The Center for Ecological-Noosphere Studies, NAS, Yerevan, Republic of Armenia

**Nuriya R. Ayupova** Institute of Mineralogy, Ural Branch of Russian Academy of Science, Miass, Russia; South Urals State University, Miass branch, Miass, Russia

**Kseniia A. Bakhmatova** Saint Petersburg State University, Saint Petersburg, Russia

**Irina V. Bardamova** Geological Institute of SB RAS, Ulan-Ude, Russia

**Gulnaz Ya. Biktimerova** State Autonomous Scientific Institution, "Institute of Regional Research of the Republic of Bashkortostan", Sibay, Russia

**Natalia M. Boeva** IGEM of the Russian Academy of Sciences, Moscow, Russia

**Nikolay S. Bortnikov** IGEM of the Russian Academy of Sciences, Moscow, Russia

**Aleksandr B. Brik** Semenenko Institute of Geochemistry, Mineralogy and Ore Formation of NAS of Ukraine, Kiev, Ukraine

**Marina V. Charykova** Saint Petersburg University, Saint Petersburg, Russia

**Svetlana Y. Chazhengina** Institute of Geology, Karelian Research Centre, Russian Academy of Sciences, Petrozavodsk, Russia

**Olga Ya Cherviatsova** Shulgan-Tash State Nature Reserve, Irgizly Village, Republic of Bashkortostan

**Ekaterina S. Chikanova** Omsk F.M. Dostoevsky State University, Omsk, Russia

**Natalia P. Chizhikova** V.V. Dokuchaev Soil Science Institute, Russian Academy of Sciences, Moscow, Russia

**Leonid V. Danyushevsky** Centre for Deposit and Exploration Studies, University of Tasmania, Hobart, Australia

**Irina G. Dobretsova** Polar Marine Geological Prospecting Expedition, St. Petersburg—Lomonosov, Russia

**Andrey V. Dolgikh** Institute of Geography, Russian Academy of Sciences, Moscow, Russia

**Svetlana G. Doroshkevich** Geological Institute of SB RAS, Ulan-Ude, Russia

**Oleksii Dubok** Institute for Problems of Material Science of NAS of Ukraine, Kiev, Ukraine

**Viktoria Ershova** Institute of Earth Sciences, Saint Petersburg State University, Saint Petersburg, Russia

**Alexey Filimonov** St. Petersburg State Polytechnical University, St. Petersburg, Russia

**Olga V. Frank-Kamenetskaya** Institute of Earth Sciences, Saint Petersburg State University, Saint Petersburg, Russia

**Alexandr S. Frid** V.V. Dokuchaev Soil Science Institute, Russian Academy of Sciences, Moscow, Russia

**Irina F. Gablina** Geological Institute RAS, Moscow, Russia

**Akhmet A. Galeev** Institute of Geology and Petroleum Technologies, Kazan Federal University, Kazan, Russia

**Lyudmila M. Gerasimenko** Institute of Microbiology RAS, Moscow, Russia

**Svetlana A. Gerk** Omsk F.M. Dostoevsky State University, Omsk, Russia

**Stepan P. Glavatskikh** Institute of Geology and Geochemistry, Ekaterinburg, Russia

**Olga A. Golovanova** Omsk F.M. Dostoevsky State University, Omsk, Russia

**Alla V. Golovchenko** Faculty of Soil Science, Moscow State University, Moscow, Russia

**Anna A. Gorbushina** Department of Earth Sciences, Department of Biology Chemistry Pharmacy, Freie Universität Berlin, Berlin, Germany; Federal Institute of Materials Research and Testing (BAM), Berlin, Germany

**Sergey V. Goryachkin** Institute of Geography, Russian Academy of Sciences, Moscow, Russia

**Valeriy M. Grishkin** Saint Petersburg State University, Saint Petersburg, Russia

**Dmitry E. Himelbrant** Saint Petersburg State University, Saint Petersburg, Russia; Komarov Botanical Institute, Saint Petersburg, Russia

**Lidia I. Inisheva** Tomsk State Pedagogical University, Tomsk Oblast, Russia

**Alina R. Izatulina** Saint Petersburg State University, Saint Petersburg, Russia

**Rinat R. Izmailov** Omsk F.M. Dostoevsky State University, Omsk, Russia

**Elena A. Kalinichenko** Semenenko Institute of Geochemistry, Mineralogy and Ore Formation of NAS of Ukraine, Kiev, Ukraine

**Nikolay S. Kasimov** Department of Landscape Geochemistry and Soil Geography of the Geography Faculty, Lomonosov Moscow State University, Moscow, Russia

**Anna V. Kazanova** The Russian Museum, Saint Petersburg, Russia

**Irina Yu. Kirtsideli** Komarov Botanical Institute, Saint Petersburg, Russia

**Tatiana M. Kisselyova** Department of Landscape Geochemistry and Soil Geography of the Geography Faculty, Lomonosov Moscow State University, Moscow, Russia

**Natalia N. Kolotilova** Moscow University, Moscow, Russia

**Valery V. Kononov** Russian Academy of Sciences, Far Eastern Geological Institute, Vladivostok, Russia

**Vyacheslav V. Korolkov** Omsk F.M. Dostoevsky State University, Omsk, Russia

**Natalia E. Kosheleva** Department of Landscape Geochemistry and Soil Geography of the Geography Faculty, Lomonosov Moscow State University, Moscow, Russia

**Alexander M. Kovshov** Saint Petersburg State University, Saint Petersburg, Russia

**Vladimir G. Krivovichev** Saint Petersburg University, Saint Petersburg, Russia

**Marina V. Kuimova** National Research Tomsk Polytechnic University, Tomsk, Russia

**Liudmila Yu Kuzmina** Institute of Biology, Ufa Scientific Center of RAS, Ufa, Russia

**Ekaterina S. Kuznetsova** Saint Petersburg State University, Saint Petersburg, Russia; Komarov Botanical Institute, Saint Petersburg, Russia

**Oksana A. Kuznetsova** Saint Petersburg State University, Saint Petersburg, Russia

**Mariya A. Kuz'mina** Saint Petersburg State University, Saint Petersburg, Russia

**Pavel A. Lazarev** The Russian Museum, Saint Petersburg, Russia

**Liubov V. Leonova** Institute of Geology and Geochemistry, Ekaterinburg, Russia

**Sofia N. Lessovaia** Institute of Earth Sciences, Saint Petersburg State University, Saint Petersburg, Russia; St. Petersburg State Polytechnical University, Saint Petersburg, Russia

**Tatiana V. Litvinova** Geological Institute of the Russian Academy of Sciences, Moscow, Russia

**Hannu Luodes** Geological Survey of Finland, Espoo, Finland

**Alexey V. Lupachev** Institute of Biological and Physico-Chemical Problems of Soil Science, Puschino, Russia

**Valeriy V. Maslennikov** Institute of Mineralogy, Ural Branch of Russian Academy of Science, Miass, Russia; South Urals State University, Miass branch, Miass, Russia

**Svetlana P. Maslennikova** Institute of Mineralogy, Ural Branch of Russian Academy of Science, Miass, Russia

**Natalia N. Matinian** Saint Petersburg State University, Saint Petersburg, Russia

**Pavel V. Medvedev** Institute of Geology, Karelian Research Centre, Russian Academy of Sciences, Petrozavodsk, Russia

**Nikita S. Mergelov** Institute of Geography, Russian Academy of Sciences, Moscow, Russia

**Elena A. Miklashevich** Kemerovo State University, Kemerovo, Russia

**Andrey V. Mokhov** IGEM of the Russian Academy of Sciences, Moscow, Russia

**Tatiana S. Mylnikova** National Research Tomsk Polytechnic University, Tomsk, Russia

**Vladimir. K. Orleanskiy** Institute of Microbiology RAS, Moscow, Russia

**Natalia G. Osmolovskay** Saint Petersburg State University, Saint Petersburg, Russia

**O.M. Osmolowskaya** Saint Petersburg State University, Saint Petersburg, Russia

**Elena G. Panova** Saint Petersburg State University, Saint Petersburg, Russia

**Ivan Yu. Parnikoza** Institute of Molecular Biology and Genetics, National Academy of Sciences of Ukraine, Kyiev, Ukraine

**Tatijana V. Pashkovskaya** The Russian Museum, Saint Petersburg, Russia

**L.M. Pavlova** Institute of Geology and Nature Management Far Eastern Branch, Russian Academy of Sciences, Blagoveshchensk, Russia

**Tatevik E. Poghosyan** The Center for Ecological-Noosphere Studies, NAS, Yerevan, Republic of Armenia

**Yury Polekhovsky** Institute of Earth Sciences, Saint Petersburg State University, Saint Petersburg, Russia

**Evgenia A. Popova** I.S. Gramberg VNIIOkeangeologia, Saint Petersburg, Russia

**Tatijana A. Popova** Herzen State Pedagogical University, Saint Petersburg, Russia

**Valentin V. Radchuk** Institute for Telecommunications and Global Information Space of NAS of Ukraine, Kiev, Ukraine

**V.I. Radomskaya** Institute of Geology and Nature Management Far Eastern Branch, Russian Academy of Sciences, Blagoveshchensk, Russia

**Yuliya S. Rafikova** State Autonomous Scientific Institution, "Institute of Regional Research of the Republic of Bashkortostan", Sibay, Russia

**Alexey V. Rusakov** Saint Petersburg State University, Saint Petersburg, Russia

**Alena S. Ryabova** Institute of Biology, Ufa Scientific Center of RAS, Ufa, Russia

**Dmitrii S. Rybakov** Institute of Geology, Karelian Research Centre, RAS, Petrozavodsk, Russia

**Sergei A. Sadykov** Institute of Mineralogy, Ural Branch of Russian Academy of Science, Miass, Russia

**Katerina V. Sazanova** Saint Petersburg State University, Saint Petersburg, Russia; Komarov Botanical Research Institute, Saint Petersburg, Russia

**Sergei M. Schiparev** Saint Petersburg State University, Saint Petersburg, Russia

**Irina N. Semenova** State Autonomous Scientific Institution, “Institute of Regional Research of the Republic of Bashkortostan”, Sibay, Russia

**Anastasiia A. Sheshukova** Saint Petersburg State University, Saint Petersburg, Russia

**Stanislav B. Shigorets** Saint Petersburg State University, Saint Petersburg, Russia

**Ilya G. Shorkunov** Institute of Geography, Russian Academy of Sciences, Moscow, Russia

**Yulia S. Simakova** Institute of Geology of the Komi Science Center, Urals Branch of RAS, Syktyvkar, Russia

**Anatoly D. Slukin** IGEM of the Russian Academy of Sciences, Moscow, Russia

**Olga P. Smyshlyaeva** Russian Academy of Sciences, Far Eastern Geological Institute, Vladivostok, Russia

**Olga I. Sorokina** Department of Landscape Geochemistry and Soil Geography of the Geography Faculty, Lomonosov Moscow State University, Moscow, Russia

**Irina S. Stepanchikova** Saint Petersburg State University, Saint Petersburg, Russia; Komarov Botanical Institute, Saint Petersburg, Russia

**Dmitry L. Sumin** Network of Independent Scientists Educators and Experts in Biopolitics (NISEEB), Moscow, Russia

**Evgeniya L. Sumina** Geological Faculty, Department of Paleontology, M.V. Lomonosov MSU, Moscow, Russia

**Olga I. Sumina** Faculty of Biology, Saint Petersburg State University, Saint Petersburg, Russia

**Yalil T. Suyundukov** State Autonomous Scientific Institution, “Institute of Regional Research of the Republic of Bashkortostan”, Sibay, Russia

**Sergey A. Svetov** Institute of Geology, Karelian Research Centre, Russian Academy of Sciences, Petrozavodsk, Russia

**Lech W. Szajdak** Institute for Agricultural and Forest Environment, Polish Academy of Sciences, Poznan, Poland

**Victor O. Targulian** Institute of Geography, Russian Academy of Sciences, Moscow, Russia

**Ivan V. Timofeev** Department of Landscape Geochemistry and Soil Geography of the Geography Faculty, Lomonosov Moscow State University, Moscow, Russia

**Galina T. Ushatinskaya** Paleontological Institute RAS, Moscow, Russia

**Aleksei D. Vlasov** Herzen State Pedagogical University of Russia, Saint Petersburg, Russia

**Alexey D. Vlasov** Herzen State Pedagogical University, Saint Petersburg, Russia

**Dmitry Yu. Vlasov** Saint Petersburg State University, Saint Petersburg, Russia; Komarov Botanical Research Institute, Saint Petersburg, Russia

**Lyubov V. Zaytseva** PIN of the Russian Academy of Sciences, Moscow, Russia

**Elya P. Zazovskaya** Institute of Geography, Russian Academy of Sciences, Moscow, Russia

**Marina S. Zelenskaya** Saint Petersburg State University, Saint Petersburg, Russia

**Mihael E. Zelenski** Institute of Experimental Mineralogy, Chernogolovka, Moscow, Russia

**Alexey P. Zhabko** Saint Petersburg State University, Saint Petersburg, Russia

**Elena A. Zhegallo** PIN of the Russian Academy of Sciences, Moscow, Russia

**Anatoly P. Zhukhlistov** IGEM of the Russian Academy of Sciences, Moscow, Russia

# About the Properties of Relations Between Living and Nonliving

**Evgeniya L. Sumina and Dmitry L. Sumin**

**Abstract** Despite a lack of evidence, pointing the ability of appearance of living from nonliving (inert), the problem is set according to this event sequence. The defining influence of the ambient environment on living processes has not been proven because the composition of all external surfaces is defined by the living activity. Based on the observed facts, this chapter presents a thesis about the primacy of biogenic to abiotic processes.

**Keywords** Origin of life · Living · Nonliving · Biosphere · Entropy · Universe

*Living* and *nonliving* are the most different conditions of a substance. At the present moment, there are no unambiguous definitions of these conditions or a univocal understanding of the relationship between them.

In the most general way, we attribute the condition of *living* to objects that purposefully reduce their entropy and *nonliving* (inert) to objects in which entropy increases spontaneously.

The problem of the relationship between living and nonliving substances in the present moment seems to be solved. It is observed only within the framework of the origin of living from nonliving, independent development of nonliving, and mainly, its defining influence of living development. In the other words, it is considered that living is the higher phase of independent evolution of nonliving, and all the evolutionary changes of living are the direct consequence of the ambient environment.

At the same time, one of the cardinal achievements of modern science is the refutation of the position that includes the ability of the appearance of living from nonliving, based on the experiences of F. Redi and, later, L. Paster.

E.L. Sumina (✉)
Geological Faculty, Department of Paleontology, M.V. Lomonosov MSU, Moscow, Russia
e-mail: stromatolit@list.ru

D.L. Sumin
Network of Independent Scientists Educators and Experts in Biopolitics (NISEEB), Moscow, Russia

 
O.V. Frank-Kamenetskaya et al. (eds.), *Biogenic—Abiogenic Interactions in Natural and Anthropogenic Systems*, Lecture Notes in Earth System Sciences,
DOI 10.1007/978-3-319-24987-2_1

Because no facts about the origin of living from nonliving have been determined yet, scientists believe it is necessary to return to observing the fundamental properties of the relationship between living and nonliving, using modern methods.

With natural history methods, the time and conditions of the expected origin of life on Earth are not available to research and may be discussed only using theoretical assumptions. However, in the present time, the sequence of the latest events of biosphere evolution is well known; on this basis, a geochronological scale—the most general representation of the latest phases of geological time—was constructed. However, the geochronological scale does not include any geological (i.e., abiotic or "nonliving") events. The periodization of time in geological scales is possible only by using biotical events. This content of the scale is defined by the recurrence of abiotic processes (at the general directivity), and biotical processes are always directed. The recurrence of geological events does not affect the directivity of biotical events. Therefore, in the geological scales of time, biotical events occur independently from geological events.

In addition to the development of life, which occurs independently from ambient changes, the whole life cardinally changes the environment. This can be observed in the three external Earth surfaces—the atmosphere, hydrosphere, and lithosphere. Particularly undeniable in the present time is the biogenic origin of oxygen, comprising the atmosphere and hydrosphere. The phase of the exact saturation of these surfaces by oxygen has to precede the phase of oxidation of the main mass of originally located reducing agents in the structure. At the same time in the lithosphere, a proportional amount of reduced carbon was buried, which has to be fully converted, even in the lithosphere (Trofimuk et al. 2000). Therefore, life not only keeps its highly organized condition independently from ambient conditions, but it increases the organization of ambient conditions.

In modern processes, the defining influence of life on the environment's composition is directly shown, including through the "past living substance" (Ostroumov 2011; Ermakov et al. 2012).

On the observed segment of development life is out of intervals of organization level of nonliving. Life is much more complicated than any nonliving substance. Therefore, the increase of life's organization level cannot be a consequence of the entropy of nonliving substances; moreover, living activity leads to a decrease of nonliving entropy. In general, it can be said that all three contact the biosphere's nonliving planet surfaces based on the way their properties are shown and converted by life—closer not to the external for the living environment (as for the biosphere in general), but to the internal.

Therefore, in the geologically documented phases of Earth's history, the primary processes are those connected to life, which purposefully decrease entropy. The nonliving processes are secondary and subordinated, inbuilt in the general decrease of entropy.

As currently understood, the origin of life is connected to the origin of organic molecules (molecular organisms), and life on its own is perceived as the function of these molecules, in an abiogenetic way. This scheme assumes the long timeframe of molecular evolution, previous to the formation of cellular organisms. As stated by

Rozanov (2006, 2009), according to a comparison of available data (Dobretsov et al. 2006), the time of appearance of a significant amount of water (4 billion years) and the oldest discovered prokaryotes (3.8 billion years) indicates a lack of time in Earth's history for a molecular stage of life evolution. Accordingly, Rozanov believes that the probability of life's appearance on Earth is extremely low.

However, the planets' formation is the most important phase of decreased substance entropy, for which it is possible to assume either spontaneous reasons or the impact of more organized components. With regard to the planets' formation, the assumption of their biogenic structures has already been stated and justified (Snytnikov et al. 1996a, b; Snytnikov 2006). If it is correct, the deep conversion of Earth's surface and the formation of planets are generally not biogenic and abiotic processes. Instead, they would be different stages of the same biogenic process, which have a lot of consequences in the nonliving substance cycle.

The scheme offered by V.N. Snytnikov solves the problem of the time and place of molecular evolution; however, life is no longer to be observed as the function of organic molecules. On the contrary, the appearance and complication of organic molecules has to be observed as the manifestation of life, in the same way as following the appearance and complication of cellular organisms.

Therefore, everything known in the present time about the appearance and development of life on Earth assumes the primacy of living processes with regard to nonliving ones.

To what degree is this circumstance unique? Against the backdrop of the common submissions of general entropy increase in the universe (its "thermal death"), the planet—where the decrease of entropy is possible to the stage of life's appearance—seems to be a rare exclusion, a local fluctuation. However, the absolute majority of objects seen by us in the universe do not have a decay of substance; they have its synthesis. At the same time, they radiate energy in the surrounding space, decreasing even its entropy. The expansion of the universe in general with acceleration (Riess et al. 1998) shows that the entropy falls not only on the level of single bodes, but on the level of the universe in general.

Therefore, the local manifestations of life in the Universe—if not in the chemical but in the physical point—are not the exclusions or a rare event. Rather, they are the manifestations of the general decrease of matter entropy. The idea of the secondary nature and locality of the living with regard to the nonliving is a direct consequence of the model of the "cooling" universe. As the facts demonstrate in the use of the model of a "heating" universe, living processes should be observed as the primary and general processes, with nonliving processes being secondary and local.

**Acknowledgments** We are sincerely grateful to A Kharitonov for discussing the manuscript.

## References

Dobretsov N, Kolchanov N, Suslov V (2006) On the early stages of the evolution of the geosphere and biosphere. Paleontol Suppl J 40(4):S407–S424

Ermakov V, Karpova E, Korzh V, Ostroumov S (2012) Innovational aspects of biogeochemistry. GEOCHI RAS, Moscow (in Russian)

Ostroumov S (2011) About the typology of the basic types of substance in biosphere. Ecol Chem 20(3):179–188 (in Russian)

Riess A, Filippenko A, Challis P (1998) Observational evidence from supernovae for an accelerating universe and a cosmological constant. Astron J 116:1009–1038

Rozanov A (2006) Precambrian geobiology. Paleontol Suppl J 40(4):S434–S443

Rozanov A (2009) The conditions of life on the early Earth after 4.0 billions years ago. In: Problems of life origin, PIN RAS, Moscow (in Russian)

Snytnikov V (2006) Astrocatalysis as the start phase of geobiological processes. Life creates planets? In: Evolution of biosphere and biovariety, The community of scient. Publications KMK, Moscow (in Russian)

Snytnikov V, Vshivkov V, Parmon V (1996a) Solar nebula as a global reactor for synthesis of prebiotic molecules. In: 11th international conference on the origin of life. Orleans, p 65

Snytnikov V, Vshivkov V, Parmon V (1996b) Mathematical modeling of non-stationary physico-chemical processes in natural protoplanetary catalytic reactor. In: 13th international conference on chemical reactors. Novosibirsk, pp 226–227

Trofimuk A, Molchanov V, Paraev V (2000) Biogenic oxygen of atmosphere—the equivalent of carbon surface in the interaction of external geospheres. Her Off Sci Earth RAS 3:58–69 (in Russian)

# Part I
# Biomineralization in Geosystems

# Biogenic Weathering of Mineral Substrates (Review)

**Natalia P. Chizhikova, Sofia N. Lessovaia and Anna A. Gorbushina**

**Abstract** A biological impact on weathering was recognized already at the beginning of the twentieth century, when biochemical influence of the lichen growth on rocks was convincingly demonstrated. Later it was shown that the progress of solid rock weathering initiated by biological colonization was affected by the initial porosity system and sensitivity of mineral association. In the meantime a considerable amount of diverse scientific data confirm the importance of biological rock colonizers (lichens and free-living rock biofilms) in mineral material dynamics as they occur at the atmosphere-exposed rock surfaces on local as well as global scale. Subaerial rock biofilms—microbial ecosystem including free-living heterotrophic and phototrophic settlers of bare rock surfaces—are characteristic for the first stage of primary succession of terrestrial ecosystems on mineral substrates. These cultivable and free-living communities are dominated by fungi and set the stage for the later development of a lichen cover, but in comparison to lichens also represent a new tool for laboratory experimentation and thus open a new stage of work in geomicrobiology. The minerals sensitivity to microbially induced biological weathering can be demonstrated by studies of natural samples as well as by the laboratory mesocosm experiments.

**Keywords** Lichens · Biofilms · Minerals' transformations

N.P. Chizhikova (✉)
V.V. Dokuchaev Soil Science Institute, RAS, Moscow, Russia
e-mail: chizhikova38@mail.ru

S.N. Lessovaia
Institute of Earth Sciences, St. Petersburg State University, St. Petersburg, Russia
e-mail: s.lesovaya@spbu.ru

A.A. Gorbushina
Department of Earth Sciences, Department of Biology Chemistry Pharmacy, Freie Universität Berlin, Berlin, Germany
e-mail: anna.gorbushina@bam.de

A.A. Gorbushina
Federal Institute of Materials Research and Testing (BAM), Berlin, Germany



O.V. Frank-Kamenetskaya et al. (eds.), *Biogenic—Abiogenic Interactions in Natural and Anthropogenic Systems*, Lecture Notes in Earth System Sciences,
DOI 10.1007/978-3-319-24987-2_2

# 1 Introduction and Background

Mineral weathering is a sum of Earth-forming processes that start to play a role at the scale of a single mineral grain but finally express their effects at the macro-level such as ecosystem and landscape scales. A biological impact on weathering was recognized already at the begging of twentieth century (Elenkin 1901). A. Elenkin was one of the first who published a manuscript devoted to changes of a solid rock affected by lichens. Further a mechanical effect of lichens was confirmed by micro-morphological study of the rocks (sericitizated plagioclase and hornblende) in thin sections. Disintegration of the rocks in the zones of their contact with lichen hyphae has been demonstrated (Parfenova 1950). Besides that Polynov (1945) was the first who assumed a biochemical influence of the lichens on rocks based on study of granitic gneiss and miascite affected by lithophilous lichens. Not only a mechanical disintegration of rocks but also a chemical difference was identified in the studied rocks. At the same time (1940s) a selective effect of lichens on minerals in the rock was also shown. On the example of *Rhizocarpon* it was found that hyphae penetrated deeply into the grains of mica (that can be a source of nutrients, initially potassium) avoiding quartz, which does not contain nutrients (Yarilova 1947).

After these detailed first reports that were scattered over almost 5 decades, these arguments have been further developed using the advanced techniques that become available. Lichen-induced accelerated degradation of minerals has again moved the focus of attention and lichens were recognized as ideal natural microcosms in which microbially mediated mineral weathering can be studied in details (Banfield et al. 1999). Banfield et al. (1999) systematically divided the contact area between lichens and minerals into four zones: (i) zone of unweathered rock and rock incipiently weathered by inorganic reactions; (ii) zone where reactions are accelerated by dissolved organic molecules but cells are not in direct contact with reacting mineral surfaces; (iii) zone of direct and intensive contact between microbial cells, biogenic organic matrix, and mineral surfaces; (iv) zone where photosynthetic members of the symbiosis generate fixed carbon and where crystalline lichen acids precipitate. Generally at present the following zones between thallus and rock unaffected by lichens were identified: (i) a biochemical weathering where substrate is colonized by microorganisms and (ii) biophysical and biochemical weathering where thallus fixes to the substrate and contact with microorganisms (de los Ríos and Ascaso 2001; de los Ríos et al. 2002).

## 2 Rock Weathering Initiated by Lichens—Model Studies with Natural Samples

The presence of fungal hyphae creates very special and significantly modified physicochemical conditions—as it is known for mycorrhizal root, e.g., pH value around colonized roots is significantly shifted (Rigou et al. 1995). A role of lichens as a weathering agent, which is characterized by one of the strongest effect on the substrate (Chen et al. 2000; Torbjørg and Thorseth 2002) was demonstrated in a detailed manner. It was shown that the effect of lichens leads to transformation of the rock silicates and iron extraction from silicates induced by organic acids (Wilson and Jones 1983; Prieto et al. 1994; Eick et al. 1996). In basalts, clay minerals enriched by iron disappear from the rock affected by lichens; the grains of ferromagnesian minerals and Ca feldspars are also highly sensitive to weathering based on their morphology (Jackson and Keller 1970; Jones et al. 1980). In granites the most sensitive mineral is mica, especially biotite (Varadachari et al. 1994; Ascaso et al. 1995; Silva et al. 1997). These data are in good agreement with the selective effect of lichens on rocks that was mentioned previously.

The discrepancy of data on the rate of hard rock disintegration affected by lichens is observed (Prieto et al. 1994; Sancho et al. 2003). A scheme to understand the role of a biological weathering in the Holocene accumulation of the fine earth was suggested. The findings were based on the study of rock weathering affected by pedogenesis in shallow soils mostly under lichens from East Fennoscandia (Lesovaya et al. 2008). A positive feedback was demonstrated between (i) the initial accumulation of the fine earth in the fissures and cavities and (ii) increase of biological and biochemical "impact" initiated by the predominant development of biota in these zones. As opposed to them the massive bedrock surface between the fissures is subjected by weaker action of lichens and mosses.

Additionally natural pore system in a solid rock could be considered as the prerequisite of the biological weathering processes. The interactions between primary minerals and solutions often occurring within confined or semi-confined microenvironments, such as pores, rather than in a bulk solution have been reported (Velde and Meunier 2008). Coexisting biotic and abiotic processes of rock weathering were demonstrated even for the extremely cold subarctic conditions of King George Island, Antarctica, where the initial micro-pores system in the solid rocks sets conditions for penetration of water and lichen hyphae inside the rock. Cryogenic physical weathering here is accompanied by local biochemical processes due to iron accumulation in the micro-zones of rock and lichen contacts and along the cranks (Glazovskaya 2002). In "cold" environment, freezing and thawing cycles are an additional factor of physical weathering, which results in the formation of fresh mineral surfaces that are highly susceptible to a chemical weathering (Arnaud and Whiteside 1963; Allen 2002). It can be explained by the findings that the element release from rocks depends on progress of chemical weathering on microstructure properties, characterized by nearly closed, semi-open and completely open micro-systems which are interconnected by fractures or pores (Meunier et al. 2007).

Generally the dissolution of the primary minerals in a rock creates new voids which are filled only partly by newly formed minerals. Such pore systems, which are relics of previous fluid–rock interactions, can promote further chemical weathering (Navarre-Sitchler et al. 2009). Thus, the progress of solid rock weathering initiated by lichen colonization is affected by the initial porosity system and sensitivity of mineral association.

While lichens do offer an ideal system to study rock/biota interface, these systems cannot be cultivated in the laboratory, thus limiting available experimental approaches. This drawback can be compensated for if we work with microbial associations in the form of biofilms. As biofilms were the first settlers of bare mineral surfaces at times, when the earth was inhabited only by microorganisms (while lichens appeared only in the Devonian); these communities also offer an additional advantage of an evolutionary more ancient approach.

## 3 Weathering Initiated by Microorganisms—A Chance to Perform Experimental Geomicrobiology

A functional role of microbes in the biosphere is based on their role in the elements' cycling. They dissolve the mineral from a colonized substrate and a set of elements such as K, Na, P, Fe, and Mg are transformed into forms that are easy available for other inhabitants of earth (Tsyurupa 1973). Several mechanisms of their effect were described: (i) oxidation of the elements with variable active valence, (ii) effect initiated by biogenic acids and alkalis, and (iii) chelation and biosorption (Zvyaginzev et al. 1999). Besides the electrochemical model of bacterial oxidation of minerals in the zones of cell / mineral contact has been reported (Ehrlich 1996; Yakhontova and Zvereva 2000). One of the results of the microbial activities can be expressed as destruction of silicates from the substrate due to metabolites effect, which is classified as chemically affected indirect process. Experimental findings showed that bacterially induced dissolution of quartz was much more active than initiated by chemical processes (Karavayko 2004). Based on experimental findings, the growth of bacterial biomass can be stimulated by adding of supplementary portions of substrate. That was shown on the example of *Rhodovulum* that the addition of volcanic ash led to formation of hard-grained micro-colonies (Naimark et al. 2009). Microbially induced biochemical transformation of minerals (Alekseeva et al. 2009) as well as mechanical disintegration of minerals (Ivanova 2013) has been reported. The latter coexists with chemical processes namely removal of elements such as K, Al, and M, shown on the example of vermiculite from Kovdor deposit.

Although most studies of rock colonization have been naturalistic or descriptive, sufficient information has been gained to allow the selection of a few typical "rock settlers" and to study their joint development in a simple laboratory system. The sensitivity of minerals to biological weathering initiated by associations of actinomycete and cyanobacteria as well as their individual cultures was investigated experimentally (Ivanova 2013). Minerals of mica group, which are significantly different based on their stability in soil environment; Gumbrin and Oglanlinskaya bentonite clays were used as a growth medium. The growth of the cultures on the trioctahedral mica (biotite) caused to disintegration of its particles and transformation of biotite into vermiculite through mixed-layer minerals with different tendencies of layers interstratifying in crystallites. But no changes were detected in dioctahedral mica (muscovite/sericite) (Ivanova et al. 2012). Gumbrin, which is Cenomanian bleaching clay from Georgia, composed of individual smectite and cristobalite, and quartz affected by the microorganisms associations demonstrates the significantly increase of amorphous phase (based on XRD data) as a result of smectite dissolving and residual accumulation of quartz. After the experiment in Oglanlinskaya bentonite clay, which is composed of individual smectite and zeolite of clinoptilolite type, zeolite was totally collapsed. So, in zeolite a more favorable habitat for microbial communities was created, which might be explained by the frame structure of zeolite. Additionally the structure of zeolites is characterized by important property such as exchange of freely associated cations, and high content of essential nutritional elements necessary for microbes (Chizhikova et al. 2009; Zenova et al. 2009; Ivanova et al. 2009; Ivanova 2013).

There have only been rare attempts at quantification of biological weathering usually based on insufficient ecological and mineralogical background. Fewer investigations have been carried out on the qualitative and quantitative changes that rock-weathering microbial communities undergo under global climate change.

To accelerate and study weathering under laboratory conditions an in vitro model of a rock-colonizing community was suggested. The selection of organisms for the model lab system was based on the assumption that a functional rock biofilm ecosystem has to possess separate but interacting components responsible for different tasks including weathering. One model fungus is combined with a model cyanobacterium, and this dual system is used to examine the continuum of microbial interactions that leads to stable rock colonization (Gorbushina and Broughton 2009) and rock weathering. The model system used here was already demonstrated as an efficient weathering agent (Seiffert et al. 2014) and includes:

- the oxygenic photoautotrophic cyanobacterium *Nostoc punctiforme* that possesses multiple developmental states and can form nitrogen-fixing symbioses with a variety of terrestrial plants and fungi;
- a model rock inhabiting Ascomycete *Knufia petricola* (Nai et al. 2013) known for its protective pigmentation, restricted colony growth and capacity to form close inter- and intra-cellular contacts with phototrophic cells.

Morphologically simple and microbially dominated ecosystems termed "biofilms" are coupling agents between the lithosphere and atmosphere. They are interesting geobiologically as well as ecologically existing and prevailing on earth for a remarkably long period of biosphere evolution. A microbial ecosystem including heterotrophic and phototrophic settlers form on bare rock surfaces under atmospheric (subaerial) exposure. Such systems form at interfaces of solid materials with gas or liquid phases and represent a unique structure capable of multiple impacts on substrate and element cycles (Burford et al. 2003; Gorbushina 2007). These microbial structures have occurred on virtually every rock surface throughout the entire geological history of the earth and have actively participated in rock weathering. Lithobiontic communities are the primary settlers on lava following volcanic eruptions and on rocks after the retreat of glaciers. Subaerial rock biofilms represent the first stage of primary succession of terrestrial ecosystems on mineral substrates. Through occupying the fractal dimension of the rock surface, subaerial biofilms present an immense reactive surface in contact both with the atmosphere and with the underlying mineral materials. Life at the rock / atmosphere interface influences both the rock substrate and the microclimate zone above and around it.

Summarizing the microbial influence on minerals (Sokolova 2011; Gorbushina 2007) can be split into four types: (i) direct one realized in the zones of mineral and microbe contact based on reaction of anode—cathode interaction; (ii) indirect one affected by products of biological metabolism including acids, alkalis, and chelates; (iii) direct and intensive contact between microbial cells, biogenic organic matrix, and mineral surfaces where physical influences as growth pressure and desiccation/humidification movements play a role and (iv) biogenic mineral formation that results for a powerful biomineralization activity of fungi that can create new biogenic minerals from oxalates to carbonates and from oxides to sulfides (Gadd 2007). This complete spectrum of microbial influences can be most efficiently addressed with the help of reverse genetics and a model rock biofilm (Gorbushina and Broughton 2009; Seiffert et al. 2014; Noack-Schönmann et al. 2014).

**Acknowledgments** This study was supported by the Russian Foundation for Basic Research (14-04-00327), St. Petersburg State University (18.38.418.2015) and grant of Freie Universität Berlin and Saint Petersburg State University (18.60.468.2015).

## References

Alekseeva TV, Alekseev AO, Sapova EV, Gerasimenko LM (2009) Transformation of clay minerals caused by an alkaliphilic cyanobacterial community. Microbiology (Mikrobiologiya) 78(6):776–784

Allen ChE (2002) The influence of schistocity on soil weathering on large boulder tops, Kärkevagge, Sweden. Catena 49:157–169

Arnaud RJST, Whiteside EP (1963) Physical breakdown in relation to soil development. J Soil Sci 14:267–281

Ascaso A, Wierzchos J, De Los Ríos A (1995) Cytological investigations of lithobiontic microorganisms in granitic rocks. Botanica Acta 108:474–481

Banfield JF, Barker WW, Welch SA, Taunton A (1999) Biological impact on mineral dissolution: application of the lichen model to understanding mineral weathering in the rhizosphere. PNAS 96:3404–3411

Burford EP, Fomina M, Gadd GM (2003) Fungal involvement in bioweathering and biotransformation of rocks and minerals. Mineral Mag 67:1127–1155

Chen J, Blume H-P, Beyer L (2000) Weathering of rocks induced by lichen colonization—a review. Catena 39:121–146

Chizhikova NP, Omarova EO, Lobakova ES, Zenova GM, Manucharov AS (2009) Crystallochemical transformation of phyllosilicates under the impact of cyanobacteria and actinomycetes. Eurasian Soil Sci 42(1):69–74

Ehrlich HL (1996) How microbes influence mineral growth and dissolution. Chem Geol 132:5–9

Eick MJ, Grossl PR, Golden DC, Sparks DL, Ming DW (1996) Dissolution of lunar simulant as affected by pH and organic anions. Geoderma 74:139–160

Elenkin AA (1901) Lichens and soil (Lishainiki i pochva). Soil Sci (Pochvovedenie) 4:319–324 (in Russian)

Gadd GM (2007) Geomycology: biogeochemical transformations of rocks, minerals, metals and radionuclides by fungi, bioweathering and bioremediation. Mycol Res 111:3–49

Glazovskaya MA (2002) Biogeochemical weathering of andesitic volcanic rocks in subantarctic periglacial conditions. Izvestiya RAN, seriya geograficheskaya 3:39–48 (in Russian)

Gorbushina AA (2007) Life on the rocks. Environ Microbiol 9:1613–1631

Gorbushina AA, Broughton WJ (2009) Microbiology of the atmosphere-rock interface: how biological interactions and physical stresses modulate a sophisticated microbial ecosystem. Annu Rev Microbiol 63:431–450

Ivanova EA, Chizhikova NP, Zenova GM, Omarova EO, Manucharov AS (2009) Biodegradation of clay minerals by the influence of cyanobacterial-actinomycetal associations. Moscow Univ Soil Sci Bulletin 3:12–16 (in Russian)

Ivanova EA (2013) Model associations of cyanobacteriae Anabaena variabilis and actinomycete and their role in the changes of clay minerals' structure. (Model'nye assotsiatsii tsianobacterii Anabaena variabilis i actinomitsetov i ikh rol' v izmenenii struktury glinistykh mineralov). Extend abstract of PhD Dissertation in Biology, Moscow State University (in Russian)

Ivanova EA, Chizhikova NP, Zenova GM (2012) The transformation of micas of various crystal chemistry under the cyanobacterial-actinomycetes associations influence. Moscow Univ Soil Sci Bull 3:19–22 (in Russian)

Jackson TA, Keller WD (1970) A comparative study of the role of lichens and "inorganic" processes in the chemical weathering of recent Hawaiian lava flows. Am J Sci 269:446–466

Jones D, Wilson MJ, Tait JM (1980) Weathering of a basalt by Pertusaria corellina. Lichenologist 12:277–289

Karavayko GI (2004) Microbial destruction of silicate minerals (Mikrobnaja destrukcija silikatnyh mineralov). Trudy Instituta mikrobiologii im.S.N. Vinogradskogo: Jubilejnyj sbornik k 70-letiju instituta. V. XII, Nauka, Moscow (in Russian)

Lesovaya SN, Goryachkin SV, Pogozhev EYu, Polekhovskii YuS, Zavarzin AA, Zavarzina AG (2008) Soils on hard rocks in the northwest of Russia: chemical and mineralogical properties, genesis, and classification problems. Eurasian Soil Sci 41:363–376

Meunier A, Sardini P, Robinet JC, Prêt D (2007) The petrography of weathering processes: facts and outlooks. Clay Miner 42:415–435

Nai C, Wong HY, Pannenbecker A, Broughton WJ, Benoit I, de Vries RP, Gueidan C, Gorbushina AA (2013) Nutritional physiology of a rock-inhabiting, model microcolonial fungus from an ancestral lineage of the Chaetothyriales (Ascomycetes). Fungal Genet Biol 56:54–66

Naimark EB, Kompantseva EI, Komova AV (2009) Interaction between anoxygenic phototrophic bacteria of the genus *Rhodovulum* and volcanic ash. Microbiology (Mikrobiologiya) 78 (6):747–756

Navarre-Sitchler A, Steefel CI, Yang L, Tomutsa L, Brantley SL (2009) Evolution of porosity and diffusivity associated with chemical weathering of a basalt clast. J Geophys Res 114:14. doi:10.1029/2008JF001060art.no.F02016

Noack-Schönmann S, Bus T, Banasiak R, Knabe N, Broughton WJ, Den Dulk-Ras H, Hooykaas PJJ, Gorbushina AA (2014) Genetic transformation of Knufia petricola A95—a

model organism for biofilm-material interactions. AMB Express 4: 80. http://www.amb-express.com/content/4/1/80
Parfenova EI (1950) Study of primitive mountainous soils on diorite of mountain ridge Magisho (the North Caucasus). (Issledovanie primitivnykh gorno-lugovykh pochv na dioritakh khrebta Magisho (Severnyi Kavakaz)). Bulletin of V.V. Dokuchaev Soil Science Institute. Moscow-Leningrad AS SSSR 34:49–109 (in Russian)
Prieto B, Rivas T, Silva B (1994) Colonization by lichens of granite dolmens in Galicia (NW Spain). Int Biodeterior Biodegradation 34:47–60
Polynov BB (1945) Pervye stadii pochvoobrazovaniya na massivno-kristaalocheskich porodack (The first stages of pedogenesis on massive crystalline rocks). Soil Sci (Pochvovedenie) 7:327–339 (in Russian)
de los Ríos A, Ascaso C (2001) Preparative techniques for transmission electron microscopy and confocal laser scanning microscopy of lichens. In: Kranner I, Beckett RP, Varma AK (eds) Protocols in lichenology. Springer, Berlin
de los Ríos A, Wierzchos J, Ascaso C (2002) Microhabitats and chemical microenvironments under saxicolous lichens growing on granite. Microbial Ecol 43:181–188
Rigou L, Mignard E, Plassard C, Arvieu JC, Remy JC (1995) Influence of ectomycorrhizal infection on the rhizosphere pH around roots of maritime pine (Pinus pinaster Soland in Ait.). New Phytol 130:141–147
Sancho C, Fort R, Belmonte A (2003) Weathering rates of Historic Sandstone structure in semiarid environments (Ebro Basin, NE Spain). Catena 53:53–64
Seiffert F, Bandow N, Bouchez J, von Blanckenburg F, Gorbushina AA (2014) Microbial colonization of bare rocks: laboratory biofilm enhances mineral weathering. Procedia Earth Planet Sci 10:123–129
Silva B, Prieto B, Rivas T, Sanchez-Biezma MJ, Paz G, Carballal G (1997) Rapid biological colonization of a granitic building by lichens. Int Biodeterior Biodegradation 40:263–267
Sokolova TA (2011) The role of soil biota in the weathering of minerals: A review of literature. Eurasian Soil Sci 44(1):56–72
Torbjørg B, Thorseth IH (2002) Comparative studies of lichen—rock interface of four lichens in Vingen, western Norway. Chem Geol 192:81–98
Tsyurupa IG (1973) The role of microorganisms in weathering of aluminosilicates and formation of mobile, easy migrating compounds (Rol' microorganizmov v vyvetrivanii alumosilikatov i obrazovanii podvizhnykh, legkomigriruiushchekh soedinenii) Kora vyvetrivaniya 13:3–38 (in Russian)
Varadachari C, Barman AK, Ghosh K (1994) Weathering of silicate minerals by organic acids: II. Nature of residual products. Geoderma 61:251–268
Velde B, Meunier A (2008) The origin of clay minerals in soils and weathered rocks. Springer, Berlin
Wilson JM, Jones D (1983) Lichen weathering of minerals: implications for pedogenesis. In: RCL Wilson (ed) Residual deposits: surface related weathering processes and material. Special Publication of the Geological Society. Blackwell, London
Yakhontova LK, Zvereva VP (2000) Basic foundation mineralogy of hypergenesis: teaching aid. (Osnovy mineralogii gipergeneza: uchebnoe posobie). Dal'nauka, Vladivostiok (in Russian)
Yarilova EA (1947) The role of lithophilous lichens in the weathering of solid rocks (Rol' litofil'nykh lishainikov v vyvetrivanii massivno-kristallicheskekh porod) Soil Sci (Pochvovedenie) 9:533–548
Zenova GM, Chizhikova NP, Omarova EO, Ivanova EA (2009) Ecological role of cyanobacterial-actinomycetes associations in change of clay minerals structure. Prob Agric Ecol (Problemy agrokhimii i ekologii) 4:25–32 (in Russian)
Zvyaginzev DG, Dobrovol'skaya TG, Bab'eva IP, Zenova GM, Lysak LV, Marfenina OE (1999) Role of microorganisms in biogeocenosis functions of soils. (Role microorganizmov v biogeotsenoticheskikh funktsiyakh pochv). In: Structure—functional role of soil in biosphere (Strukturno-funktsional'naya rol' pocgvy v biosfere). GEOS, Moscow (in Russian)

# Biomineralization of Precious Metals

**L.M. Pavlova and V.I. Radomskaya**

**Abstract** In model tests, the kinetics and the possibility of concoction and biogenesis of ultradisperse forms of Au, Pd, and Pt in different physical and chemical conditions were examined. Studied biosorbents included peat, humic acids, nonhydrolyzed residue, and microscopic fungi in whole extract gold at 86–99 %, palladium at 89–95 % from chloride solutions, and platinum at 73–87 %. The maximum rate of extraction of Pd and Pt from solution by *Penicillium canescens* Sopp. biomass at different pH values was observed in sour solutions (pH 1–2) to be 16 and 47 mg/g of an element, respectively. The maximum rate of extraction of Au (to 24 mg/g) by fungic biomass was observed in subacidic solutions (pH 4–5). Palladium and platinum from chloride solutions were extracted from microfungi at rate of 87–95 % at optimum acidity of solution, with gold to 99 %. Using scanning electronic microscopy, the Pd and Pt distribution over the entire mycelium surface was established. The sorption process of precious metals by a fungic cell wall proceeded differently: the interaction process of ionic Au with an organic matrix continues up to the formation of metal nanoparticles, on which gold units are further formed, whereas the interaction of Pd and Pt stops at the sorption stage. It is thought that the biomineralization mechanisms of precious metals are caused by several reactions, including ion exchange, complex formation, and oxidation–reduction.

**Keywords** Biomineralization · Precious metals · Mechanisms of sorption · Fungus biomass · Organic sorbents · Nanoparticles

Biomineralization is the insulation process of mineral formation by living organisms or biomolecules that are preserved in the environment after dying off, causing biogeochemical cycles of chemical element circulation in the crust. The main evolution stages of both the biological and mineral worlds have already modified and are still changing the planet's surface, and they are directly related to the

L.M. Pavlova (✉) · V.I. Radomskaya
Institute of Geology and Nature Management Far Eastern Branch, Russian Academy of Sciences, Blagoveshchensk, Russia
e-mail: pav@ascnet.ru

 
O.V. Frank-Kamenetskaya et al. (eds.), *Biogenic—Abiogenic Interactions in Natural and Anthropogenic Systems*, Lecture Notes in Earth System Sciences,
DOI 10.1007/978-3-319-24987-2_3

biomineralization process. Carbon formations (peat, sapropel, coal, black shales) and living organisms (plants, fungi, bacteria, etc.) involved in organogenic formation processes play a major role in the biogenic concentration of elements. Living matter actively influences the geochemical environment, differentiation, migration of chemical elements through the food chain, metabolites, and post-mortem residues, providing for the migration of biogenic forms of chemical elements and the biogenic origin of some deposits of gold, uranium, iron, phosphorus, and others. Microorganisms play a significant role in the concentration process of chemical elements in the weathering crust. This is due not only to a huge variety of occupied ecological niches, but also to the great potential of the enzyme system. Moreover, bacteria have the highest sorption capacity of all life forms, as they have the highest surface area-to-volume cell ratio and a negatively charged surface related to the environment (Trudinger and Swaine 1979; Beveridge 1989). For example, the gold concentration by microorganisms relative to the environment reaches $10^4$–$10^5$ degrees (Mann et al. 1987), which is a very high concentration factor considering the low abundance of Au in nature (2–6 mg/t).

There are a few studies of the role of biogenic factors in the genesis of nanodisperse forms of precious metals in high-carbon formations. The biosynthesis of nanoparticles of Au, Ag, Au–Ag, Pt, Pd, Ti, and other elements by bacteria, fungi, and viruses is established in experimental studies (Konishi et al. 2007; Thakkar et al. 2010); the mechanisms of the biogenic concentrations of Cu, Zn, Ag, and Au are actively discussed (Bernatonis 1990; Belyaev and Pedash 1989; Yudovitch and Ketris 2004; Arbuzov et al. 2004). The possibility of the precious metal reduction of palladium has also been pointed out, up to crystalline nanoparticles of 80–100 nm by tropical plant extracts such as *Terminalia chebula* (Kumar et al. 2013), *Annona squamosa* (Roopan et al. 2012) and extracts of coffee and tea (Nadagouda and Varma 2008) due to the presence of significant amounts of aldehyde and hydroxyl groups in aqueous extracts. $PdCl_2$ reduction by the aqueous extract of cinnamon powder (Sathishkumar et al. 2009) up to 15–20 nm has been noted. The main role in the migration of platinum group elements in natural conditions at low temperatures has been found to belong to the organic matter present in groundwater and surface waters (Bowles et al. 1994; Wood 1996); in addition, a biochemical mechanism causes a selective concentration of Pt by Fe–Mn nodules in the ocean floor (Zhmodik et al. 2009).

To clarify the regularities and mechanisms of precious metal (PM) mineralization by natural organic matter, laboratory model experiments on the concentration ions of Au(III), Pd(II), and Pt(II) by microfungal biomass, peat organic matter, and its derivatives, as well as low ash preparations of humic acids (HA) and nonhydrolyzed residue (NHR) fraction, were conducted. These organic matter groups are the most inactive and may be the main concentrators of elements.

# 1 Methodology

In the capacity of a natural organic biosorbent, the lowland peat of Zeya River's second bottom (the Amur region) was used. Peat samples were selected at 2.7 km from the river bed from the 2-m reservoir layer by layer: 0–10 cm, 50–60 cm, 100–110 cm, and 170–180 cm. In model experiments of PM biomineralization, only peat from the 100–110 cm layer was used. In a preliminary study (Kuimova et al. 2011), the maximum HA content (51 %) and an adequate content of NHR (34 %) were found in a substance from this horizon. The HA was obtained from an air-dried quantity of milled peat after the bitumen was preliminary removed by hexane extraction.

Debituminous peat was triple-treated with a 0.1 M solution of NaOH, separating the nonhydrolysable residue by centrifugation. Humic acid was precipitated by coagulation within acidification by 5 % HCl solution up to pH 1.0–1.5. All organic substances that remained in the acid solution after HA was discharged were considered to be fulvic acids (FA). HA was centrifuged, transferred to a filter, washed with distilled water to neutral pH, then dried in a vacuum oven at 60–70 °C to a constant weight. The obtained HA product was demineralized with 5 % HF solution in a water bath. The ashless HA precipitate was repeatedly washed and air-dried.

Nonhydrolyzed residue (NHR) obtained after HA removal was washed from alkali, centrifuged, and dried to a constant weight (Radomskaya et al. 2014).

Microfungal biomass *Penicillium canescens* Sopp., previously isolated from brown coal from a Raichikhinskiy deposit, was obtained by cultivation in liquid Czapek's medium at 20 °C for 7 days. Then, the biomass was centrifuged and dried to a constant weight (Kuimova et al. 2012a, b).

The kinetics of biosorption ions Au, Pd, and Pt were studied using a "limited value" method under static conditions at 25 °C, constant pH of 1.3, and periodic agitation. In a container with a certain quantity of organic sorbent, the chloride solution of the precious metal was added, varying the solution and biosorbent ratio between 100 and 1000 ml/g. At certain intervals, the solution aliquot was collected to determine the amount of adsorbed metal ions, wherein the total value of samples did not exceed 5–7 % of the original. After the sorption equilibrium was established, the biosorbent was filtered from the solution.

The acid–base environmental characteristics actively influenced the elements' migration and concentration. Therefore, in the second set of experiments, the bioaccumulation degree of the PM was investigated by pH gradient.

Original solutions of Au(III), Pd(II) and Pt(II) with concentrations of 1000 mg/ml in 1 M HCl were prepared by dissolving a certain quantity of metal in chlorazotic acid, followed by evaporation to humid salts. This was followed by additional treatment with concentrated HCl, repeated evaporation, and residue dissolving in 1 M HCl (Radomskii and Radomskaya 2004). On the experiment day, the working solutions were prepared by diluting the original solutions in distilled water.

The content of Au and Pt in the solutions was determined using an atomic absorption method on the first class spectrophotometer (Hitachi 180-50). The content of Pd was determined using a voltamperemetric analyzer TA-4 and two-electrode unit (SIE TomAnalit, Tomsk). The work electrode was graphite, and the comparison electrode was a saturated silver-chloride electrode. Quartz cups of 20 $cm^3$ volume were used as an electrolyzer. Pd(II) was determined on 0.1 M HCl background after electrocumulation at −0.8 V; the potential of the palladium anodic peak was 0.55 ± 0.05 V. The adsorption amount (*A*), expressed in milligrams per gram of sorbent, is calculated as $A = (C_{initial} - C_{equilibrium})$ V/m, or the difference between the initial concentration and the metal concentration in the solution after sorption.

Ultrastructural analysis of the sample surface and microprobe analysis were performed using a scanning electron microscope *JSM-6490LV* with EDS *INCA Energy, X-max,* and HDS *INCA Wave.*

## 2 Results and Discussion

The kinetic curves of ion PM sorption (Fig. 1) show that the equilibrium in Au(III) distribution between the solution and organic sorbents was set for 24 h. The maximum sorption of palladium was achieved in 24 h for microfungus biomass, 36 h for peat, and 48 h for NHR and HA, whereas the maximum sorption of platinum was reached in 1 h. Microscopic fungi as well as HA low-ash products have a greater selectivity to palladium than peat and NHR. The maximum possible exchange capacities of palladium for HA, microscopic fungi, NHR, and peat were 11.7 mg/g, 10.72 mg/g, 9.6 mg/g and 6.39 mg/g, respectively. The form of the kinetic curve of palladium sorption on microfungus biomass indicates a high sorption capacity of the micromycetes cell wall surface in relation to low concentrations of palladium ions (Fig. 1b). Similar effects were obtained for gold(III) (Fig. 1a).

The highest sorption capacity of platinum among studied biosorbents was also observed for microscopic fungi. The sorption curve of Pt, which depended on phase contact time (Fig. 1c), showed that Pt(II) ion extraction accelerated several times in the first half hour compared with the sorption of palladium and gold hydroxochloride complexes. Fungus biomass adsorbs up to 87 % platinum presenting in solution per hour; the platinum extraction ratio by other organic sorbents is just below this value. This is likely stipulated by the greater propensity of Pt(II) complexes to the replacement process in the ligand coordination sphere. However, further increase of the extraction ratio in the course of time practically does not occur.

In quantitative terms, the sorption potentials of the studied organic materials for each element can be represented by the following ordered series (mg/g): for Au, fungus biomass (20–200) > HA (23–54) > NHR (2–23) > peat (2–4) (Kuimova et al. 2012a, b); for Pd, micromycetes biomass (8–41) > HA (1–25) > NHR

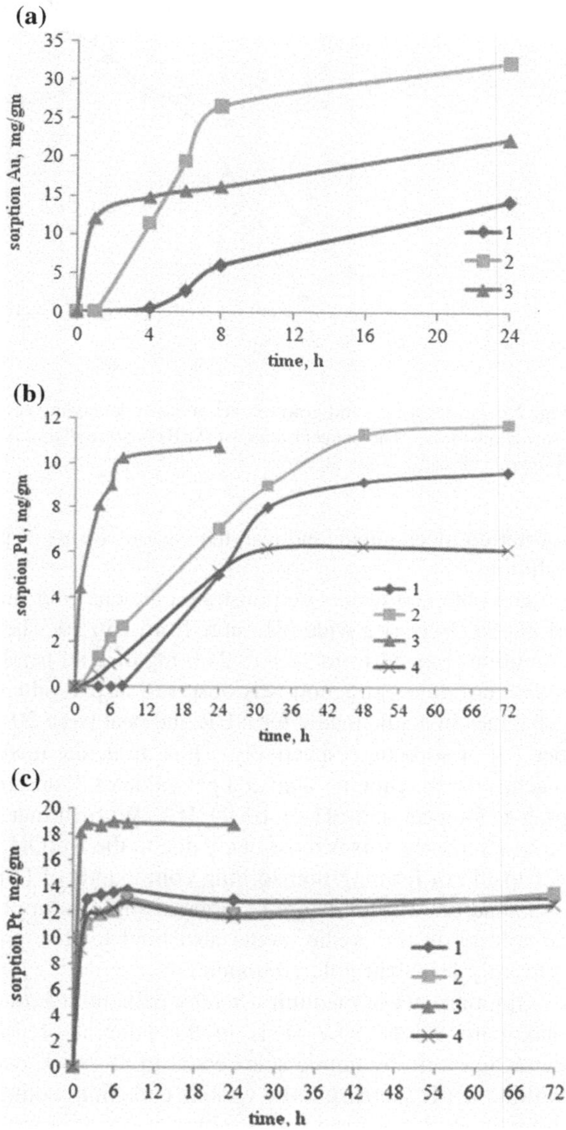


**Fig. 1** Kinetic curves of PM sorption: **a** Au; **b** Pd; **c** Pt on NHR (*1*), HA (*2*), and microfungus biomass (*3*); peat (*4*). The initial concentration of Au(III) is 65 mg/l, Pd(II) is 60 mg/l, and Pt(II) is 44 mg/l; solution volume is 20 ml, $mass_{sorb.}$ = 40 mg

(0.5–10) > peat (0.5–7); and for Pt, fungus biomass (17–19) > NHR (9–16) > peat (9–15) > HA (4–14).

Diagrams of the palladium and gold extraction ratios from solutions in a pH gradient by organic sorbents are presented as an S-shaped curve (Fig. 2) due to several factors: the heterogeneity of sorbent surface, the multifunctionality of surface groups, the implementation of the anion-exchange mechanism or complex

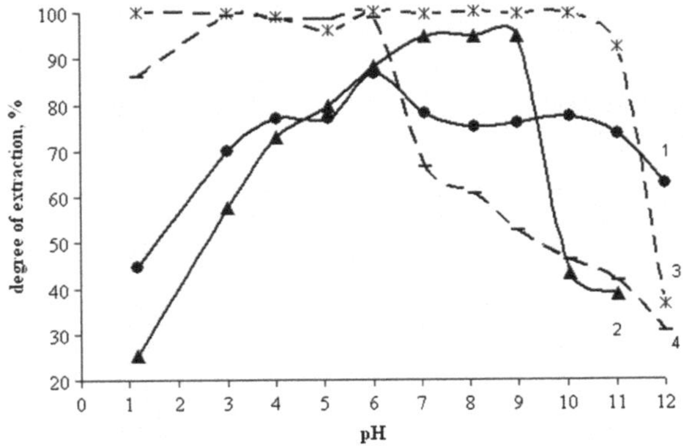


**Fig. 2** The palladium and gold extraction ratio depending on pH: *1* HA (Pd), *2* peat (Pd), *3* HA (Au), *4* peat (Au). The concentration of Pd(II) is 50 mg/l and Au(III) is 46.1 mg/l; solution volume is 10 ml, $mass_{sorb.}$ = 20–30 mg

formation mechanism, and also the various forms of palladium and gold in chloride solutions.

The obtained results demonstrate efficient gold binding by HA (extraction ratio of 86–99 %) over a wide pH range from 1 to 10. The sorption capacity under these conditions ranged from 22.8 to 24.9 mg of gold per 1 g of HA. For peat and NHR, maximum gold extraction was observed at pH values from 1 to 5–6. The sorption capacities in acidic media for NHR and peat were 20–21.2 and 19.2–22.1 mg of Au per 1 g of sorbent, respectively. Thus, in acidic media, Au(III) sorption increases, reaches the maximum value at a pH value of 3, and remains constant to a pH value of 5–6 for peat and pH of 10 for HA. With further pH increase, the Au sorption value decreases; this is most likely due to the $Au(OH)_3$ amphoteric character, which is capable of forming non-sorbing compounds of $[Au(OH)_4]^-$ type.

Palladium extract nature by organic sorbents in pH gradient differ. Its inorganic complexes in the acidic media also bind by HA and peat surface, although significantly less than gold compound.

Optimum pH of medium whereby palladium extraction ratio by peat reaches the maximum value—94.7 % is in the interval of 7–9. The maximum palladium extraction ratio by humic acids equal to 89–91 % occurred at pH of 6–7. Possibly within the pH increasing the various palladium complexes are formed those have a different ability to form bonds with functional groups such as peat, HA, and NHR.

Regarding PM sorption on fungus biomass, the maximum Au extraction ratio was observed in weakly acid solutions (pH 4–5), while the maximum Pd and Pt extraction ratio by *P. canescens* biomass was observed in acid solutions (pH 1–2) (Fig. 3). Microfungus extracted palladium and platinum from chloride solutions at 87–95 and 99.9 % gold.

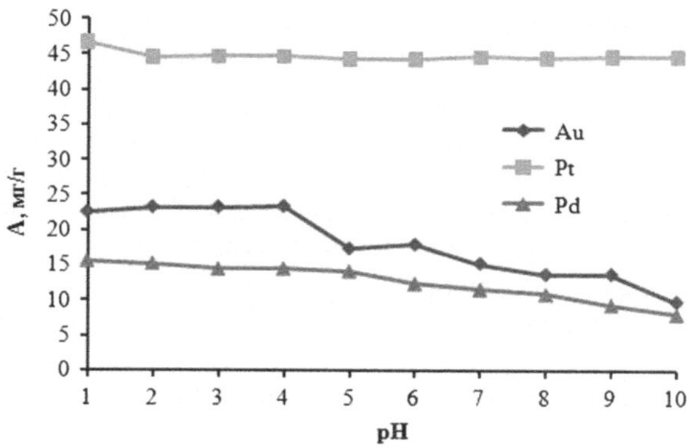


**Fig. 3** Sorption curves of Au, Pd, and Pt by microfungus biomass at different pH values. The concentration of Pd(II) is 41.2 mg/l, Au(III) is 48.9 mg/l, and Pt(II) is 99.8 mg/l; solution volume is 10 ml, $mass_{sorb.}$ = 20 mg, and sorption time is 72 h

Two main factors determine the organic sorbents' sorption in relation to precious metals—the chemical forms of metal ions' ratio in solution and the amount of functional sorbent groups.

The chemistry of precious metals in chloride solutions is difficult. They are characterized by a variety of chemical forms, the transition from one form to another, chlorine complexes addition to hydrolysis, and polymerization—the speed and completeness of which substantially depend on the initial chemical form, temperature, salt background, solution exposure time, metal concentration, chloride ions, and others.

The complexes $[AuCl_4]^-$ and $[PdCl_4]^{2-}$ are the dominant forms of gold and palladium in strongly acidic solutions. $[AuCl_3OH]^-$, $[AuCl_4(H_2O)_2]^-$, $[Pd(H_2O)Cl_3]^-$, $[Pd(H_2O)_2OH]^+$, and $[Pd(H_2O)_2(OH)_2]$ are additionally formed in weakly acidic solutions. $[PtCl_6]^{2-}$ complexes are the dominant forms of platinum in strongly acidic solutions. $[Pt(OH)Cl_5]^{2-}$ and $[Pt(OH)_6]^{2-}$ are additionally formed in weakly acidic solutions (Buslayeva and Simanova 1999; Busev and Ivanov 1973).

***Peat*** is a weakly acidic polyfunctional ion exchanger consisting of various chemical components, which are both organic and inorganic in nature. Peat contains hydrophilic substances (cellulose, hemicellulose, lignin, humic substances), hydrophobic components (bitumen, wax), and mineral and organo-mineral impurities with different degrees of dispersion. Therefore, the diversity of functional groups presented in peat covers the entire range of $pK_a$ values, from 2 to 14.

***Nonhydrolysed residue (NHR)*** is a mixture of substances, the main component of which is lignin, its nitrogen-containing derivatives, and humification intermediates.

***Humic acid (HA)*** is a hydrophobic aromatic backbone with a high content of functional groups. The structure of the humic acids has a plurality of oxygen-containing functional groups: carboxyl groups, phenolic and alcoholic hydroxyls, carbonyl groups, quinoid, lactone, and ether groups, as well as nitrogen-containing functional groups (Orlov 1990). Humic acid, by the nature of the functional groups, as a sorbent is similar to a subacid carboxyl cation exchange resin. Among the basic functional groups of HA, two types can be distinguished: carboxyl group –COOH and phenolic –OH; their main dissociation areas are characterized by $pK_a$ values of 3–5 and 8–9, respectively. Strongly acidic carboxyl functional groups and carboxyl groups of mean force at pH less than 6 will be completely dissociated (Varshal et al. 1996). The content of –COOH groups in the sample of studied HA, defined by the acetate method, was 2.57; in –OH-groups, the content was 4.41 mg-equiv/g. Thus, humic acids isolated from fen sedge peat are characterized by a predominance of the phenolic groups over carboxyl by 1.7 times.

***Microfungus cell wall*** contains even more phenolic groups: 27.93 mg-equiv/g, which exceeds the carboxyl group content (3.22 mg-equiv/g) severalfold. This stipulates the negatively charged surface of the cell wall in relation to the environment.

Therefore, the sorption isotherm is unlikely to be homogeneous for the different types of organic sorbents. Under realistic conditions, the dependencies overlapping that characterize metal ions sorption with different functional groups is possible. An organic sorbent surface could acquire a positive or negative charge depending on the pH value of the model solution.

As mentioned, at pH of 1–3, gold(III) presents as negatively charged chloro-complexes and hydrated ions; palladium (II) presents as $[PdCl_4]^{2-}$, $[Pd(H_2O)Cl_3]^-$, $[Pd(H_2O)_2OH]^+$, and $[Pd(H_2O)_2(OH)_2]$. In this pH range, the effect of electrostatic repulsion of a positively charged sorbent surface and neutral or positively charged palladium complexes and attraction of negatively charged gold complexes was observed, which stipulates a lower palladium(II) sorption in comparison with gold (III) in acidic media. With increasing pH, the surface acquires a negative charge. Therefore, the electrostatic attraction between the negatively charged sorbent surface and palladium complexes increases, which releases the sorption process.

The sorbent surface after PM ion sorption was investigated by **scanning electron microscopy**. The sorption process of precious metals by the microfungus cell wall (Fig. 4) flows differently: the ion Au interaction continues until the formation of metal nanoparticles (Fig. 4a), on which gold aggregates are then formed (Fig. 4b) (Kuimova 2004).

Scanning the fungal mycelium surface after interaction with palladium chloride solution showed the presence of the latter as part of spherical particles with phosphorus and iron in the interhypha space and on the hypha surface (Fig. 4c). However, reduced palladium forms on the fungus biomass were not found.

The interaction process of platinum ions with the micromycete cell wall also appears to stop at the sorption stage (Fig. 4d) because nanoparticles of reduced platinum were not found by electronic microscopy.

The interaction of gold with HA, NHR, and peat also lasts until metal reduction as nanosized particles (Fig. 5). The analysis of microphotographs showed an uneven distribution of gold nanoparticles on the grain sorbents' surface. The selectivity of this process appears in preferential gold localization on more carbon

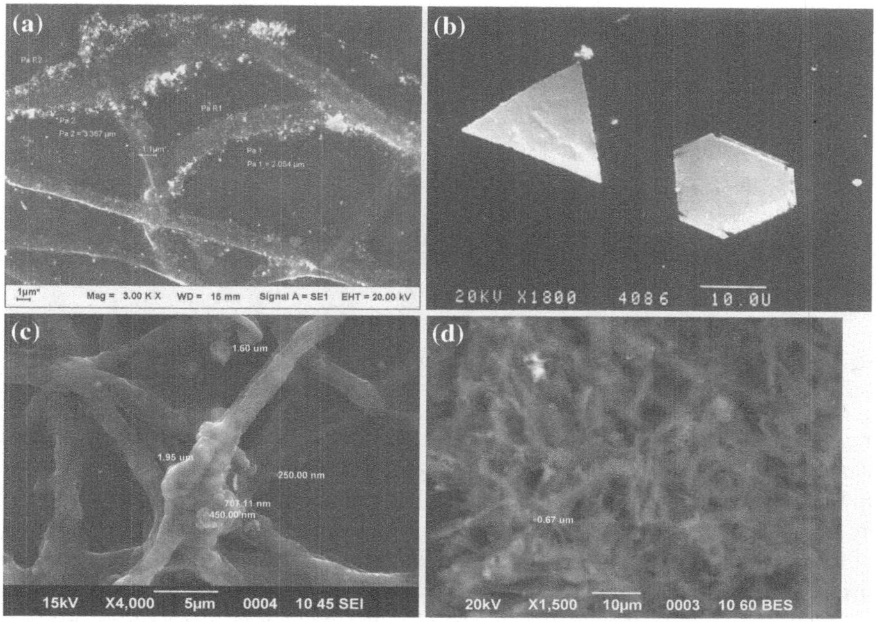


**Fig. 4** Micromycete mycelium surface with sorbed particles of Au (**a**, **b**), Pd (**c**), and Pt (**d**)

**Fig. 5** Clusters of globular complex aggregates of gold nanoparticles on peat (**a**, **b**), HA (**c**, **d**), and NHR (**e**–**g**). General view (**a**, **e**, **f**) and separate particles (**b**, **c**, **d**, **g**)

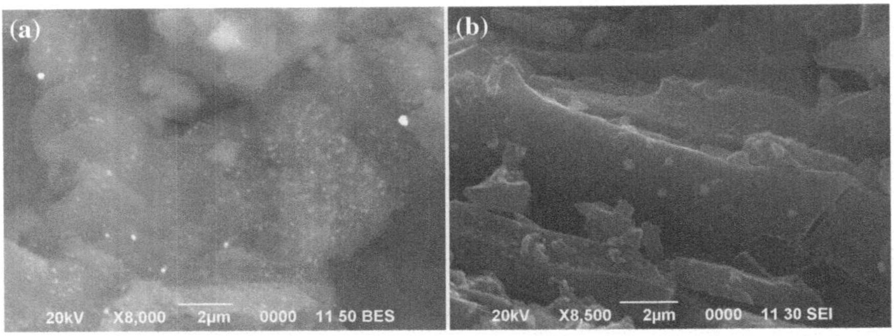


**Fig. 6** The fragment of peat surface with palladium nanoparticles (**a**) and NHR (**b**)

fragments of structurally heterogeneous substrates. The average size of gold nanoparticles is 0.2 μM. Discrete "single" gold grains of size 1–3 μM (occasionally up to 9 μM) are quite rare. In composition, the metal grains studied on peat and HA are mainly gold with low Cu impurities (up to 1–2 wt%).

In the process of palladium sorption on peat, palladium reduction was also observed. Microphotographs (Fig. 6a) show the domination of palladium nanoparticles smaller than 0.2 μM. For HA, the palladium distribution over the entire surface of the absorbing matrix was found, but the reduced palladium forms were not detected. This fact indicates that HA interaction with palladium solution stops at the sorption stage. The NHR matrix (Fig. 6b) is characterized by the presence of palladium in the metal phase, referred to as blocks, with a higher carbon content, such as palladium adsorbed phases.

Thus, the results of electron microscopic studies indicate that Au sorption by all studied natural organic sorbents (peat, HA, NHR, microfungus biomass) proceeds to the formation stage of elemental metal of nanoscale size.

Pd sorption by natural organic sorbents proceeds ambiguously. Pd reduction to nanoparticles is observed only on peat and partly on NHR. Humic acids and the microfungus cell wall adsorb Pd ions only. Pt interaction with microfungus biomass also stops at the sorption stage.

The correlation of experimental observations and oxidation–reduction potential values of PM chloride solutions indicates that the oxidation–reduction reaction can be the last stage of interaction between organic sorbents and PM complexes. The possibility of oxidation–reduction reaction behavior for precious metals in weakly acidic solutions is defined by the oxidation–reduction reactants' potential ratio. According to the literature, the standard oxidation–reduction potential for $AuCl_4^- + 3e \leftrightarrow Au + 4Cl^-$ is 1.00 V; for system $[PdCl_4]^{2-} + 2e \leftrightarrow Pd + 4Cl^-$, it is 0.62 V; and for $[PtCl_4]^{2-} + 2e = Pt + 4Cl^-$, it is 0.73 V.

Oxidation–reduction potentials of HA solutions and chloride with fungus biomass were measured. The oxidation–reduction potential of HA solution is 0.71 V, indicating that HA is able to reduce oxygenized forms of oxidation–reduction pairs with potential ≥0.71 V. The oxidation-reduction potential of chloride solution with fungus biomass is 0.74 V, indicating that the fungus cell wall is able to reduce gold only.

Thus, the comparative analysis of experimental data on precious metals sorption (Au, Pd, Pt) by organic natural sorbents (peat, HA, NHR, microfungus biomass) and the status data of PM in weakly acidic solutions lead to the conclusion that there is a complicated sorption mechanism for these systems. Possibly, ion exchange occurs at the first stage. According to our data, the solution pH value shifts to the acidic region due to adsorption, possibly because of the metal ion exchange by the carboxyl group hydrogen. At the second stage, the implantation of atoms of sorbent functional groups occurs inside the precious metal complex; for this reason, the sorption equilibrium is set for a long time and reduced gold nanoparticles contain other metal impurities, particularly copper. Organic sorbents such as peat, HA, and NHR are characterized by a hydrophobic nature and are water insoluble. The metal ions' diffusion to a solid body through the surrounding hydrodynamic film (i.e., film diffusion) and metal ions' diffusion to sorbent ligand groups inside pores (i.e., gel diffusion) precede the interaction between organic sorbents and PM.

The process of Au, Pd, and Pt sorption proceeds differently. The Au ion interaction with any organic matrix continues until elemental metal formation in the form of nanoscale particles, after which gold aggregates form. Pd sorption proceeds less clearly. Pd reduction to nanoparticles is observed only on peat and partly on NHR. Humic acids and microfungus cell walls adsorb Pd ions only. Pt interaction with microfungus biomass also stops at the sorption stage.

## 3 Conclusions

Based on precious metal sorption experiments, it has been discovered that the studied biosorbents—peat, HA, NHR, and microscopic fungi—extract gold at 86–99 %, palladium at 89–95 % from chloride solutions, and platinum at 73–87 %.

With a reduction of the acidity, the extraction ratio of precious metals increases for all sorbents, reaching a maximum in a weakly alkaline pH range and then decreasing almost to the original value. The maximum Pd and Pt extraction ratio by *P. canescens* biomass was observed in acidic solutions (pH 1–2); for Au, it was found in in weakly acidic solutions (pH 4–5).

The mechanisms of noble metal mineralization are stipulated by several reactions, including ion exchange, complex formation, and oxidation–reduction.

**Acknowledgments** This study was supported by the Russian Foundation for Basic Research (project no. 15-05-08819).

## References


Arbuzov SI, Rikhvanov LP, Maslov SG, Arkhipov VS, Pavlov ZI (2004) Anomalous gold contents in brown coals and peat in the south-eastern region of the Western-Siberian platform. Bull Tomsk Polytech Univ 307(7):25–30

Belyaev VK, Pedash ET (1989) Small elements in coal and enclosing rocks of Shubarkol deposit. Explor Conserv Mineral Resour 11:12–16

Bernatonis VK (1990) The role of organic matter in the processes of supergene gold migration. Hydrochemical prospecting of mineral deposits. Nauka, Novosibirsk

Beveridge TJ (1989) Role of cellular design in bacterial metal accumulation and mineralization. Annu Rev Microbiol 43:147–171

Bowles JW, Gize AP, Cowden A (1994) The mobility of the platinum-group elements in the soils of the Freetown peninsula, Sierra-Leone. Can Mineral 32:957–967

Busev AI, Ivanov VM (1973) Analitical chemistry of gold. Nauka, Moskow

Buslayeva TM, Simanova SA (1999) Condition of platinum metals in chloride water solutions. Palladium, platinum, rhodium, iridium. Coord Chem 25(3):165–176

Konishi Y, Tsukiyama T et al (2007) Microbial deposition of gold nanoparticles by the metal-reducing bacterium *Shewanella algae*. Electrochim Acta 53:186–192

Kuimova NG (2004) Accumulation and crystallization of gold by microorganisms isolated from ore and placer deposits. Dal'nauka, Vladivostok

Kuimova NG, Pavlova LM, Sorokin AP, Noskova LP (2011) Experimental modeling of gold concentrating processes in peat. Lithosphere 4:131–136

Kuimova NG, Pavlova LM, Radomskay VI (2012a) Biogenic concentration of precious metals. Georesources 1(12):21–24

Kuimova NG, Radomskaya VI, Pavlova LM (2012b) Concentration of palladium on the biogeochemical barriers. Mt Inf Anal Bull (Sci Tech J) 9:77–84

Kumar KM, Mandal BK, Kumar KS, Reddy PS, Sreedhar B (2013) Biobased green method to synthesise palladium and iron nanoparticles using *Terminalia chebula* aqueous extract. Spectrochim Acta Part A Mol Biomol Spectrosc 102:128–133

Mann H, Tasaki K, Fife WS et al (1987) Cellular precipitation and heavy-metal sorption in *Euglena* sp.: implication for biomineralization. Chem Geol 63(1):39–43

Nadagouda MN, Varma RS (2008) Green synthesis of silver and palladium nanoparticles at room temperature using coffee and tea extract. Green Chem 10(8):859–862

Orlov DS (1990) Soil humic acids and general humification theory. MSU, Moscow

Radomskaya VI, Noskova LP, Pavlova LM (2014) Interaction of chloride complexes of gold and palladium with humic acids. Bull Irkutsk State Tech Univ 8(91):62–68

Radomskii SM, Radomskaya VI (2004) Non-equilibrium sorption of silver, platinum, gold from solutions with synthetic fibrous sorbents. Chem Technol 7:21–25

Roopan SM, Bharathi A, Kumar R, Khanna VG, Prabhakarn A (2012) Acaricidal, insecticidal, and larvicidal efficacy of aqueous extract of *Annona squamosa* as biomaterial for the reduction of palladium salts into nanoparticles. Colloids Surf B 92:209–212

Sathishkumar M, Sneha K, Kwak IS, Mao J, Tripathy SJ, Yun YS (2009) Phyto-crystallization of palladium through reduction process using *Cinnamom zeylanicum* bark extract. J of Hazard Mater 171:400–404

Thakkar K, Mhatre S, Parikh R (2010) Biological synthesis of metallic nanoparticles. Nanomed Nanotechnol Biol Med 6:257–262

Trudinger PA, Swaine DG (1979) Biogeochemical cycling of mineral forming elements. Elsevier, Amsterdam

Varshal GM, Koscheeva IYa, Velukhanova TK, Chkhetia DN, Tyutyunnik OA, Grinevskaja ZhM (1996) Sorption of heavy metals and isotope carriers of long-lived radionuclides on Humic Acids: message I. Sorption of Cesium (I), Strontium (II), Cerium (III), Ruthenium (IV). Geochemistry 11:1107–1112

Wood SA (1996) The role of humic substances in the transport and fixation of metals of economic interest (Au, Pt, Pd, U, V). Ore Geol Rev 11:1–31

Yudovitch YaE, Ketris MP (2004) Gold in coals. In: Lithogenesis and geochemistry of sedimentary formations of the Timan-Ural Region. Proceedings Institute of Geology of the Komi Science Center, Urals Branch of RAS, vol 116(5), pp 80–109

Zhmodik SM, Belyanin DK, Mironov AG, Parkhomenko VS, Titov AT, Teplyakova TV, Zimbalist VG, Tatarinov AV (2009) Role of the biogenic factor in platinum accumulation by oceanic ferromanganese nodules. Doklady RAS 426(5):658–663

# Modifications of Selected Clay Minerals Due to Activity of Filamentous Alkaline Cyanobacteria

**Andery O. Alekseev, Tatiana V. Alekseeva, Lyudmila M. Gerasimenko, Vladimir. K. Orleanskiy and Galina T. Ushatinskaya**

**Abstract** The aim of the study was to investigate the transformations of clay minerals at the laboratory experiments under the growth and fossilization of alkaline cyanobacteria. The clays incubation with cyanobacteria resulted in different trends of their transformations. The direction and intensity of transformation depends on type of clay mineral. The observed processes were fast and completed within the first 10–60 days of experiments. Cyanobacteria most actively influenced the processes of mineral dissolution and the transformations during the stage of their photosynthesis. Formation of carbonate in the experiments with palygorskite, bentonite, and kaolinite was observed at the stage of cyanobacteria fossilization.

**Keywords** Alkaliphilic cyanobacteria · Clay minerals · Palygorskite · Bentonite

The interactions between the geosphere and the biosphere are central point in the environmental and geological research. Minerals and rocks are the most fundamental earth materials with which microbes interact at all scales. Microbial interactions with mineral surfaces are widespread in sediments and soils. The recent studies (Gerbersdorf and Wieprecht 2015 ) have revealed an enormous role of bacteria, including cyanobacteria, in the processes of ancient sedimentogenesis and biostabilization of sediments. The residues of fossilized microorganisms, including cyanobacteria, are an evidence of their communities being the most important factor of sedimentation on the earth's surface from the Archaean time. Ancient microbial communities were involved in both formation and transformation of many minerals. Cyanobacteria have played

A.O. Alekseev (✉) · T.V. Alekseeva
Institute of Physicochemical and Biological Problems in Soil Science RAS, Pushchino, Russia
e-mail: alekseev@issp.serpukhov.su

L.M. Gerasimenko · Vladimir.K. Orleanskiy
Institute of Microbiology RAS, Moscow, Russia

G.T. Ushatinskaya
Paleontological Institute RAS, Moscow, Russia

 
O.V. Frank-Kamenetskaya et al. (eds.), *Biogenic—Abiogenic Interactions in Natural and Anthropogenic Systems*, Lecture Notes in Earth System Sciences,
DOI 10.1007/978-3-319-24987-2_4

fundamental and important roles in biogeochemical cycling of elements and in the oxygenation of the biosphere. Cyanobacteria or blue-green algae are oxygenic photoautotrophic organisms. Silicified microfossils of cyanobacteria (stromatolites) are found in deposits beginning from the age of 3.5 Ga years (Rozanov 2003). In our days, cyanobacteria create the relict communities in the extreme ecological conditions—regions with the hydrothermal activity, hyper-saline reservoirs or alkaline lakes (Dubinin et al. 1992; Zavarzin 1993; Schultze-Lam et al. 1995; Gerasimenko et al. 1996; Tazaki 1998; Phoenix et al. 2000; Tazaki et al. 2003).

Clay minerals, i.e., aluminosilicates with layer structure, are the most widespread minerals of sedimentation (marine and continental) basins, residuum, and soils. Frequent occurrence in their structure of the ions with variable valence makes clay minerals extremely sensitive to ambient conditions and, first of all, to variation of pH and redox conditions of the medium, so that they can be used as indicators of state of the environment. The interactions between clay and cyanobacteria can occur at several levels. Clays may directly influence the growth of cyanobacteria cells, e.g., by shading and flocculation. Microorganisms modify the rates and mechanisms of chemical and physical weathering and clay growth, thus playing fundamental roles in the dissolution of silicate structure in the rock weathering process, in the genesis of clay minerals, soils and sediments formation (Banfield et al. 1999). The presence of clay minerals is a typical symptom of biogeochemically weathered rocks (Barker et al. 1998). There are many reasons for microorganisms to colonize mineral surfaces, essential macro- or micronutrients can be found in many minerals (Bennett et al. 1996; Bennett at al. 2001; Grantham and Dove 1996; Kalinowski et al. 2000). Microbe-mediated mineral dissolution is usually incongruent, which means that certain elements are removed preferentially relative to others (Dong 2012). When individual bacteria discover the advantages of living on mineral surfaces because of sufficient supplies of nutrients and protection from lethal environmental stress, they tend to produce signal molecules to attract other organisms to gradually build a community called "biofilm" (Harrison et al. 2005). Biofilms may contain proteins, lectins, and polysaccharides, which help microorganisms to attach to mineral surfaces as Thorseth et al. (1995) and Barker et al. (1998) have found. Once a biofilm community is developed, minerals and rocks undergo weathering, and some minerals dissolve, other precipitate, or undergo transformations. Weathering of minerals usually occurs under the influence of organic acids and other ligands (such as siderophores) of the microbial community, oxidation of reduced minerals, and reduction of oxidized minerals. Besides the organic matter, bacterial mats usually contain mineral components—carbonate, silica, Fe and Mn oxides/hydroxides, clay minerals, phosphates, sulfides (Zavarzin 1993; Tazaki 1998). The recent findings show that microorganisms also participate in the synthesis of layer silicates, e.g., smectite-seladonite, glauconite-nontronite (Geptner and Ivanovskaya 1998; Ueshima and Tazaki 2001), imogolite-allophane-chamosite (Kawano and Tomita 2002), saponite (Geptner et al. 2002). Basing on electron microscopy observations of fresh water microorganism communities, Konhauser and Urrutia (1999) concluded that these communities are mineralized and are associated with amorphous (Fe, Al)-silicates with the composition close to chamosite/bertherine. Authors concluded that this phenomenon has the global distribution.

Mineralization processes in the presence of cyanobacteria are extensively studied for the hot spring communities (Phoenix et al. 2000; Likhoshway et al. 2006). But not many publications exist about mineralization in the presence of alkaline species (Ushatinskaya et al. 2006), and there are limited data on the effects of alkalophilic cyanobacteria and their communities on clay minerals (Alekseeva et al. 2008, 2009).

The goal of this work was to study the changes in mineral composition of different clays under the action of productive and destructive processes of alkalophilic cyanobacterial community.

## 1 Objects and Methods

In our experiments, we used a culture of alkalophilic filamentous cyanobacteria *Microcoleus chthonoplastes*, a unique member of laminated benthic community, isolated from soda Lake Khilganta, Buryatia (Russia) (collection of the Institute of Microbiology, RAS). In addition, in experiments with palygorskite a culture of alkalophilic filamentous cyanobacteria *Oscillatoria terebriformis* from collection of the Institute of Microbiology, RAS was used. The stimulating growth of cells was done using «S» medium (g/l): KCl—1.0; $K_2SO_4$—1.0; $NaHCO_3$—16.8; $K_2HPO_4$—0.5; $NaNO_3$—2.5; NaCl—30.0; $MgSO_4 \cdot 7H_2O$—0.2; $CaCl_2$—0.04; $FeSO_4$—0.01; A5-1 ml (Gerasimenko et al. 1996). pH of the medium was 8.5. The cultures were grown autotrophically under continuous illumination (white fluorescent light) at approximately 3000 lx (10000E $sm^{-2}$ $s^{-1}$) and 28 °C with $CO_2$ provided by bubbling air. After 4 days of stimulating growing 4 ml of suspension containing cyanobacteria was added to bottles which contained: clay (0.5 g), NaCl (2.0 g); $NaHCO_3$ (1.2 g), and sterile distilled water (100 ml). Cyanobacteria were incubated for 7, 14, 28, and 60 days at 28 °C in the light. Control bottles without cyanobacteria were incubated at the same conditions. All experiments have been done in duplicate.

It is well known that the intensity of the weathering and their trends are determined by the mineral's structure and composition. We studied the influence of *Microcoleus chthonoplastes* on the properties of different clays: 2:1 swelling (bentonite); 1:1 (kaolin), 2:1 non-swelling mineral (illite) (Alekseeva et al. 2008, 2009). The last study particularly was connected with possible transformations palygorskite at the laboratory experiments under the growth and fossilization of cyanobacteria *Microcoleus Chthonoplastes* and *Oscillatoria terebriformis*. Palygorskite, a magnesium-rich chain silicate (ideal formula $[(Mg, Al)_4Si_8O_{20}(OH)_2(OH_2)_4\_4H_2O]$) is a clay mineral with fibrous habit which consists of modulated 2:1 layers (for a review on their structure, mineralogy, and properties, see Singer 1989). The palygorskite was sampled from Late Carboniferous deposits in Moscow region (Russia). The information about clays and some of their properties are given in Table 1.

Element concentrations in clays were determined with the «SPECTROSCAN MAKC-GV» XRF crystal diffraction scanning spectrometer. Mineralogical

**Table 1** Information about clays and some of their properties

| Sample | Mineralogy | Content (%) | | CEC, C mol/kg | Magnetic susceptibility, (χ) ($10^{-8}$ $m^3$ $kg^{-1}$) | OC (%) |
|---|---|---|---|---|---|---|
| | | $Fe_2O_3$ | $K_2O$ | | | |
| Palygorskite (Moscow region, Russia) | Palygorskite, quartz | 4.35 | 1.12 | 20.50 | 11.4 | 0.1 |
| Bentonite SWy-2 | Montmorillonite, quartz, feldspars, calcite | 3.10 | 0.50 | 94.53 | 10.47 | 0.86 |
| Illite (Hungary) | Illite, quartz, microcline | 0.60 | 6.20 | 12.84 | 0 | 0.15 |
| Kaolin (Vimianzo, Spain) | Kaolinite, muscovite, quartz | 0.69 | 0.92 | 2.25 | 0.9 | 0.24 |
| Smectite–zeolite (Russia) | Montmorillonite, clinoptilolite, quartz, muscovite, calcite | 1.53 | 1.40 | 34.6 | 2.73 | 0.04 |

composition of clays was analyzed by X-ray diffraction (CuKα radiation). Oriented specimens were prepared by sedimentation on glass slides and step-scanned in steps 0.1° 2θ and 10 s/step counting time. Mg- and K-saturated clays were examined at room temperature, after ethylene glycol salvation and after heating to 300 and 550 °C. Concentrations of Fe, Si, Al in solutions were measured by AAS and K by flame photometry. The fossilization process was studied using scanning electron microscope (SEM). For the observation of dynamics of *Microcoleus* growth in the presence of clays, SEM-EDAX (CamScan with Link-860 and TESCAN VEGA 3 LMU with EDS and AZtec software) were used. FTIR spectra were obtained on a Thermo Scientific Nicolet 6700 FT-IR Spectrometer. In addition to the traditional transmission method (KBr-pellet techniques), modern reflectance techniques such as diffuse reflectance (DRIFT) for near-infrared (NIR) range was used. Spectra manipulations were performed using the Thermo Scientific OMNIC TM software package. Instead of the more commonly used peak-fitting procedures of FTIR spectra, the second derivative formalism (with OMNIC defaulted parameters) was employed to enhance the resolution of the sharp features for NIR range.

The content of chlorophyll-a was detected by spectrophotometric analysis at 665 nm wavelength after extraction in ethanol (Gerasimenko et al. 1989). The velocity of oxygen flow was determined by polyarographic method using Clark—electrode and 1 mm cell (Trebst 1980). Organic matter content in clays before and after incubation with cyanobacteria was studied on an Elementar Vario EL III CNS analyzer. The release of elements during the incubation period was detected with ICP-OES. The dissolved Si and other elements in the aqueous phase were used to monitor the dissolution of minerals.

## 2 Results and Discussion

Biological Activity of *Microcoleus chthonoplastes* in the presence of palygorskite.

For all experiments with palygorskite, the pH values were from 8.6 to 9.2–9.5 and similar for the systems with cyanobacteria and without (controls). Figure 1b shows the dynamics of chlorophyll-ɑ concentration and the oxygen flow for *Microcoleus chthonoplastes* in the presence of clays. The maximum of both parameters is observed for the first 7 days of experiments. After that, the gradual decrease of biological activity took place (Fig. 1b, c). During the first week, the growth of cyanobacteria was observed on palygorskite and their biomass increased near threefold. Photosynthetic activity was rather high (≈0.6 μmol $O_2$ mg $chl^{-1}$ $h^{-1}$), during further cultivation, these values decreased, smoothly for chlorophyll-ɑ and abruptly for oxygen. On day 14, photosynthetic activity was still rather high. On day 30, the drop of photosynthetic activity was especially abrupt. The cells remained viable for the first 14 days and then began to die. 28 days of incubation with cyanobacteria lead to the decrease of organic carbon and nitrogen content with the decline of biological activity. The obtained results allow distinguishing the

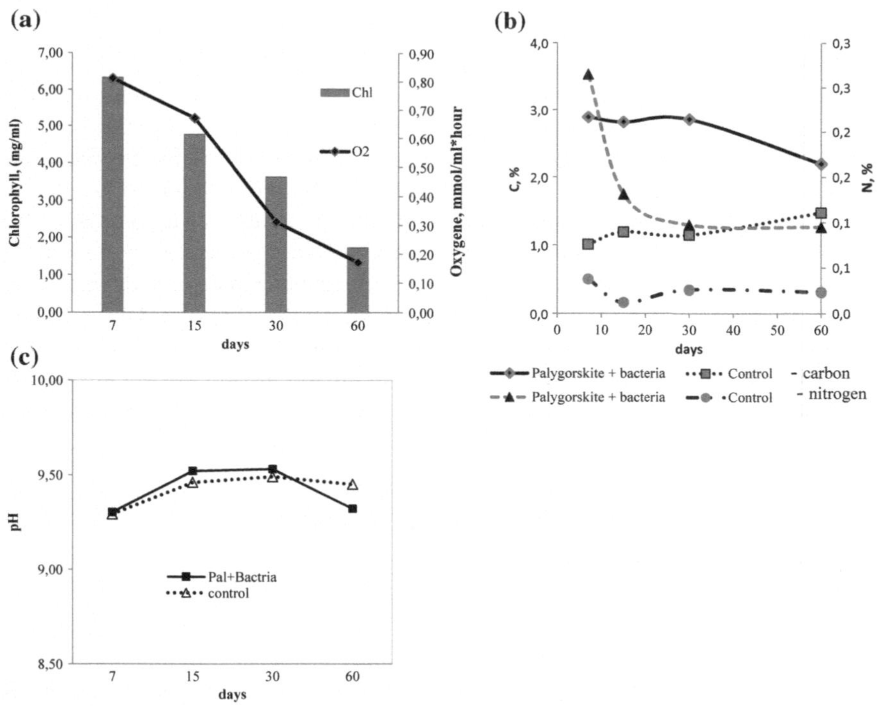


**Fig. 1** Biological activity of *Microcoleus chthonoplastes* on the palygorskite: **a** dynamics of chlorophyll ά concentration (mg/ml) and oxygen (mmol/ml/h); **b** dynamics of organic carbon an nitrogen content, **c** dynamics of pH

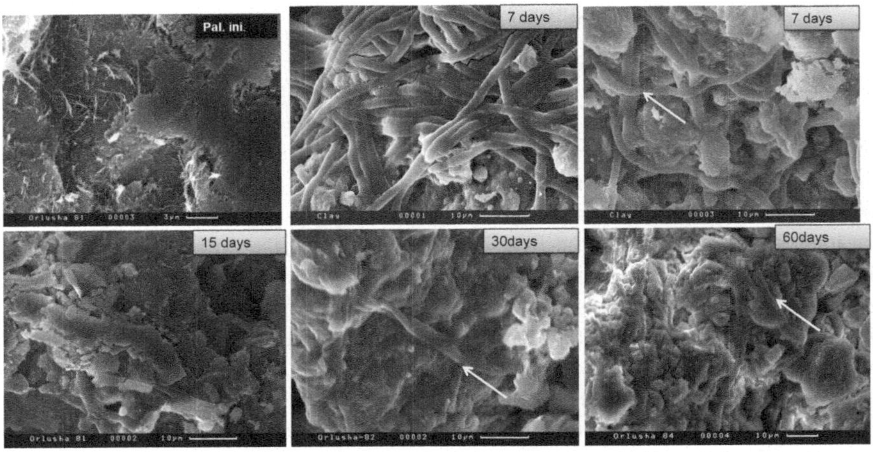


**Fig. 2** Interaction of cyanobacteria *Microcoleus chthonoplastes* with palygorskite after day 7, 14, 28, and 60 of the experiments (cyanobacterial filaments marked with *arrows*)

period of culture growth (7–30 days) and destruction of organic substances (after 14–28 days).

SEM micrographs show that during the first 14 days, the filaments of cyanobacteria actively developed on the clay surfaces and twined round the clay lumps maintaining normal morphology. After 2 weeks, a dense layered mat was formed (Fig. 2a) shows the layers of cyanobacteria with normal morphology and the layers of clay particles with fragmented cyanobacterial filaments shown in Fig. 2b. The interaction of cyanobacteria with clay minerals was accompanied by abundant glycocalyx in a form of extracellular polysaccharide sheath formation, which explains good structural integrity of the formed mat. On day 28, the filaments on the mat surface were covered with the particles of clay (Fig. 2c). On 60th day, trichomes became lithified and were hardly recognized on the surface of clays (Fig. 2d). The similar processes were observed in experiments of incubation of cyanobacteria *Oscillatoria terebriformis* in the presence of palygorskite (Fig. 3).

## 2.1 Changes in Clay Minerals After Incubation with Cyanobacteria

The experiments with palygorskite and *Microcoleus chthonoplastes* showed that in the course of interaction of bacteria with clay it is not noted essential changes in structure of a palygorskite according to a XRD (Fig. 4). Only partial congruent dissolution of a mineral was observed. This fact was confirmed by the decrease of macroelements (Si, Al, Fe) in the mixture of clay with bacteria during the all experiment in a comparison with control experiments (Fig. 5). Bacteria, as have

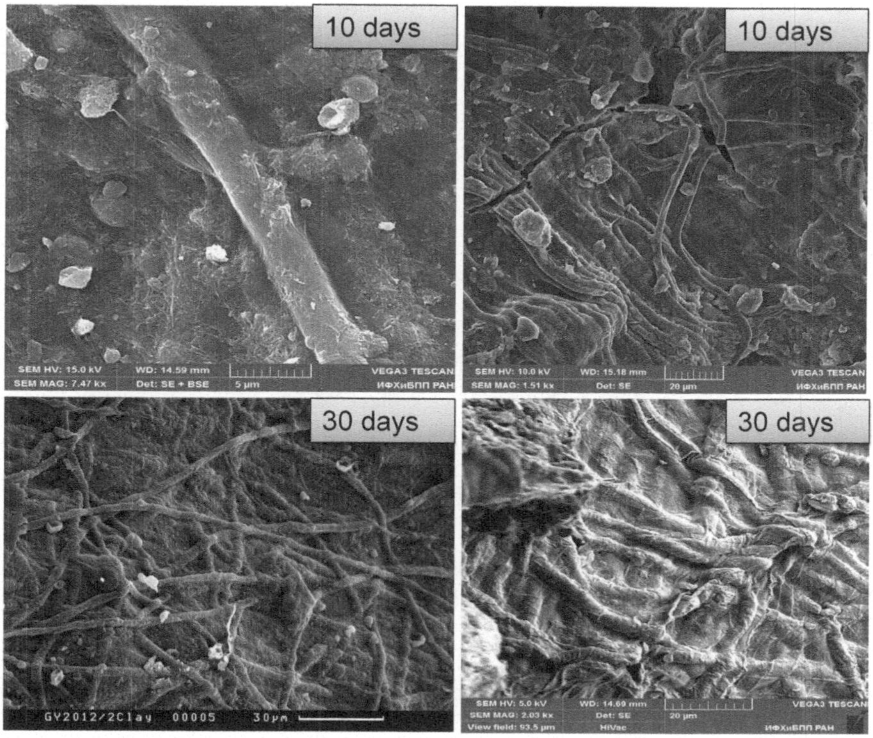


**Fig. 3** Interaction of cyanobacteria *Oscillatoria terebriformis* with palygorskite after day 10 and 30 of the experiments (cyanobacterial filaments marked with *arrows*)

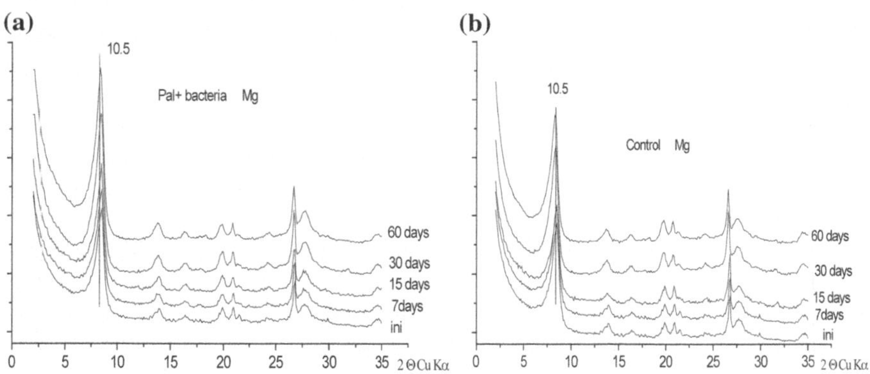


**Fig. 4** XRD patterns of palygorskite after day 7, 14, 28, and 60 of the experiment with cyanobacteria *Microcoleus chthonoplastes*: **a** samples with cyanobacteria, Mg-saturated; (spacing values are given in Å) **b** control samples Mg-saturated

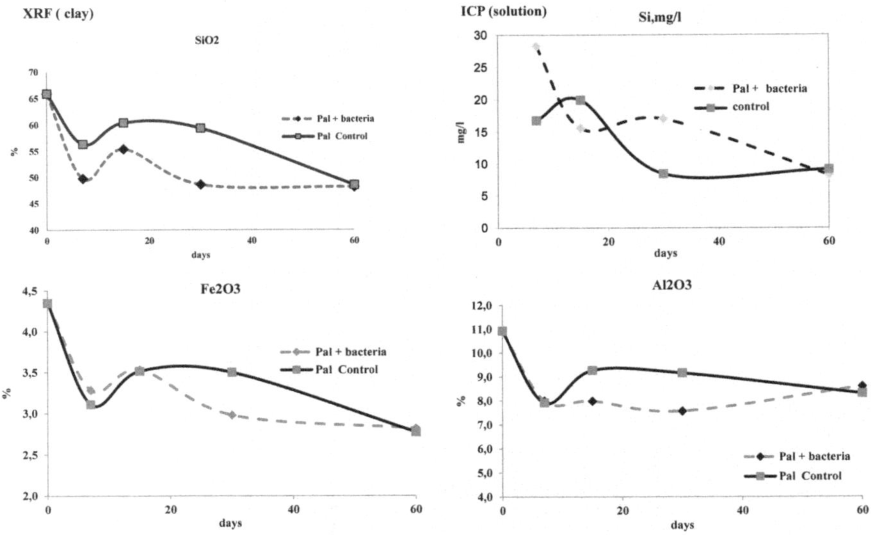


**Fig. 5** The content of elements in palygorskite after the interaction of clays with cyanobacteria: XRF data for clay samples after interaction with bacteria, 0 days—initial palygorskite. ICP data for solutions

been shown, accelerate the dissolution of silicates in some cases by the production of hydroxyl anion, extracellular polysaccharides (EPS) (Hiebert and Bennett 1992; Vandevivere et al. 1994). The organisms affect mineral weathering reactions by producing $CO_2$ and the formation of the weak carbonic acid ($H_2CO_3$) can also result in solubilization of silicates (Gadd 2010). Alkaline conditions can mobilize silicon from silicates (Ehrlich and Newman 2009).

FTIR was used to get the information about structural changes, which might occur with palygorskite lattice under the impact of the *Microcoleus chthonoplastes*. The analysis of FTIR spectra also demonstrates the possible congruent dissolution of palygorskite. This finding is confirmed by the decrease of intensity of the bands at ca. 1125, 1090, 1020, and 975 $cm^{-1}$ attributed to the asymmetric stretching modes of the tetrahedral silicate sheet (Si–O–Si, the strongest Si–O stretching mode at 975 $cm^{-1}$) (not shown). Summary of the proposed near- and mid-infrared assignments bands for palygorskite are given in Gionis et al. (2006). In addition to the spectra obtained in the MIR region, the spectra measured in the NIR region also provide the useful information about clay minerals since the observed bands related to the vibrations of OH groups are sensitively affected by the variations in the mineral structure (Madejová et al. 2011). The application of NIR allowed to get information on structural OH-groups at 7093 $cm^{-1}$, and a multiplet at 4503, 4320, 4280, 4102, and 4050 $cm^{-1}$ (the combination modes of hydroxyls in Mg–OH, Al–OH, and Fe–OH). No changes in the positions of these hydroxyls were observed after incubation with *Microcoleus chthonoplastes* (Fig. 6). Under the influence of

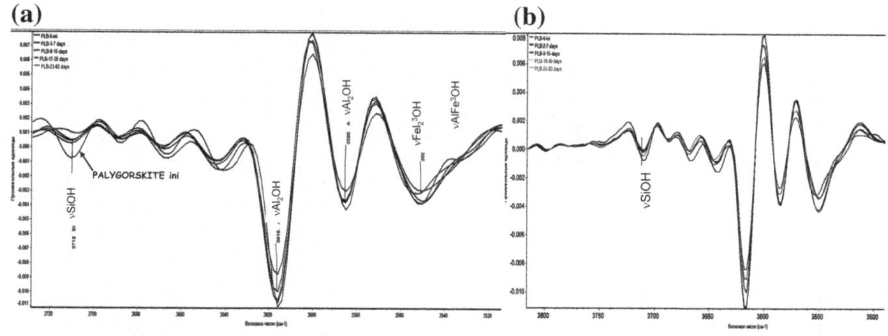


**Fig. 6** Near-infrared spectra (FTIR) of palygorskite, their second derivatives: **a** Samples with cyanobacteria after day 7, 14, 28, and 60 of the experiment, **b** control samples after day 7, 14, 28, and 60 of the experiment

cyanobacteria, we observed the decrease of the intensity of silanol groups (Si–OH outer surface at 3712 $cm^{-1}$) of palygorskite. It is necessary to take into account that according to Kuang et al. (2004) ~8 % of Si nuclei in palygorskite exist as Si–OH.

The experiments with palygorskite and *Oscillatoria terebriformis* demonstrated no changes in structure of a palygorskite according to a XRD data (Fig. 7). But in

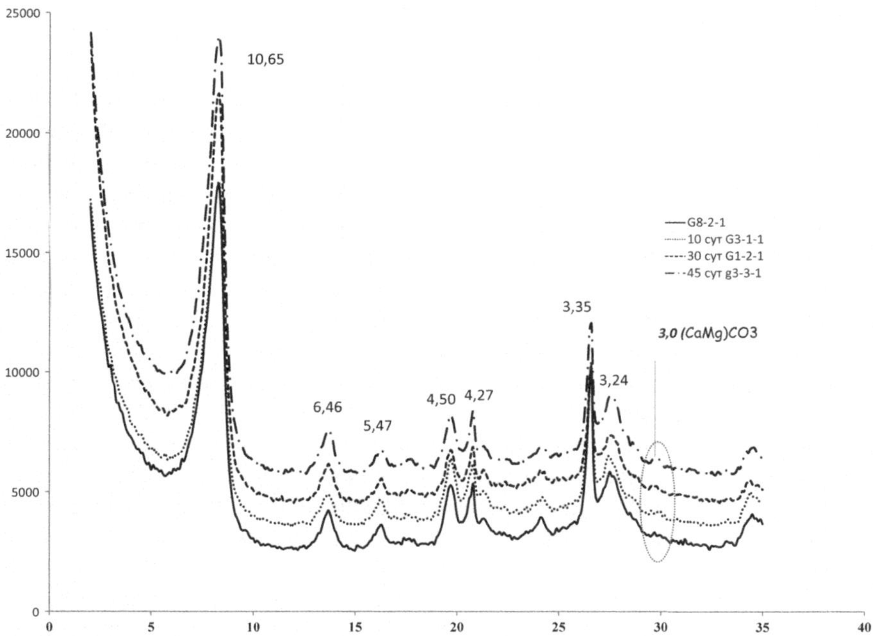


**Fig. 7** XRD patterns of palygorskite after day 7, 14, 28, and 60 of the experiment with cyanobacteria *Oscillatoria terebriformis* (spacing values are given in Å.)

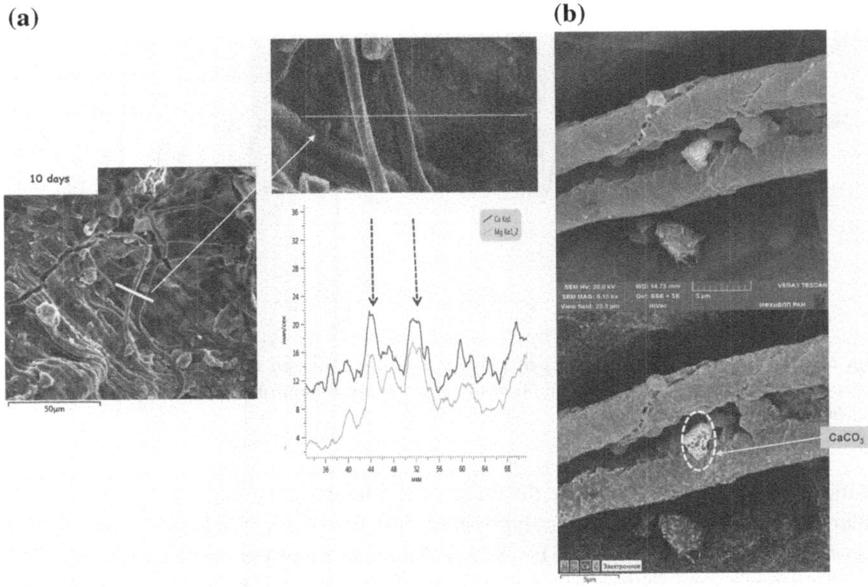


**Fig. 8** Scanning Electron Microscope—SEM/EDS microanalysis data for palygorskite after day 10 and 30 of the experiment with cyanobacteria *Oscillatoria terebriformis*. **a** Microgeochemical profiling of elements distribution (Ca, Mg), 10 days. **b** Map of elements distribution with carbonate microconcretions (30 days of experiments)

comparison with *Microcoleus chthonoplastes* at a stage of a bacteria fossilization (30–45 days) the formation of Mg-calcite is observed. Microgeochemical profiling of elements distribution obtained by SEM–EDS also showed the increased concentrations of Ca and Mg at the contact of bacterian trichomes with clay (Fig. 8a). Additionally, carbonate microconcretions were found on trichomes of *Oscillatoria terebriformis* (Fig. 8b). FTIR spectroscopy also demonstrates the formation of calcium carbonate in the experiments of palygorskite, bentonite, and kaolinite after incubation with *Microcoleus chthonoplastes* (Fig. 9).

The described processes of clays transformations were noted in our previous studies of the influence of *Microcoleus chthonoplastes* on the properties of different clays: kaolin, illite, bentonite, and zeolite (Alekseeva et al. 2008, 2009). The obtained results showed that kaolin and bentonite exhibit the maximum stability with the insignificant transformations of their mineralogical composition besides the formation of Mg-calcite in the presence of kaolin after 28 days of incubation (at the fossilization stage). XRD data showed that transformation of smectite–zeolite exhibited in the formation of a defect "island" layer in the interlayer space of montmorillonite on day 28 of incubation and is probably associated with a potassium substitution by the molecules of hydroxides (e.g., A1) or their complexes with organic anions. The defect layer was not developed in the control experiments

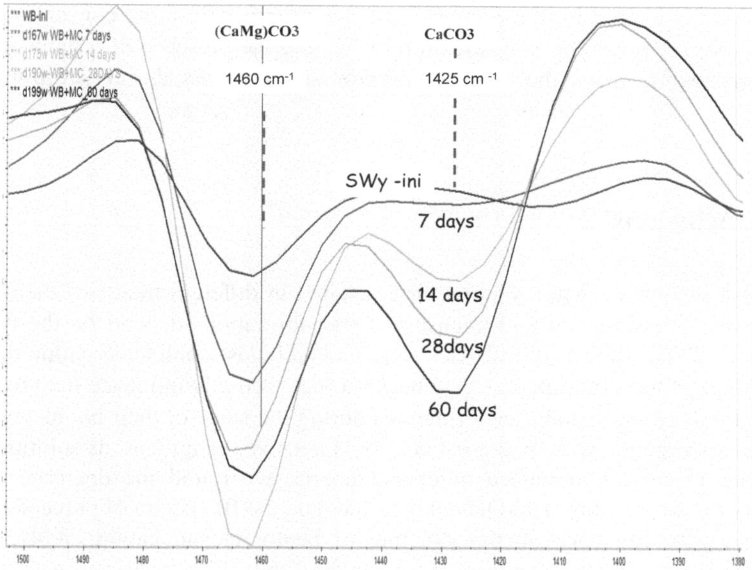


**Fig. 9** Example of FTIR spectra of Bentonite SWy-2. Samples with cyanobacteria after day 7, 14, 28, and 60 of the experiment of incubation *Microcoleus chthonoplastes*

(without cyanobacteria). Previously it was shown (Ransom et al. 1999) that the swelling minerals of the smectite type in biological systems can act as a "catch pit" for the accumulation of ammonium and other metabolic products. The most significant changes in the mineralogical composition were found in the experiments with illite. Dioctahedral illite from the control experiments is characterized by the following series of XRD peaks: 10.5–10.6; 5.07; and 3.34 Å. After treatment with ethylene glycol, some layers swelled to 11.33 Å, which corresponds to the content of the smectite phase within the range of 17–26 %. Thus, cyanobacteria contributed to the release of a part of potassium from the mica lattice and increase the number of swelling smectite-type layers, which was registered on days 7 and 14 by the appearance of an additional XRD peak at 17.33 Å after ethylene glycol saturation of a specimen.

As a rule, clay minerals possess paramagnetic properties. Magnetic properties of clays are caused by impurities of iron oxide minerals that have ferromagnetic properties (dispersed forms of magnetite or maghemite, hematite, and iron hydroxides). *Microcoleus chthonoplastes* visibly influenced the properties of iron compounds. Both dissolution and precipitation processes were observed. Kaolin and smectite–zeolite demonstrated the increasing of magnetic susceptibility and magnetization values which connected with the precipitation of Fe in the form of metastable ferrihydrite ($5Fe_2O_3 \cdot 9H_2O$) following by the formation of more stable goethite (α-FeOOH). In our experiments, the formation of poorly ordered ferrihydrate is caused by the properties of the system with large concentration of oxygen that promotes the fast oxidation of Fe, and the presence of organic matter, which, probably, plays the role of inhibitor and prevents the crystallization of goethite.

## 3 Conclusions

The clays incubation with cyanobacteria resulted in different trends of their transformations. The direction and intensity of transformations depend on the type of clay mineral. The observed processes were relatively fast-completed within the first 10–60 days of the experiment. Cyanobacteria most actively influence the processes of mineral dissolution and transformations during the stage of their photosynthesis. In the experiments with palygorskite, the intensive congruent dissolution was observed. Under the influence of cyanobacteria, we found the decrease of the intensity of sialon groups (Si–OH outer surface) on the IR spectra of palygorskite. It had no visible influence on the structure of bentonite and kaolin. Whereas the development of extra smectitic layers within the illite matrices was found. Cyanobacteria have the visible influence on the properties of iron compounds, which are present in all the studied clays as impurities. We may conclude that *Microcoleus chthonoplaste* intensifies process of dissolution in iron hydroxides on the one hand and affects the low temperature autigenic mineral formation on the other hand. Formation of carbonate in the experiments with palygorskite, bentonite, and kaolinite was observed at the stage of cyanobacteria fossilization.

**Acknowledgments** The investigation was supported by Program N28 of the Russian Academy of Sciences «Biosphere Origin and Evolution».

## References

Alekseeva T, Gerasimenko L, Sapova E, Alekseev A (2008) Transformational changes of clays due to cyanobacteria (Chap. 17). In: Dobretsov N et al (eds) Biosphere: origin and evolution. Springer, Berlin, pp 227–238

Alekseeva T, Sapova E, Gerasimenko L, Alekseev A (2009) Transformation of clay minerals caused by an alkaliphilic cyanobacterial community. Microbiology 78(6):776–784

Banfield JP, Barker WW, Welch SA, Taunton A (1999) Biological impact on mineral dissolution: application of the lichen model to understanding mineral weathering in the rhizosphere. Proc Nat Acad Sci USA 96:3404–3411

Barker WW, Welch SA, Chu S, Banfield JF (1998) Experimental observations of the effects of bacteria on aluminosilicate weathering. Am Mineral 83:1551–1563

Bennett PC, Hiebert FK, Choi WJ (1996) Microbial colonization and weathering of silicates in a petroleum-contaminated groundwater. Chem Geol 132:45–53

Bennett PC, Rogers JR, Choi WJ, Hiebert FK (2001) Silicates, silicate weathering, and microbial ecology. Geomicrobiol J 18:3–19

Dong H (2012) Clay–microbe interactions and implications for environmental mitigation. Elements 8:113–118

Dubinin AV, Gerasimenko LM, Gusev MV (1992) Physiological characteristics of Microcoleus chthonoplastes from hyper saline lake. Microbiology 61(1):63–69 (in Russian)

Ehrlich HL, Newman DK (2009) Geomicrobiology, 5th edn. CRC Press, Taylor & Francis, Boca Raton, FL

Gadd GM (2010) Metals, minerals and microbes: geomicrobiology and bioremediation. Microbiology 156:609–643

Geptner AR, Ivanovskaya TA (1998) Biochemogenic genesis of the glauconite-nontronite series minerals in present-day sediments of the Pacific ocean. Lithol Min Resour 33(6):503–517

Geptner A, Kristmannsdottir H, Kristjansson J, Marteinsson V (2002) Biogenic saponite from an active submarine hot spring, Iceland. Clays Clay Miner 50(2):174–185

Gerasimenko LM, Nekrasova VK, Orleansky VK, Venetskaya SL, Zavarzin GA (1989) The primary production of halophilic Cyanobacterium cenoses. Microbiology 58(3):507–514 (in Russian)

Gerasimenko LM, Dubinin AV, Zavarzin GA (1996) Alkaliphilic cyanobacteria from soda lakes of Tuva and their ecophysiology. Microbiology 65(6):736–740

Gerbersdorf SU, Wieprecht S (2015) Biostabilization of cohesive sediments: revisiting the role of abiotic conditions, physiology and diversity of microbes, polymeric secretion, and biofilm architecture. Geobiology 13:68–97

Gionis V, Kacandes GH, Kastritis ID, Chryssikos GD (2006) On the structure of palygorskite by mid- and near-infrared spectroscopy. Am Mineral 91(7):1125–1133

Grantham MC, Dove PM (1996) Investigation of bacterial–mineral interactions using fluid tapping mode atomic force microscopy. Geochim Cosmochim Acta 60:2473–2480

Harrison J, Turner RJ, Marques L, Ceri H (2005) A new understanding of these mcirobial communities is driving a revolution that may transform the science of microbiology. Am Sci 93:508–515

Hiebert FK, Bennett PC (1992) Microbial control of silicate weathering in organic-rich ground water. Science 258:278–281

Kalinowski BE, Liermann LJ, Brantley SL, Barnes A, Pantano CG (2000) X-ray photoelectron evidence for bacteria enhanced dissolution of hornblende. Geochim Cosmochim Acta 64:1331–1343

Kawano M, Tomita K (2002) Microbiotic formation of silicate minerals in the weathering environment of a pyroclastic deposit. Clays Clay Miner 50(1):99–110

Konhauser KO, Urrutia MM (1999) Bacterial clay authigenesis: a common biogeochemical process. Chem Geol 161: 399–413

Kuang W, Glenn AF, Detellier C (2004) Dehydration and rehydration of Palygorskite and the influence of water on the nanopores. Clays Clay Miner 52: 635–642

Likhoshway EV, Sorokovikova EG, Bel'kova NL, Belykh OI, Titov AT, Sakirko MV, Parfenova VV (2006) Silicon mineralization in the culture of cyanobacteria from hot springs. Doklady Biol Sci 407:201–205

Lowenstam HA (1981) Minerals formed by organisms. Science 211:1126–1130

Madejová J, Jankovič L, Pentrák M, Komadel P (2011) Benefits of near-infrared spectroscopy for characterization of selected organo-montmorillonites. Vib Spectrosc 57(1):8–14

Phoenix VR, Adams DG, Konhauser KO (2000) Cyanonacterial viability during hydrothermal biomineralisation. Chem Geol 169:329–338

Ransom B, Bennet RH, Baerwald R, Hulbert MH, Burkett P-J (1999) In situ conditions and interactions between microbes and minerals in fine-grained marine sediments: a TEM microfabric perspective. Am Mineral 84:183–192

Rozanov AY (2003) Fossil bacteria, sedimentogenesis, and the early biospheric evolution. Paleontol Zh 39(6):41–49 [Paleontol J (Engl. Transl.) 39(6):600–609]

Singer A (1989) Palygorskite and sepiolite group minerals, in Dixon JB , Weed SB (eds.) Minerals in soil environments, Soil Science Society of America, Madison, WI, 2nd Ed., p. 829–872

Schultze-Lam S, Forris FG, Konhauser KO, Wiese RG (1995) In situ silicification of an Icelandic hot spring microbial mat: implications for microfossil formation. Can J Earth Sci 32:2021–2026

Tazaki K (1998) A New world in the science of biomineralization—environmental biomineralization in microbial mats in Japan. The Science Reports of Kanazawa University, Japan. XLII (N 1, 2)

Tazaki K, Okrugin V, Okuno M, Belkova N, Islam AR, Chaerun SK, Wakimoto R, Sato K, Moriichi S (2003) *Heavy metallic concentration in microbial mats found at hydrothermal systems, Kamchatka, Russia.* Science reports of Kanazawa University, Japan. 47 (N 1–2)

Thorseth IH, Furnes H, Tumyr O (1995) Textural and chemical effects of bacterial activity on basaltic glass: an experimental approach. Chem Geol 119:139–160

Trebst A (1980) Inhibitors in electron flow: tools for the functional and structural localization of carriers and energy conversation sites. Methods Enzymol 69:675–715

Ueshima M, Tazaki K (2001) Possible role of microbial polysaccharides in nontronite formation. Clays Clay Miner 49(4):292–299

Ushatinskaya GT, Zaitseva LV, Orleansky VK, Gerasimenko LM (2006) Significance of bacteria in natural and experimental sedimentation of carbonates, phosphates and silicates. Paleontol J 40(Suppl 4):524–531

Vandevivere P, Welch SA, Ullman WJ, Kirchman DL (1994) Enhanced dissolution of silicate minerals by bacteria at near-neutral pH. Microb Ecol 27:241–251

Zavarzin GA (1993) Development of microbial community in the history of Earth. In: Problems of before anthropogenic evolution of biosphere. Nauka, Moscow, pp 212–222 (in Russian)

# Biomineralization Processes During the Formation of Modern Oceanic Sulfide Ore and Ore-bearing Sediments

**Irina F. Gablina, Irina G. Dobretsova and Evgenia A. Popova**

**Abstract** Sulfide ores were investigated along with ore-bearing and metalliferous sediments of the hydrothermal fields in the northern near-equatorial Mid-Atlantic Ridge (MAR) zone: Semenov (13°30–31′N), Ashadze-1 (12°58′N), Zenit-Victoria (20°08′N), and Peterburgskoe fields (19°52′N), discovered during legs 26, 32, and 33 of the R/V *Professor Logachev* FSUE PMGE. Biogenic carbonate and background sediments of this region were also examined. Lithological, biostratigraphic, and geochemical physical-chemical investigations methods were used. Mineragraphic and precision structural and chemical research of typomorphic minerals were carried out at various stages of lithogenesis. It was found out that most sulfide constructions in the Zenit-Victoria and Peterburgskoe fields, as well as the eastern field of the Semenov cluster, are located in biogenic carbonate sediments of the Holocene and Late Pleistocene ages and represent a new type of sulfide mineralization, unknown earlier in the MAR zone. This mineralization was formed by metasomatic replacement of biogenic carbonate sediments by ore minerals, simultaneously with diffuse percolating of hydrothermal solutions through the sediments.

**Keywords** Ocean · Mid-Atlantic ridge · Modern sulfide ores · Ore-bearing sediments · Biomineralization · Hydrothermal · Metasomatic

I.F. Gablina (✉)
Geological Institute RAS, Moscow, Russia
e-mail: gablina@ilran.ru

I.G. Dobretsova
Polar Marine Geological Prospecting Expedition, St. Petersburg—Lomonosov, Russia
e-mail: ocean-party@peterlink.ru

E.A. Popova
I.S. Gramberg VNIIOkeangeologia, St. Petersburg, Russia

 
O.V. Frank-Kamenetskaya et al. (eds.), *Biogenic—Abiogenic Interactions in Natural and Anthropogenic Systems*, Lecture Notes in Earth System Sciences,
DOI 10.1007/978-3-319-24987-2_5

# 1 Introduction

Massive sulfide ores from the ocean are chimneylike ore mounds (black smokers) formed under highly forced hydrothermal fluids, which are vented on the seafloor in various geological settings, mainly in spreading zones of the mid-oceanic ridges. Sulfide mineralization hosted in recent sediments has been discovered near continents and in intracontinental rifts, such as in the Red Sea.

It is suggested that most ore material is supplied into marine sediments from hydrothermal plumes, which form in the ocean above active hydrothermal vents. The effect of hydrothermal solutions on sediments while the solutions seep through them has been much less studied.

Ore-bearing sediments formed under the effect of hydrothermal solutions have been studied most completely in the Red Sea rift zone (Baturin 1971; Pushelt and Laschek 1984; Bogdanov et al. 1986; Butuzova 1998, 2003) and in the rifts adjoining the North American continental block in the west: the Guaymas Basin in the Bay of California, the Middle Valley of the Juan de Fuca Ridge, Escanaba Trough, etc. in the Pacific Ocean (Curray et al. 1982; *Gidrotermal'nye* ... 1993; Goodfellow and Franklin 1982; Kurnosov et al. 1994; Zierenberg et al. 1993; Lisitsyn et al. 1989; *Geologicheskoe* ... 1990; Bogdanov et al. 1989, 2006). These processes have remained poorly studied until now in mid-ocean ridges, which are distant from continents and covered only with thin sedimentary cover.

We believe that the main reason for this situation is the generally agreed opinion that the formation of ore-bearing sediments predominantly occurs by sedimentation (due to destruction of surface hydrothermal structures and precipitation of the ore substance from suspensions of hydrothermal plumes). Another cause is the unilateral approach to studying ore-bearing and metalliferous sediments: commonly, specialists (lithologists, paleontologists, geochemists) explore their own field of expertise, applying methods used in that field. These methods do not always allow them to determine the nature of mineral components. These components can be hydrothermal-sedimentary (precipitating from hydrothermal plumes or formed by destruction of hydrothermal mounds and vents), authigenic (appearing in situ as products of postsedimentation transformations, including hydrothermal metasomatic alteration), terrigenous (coming from continents), and edaphogenic (products of ocean bed destruction).

Massive sulfide deposits—ancient analogues of the contemporary oceanic massive sulfide ores—are thick stratiform bodies embedded, as a rule, in sedimentary and volcano-sedimentary rocks (Maslennikov 2006). This is similar to other type of deposits (cuprous sandstones and shales) formed under underground waters that emanated to submarine bottom sediments (Hoyningen-Huene 1963; Lur'e and Gablina 1972; Lur'e 1988; Rose 1976; White 1971). Both types are characterized by high concentrations of copper, lead, zinc, and rare elements and are the main sources for them. Enrichment in metals is explained by protracted circulation of ore-bearing fluids in sediments that serve as a barrier for metals (Gablina 1997), whereas venting of hydrothermal ore-bearing fluids in seawater results in their dispersion. Less than 5 % in the initial stage and up to 50 % in further periods are precipitated in sulfide mounds, which are formed on the seafloor near hydrothermal vents (Rona 1986; Gurvich 1998). The destruction of unstable massive sulfide ores under seawater began just after their formation (*Gidrotermal'nye*... 1993; Gablina et al. 2000).

In recent years, we and other researchers (Gablina et al. 2010, 2011, 2012; Dobretsova and Laiba 2009; Rusakov et al. 2013) obtained new data about the metasomatic effect of hydrothermal solutions on modern biogenic carbonate sediments of the Ashadze-1, Eastern (Semenov ore cluster), Zenit-Victoria, and Peterburgskoe ore fields of the Mid-Atlantic ridge (MAR), which led to the substitution of microfaunal shells of sediments by sulfides and other hydrothermal minerals, as well as to the formation of sulfide structures in them. This information is very important from both scientific and practical points of view, as the seepage of hydrotherms through the sediments might evoke precipitation of a larger part of the net load as well as the formation and preservation from destruction of a rich accumulations of sulfide ores.

## 2 Materials and Methods

Sulfide ores were investigated along with ore-bearing and metalliferous sediments of hydrothermal fields in the northern near-equatorial MAR zone: Semenov (13°30–31′N), Ashadze-1 (12°58′N), Zenit-Victoria (20°08′N), and Peterburgskoe fields (19°52′N), discovered during legs 26, 32, 33 of the R/V *Professor Logachev* FSUE PMGE, and carbonate, background sediments of this region. The investigations are based on a systematic approach to studies of the conditions of mineral and ore formation, which consists of a combination of lithological, biostratigraphic, and geochemical investigations. Mineragraphic, precision structures, and chemical research of typomorphic minerals, ore inclusive, were conducted at various stages of lithogenesis. The findings were compared with the data of direct measurements of Eh-pH parameters of ore-bearing and background sediments obtained during the leg. The main object of the research is the biogenic material of sediments, the mineral composition of which is uniform at the stage of sedimentogenesis within the ocean zone under consideration; these are calcite shells of dead plankton and benthos. Their different transformations may be associated with various conditions of diagenesis or superimposed processes.

**Fig. 1** The studied area. The studied subjects are earmarked

## 3 Characterization of the Work Region

Ore fields are situated at the boards of the MAR rift valley in the area from 12° 58.4’с.ш (Ashadze-1 field) to 20°08′N (Zenit-Victoria field) at depths from ~2400 to 4200 m (Fig. 1). Ore fields are composed of basalts, gabbro, peridotites, carbonate, metalliferous, and ore-bearing sediments with inclusions of sulfide ores.

The considered ore fields occupy a specific tectonic position: they are all situated in rift valley walls, with the Peterburgskoe field being found in the second ridge of the wall. Hydrothermal fields are characterized by developing metalliferous ($Fe_{cfs}$[1] $\geq 10\ \%$, $(Cu + Zn)_{cfs} < 0.25\ \%$), ore-bearing ($Fe_{cfs} \geq 10\ \%$, $(Cu + Zn)_{cfs} \geq 0.25\ \%$), and carbonate ($Fe_{cfs} < 10\ \%$, $(Cu + Zn)_{cfs} < 0.25\ \%$; $CaCO_3 \geq 50–70\ \%$) sediments. The thickness of carbonate sediments in this MAR sector does not exceed 1.5–3 m; ore-bearing sediments are no more than 0.5–0.6 m. Unaltered sediments are represented by calcite shells of foraminifers, coccolithes, and pteropods with admixture of clay and edaphogen material. According to the

[1]/cfs—based on carbonate-free substance.

fauna assemblage, the age of the sediments is Holocene-Late Pleistocene (0–128 ka) (Gablina et al. 2011; Rusakov et al. 2011; Shilov et al. 2012).

The Ashadze-1 hydrothermal field is represented by two ore bodies with relict and active hydrothermal constructions and ore-bearing sediments. The Semenov ore cluster contains both active constructions (North-Western hydrothermal field) and massive ores embedded in sediments (Eastern hydrothermal field). No active constructions have been revealed in the Zenit-Victoria and Peterburgskoe hydrothermal fields, whereas sulfide formations were recorded in topography in the form of small hills. The majority of the investigated sites are characterized by hydrophysical anomalies that indicate continuing hydrothermal activity. The age of sulfide ores, dated by the isotopic Th/U metod, ranges from <2–60 ka (Zenit-Victoria) to 37–124 ka (Semenov) (Shilov et al. 2012; Rusakov et al. 2013).

Most sulfide constructions in the Zenit-Victoria, Peterburgskoe, and Eastern fields of the Semenov cluster are located in sediments and represent a new type of sulfide mineralization, unknown earlier in the MAR zone. The mineralization was formed by metasomatic replacement of organic carbonate sediments by hydrothermal minerals, simultaneously with diffuse percolating of hydrothermal solutions through the sediments (Gablina et al. 2012; Rusakov et al. 2013).

# 4 Results

Sulfide ores have been most completely studied in Zenit-Victoria and Peterburgskoe hydrothermal fields. Here the ores are represented by scattered sulfide constructions composed of a porous pyrite-marcasite or sphalerite-marcasite (in Cu–Zn–Fe ores) framework. Chalcopyrite, sphalerite, copper sulfides, and, in rare cases, bornite develop in the most porous places of the framework. Opal, barite, and Fe–Si gel are most abundant among nonmetallic hydrothermal minerals.

Investigations have shown that sulfide constructions in sediments of the Zenit-Victoria and Peterburgskoe fields have certain features that distinguish them from tubular bodies of "black smokers" described in the literature, which were generated on the surface of the ocean floor (Gablina et al. 2012). Those features include the following:

1. Unusual shape: streamlined pear- and mushroom-shaped, platy stratified (Fig. 2).
2. Multichannel inner structure characteristic of "diffusers" (Fig. 2b).
3. Hardening crusts in the roof and on the walls of structures—lithified sediments, commonly situated at channel outputs (Fig. 2a, c–f).
4. Presence of relics of sedimentary layering in platy structures (Fig. 2f).
5. Widely spread relics of microfossils replaced by ore minerals (Fig. 3).
6. Conformity of the height of structures with the thickness of enclosing sediments.

**Fig. 2** Morphology and the inner structure of sulfide edifices. Zenit-Victoria field: **a** a pear-shaped construction with a crust of lithified sediments in the roof, st. 33L70; **b** a multichannel sulfide construction ("diffuser"), st. 33L68; **c** a fragment of platy stratified edifice, st. 33L69. Peterburgskoe field, st. 33L159: **d** a large construction of stratiform type with a channel in the floor and a crust of lithified sediments in the roof; **e** a small construction with a crust of lithified sediments in the roof; **f** a fragment of sulfide construction of stratiform type with relics of the stratified texture and with a crust of lithified sediments

7. The presence of aragonite crusts with Fe and Mn hydroxides on the surface of sediments that contain sulfide formations or aragonite zones at the contact of sediments with sulfide structures (Fig. 4).

Within hydrothermal fields, metalliferous and ore-bearing sediments are spread in the form of spots, which is not consistent with the idea of a uniform precipitation of ore particles from hydrothermal plumes. They differ from background ones in their physical-chemical characteristics, by elevated content of ore components (Fe, Cu, Zn, and Mn), Si, Mg, metasomatic replacement of calcite tests by hydrothermal minerals, and development of near-ore mineral-geochemical zones.

Lateral and vertical zonation was established in the distribution of major rock- and ore-forming elements and secondary (hydrothermal) minerals in the sediments of the Ashadze-1 field. Hydrothermal minerals are represented by silicates (tremolite-actinolite, hornblende, serpentine, Fe–Mg-smectite, and, probably, polygorskite [sepiolite]), carbonates (Mg–Mn-bearing siderite), sulfides (pyrite, isocubanite), sulphates (millerite, barite, celestine), chlorides (atacamite, halite), and Fe and Mn hydroxides, which metasomatically substitute calcite in microfossil tests and form porous and crustified cement of the sediments. Pyrite, siderite, and Fe hydroxide and sulphate are most widespread newly formed minerals.

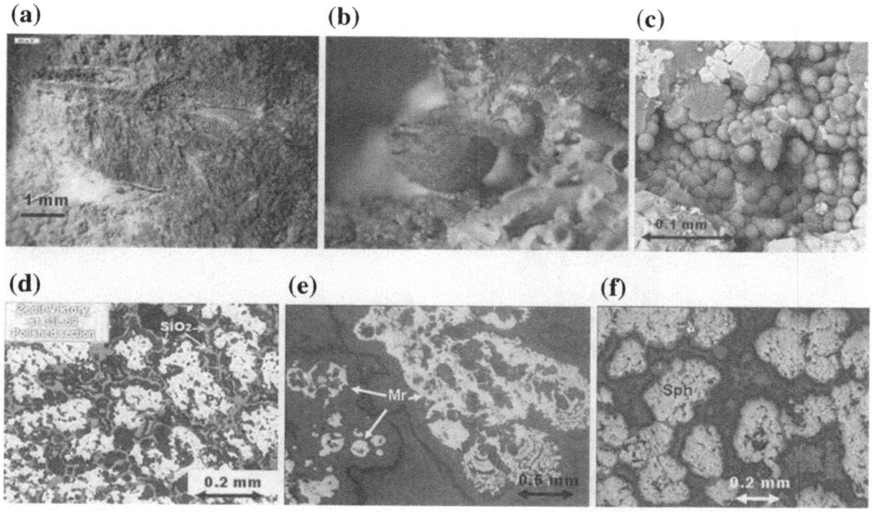


**Fig. 3** Relics of microfossils replaced by ore minerals. Microphotograph of the samples under the binocular: **a** pteropod imprints in chalcopyrite-sphalerite ores, Zenit-Victory hydrothermal field, st. 33L69; **b** opal pseudomorph after pteropod shell at the channel mouth, white below on the *right*—unaltered calcite shells, pyrite-chalcopyrite ores with Cu-sulfides, opal, Peterburgskoe field, st. 33L159. Zenit-Victory hydrothermal field, st. 33L69: **c** accumulations of opal pseudomorphs after foraminifers in the cavities of pyrite-chalcopyrite ores, electron microscopy; **d** microphotograph of a polished section of pyrite (Py) pseudomorphs after foraminiferas opal was crusted over; **e** microphotograph of a polished section of marcasite (Mr) pseudomorphs after foraminiferas in $SiO_2$ in pyrite-marcasite ores; **f** microphotograph of a polished section of sphalerite (Sph) pseudomorphs after foraminiferas in $SiO_2$ in pyrite-chalcopyrite-sphalerite ores

The following zones are identified according to the composition of the dominant secondary minerals:

(1) A sulfide zone, which coincides with ore bodies
(2) A zone of minerals with elevated magnesia content, which partly coincides with ore body 1, going beyond its contours in the west and southwest
3) A zone with developing Fe–Mn crusts along the periphery of ore bodies 1 and 2.

Vertically, sulfide paragenesis is substituted by the hydroxide-ferruginous. Farther away from the ore bodies, the processes of dissolution and replacement of carbonate shells by secondary minerals in sediments fade (Gablina et al. 2010, 2011).

The sediments of the Semenov field also exhibit metasomatic substitution of foraminiferal tests by Fe and Cu sulfides and near-ore hydrothermal alterations. Vertical zonation is also recorded in the distribution of hydrothermal minerals in host rocks. In the upward direction, from the roof of the sulfide deposit toward the surface of the ocean floor, the sulfide-barite association progressively gives

**Fig. 4** Aragonite genesis. **a**, **b** Crusts covered with a shade of Mn hydroxides on the surface of the sediment, which contains sulfide edifices (Peterburgskoe field, st. 33L159). **c**, **d** Zone at the contact of pyrite (Py) ores and lithified sediment in the roof of edifice (Semenov field, st. 32L357)

way to the hydroxide-ferruginous-sulfide-barite, barite-hydroxide-ferruginous, and atacamite-hydroxide-ferruginous associations (Rusakov et al. 2013). This zonation reflects an increase in the oxidation potential of porous solutions of sediments up the section as a result of the growing influence of seawater.

Ore-bearing sediments and host sulfide constructions of the Zenit-Victoria and Peterburgskoe fields were intensely altered, which is reflected in their lithification at the contact with ores, the development of secondary minerals in them (Fig. 5), a sharp decrease in Eh and pH values with respect to background, and an increase of porous water temperature. Simultaneously, the oxidation potential decreases (to negative values) while the acidity increases down the section, clearly indicating that acid-reducing hydrothermal fluids arrive to the sediments from below (Gablina et al. 2012).

In general, the following characteristics are typical of the studied objects:

1) Reduction in the thickness of sediments occurring above sulfide constructions or enclosing them (probably due to dissolution of primary biogenic calcite).

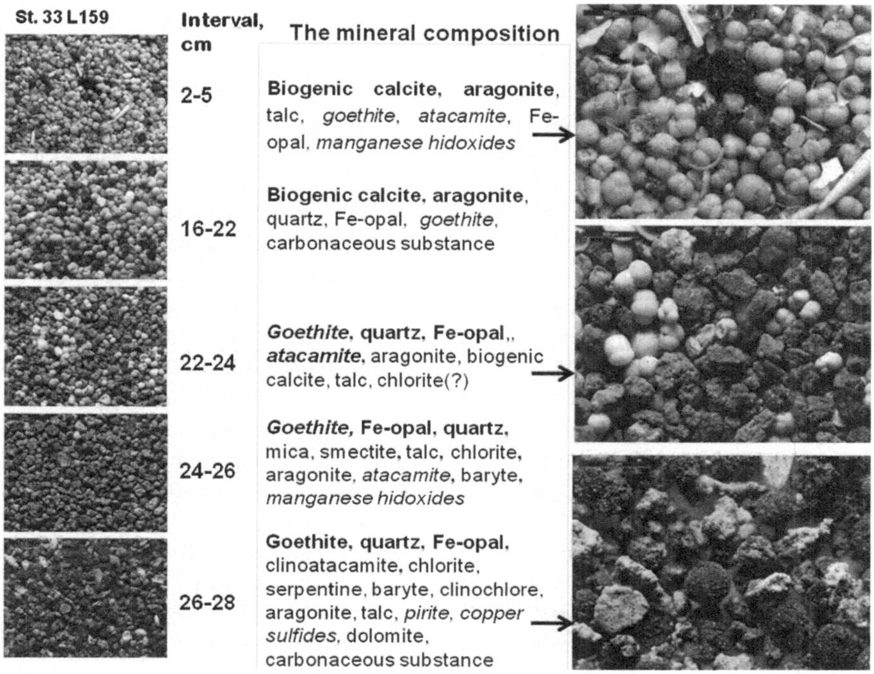


**Fig. 5** Vertical mineral zonation of biogenic carbonate sediments in the zone of near-ore alterations in the Peterburgskoe hydrothermal field (core st. 33L159). A fraction contains 0.25–0.5 mm of sediments. The mineral composition is based on the data of mineragraphic, electron microscopic, electron microprobe, and X-ray investigations. *Left* Microphotograph of the samples under the binocular. *Right* Their enlarged fragments. Predominant minerals are given in *bold type*, with mineral indicators of the physical-chemical environment shown in *italics*

2) Development of pseudomorphs of ore and other hydrothermal minerals after microfauna tests in ore-bearing sediments (Gablina et al. 2011).
3) Vertical and lateral mineral-geochemical zonation in ore-bearing sediments around ore bodies; the sedimentary section, from bottom to top, reflects an increase in the oxidation potential and pH up the sedimentary column and at the flanks of the ore fields (Fig. 5).
4) A sharp difference of ore-bearing sediments from background sediments revealed by the physical and chemical characteristics of their interstitial waters; Eh and pH of ore-bearing sediments are close to the physical-chemical parameters of hydrothermal solutions (Gablina et al. 2012).

# 5 Conclusions

The described shapes and the inner structure of sulfide constructions ("diffusers") suggest that their formation occurred under incomplete permeability of the environment, which hampered the spread of fluids (i.e., the diffusion mechanism of flow promotion). In sulfide ores, the presence of relics of the layered sedimentary structure and pseudomorphs of ore minerals after microfauna tests suggest the metasomatic origin of ore constructions. The biogenic calcite that composed the sediments dissolved during the test and was replaced by hydrothermal minerals. The ore-forming process can be associated with the influence on sediments of acid metalliferous low pressure fluids of diffusion type, which arrive from below through permeable zones in basalts. Mineralization occurred on the geochemical barrier in the zone of interaction among metalliferous fluids, seawater saturating the sediments, and biogenic sedimentary calcite that is unstable in the acidic environment. Carbonate bottom mud served both as a geochemical barrier (with the presence of reactive carbonate material) and a physical barrier, making rapid migration of ore-bearing solutions difficult and encouraging precipitation and accumulation of useful components. At the initial stages or at low pressure of fluids (e.g., in the "exocontact" of the already formed constructions), metasomatic replacement of isolated tests leads to the formation of ore-bearing and/or metalliferous sediments. The prolonged influence of hydrotherms brings about the complete replacement of sedimentary biogenic material and the formation of sulfide constructions in its place. The recycling of dissolved biogenic calcite occurs on the surface of sediments (the sea floor) in aragonite crusts, enriched in Fe and Mn hydroxides, or at the contact of sediments with sulfide structures (Fig. 4). The presence of hardening crusts of sediments in the roof and on the walls of the sulfide constructions indicates the increased temperature of leaking fluids. The occurrence of isocubanite among the newly formed hydrothermal minerals indicates a high temperature (>260 °C) of hydrothermal solution supplied to sediments (Vaughan and Craig 1978).

Our conclusions concerning superimposed metallization on Holocene-Late Pleistocene (0–128 ka) bottom sediments do not contradict the data on the absolute age of sulfide ores (from less than 2 to ~124 ka)[2] and the availability of recent hydrophysical anomalies above the Zenit-Victoria ore field, which indicate ongoing hydrothermal activity there (Shilov et al. 2012).

**Acknowledgments** The material used in our studies was acquired during cruises of the R.V. *Professor Logatchev*, organized by the Polar Marine Geological Prospecting Expedition and financed by the Federal Agency of Subsurface and Natural Resources and Ecology of the Russian Federation. This work was financially supported by RFBR grants 08-05-00,799, 11-05-01117, and 14-05-00480.

[2]The dating in the Petersburgskoe ore field of 176.2 ± 59 ka (Shilov et al. 2012) needs to be clarified.

## References


Baturin GN (1971) Deepwater ore sediments of hydrothermal genesis. In: Zenkevich NA (ed) History of the World Ocean. Nauka, Moscow (in Russian)

Bogdanov YuA, Gurvich EG, Butuzova GYu et al (1986) Metalliferous sediments of the Red Sea. Nauka, Moscow (in Russian)

Bogdanov YuA, Khvorova IV, Serova VV, Gorbunova ZN (1989) Sedimentation in rift zone of the Juan de Fuca Ridge. Izv Akad Nauk USSR Ser Geol 5:26–35 (in Russian)

Bogdanov YuA, Lisitsyn AP, Sagalevich EG, Gurvich EG (2006) Hydrothermal ore formation on oceanic bottom. Nauka, Moscow (in Russian)

Butuzova GYu (1998) Hydrothermal-sedimentary ore formation in rift zone of the Red Sea. GEOS, Moscow (in Russian)

Butuzova GYu (2003) Hydrothermal-sedimentary ore formation in World Ocean. GEOS, Moscow (in Russian)

Curray JR, Moore DC et al (1982) Initial reports of DSDP, D.C.: U.S. Gov. Print. Off., vol. 64. Washington

Dobretsova IG, Laiba AA (2009) Formation of mineral carbonate in hydrothermal sulfide ore in MAR. In: Lisitsyn AP (ed) Geologiya morei i okeanov: Proceedings of the 18th International. Scientific Conference (School) on Marine Geology. 2:155–159. GEOS, Moscow (in Russian)

Gablina IF (1997) Formation condition of largy cupriferrous sandstone and shale deposits. Geol Ore Dep 39, 4: 320–333

Gablina IF, Mozgova NN, Borodaev, Yu S et al (2000) Copper sulfide associations in recent oceanic ores of the Logachev hydrothermal field (Mid-Atlantic Ridge, 14° 45′N). Geol Ore Dep 42, 4: 296–316

Gablina IF, Popova EA, Sadchikova TA et al (2010) Hydrothermal alterations of modern organic sediments at the Ashadze-1 hydrothermal field, Mid-Atlantic Ridge, 13° N. Dokl. Earth Sci 433(2):998–1002

Gablina IF, Demina LL, Dmitrenko OB et al (2011) Composition and secondary alterations of microfossils in sediments of the Ashadze-1 hydrothermal field (tropical Mid-Atlantic Ridge). Oceanology 51(3):476–490

Gablina IF, Dobretsova IG, Bel'tenyov VE et al (2012) Peculiarities of presentday sulfide mineralization at 19°15′–20°08′N, Mid-Atlantic Ridge. Dokl Earth Sci 442, 2: 163–167

*Geology and Hydrothermal Activity of the Juan de Fuca Ridge* (1990). In: Lisitsyn AP (ed) Nauka, Moscow (in Russian)

*Hydrothermal Systems and Sedimentary Formations in Atlantic Mid-Ocean Ridges* (1993). Lisitsyn AP (ed), Nauka, Moscow (in Russian)

Goodfellow WD, Franklin JM (1982) Geology, mineralogy, and chemistry of sediment-hosted clastic massive sulfides in shallow cores, Middle Valley, Northern Juan de Fuca Ridge. Econ Geol 88:2037–2068

Gurvich EG (1998) Metalliferrous Sediments of the Ocean. Nauchnyi Mir, Moscow (in Russian)

Hoyningen-Huene E (1963) Zur Paläohidrologie des Oberrotligenden und Zechsteins in Harzvorland. Ber Geol Ges DDR S.-H. 1:201–220

Kurnosov V, Murdmaa I, Rosanova T et al (1994) Mineralogy and hydrothermally altered sediments and igneous rocks at Sites 856858, Middle Valley, Juan de Fuca Ridge, Leg 139. Proc ODP Sci Rep 139:113–131

Lisitsyn AP, Bogdanov YuA, Zonenshain LP et al (1989) Black smokers in the Bay of California. Izv Akad Nauk USSR Ser Geol 5:3–18 (in Russian)

Lur'e AM (1988) Genesis of Cuprous Sandstones and Shales. Nauka, Moscow (in Russian)

Lur'e AM, IF Gablina (1972) Sources of copper for the formation of mansfield-type deposits in the Western Ural Region. Geochem Int 10, 1:75–88

Maslennikov VV (2006) Lithogenesis and massive sulfide formation. IMin UB RAS, Miass (in Russian)

Pushelt H, Laschek D (1984) Marine Erzvorkommen in Roten Meer, Fridericiana. Zeitschrift der Universitet at Karlsruheno 34:3–17

Rona P (1986) Hydrothermal Mineralization in Seafloor Spreading Centers. Earth. Science Reviews, Mir, Moscow 20(1):1–104 (in Russian)

Rose AW (1976) The effect of cupriferous chloride complexes in the origin of red-bed copper and related deposits. Econ Geol 71:1036–1048

Rusakov V, Shilov VV, Dobretzhva IG et al (2011) Litochemostratigraphical horisons of upper pleistocene-holocene sediments of the Semenov ore cluster. In: Main Results in Russian Study of the Mid-Atlantic Ridge Processes in the First Decade of XXI. Russian Ridge' 2011. IGEM RAS. Moscow, pp 68–70

Rusakov VYu, Shilov VV, Ryzhenko BN et al (2013) Mineralogical and geochemical zoning of sediments at the Semenov cluster of hydrothermal fields, 13°31′–13°30′ N, Mid-Atlantic Ridge. Geochem Int 51(8):646–669

Shilov VV, Beltenev VE, Ivanov VN et al (2012) New hydrothermal fields on Mid-Atlantic Ridge: Zenith-Victory (20°08′ с.ш.) and Peterburgskoe (19°52′ с.ш.). Doklady Earth Sci 442(1):63–69

Vaughan DJ, Craig JR (1978) Mineral chemistry of metal sulfides. Cambridge University Press, Cambridge

White WS (1971) A paleohydrologic model for mineralisation of the White Pine Copper Deposit, Norten Michigan. Econ Geol 66(1):1–13

Zierenberg R, Koski RA, Morton JL, Shanks WC (1993) Genesis of massive sulfide deposits on a sediment-covered spreading center, Escanaba Trough, Southern Gorda Ridge. Econ Geol 88 (8):2069–2098

# Biogenic–Abiogenic Interactions in Stromatolitic Geosystems and Their Mineralization

**Tatiana V. Litvinova**

**Abstract** Investigation of biogenic ultramicrotextures in stromatolites is a newer research avenue, which makes it possible to reconstruct this process on the basis of factual evidence and significantly improve current understanding of the stromatolitic system as a whole. New techniques utilizing the electron microscope allowed to establish a variety of mineralized biogenic ultramicroformations, responsible for occurrence of a particular rock microstructure and to determine their elementary composition. Biota significantly affects the development of stromatolitic reefs and their architecture. The cyanobacterial community is a self-learning biological system, flexibly adapting to climatic, geological, atmospheric, and other natural environments. The obtained results present stromatolites as a complexly organized geosystem, with various interacting components constantly exchanging matter, energy, and information.

**Keywords** Stromatolite reefs · Geosystem · Bacterial biofilms · Morphotypes of cyanobacteria · Construction

## 1 Introduction

While investigating stromatolitic edifices as geosystems, the following sections should be considered:

- Morphotypes of cyanobacteria in stromatolites.
- Stromatolite reef as a geosystem and its components, their direct and backward relations (Zavarzin and Rozhnov 2011).
- Evolution of the ecological community in stromatolites, its characteristics and stability (Awramik et al. 1983) and other.

T.V. Litvinova (✉)
Geological Institute of the Russian Academy of Sciences, Moscow, Russia
e-mail: litvinova-geo@rambler.ru

 
O.V. Frank-Kamenetskaya et al. (eds.), *Biogenic—Abiogenic Interactions in Natural and Anthropogenic Systems*, Lecture Notes in Earth System Sciences,
DOI 10.1007/978-3-319-24987-2_6

- Spatiotemporal setting of stromatolite geosystems as global structural systems on the planet and the role of these systems in its evolution (Semikhatov and Raaben 1994, 1996 and other).

Items 2–4 have been addressed in the literature in more or less detail, but the morphotypes of stromatolite builders have never been studied so far, except for few publications (Litvinova 2013, 2014; Naugolnych and Litvinova 2014). However, biota has a profound effect on the formation of the stromatolite reef as a whole and plays a primary role in the creation of the edifice architecture (Bertrand-Sarfati 1972; Zavarzin and Rozhnov 2011). In this paper the author tried to depict them and offer their provisional classification based on evidence obtained from stromatolites of various ages and setting.

## 2 Discussion

Recall that stromatolites are thin-bedded rocks, generated as a result of interaction between cyanobacteria and sedimentation. They appeared 3.5 billion years ago, were widespread in the Precambrian, encountered in the Phanerozoic; their formation is still going on. Their structure and distribution were studied, their formal classification was carried out on the basis of the morphology of constructions and their microstructure; however, the mechanism of their formation can be understood if the contribution of stromatolite builders will also be taken into account. Stromatolites are classified into stratiform, columnar, and nodular (Krylov 1975; Semikhatov 1974), with these types usually occurring in this exact order in the vertical sections of the edifices (listed in order from bottom to top). Their structure and distribution were studied, formal classification based on the morphology and microstructure of the buildings was proposed, though the mechanism of formation of these rocks can be understood only if we evaluate the contribution stromatolite builders. Nonetheless, the mechanism forming these rocks can be elucidated only by estimating the contribution of stromatolite builders. It is proposed that the only reason for the diversity of stromatolite edifice morphologies is environmental conditions not the activity of various living organisms (Grotzinger and Knoll 1999). However, this is hard to agree with this taking into consideration that formal stromatolite types were the first stratigarphic benchmarks of the Precambrian, and each of them marked an exact stage of geological time.

The application of the new technique, involving the scanning electron microscope TesScan MV 2300, allowed us to distinguish diverse mineralized biogenic ultramicrounits that were responsible for the formation of certain microtextures of the rock and to determine their chemical composition. These were distinguished by their morphologies, which were different from those of mineral aggregations and, hence, could be considered as a identification guide in a fossil state (Zavarzin 1993). The application of the energy dispersion spectrometer (EDS) Inca 200 for pollination of samples with gold makes it possible to reveal elevated contents of carbon in them and the presence of some biophile elements, which differentiate organisms from their host rock, mostly composed of pelitomorphic carbonate or silico-carbonate material and thus eliminates any doubt concerning their genesis. Below we consider 14 morphotypes of biogenic ultramicrotextures, which were recognized by the author when studying stromatolites of various ages.

1. Bacterial biofilms, complex and simple, with wrinkled surface (Fig. 1a) and complex (Fig. 1b, d) bacterial slimes. Their crumpled surfaces appeared due to desiccation during sediment lithification. Such formations are widespread in stromatolites of different age.
2. Lower biofilms are often associated with coccoid bacteria (Fig. 1c).
3. Filamentous bacteria (Fig. 1i), which formed ultramicrostructures due to the substitution by silica (Fig. 1f) and their tracks (Fig. 2a).
4. Fairly large endolithic algae (Fig. 2b–d) and formed with their contribution microtextures (Fig. 2i, f). They introduced into the calcareous sediment and due to the release of organic acids, dissolved it. They are found only in Lower Proterozoic stromatolites of Karelia.
5. Rod-shaped organisms (Fig. 3a).
6. Tubes, which often have an aperture inside (Fig. 3b).
7. Geometrically regular spherules (Fig. 3c).
8. Single relatively large coccoid bacteria (Fig. 3d).
9. Colonies of small coccoid bacteria (Fig. 3I, f).
10. Carbon clots (Fig. 4a), the origin of which is not quite clear. Such particles are often found in Precambrian stromatolites. They are almost entirely composed of carbon.
11. Complex formations of diverse forms (Fig. 4b–f), each is confined to stromatolites of a definite age (see Captions)
12. Arrow-shaped constructions (Fig. 5a) in which one of the spikes enters the round hole of the other. Such construction can include up to eight elements-arrows (Fig. 5c).
13. Codweb-shaped coccoid bacteria preserved in some of their round holes (Fig. 5b, i) 12–13 morphotypes were known from the Permian time.
14. Iron-depositing bacteria form colonies and never occur as individual organisms (Fig. 5d, f). They might occur in stromatolites of various ages, though commonly they are spread locally within one sedimentation basin.

The obtained results present stromatolites as a complexly organized geosystem, with various interacting components constantly exchanging matter, energy, and information. The changing composition of the community directly influenced the architecture of the reefs. But the construction itself, that is, the geographical environment in which they were formed, hydrodynamics of seawater from which to extract calcium, carbon, trace elements, and their concentration in the pool, bottom topography, lighting mode, changing water levels—all this created geochemical conditions that can affect the development of the community and its biodiversity.

A thin mucous biofilm with diverse green-blue algae was formed on the mineral sediment. It rapidly grew and covered an increasingly wider area. The community

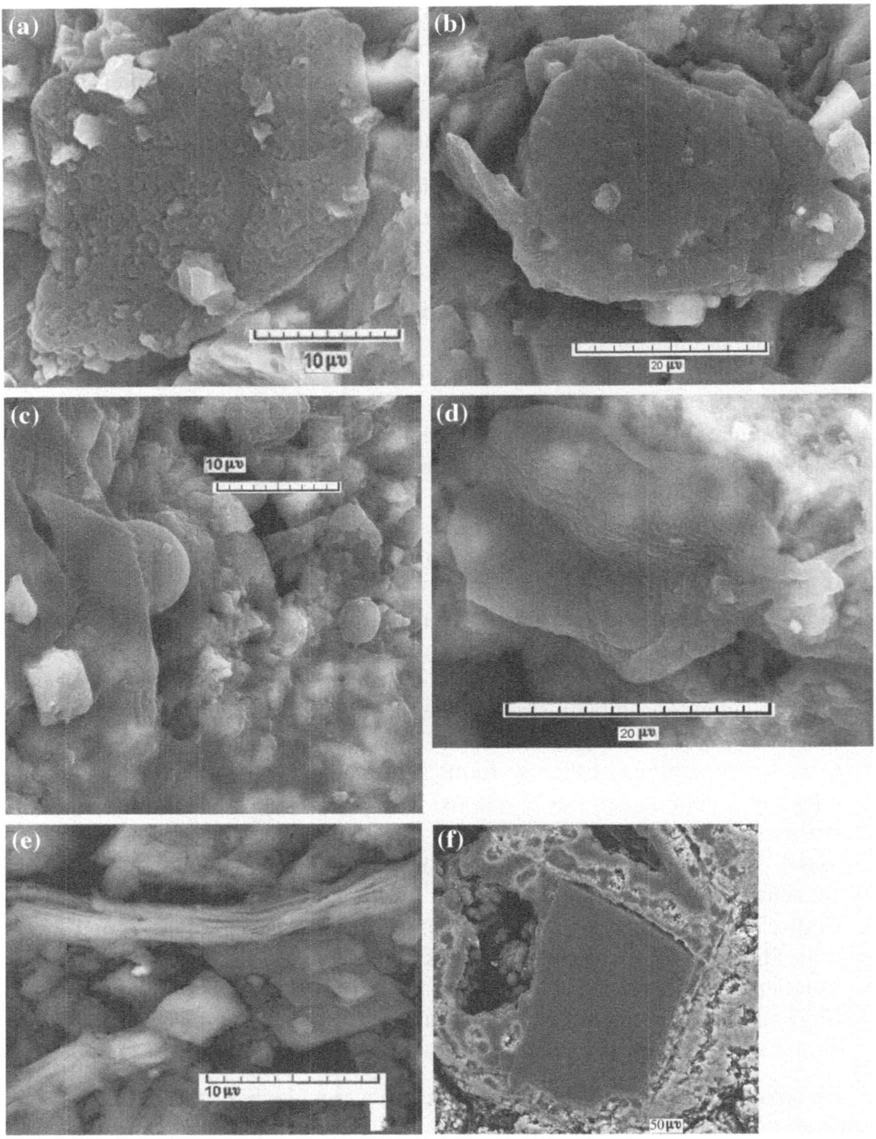


**Fig. 1** *Morphotype 1*: **a** (Permian fore-Urals, Middle Permian), **b** (Karelia, Paleoproterozoic), **d** (Shark Bay, Australia, modern)—simple and complex bacterial films with a crumpled surface; *morphotype 2*: **c** blanket biofilm with coccoid bacterium (Kerch Peninsula, Early Neogene); *morphotype 3*: **i** filamentous bacteria (Northern Anabar region, Early Riphean), **f** ultramicrostructure, formed filamentous bacteria (Karelia, Paleoproterozoic)

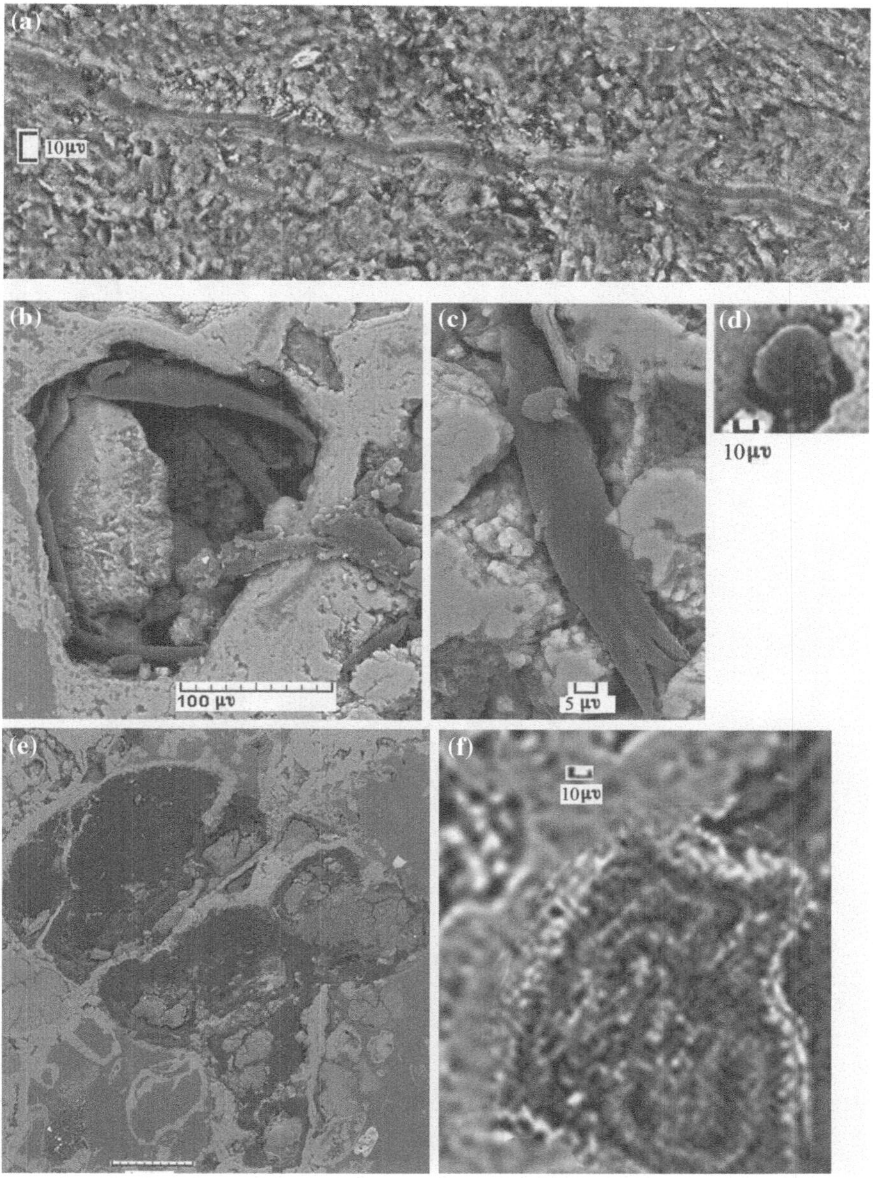


**Fig. 2** *Morphotype 3*: **a** traces of filamentous bacteria (Northern Anabar region, Early Riphean); *morphotype 4:* **b–d** algae (Karelia, Paleoproterozoic); **i**, **f** ultramicrostructures formed by algae (Karelia, Paleoproterozoic)

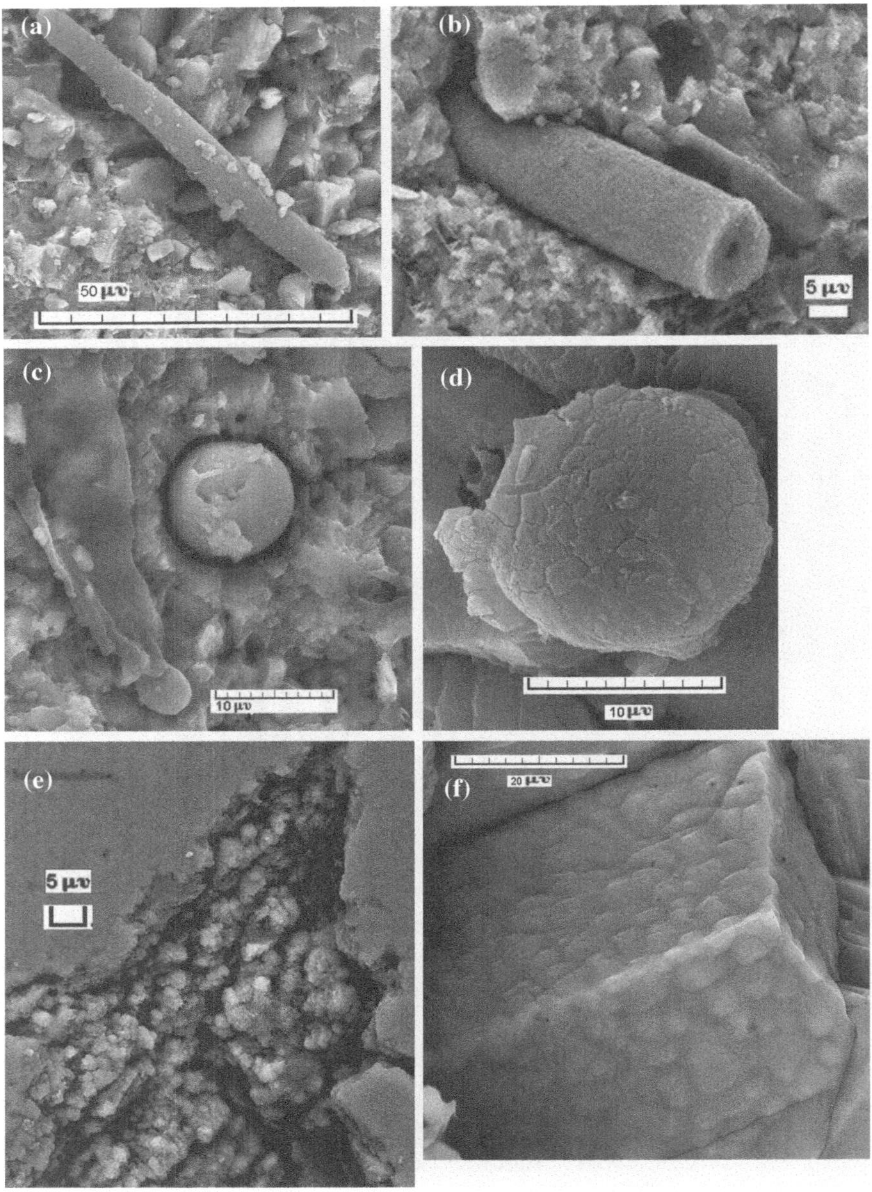


**Fig. 3** *Morphotype 5*: **a** rod-shaped organisms (Olenek, Yakutia, Early Cambrian); *morphotype 6*: **b** tube with aperture inside (Kerch Peninsula, Early Neogene); *morphotype 7*: **c** regular balls (Kerch Peninsula, Early Neogene); *morphotype 8*: **d** coccoid bacterium (Karelia, Paleoproterozoic); *morphotype 9*: **i**, **f** colonies of coccoid bacteria (Karelia, Paleoproterozoic)

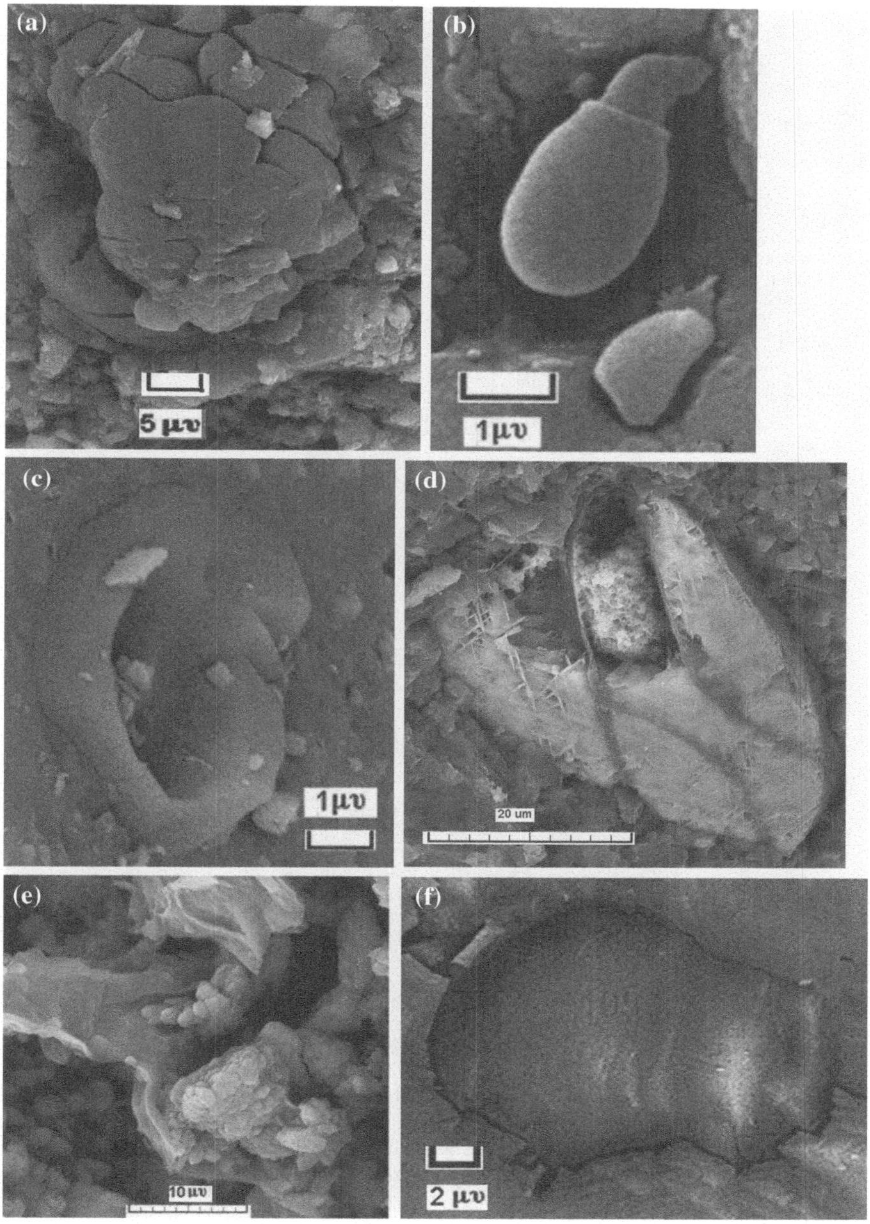


**Fig. 4** *Morphotype 10*: **a** carbon clots (Northern fore-Urals, Early Permian); *morphotype 11*: **b** (Southern Ural, Early Riphean), **c**, **f** (Northern Anabar region, Early Riphean), **d** (Northern fore-Urals, Early Permian), **i** (Kerch Peninsula, Early Neogene)—single complex formation of various forms

**Fig. 5** *Morphotype 12*: **a** arrows, **c** construction, where a spike of one of them enters in other (Permian fore-Urals, Middle Permian); *morphotype 13*: **b** (Kerch Peninsula, Early Neogene), **i** (Northern fore-Urals, Early Permian)—webs with coccoid bacteria; *morphotype 14*: **d** (Permian fore-Urals, Middle Permian), **f** (Karelia, Paleoproterozoic)—iron-depositing bacteria

extracted carbon and other biophile elements from the environment, thus providing mutual penetration of components of the geosystem. When colony reached a maximal size, the substrate was depleted, and the community lost its nutrient medium. The concentration of macro- and microelements, necessary for cell growth, without which organisms could not perform their essential life functions,

**Fig. 6** **a** Stromatolitic reef (Northern Anabar region, Early Riphean); **b** embedded stromatolites turns into columnar ones; **c** columns branching (Permian fore-Urals, Middle Permian); **d** nodular stromatolite (Karelia, Paleoproterozoic); **i** oncolites (Kazakhstan, Small Kara-Tau, Early Cambrian)

decreased; their growth and reproduction ceased. Withdrawal of carbon dioxide by organisms increased the importance of water pH, so accelerating the sedimentation process. Direct connection—the vital activity of cyanobacteria—disturbed its dynamic equilibrium, while the feedback connection, through precipitation, weakened the impact of organisms on the geosystem and stabilized the situation for a certain time (Prigogine and Stengaers 2014). Therefore, not only direct, but also feedback connection existed between components, the latter amplifying the cyclical process and initiating the self-regulation of the geosystem.

During sediment precipitation, the gliding trichomes rapidly moved to a new feeding substrate and established a colony on its surface. The biocenosis evolved by selecting and transforming organisms, which allowed them to survive under varying external conditions. The colony accumulated and used its earlier experience, but the physical and chemical characteristics varied within limited ranges (Zavarzin and Rozhnov 2011). A successive change in the reproduction of the bacterial films and precipitation of mineral sediment led to the formation of thin-bedded rocks. Such stromatolitic reefs may extend for several kilometers (Fig. 6a). The occurred unfavorable conditions prevented further integration of the community: the bacterial film, which had initially covered the entire surface of the mineral sediment, becomes unable to unite. Organisms colonized «islands», which are separated by variable amounts of sedimentary material. This process in the sequences manifested itself as the replacement of stratiform stromatolites by their columnar forms. Certain differentiation of edifices may be observed within a single sample (Fig. 6b). At the regressive stage of the basin existence, the volume of constructions and sedimentation intensity notably decreased; there appeared formations not attached to the substrate, which were capable of passive dislocation, and which further generated oncolites (Fig. 6i). The evolutionary process worked out incredible flexibility of the stromatolite geosystem and its components, which ensured a relative autonomy of the boundaries and strengthened internal connections. The unity and interaction of biogenic and abiogenic components provided the independence of the system, formed its integrity and resistance of the construction to the influence of the environment and secured the spatial-time arrangement of its components, their order, and mutual arrangement. The community of stromatolite reefs is a complex biosystem, whose texture and shape are mainly formed by organisms (edificators).

When outer impacts disturbed the vertical and lateral structures of the geosystem and the interconnections among their components, it was gradually losing its ability to recover and degraded.

## 3 Conclusion

1. Stromatolite reefs were complicated geosystems, which included biogenic and abiogenic components. Their unity and interaction ensured the independence of the system, its integrity, resistance to environmental influences, and maintained

the spatiotemporal organization of its components, their ordering and mutual arrangement.

2. The geosystem had numerous direct and reverse between its components at various levels, which alternately lead to the stabilization of the situation in the basin and the growth of the bacterial film, or to a disturbance of the dynamic equilibrium, activation of sedimentation, and consequently, burial of the community. This led to the accumulation of finely intercalating biogenic and mineral sediments, transformed by lithification and later secondary processes into a rock that shows the typical texture-structure features.
3. The long-lasting stable evolution of the biocenosis within the reef was secured by selection and replacement of organisms in subsequent generations and their adjustment to changing outer conditions. The cyanobacterial community as a whole was a "learning" and an adaptable system, constantly exchanging its material, energy, and information with the environment. This enabled this system to flexibly integrate itself into the climatic, geological, atmospheric, and other natural conditions.
4. Stromatolite reefs, which were widespread in the Precambrian, were a geomorphologic barrier that was expressed in the seafloor topography and was affected by the biota. This barrier separated zones of different depth. These geosystems acted as sedimentary and geochemical traps and controlled the carbon dioxide cycle; carbon dioxide was partly incorporated into biogenic carbonate, and the rest of the carbon was released into the atmosphere; this not only affected the climate but also substantially contributed to the evolution of the atmosphere on our planet.

## References

Awramik SM, Schopf JW, Walter MR (1983) Filamentous fossil bacteria from the Archean of Western Australia. Precambr Res 20:357–374

Bertrand-Sarfati J (1972) Stromatolitescolumnaires of Précambriensupérieur Sahara nordoccidental. Arid Res Center Ser Geol Soc Notes 14:245 (In France)

Grotzinger JP, Knoll AH (1999) Stromatolites in precambrian carbonates: evolutional columns or environmental indexes? In: *Science* (ed). Annular review *Earth and Planetary*, vol. 27, pp 313–358

Krylov IN (1975) Stromatolites of the Riphean and Phanerozoic in the USSR. Science, Moscow (In Russian)

Litvinova TV (2013) Stromatolite builders and their importance in the formation of stromatolite structures in Karelia. In: Paleontological Institute RAS (ed) Algae in the evolution of the biosphere. Moscow (In Russian)

Litvinova TV (2014) Ultramicrostructures of the stromatolite reef in the Northern Anabar Region and their origin. Lithol Miner Resour Soc Notes 5:416–426 (In Russian)

Naugolnych SV, Litvinova TV (2014) Stromatolites from the Permian deposits of the Perm Cis-Urals: a new form-genus *Alebastrophyton* Naug. et Litvinova, Gen. nov. In: Media-Grand (ed) Palaeontology in the museum practice. Collection of the scientific articles. Moscow (In Russian)

Prigogine IP, Stengaers E (2014) Order from chaos: a new dialog between humans and nature. Editorial, Moscow (In Russian)
Semikhatov MA (1974) Stratigraphy and geochronology of Proterozoic. Science, Moscow (In Russian)
Semikhatov MA, Raaben ME (1994) Dynamics of the global diversity of Proterozoic stromatolites. Article 1: Northern Eurasia, China, and India. Strat Geol Correl 2:492–513
Semikhatov MA, Raaben ME (1996) Dynamics of the global diversity of Proterozoic stromatolites. Article II: Africa, Australia, North America, and general synthesis. Strat Geol Correl 4:24–50
Zavarzin GA (1993) Development of microbial communities in through Earth's history. In: *Science* (ed) Problems of preanthropogenic evolution of the biosphere. Moscow (In Russian)
Zavarzin GA, Rozhnov SV (2011) Reefs in evolution of geo—biological systems. Problems formulation. In: Paleontological Institute RAS (ed) Reef—forming system and reefs in the evolution of the biosphere. Moscow (In Russian)

# Biomineralization in Bauxitic Laterites of Modern and Paleotropics of Earth

**Anatoly D. Slukin, Nikolay S. Bortnikov, Elena A. Zhegallo, Lyubov V. Zaytseva, Anatoly P. Zhukhlistov, Andrey V. Mokhov and Natalia M. Boeva**

**Abstract** The top part of a lateritic profile including laterites *in sensu stricto*, bauxites and even zones of clay minerals (кaolinite, halloysite, montmorillonite), contains plentiful traces and products of interaction of a substratum with live organic substance and mortmasses. The surface of laterites is hidden under the cover of trees and bushes and the dense felt like a film weaved from filamentous fungus, roots of vegetation, and fossilized microorganisms. Dying off, they turned into biofilms, which is replaced with biominerals such as gibbsite-Al $(OH)_3$, goethite-$HFeO_2$, hematite-$\alpha$-$Fe_2O_3$, and halloysite-$Al_4(OH)_{10}[Si_4O_8](OH)_2 \cdot 4H_2O$. A zone of a biological pedoturbation is completely processed by the burrowing organisms. Earlier it was represented that they make only mechanical impact on rocks. We revealed that burrowing organisms, including worms, mill all minerals of a substratum, passing it through the digestive path, and satiate it with biochemically active substances. It causes dissolution of all minerals except rutile—$TiO_2$. Gibbsite, when passed through a digestive path, recrystallizes and forms surprisingly beautiful and perfect idiomorphic crystals filling the burrows of the worms and covering their walls. We managed to find the paradoxical phenomenon: passing of worms through large (2–4 cm) quartz crystals. The entrances are surrounded with a biofilm similar to slimes and turned into units of crystals of hematite. The described phenomena are established on numerous bauxite deposits of India, Guinea, and Brazil of the modern tropic and on the ancient deposits of Siberia.

**Keywords** Laterite · Bauxite · Microorganisms · Biological pedoturbation · Biomorphs · Gibbsite · Hematite · Quartz

---

A.D. Slukin · N.S. Bortnikov · A.P. Zhukhlistov · A.V. Mokhov · N.M. Boeva (✉)
IGEM of the Russian Academy of Sciences, Moscow, Russia
e-mail: boeva@igem.ru

E.A. Zhegallo · L.V. Zaytseva
PIN of the Russian Academy of Sciences, Moscow, Russia

 
O.V. Frank-Kamenetskaya et al. (eds.), *Biogenic—Abiogenic Interactions in Natural and Anthropogenic Systems*, Lecture Notes in Earth System Sciences,
DOI 10.1007/978-3-319-24987-2_7

# 1 Introduction

At present, the concept of laterite or cuirasse may be used to describe the final product of weathering rocks under warm tropical climate with seasonal rainfall. The weathering rocks, including laterites, occupy extensive territories within the boundaries of modern and ancient tropical and partly subtropical belts. They have many specific geological, mineralogical, geomorphological, and some other peculiarities. Logically, they must be considered as the fourth major rock group besides the groups of the igneous, metamorphic, and sedimentary rocks (Bardossy and Aleva 1990; Bortnikov et al. 2011).

The weathering rocks usually have a zonal profile. We have studied the profiles of modern tropics in India, Guinea, and Brazil and the Cretaceous-Paleogene profiles in Siberia. Thicknesses of lateritic profiles vary from some meters to 600 m (Slukin 1973). For all these, the thickness of the bauxite zone reaches 51 m (Slukin et al. 1989) or 54 m (Bardossy and Aleva 1990). Thicknesses of lateritic profiles depend on the geomorphology and the tectonic position. The mineral and chemical compositions of the weathering profile depend on the composition of the parent rock. For instance, the parent rocks, containing aluminous silicates, produce the complete lateritic profile, consisting of the zones: disintegration, saprolite, bauxite, duricrust, and residual tropical soil. Saprolite is composed of clay minerals such as кaolinite, $Al_4(OH)_8[Si_4O_{10}]$, halloysite, $Al_4(OH)_{10}[Si_4O_8](OH)_2 \cdot 4H_2O$, montmorillonite, $m\{Mg_3[Si_4O_{10}](OH)_2\} \cdot p \{(Al, Fe)_2[Si_4O_{10}](OH)_2\} \cdot nH_2O$, and allophane, $mAl_2O_3 \cdot nSiO_2\ pH_2O$. Weathering coarse-grained quartz-bearing rocks, e.g., granite and khondalite, contain also quartz. Bauxite can be almost a monomineral aluminum ore, consisting of gibbsite or a bimineral paragenous association between gibbsite, Al $(OH)_3$, and boehmite—$\gamma AlO(OH)$. The most common impurities are goethite, $\alpha$ FeO(OH), hematite, $\alpha$ $Fe_2O_3$, rutile, $TiO_2$, anatase, $TiO_2$, and relics of the enumerated clay minerals and quartz, $SiO_2$. The duricrust is the zone of the greatest accumulation of the lateritic substances owing to an interaction of processes such as evapotranspiration, oxidation, absorption, adsorption, hydrolysis, and biological activity and decomposition of organic matter. These processes led to the enrichment of hematite, кaolinite, halloysite, and amorphous Al–Si, Al–Si–Fe, Al–Fe, Fe, and Mn compounds.

The role of organic substances in the processes of lateritization was a long-disputed problem. Many authors believed that the development of laterites and bauxites occurred without the participation of living organisms and several authors considered them as an important factor. Finally, the genus *Bacillus* has been revealed in some lateritic bauxites at the present tropic regions (Heydeman et al. 1983). Using electronic microscopes, we studied the Indian lateritic bauxites and discovered biohematite (Slukin et al. 1989). This mineral has morphological features typical for ferrihydrite formed with the participation of the iron-bacteria *Gallionella* and *Leptothrix*. Research on bauxites and the Fe and Mn ores associated with weathering rocks under electronic microscopes (SEM and TEM) discovered in them a great number of various biomorphes.

## 2 Methods

We undertook a more detailed study of bauxites and laterites of various ages and added the previous results with the new data. The main morphological and structural peculiarities were studied using scanning (SEM) CamScan-4 (Cambridge) and TESCAN VEGA IIXMU (Tescan), and transmitting (TEM) JEM 2100 (JEOL, Japan) microscopes. Microanalysis with TEM was performed using a device for X-ray energy-dispersive analysis X-Max (Oxford Instruments, United Kingdom).

## 3 Geological Setting

The top part of a lateritic profile including laterites *in sensu stricto*, bauxites, and even zones of clay minerals (кaolinite, halloysite, montmorillonite), contains abundant traces and products of interaction of a substratum with live organic substance and mortmasses.

Laterites were broken by cracks and penetrated by roots of trees and bushes of depth more than 10 m that provides deep penetration of microorganisms. The surface of laterites is hidden by a cover of trees and bushes and the dense felt like a film weaved from filamentous fungus, roots of vegetation, and fossilized microorganisms. Dying off, they turned into plant-shaped concretions, replaced with such biominerals, as gibbsite, goethite, hematite, and halloysite. They formed poorly crystallized masses and idiomorph crystals in cavities and pores.

The top 100 cm of laterites called by a zone of a biological pedoturbation, are completely processed by burrowing organisms: worms, termites, ants, nematodes, and others (Schaetzl 2008). Earlier it was represented that they make only mechanical impact on rock (Robert and Berthelin 1986). By us it is shown that plants and burrowing organisms, including worms, completely mill all minerals of a substratum, passing it through the digestive path and satiate it with biochemical active substances (Slukin et al. 2014). In particular, we studied the lateritized quartz-muscovite-feldspar slates of the Chadobets uplift (Slukin 1973). The thickness of lateritic bauxites is 8.5 m. They preserved the relict texture and bedding of the slates. Below 1 m, the bauxites are porous pseudomorphous rocks consisting of well-crystallized pseudohexagonal plates of gibbsite. The bauxite contains 62 % $Al_2O_3$ and 0.6 % $SiO_2$. In the biological pedoturbation zone every crystal has been disintegrated. The bauxites preserved the relict bedding but they lost the relict texture and became cavernous owing to numerous tunnels, burrowing by different macroorganisms. The tunnels are a fraction of millimeter to 1 mm in diameter and up to 5 cm long. They are filled with continuous uneven chains of stuck together by mineral lumps ejected from the digestive tract and resembling the product of recent life activity of earthworms (Fig. 1a). The tunnels are enveloped by biomineral films variably covered by gibbsite crystals. Gibbsite, passed through a digestive path, recrystallized, and formed surprisingly beautiful and perfect

**Fig. 1** Gibbsite: **a** general view of the worm burrow (SEM); **b** gibbsite in the worm burrow (SEM)

idiomorphic crystals filling the burrows of worms and covering their walls (Fig. 1b). The contents of burrows reworked by macroorganisms may be subdivided into two types: (1) partly recrystallized with preserved remains of reworked mineral mass with occasional newly formed gibbsite crystals at its surface; and (2) entirely recrystallized and transformed into druses of gibbsite crystals extended along the entire burrow. The first and the second types both contain rare remains of vermicular kaolinite. Kaolinite is corroded along the surfaces of polysynthetic twins and the base planes (001) of split intergrowths and subjected to intense biochemical dissolution.

## 4 Results

The finds of the products of biological activity and the new-formed biominerals raise further question about the development scale of such phenomena particularly in the present tropical belt of the Earth, which is characterized by an environment favorable for lateritization.

We managed to find the paradoxical phenomenon: pass of worms through large (2–4 cm) quartz crystals. The entrance is surrounded with a biofilm similar to a slime and turned into units of crystals of hematite, and the mineralized excrements corresponding to halloysite.

The described phenomena are established on numerous bauxite deposits of India, Guinea, and Brazil of the modern tropic and on the ancient deposits of the Chadobets uplift in Siberia, where they remain more than 30 million years after alteration of paleotropic climatic conditions.

Quartz in the process of weathering and lateritization takes special place. Its behavior caused many debatable questions. There were such opinions: (1) is quartz a stable mineral, it changes a little and remains in the final products of weathering; (2) in the presence of quartz formation of gibbsite is impossible; (3) quartz-bearing rocks: granites, gneisses, sandstones, etc., cannot be parent rocks of bauxites. The exception of these rocks of favorable sources at the look-ahead estimates of the prospecting areas was a consequence of such opinions.

Stability of quartz spoke, in particular, its high hardness and lack of cleavage (Deer et al. 1963). Actually, hardness of a mineral does not rescue it from the weathering processes. It is known that a more solid emerald is transformed into kaolinite (Vlasov and Kutukova 1960). According to the results of our last researches, the firmest rock formations—Precambrian corundum rocks of the Alag-Ula deposite of Mongolia have undergone physical weathering and chemical replacement by gibbsite and halloysite. As to lack of cleavage in quartz, it with success filled with cracking, the highest among all rock-forming minerals (Fig. 2a). Except for cracks in quartz, pits and channels of dissolution caused by defects and dislocations in crystal structure develop (Fig. 2b). Quartz, as well as some carbonates of alkaline and alkaline earth elements, being dissolved, will form empty spaces which provide progressing filtering waters and promote weathering of the rock as a whole. At last, the results of the joint impact of biological and biochemical weathering processes on quartz were most brightly shown (Huang 2008).

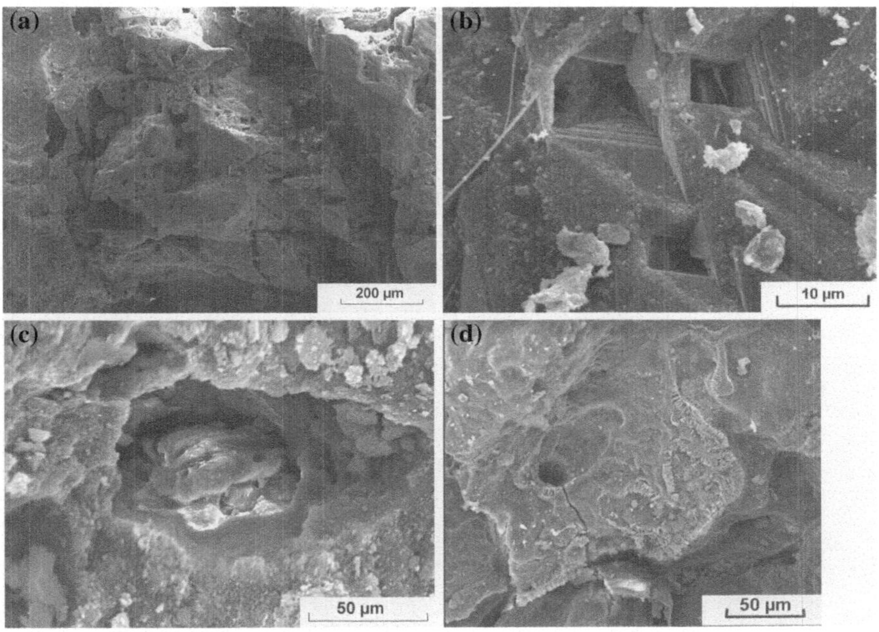


**Fig. 2** Mineralized slime (biofilm) in bauxite, India (SEM): **a** biofilm and two needle-like crystals of rutile; **b** and **c** mineralized biofilms with different crystallinity of gibbsite; **d** mixed scraps of mineralized biofilms and gibbsite crystals in the worm burrow

We managed to observe numerous forms of dissolution of quartz grains in lateritized granites, khondalites, graptolitic slates, and quartz-kaolinite clays. Under the influence of biochemical processes even round-shaped grains again acquire the forms of the quartz crystals, lost at the previous removing, crushing, and rubbing processes. Among them prism and pyramid sides are shown. The sides of prism consist of sharp parallel circles of rhombs which on fresh crystals will form shading. The sides of the pyramid are covered with smaller pyramids combined in a complicated manner.

The joint influence of biological and biochemical factors is shown in two events: first is the passing of fine grains of quartz through the digestive path of burrowing organisms and their fixation in mineralized excrements (Fig. 2c) and the second is passing of burrowing organisms through quartz veins, development of tunnels of the large section covered inside and outside of plentiful mineralized slime, transformed in tubular hematite allocation (Fig. 2d). In excrements except silica, the reworked phosphates (33 % $P_2O_5$) of rare earths (11 % Ln-REE) are found.

It is characteristic that weathered grains of quartz almost always are covered with biofilms on which abundant gibbsite crystals are formed.

Iron is the second principal chemical element in lateritic profile. Almost on all flat-topped mountains (plateau) of Siberia and India, the surfaces of laterites are covered by constants or temporary lakes or marshes. In the lakes of Siberia, water is pure and transparent, but in freshets friable yellow hydroxides of iron immediately are allocated. They are X-ray-amorphous and contain up to 19 % of $SiO_2$, being fixed on herbs, they will form uncountable phytomorphs, but without appreciable presence of microorganisms. Phytomorphs in laterites and bauxites are always accompanied plentiful biomorphs (Fig. 3). In the southeast part of the Siberian platform, among lateritized rocks the algal horizon has met. It was completely replaced with goethite, having well saved the relict algae structure (Fig. 3a).

It was interesting to us to track products of weathering ferruginous aluminosilicates. It was found out that at the Pocos de Caldas (Posus-di-Caldas, Brazil)

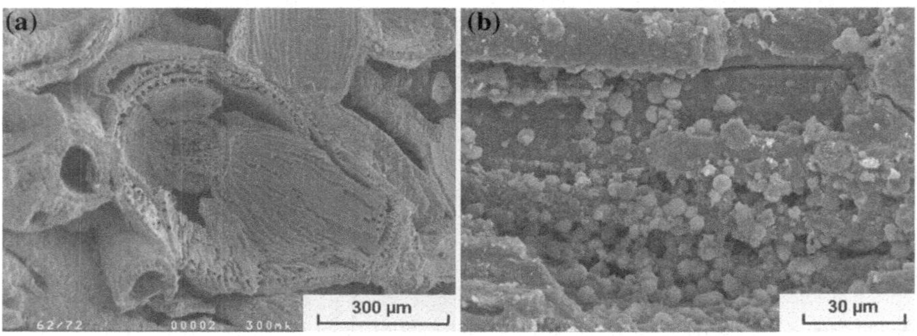


**Fig. 3** Morphology of hematite in the products of lateritization (SEM): **a** phytomorphs of hematite after algae, south-east part of Siberian platform; **b** biomorphs of coccoidal bacteria on phytomorph after wood, west part of Siberian platform

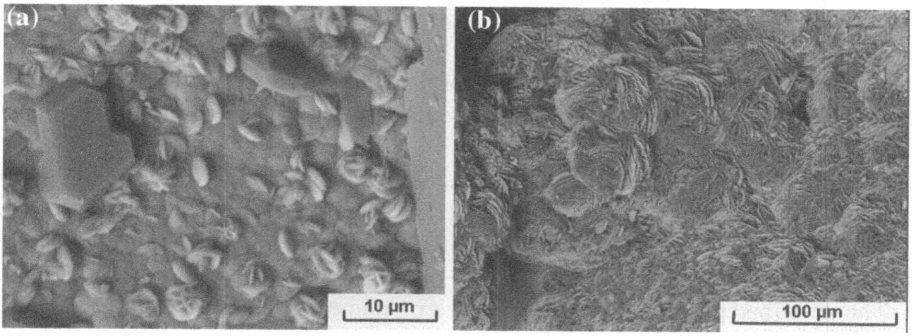


**Fig. 4** Morphology of gibbsite crystals and hematite in the products of lateritization (SEM): **a** twin and single crystals of gibbsite on the surface of the biofilm covered by dispersed hematite biomorphs; **b** the hematite biomorphs as flowers of iron

deposit weathered amphibole formed pseudomorphs, consisting of two minerals: hematite and gibbsite. Hematite is allocated in the form of the crystal framework which being crossed elements remind steel concrete structure. Inside the framework, separate idiomorphic crystals of gibbsite occur. Disseminated and continuous biomorphs of hematite are covered with edges of pseudomorphs. In laterites of the Eastern Ghats, India, almandine formed cellular hematite pseudomorphs with gibbsite. The empty spaces in laterites, pores and cavities, are covered by continuous biomorphs of hematite in the forms of separate disks, tyrelikes, stone flowers, and other fantastic combinations (Fig. 4a and b).

## 5 Conclusions

The systematic research of the role of biota in the formation of bauxites has shown that micro- and macroorganisms are ubiquitous participants of biological and biochemical interactions in the upper part of lateritic profiles. In the processes of interactions of water and rock, it is impossible to allocate purely chemical and purely mechanical natural participants. Rain waters, getting through live and dead vegetation cover, interacting with live organisms and metabolism products, essentially modify the properties. Forming organic acids, they influence frontally on the weathering rocks. Having direct contact to decaying organic substance or live organism, they are capable to cause local changes. Especially brightly it is shown as an example of contacts of burrowing organisms with quartz and gibbsite matrix. Amazing results of penetration of burrowing organisms through quartz veins find high reactivity of slime produced by them. The results of the influence on matrix which has been processed in the digestive path are not less amazing. It was recrystallized and transformed to the most perfect gibbsite crystals. In these conditions, the last remains of kaolinite decay, and the released alumina participates in formation of gibbsite. On weathering of iron-bearing minerals, pseudomorphs of the ferric hydroxides arise whose forms are substantially caused by the crystal structure of the parent minerals. Microorganisms provide huge variety of biomorphs. In all these processes, the biogeochemical processes participate in the mineralization of biofilms.

It causes dissolution of all minerals (kaolinite, gibbsite, hematite, quartz, etc.) except rutile—$TiO_2$ occurring as the thinnest needle crystals. Possibly, rutile can be used as the indicator of the inferior limit of acidity in biogeochemical reaction ($pH \geq 2$).

The described phenomena are established on numerous bauxite deposits of India, Guinea, and Brazil of the modern tropic and on the ancient deposits of the Chadobets uplift in Siberia, where they remain more than 30 million years after the alteration of paleotropic climatic conditions.

**Acknowledgments** This study was supported by the Russian Foundation for Basic Research, project nos. 13-05-00765a, 13-04-00933a.

## References

Bardossy G, Aleva GJJ (1990) Lateritic bauxites. Akademiai Kiado, Budapest

Bortnikov NS, Bugelski YuYu, Slukin AD, Novikov VM, Piloyan GO (2011) The basic aspects of the ore-bearing weathering rocks science at XXI century. Geol Ore Deposits 53:491–505 (in Russian)

Deer WA, Howie R, Zussman MA (1963) Rock-forming minerals, vol 4., Framework silicatesLongmans, London, WI

Huang PM (2008) Clay mineral alteration in soils. In: Chesworth Ward (ed) Encyclopedia of soil sciences. Springer, Berlin, pp 122–135

Heydeman MT, Button AM, Williams HD (1983) Preliminary investigation of micro-organisms occurring in some open blanket lateritic bauxites. In: Proceeding of 2nd international seminar on lateritisation processes, Sao Paulo, Brazil, pp 225–235

Robert M, Berthelin J (1986) Role of biological and biochemical factors in soil mineral weathering. In: Huang PM Schnitzer M (eds) Interaction of soil minerals with natural organisms and microbes, SSSA Special Publication 17. Soil Science Society of America, Madison, WI, pp 453–495

Schaetzl RJ (2008) Pedoturbation. In: Chesworth W (ed) Encyclopedia of soil sciences. Springer, Berlin, pp 516–520

Slukin AD (1973) Weathering rocks and bauxites of the Chadobets uplift. Publishing House Nauka, Moscow (in Russian)

Slukin AD, Boyarskaya RV, Nandi AK (1989) Mineralogy of lateritic products and bauxites of khondalites, Eastern Ghats, India. In: Zographou SA (ed) Weathering; its products and deposits, vol II. Products—Deposits—Geotechnics. Theophrastus Publications, Athens, Greece, pp 101–112

Slukin AD, Bortnikov NS, Zhegallo EA, Zhukhlistov AP, Boeva NM (2014) Gibbsite and кaolinite in the biological pedoturbation zone of the lateritic profile: a different fate (exemplified by deposits of Siberia, India, Guinea, and Brazil). ISSN 1028-334X; Doklady Earth Sciences 458(Part 2):1220–1225

Vlasov KA, Kutukova EI (1960) Izumrudny Kopi (Emerald Pits). Publishing House of Academy of Sciences of the USSR, Moscow (in Russian)

# Some Mineralogical Approaches to Study the Biocarbonate and the Carbonate-Siliceous Nodules

**Liubov V. Leonova, Akhmet A. Galeev, Yulia S. Simakova, Alena S. Ryabova, Liudmila Yu Kuzmina, Stepan P. Glavatskikh and Olga Ya Cherviatsova**

**Abstract** Nodules of different compositions from Paleozoic sedimentary rocks and those deposited by microbial communities in laboratory-scale experiments were studied by the use of electron paramagnetic resonance, X-ray diffraction, and scanning electron microscopy with energy-dispersive spectrometer. The study of the mineral composition of fossil nodules showed that they have monomineral composition of dolomite or chalcedony, mixed composition of dolomite-chalcedony or of opal-dolomite, and finally nodules can be composed of alternating opal and chalcedony layers cementing fine dispersed dolomite grains or clusters of irregular shape. Similarity in dolomite crystal lattice defects in both the nodules and the host rocks confirms their formation during synsedimentary early diagenesis. Bacterial activity during sedimentary nodules growth is evidenced by the presence of paramagnetic carbon-centered free radicals of fossilized protein substances and findings of fossil bacteria. Experimental laboratory-scale modeling of natural carbonate deposition by microbial communities confirms that bacteria can promote nodules formation.

L.V. Leonova (✉) · S.P. Glavatskikh
Institute of Geology and Geochemistry, Ekaterinburg, Russia
e-mail: lvleonova@yandex.ru

A.A. Galeev
Institute of Geology and Petroleum Technologies, Kazan Federal University, Kazan, Russia
e-mail: akhmet.galeev@kpfu.ru

Y.S. Simakova
Institute of Geology of the Komi Science Center, Urals Branch of RAS, Syktyvkar, Russia
e-mail: yulia5-07@mail.ru

A.S. Ryabova · L.Y. Kuzmina
Institute of Biology, Ufa Scientific Center of RAS, Ufa, Russia
e-mail: ljkuz@anrb.ru

O.Y. Cherviatsova
Shulgan-Tash State Nature Reserve, Irgizly Village, Republic of Bashkortostan
e-mail: kittary@yandex.ru



O.V. Frank-Kamenetskaya et al. (eds.), *Biogenic—Abiogenic Interactions in Natural and Anthropogenic Systems*, Lecture Notes in Earth System Sciences,
DOI 10.1007/978-3-319-24987-2_8

**Keywords** Nodules · Bacterial activity · Biogenic carbonates and flints · Methods of investigation · Experiments

Sedimentary and volcanic-sedimentary rocks are formed in the water-saturated upper part of the lithosphere. Water-saturated conditions can be considered to be the most favorable for biogenic–abiogenic interaction when the biota affects both upwelling gas–fluid flows from the Earth's interior and surface deposits resulting in the formation of biominerals. Current reassessment of the role of bacterial assemblages in sedimentogenesis and lithogenesis creates a new direction in the lithology called as "bacterial lithogenesis" (Antoshkina 2012). Since almost all sedimentary rocks formed in varying degrees with the participation of the microbiota, one of the tasks of lithogenesis is to develop models of sedimentation in ancient epicontinental basins taking into consideration the essential role of bacterial factor. The study of nodules can be useful for solving these tasks. The problem is that the participation of bacteria in sedimentogenesis and lithogenesis should be proved using a complex of physical research techniques.

Lithological syngenetic nodules are often found in sedimentary rocks of different geological age formed in various facies conditions. Chemical and mineral compositions of the nodules and the surrounding rock may be the same or different from one another: carbonates, phosphates, silica oxides (opal, chalcedony, quartz), oxide-hydroxides of iron and manganese, and mixed iron-manganese oxides. There are microscopic and giant, monomineralic or mixed mineral nodule, including clays and sandstones. Often, nodules are spheroid or disk-shaped but sometimes form a variety of more complex shapes. Nodules of similar shape but of different mineral composition are presented in our collection (Fig. 1a–f): a dolomite (gypsum-dolomite stratum $P_2$ $kz_2$, right bank of the Volga river, v. Klyuchischi); b vernadite-lithiophorite-quartz mineral association (jasper-silicite deposits with manganese ore interlayers $D_2$ ef, Southern Urals, Baimaksky district, v. Fayzulino); c sandstone with quartz cement (jasper-silicite deposits $D_2$ ef-zv, Southern Ural, Chelyabinsk region, v. Kizilskoe); d opal-dolomite association (gypsum-dolomite deposits $P_2$ $kz_2$, right bank of the Volga river, v. Krasnovidovo); e gray jasper (jasper-silicite deposits $D_2$ ef-zv, Southern Ural, Sibai); f quartz sandstone with vernadite-lithiophorite cement (jasper-silicite complex $D_2$ zv, Southern Ural, Sibai district, v. Khasanovo).

Outcrops of the nodules occur locally in the vertical section of a given sedimentary horizon, forming meadows in landscape views, and often traced from one layer to another. Their presence indicates local geological events in paleobasin development—localized gas/fluid flow associated with seafloor seeps. However, the question of the nodules genesis up to this time predominates the model of chemogenic consolidation in the late diagenesis and catagenesis, implying the presence of permeable media and migrating solutions with a high degree of mineralization

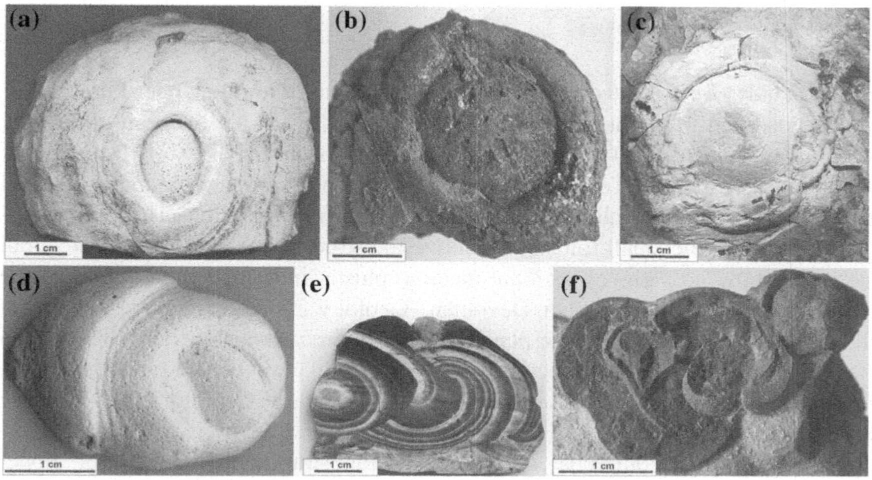


**Fig. 1** **a–f** Nodules with similar morphology and differing in mineral composition: **a** dolomite (gypsum-dolomite stratum $P_2$ $kz_2$, *right* bank of the Volga river, v. Klyuchischi); **b** vernadite-lithiophorite-quartz mineral association (jasper-silicite deposits with manganese ore interlayers $D_2$ ef, Southern Urals, Baimaksky district, v. Fayzulino); **c** sandstone with quartz cement (jasper-silicite deposits $D_2$ ef-zv, Southern Ural, Chelyabinsk region, v. Kizilskoe); **d** opal-dolomite association (gypsum-dolomite deposits $P_2$ $kz_2$, *right* bank of the Volga river, v. Krasnovidovo); **e** *gray* jasper (jasper-silicite deposits $D_2$ ef-zv, Southern Ural, Sibai); **f** quartz sandstone with vernadite-lithiophorite cement (jasper-silicite complex $D_2$ zv, Southern Ural, Sibai district, v. Khasanovo)

(Strakhov 1962). But this model cannot explain the genesis of those nodules which are the same in the mineral composition with surrounding rocks, or whose specific shapes (Fig. 1a–f) do not follow the mechanism of chemogenic consolidation.

An alternative hypothesis of the dissolved substances concentration and of forming consolidated nodules is related to the leading role of microbial communities, the part of which can promote mineral precipitation (Berner 1971; Skinner and Fitzpatrick 1992; De Craen et al. 1999; Green and Madgwick 1991; Tazaki 2000; Raiswell 1976; Stocks et al. 1999). This is supported by the findings of mineralized remains of the fossil cyanobacteria and the bacteria buried in nodules substrate. Unfortunately, in carbonate formations such finds are rare. Therefore, this paper focuses on the study of carbonate nodules. Probably, the depositions of such mineral aggregates are biochemogenic in nature and occur as a result of catalytic influence of bacterial activity. In these cases it is necessary to adapt the analytical research techniques used in the physics and chemistry of minerals, which may reveal the implicit role of bacteria in concretion formation. And of course, the experimental evidences are the best way to prove the hypothesis.

# 1 Materials

For this study, we collected 40 samples from gypsum-dolomite Upper Permian ($P_2$ $kz_2$) deposits on the right bank of the Volga river (Russia, the western part of the Republic of Tatarstan, 50–100 km downstream of Kazan: Pechischi, Krasnovidovo, Antonovka, Yashelcha, Tenishevo piers) (Fig. 2). The carbonate sediments, formed in the shallow marine-salinated basin, are dominating this area. The brachyanticlinal uplift and the small folds characterize the structural features of this area, but the rocks show no signs of regional metamorphism, inheriting the outlines of the relief on the roof of the Upper Devonian (Geology 2007). However, some local changes of the rocks due to the penetration of hydrocarbon fluids are observed in the vicinity of brachyanticlinal uplifts (Syukeevskoe, Kama-Ustyinsky, Krasnovidovskoe, Verkhneuslonsky, Sviazhsky at alias) (Geology 2007; Korolev et al. 2014).

Nodules in the outcrops occur at several stratigraphic levels of upper Kazanian horizon: Prikazanskaya, Pechishchinskaya, Verkhneuslonskaya, and Morkvashinskaya stratas (Geology 2007). They are composed of dolomite, silica, dolomite-silica (chalcedony) or dolomite-opal. Moreover, silica (chalcedony) appears in different forms: fine-dispersed, lumpy, or localized in the nucleation center of nodules (Fig. 3a). The color of the silica component of gray or black depends on ultrafine and nano-sized inclusions of pyrite (melnikovite), magnetite, and maghemite. The maximum size of up to 60 cm is typical to purely dolomite nodules, whereas the dolomite-opal nodules are of the minimum size of 1–3 cm. The dominating shapes are spheroids (Fig. 3b) or flattened spheroids (Fig. 3c), rarely—spheroids with rim (Fig. 1a).

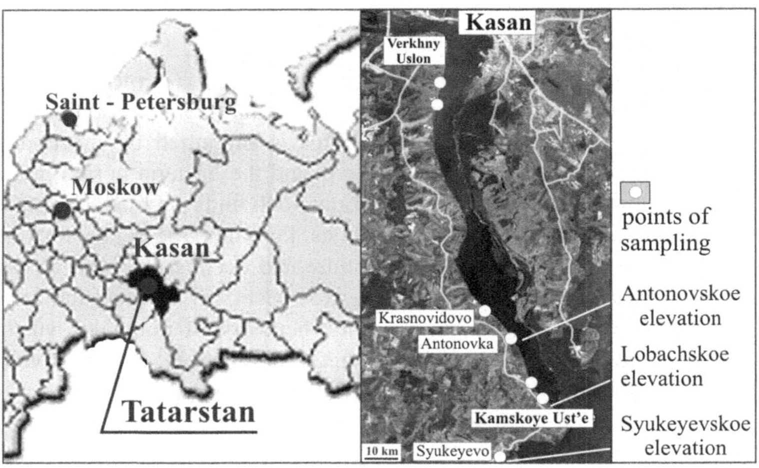


**Fig. 2** Survey maps of studied area

**Fig. 3** **a–c** Nodules from *right* bank of the Volga river, v. Krasnovidovo: **a** chalcedony component is localized in the central part of dolomite concretion; **b** spheroidal dolomite concretion in dolostone; **c** flattened-spheroidal dolomite concretion in dolostone

## 2 Methods

In this study, we used the following physical techniques:

- electron paramagnetic resonance (EPR) (spectrometer EPR-PS100.X operating at 9.772 GHz), used to reveal the remains of fossil organic matter in mineral matrices of concretions, and to study point defects in crystal structure (radiation centers) of carbonate minerals of host rocks and nodules;
- experimental laboratory-scale modeling of natural biocarbonate deposition by microbial communities;
- scanning electron microscopy (SEM) (JSM-6390LV JEOL) with energy-dispersive spectrometer (EDS—Inca Energy 450), used to study mineral aggregates and microfossils micromorphology, and to determine the chemical composition of the minerals. The fractured and polished surfaces of samples were carbon coated for microscopic purposes.
- powder X-ray diffraction (XRD) (X-ray Diffractometer Shimadzu XRD-6000, CuK$\alpha$, Ni filter, 30 kV, 20 mA, scanning in $2\theta$ range from 02° to 35° and from 15° to 55°), used to characterize the crystal structure determination of mineral composition of nodules and microbial carbonates, deposited in laboratory.

## 3 Justification of Selected Methods

A prerequisite for the use of EPR technique in our work were the previously obtained spectral characteristics of carbon radicals in organic matter (OM) for the most common fossil residues of different age, genesis, and manifestation forms in the sedimentary rocks (Votyakov et al. 2005; Murav'yev 2007; Soroka et al. 2007; Conard 1984). Depolymerization and repolymerization products of proteins, lignin, and cellulose, such as fragments of aminoacids and polysaccharides, usually contain uncompensated bonds which can be locally stabilized in the form of carbon-centered free radicals ($R_C$-org) with paramagnetic properties. Being fossilized, these radicals can be preserved within the mineral matrix for a long period of geological time (Votyakov et al. 2005). The concentration of these paramagnetic centers in fossils can be changed during the specific heat treatment in laboratory conditions (Fig. 4a). The characteristic parameters of their EPR spectra are the g-value, the line width ($\Delta H$, mT), and the line shape (mainly Lorentzian).

By measuring the paramagnetic properties and their temperature dependence, one can distinguish the following types of fossil OM:

- humus-sapropel matter metamorphized at early stages of diagenesis. EPR spectra with $g \sim 2.0031$ and $\Delta H \sim 0.5 \div 0.9$ mT are observed in the raw samples and gradually disappear when treated up to 350 °C (Fig. 4b).

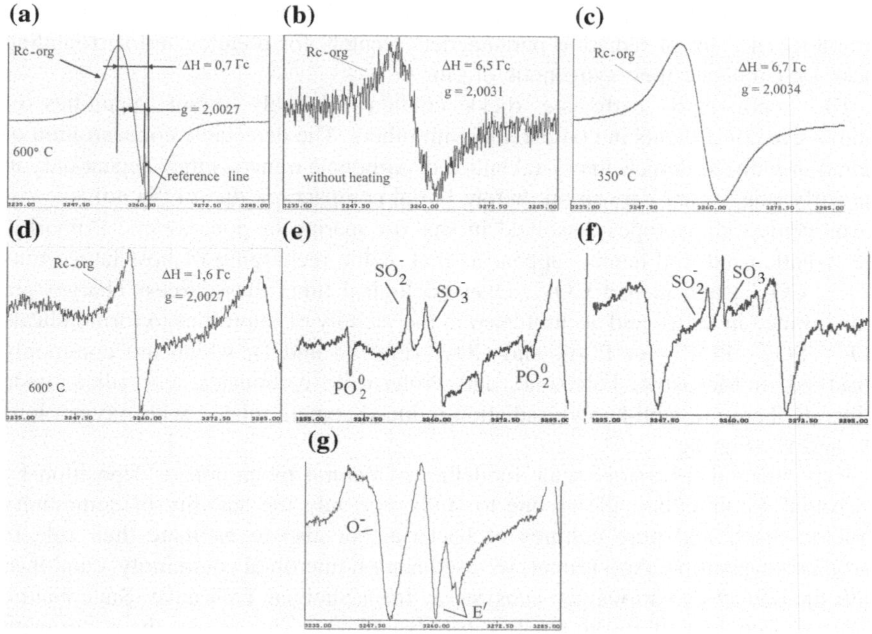


**Fig. 4** **a–g** EPR spectra in the range of radical lines: **a** characteristics of the resonance lines in spectrum of carbon-centered organic free radicals $R_C$-org (g-factor is a dimensionless parameter, defining the Zeeman splitting of spin multiplets for a given atoms, $\Delta H$—line width in mT (~10 Gs), and the upper temperature limit of stability range for paramagnetic properties of radicals); **b** typical $R_C$-org EPR spectrum ($g \sim 2.0031$, $\Delta H \sim 0.5 \div 0.9$ mT) of the humus-sapropel organic matter which is observed in the raw sample and disappears after heat treatment; **c** $R_C$-org spectrum of fossil OM of plant origin ($g \sim 2.0030 \div 2.0038$, $\Delta H \sim 0.4 \div 0.7$ mT, 350 °C); **d** typical $R_C$-org EPR spectrum of organic residues of animal, including bacterial, protein (collagen) ($g = 2.0027 - 2.0028$, $\Delta H \approx 0.5 \div 0.4$ mT, upper temperature limit of stability range is 600 °C); **e** radiation centers $SO_2^-$, $SO_3^-$, $PO_2^0$ in dolomite; **f** radiation centers $SO_2^-$, $SO_3^-$ in calcite; **g** radiation centers $O^-$ и E′ in silica components

- plant matter (lignin, cellulose). EPR spectrum with $g \sim 2.0030 \div 2.0038$ and $\Delta H \sim 0.4 \div 0.7$ mT can be observed in the raw samples and gradually increases when treated with maximum intensity at 350 °C and then disappear (Fig. 4c).
- OM of animal origin (collagen), including bacteria, is characterized by EPR spectra with—$g = 2.0027 - 2.0028$, $\Delta H \approx 0.05 \div 0.4$ mT, which disappears at treatment above 600 °C (Fig. 4d).

At high degree of natural coalification of OM, all the above-mentioned signals $R_C$-org can be observed in raw samples and disappear when treated to 350 °C or 600 °C depending on the original nature (Murav'yev 2007; Khasanov and Galeev 2008; Conard 1984).

The presence of radiation paramagnetic centers in the EPR spectra of minerals indicates that these minerals were not subjected to recrystallization because defect structure of minerals is unique for a given sediment's formation conditions. Similarity in a set of radiation paramagnetic centers for nodules and surrounding rocks may indicate their syngenetic origin.

The sedimentary carbonate rocks contain relatively minor quantities of radioactive constituents in comparison with others. The detectable concentration of radiation-induced defects in crystal lattice of carbonate minerals may appear only in the early stages after deposition, before burial compaction, due to the influence of cosmogenic radioisotopes dissolved in shallow marine or pore waters. Primarily, the radiation-induced defects appear as metastable recharging of host lattice ions $CO_3^-$, $CO_3^{3-}$, and in part $CO_2^-$. Over geological time, these excess charges are redistributed in lattice and accumulated in the vicinity of impurities to form radicals $SO_2^-$, $SO_3^-$, $PO_2^{2-}$ (or $PO_2^0$), and $CO_2^-$ (Fig. 4e and f), which are commonly observed in Mesozoic, Paleozoic, and Proterozoic carbonates. For silica oxide minerals there are well-known radiation-induced centers related with oxygen ions $O^-$ and E′ (Fig. 4g).

Experimental laboratory-scale modeling of natural biocarbonate deposition by microbial communities allows one to study not only the stability of community structure itself and pure cultures of bacteria, but also to estimate their role in carbonate deposition. As a primer, we used natural microbial community, combined with the habitat of various geo ecosystems from Shulgan-Tash cave (State Nature Reserve, Republic of Bashkortostan, Irgizly village). The aim of the experiment no. 1 was to identify the nutrient media, the most favorable for the development of bacterial communities and carbonate deposits. A Petri dish with bacterial colonies on an agar-based growth medium Variant 3 were inoculated by mineral pieces of needle-fiber (NFC), pool fingers (lake aggregates of calcite), and soil from the cave hall "Raduzhny" where gypsum crystals are growing. The most productive variants of culture media were used for the following experiments. Experiment no. 2 was conducted to study the newly formed biocarbonates and possible intermediate phases by the use of analytical research techniques: SEM, EDS, and XRD. Microbial isolates for this series of experiments were grown in medium variant 3 (Danielli and Edington 1983), g/l (modified): $KNO_3$(0.5); $Na_2HPO_4*12H_2O$ (0.25); Ca succinate (5.0) (succinic acid (3.54); $CaCl_2*2H_2O$ (3.3)); Oxoid agar (15.0); pH 7.1, and also in Grans medium (Mason-Williams 1969) g/l (modified): Ca malate (5.0) (malic acid (4.02); $CaCl_2*2H_2O$ (3.3)); $KNO_3$ (0.5); $Na_2HPO_4*12H_2O$ (0.25); Oxoid agar(15.0); pH 7.2. The isolates were grown in the growth medium peptone water g/l (modified) (Antipchuk 1979): Peptone (5.0); $K_2HPO_4$ (1.0); $KH_2PO_4$ (1.0); $MgSO_4*7H_2O$ (0.5); $CaCl_{2-3}$; NaCl a very small content, were studied only by XRD.

The samples were prepared as follows: microbial community and isolates were plated on solid culture media (the crops for XRD were duplicated in liquid media) and cultured for 30 days. The specimens were prepared from these substrates including the spent culture medium with biofilm and the crystalline phase of newly formed carbonate minerals of different dimensions.

The substrate was dried at a temperature not higher than +25 °C and then the piece cut out film with crystals was mounted on carbon adhesive tape and placed in vacuum coating systems for SEM. To study the newly formed carbonates by XRD, the bacterial biomass were scraped from the culture media and frozen in eppendorf polypropylene tubes at −30 °C. Grown in liquid media carbonates and microbial biomass were centrifuged at 5000 rpm and then frozen for transport to XRD laboratory. It is assumed that in this manner the transformation of uncontrolled hydrocarbonate into carbonate will proceed slowly that allows to reveal all the phases of the newly formed carbonates. Unoriented samples for XRD analyses were prepared by drying the gel on rectangular glass plates. Directly before the analytical procedure ethyl alcohol was added to the glass slide with the sample for accelerated drying.

## 4 Results

The study of the mineral composition of fossil nodules showed that they can have monomineral composition of dolomite or chalcedony, mixed composition of dolomite-chalcedony or of opal-dolomite, and finally nodules can be composed of alternating opal and chalcedony layers cementing fine dispersed dolomite grains or clusters of irregular shape.

## 5 Electron Paramagnetic Resonance

The results of the EPR study are present in Table 1 for the samples of most representative mineral associations:

No. 1—dolomite, no. 2—black chalcedony concentration in middle of the dolomite concretions, no. 3—opal-dolomite mineral association, and no. 4—black pure chalcedony in dolomite host rock. EPR spectra of radiation centers $SO_2^-$, $SO_3^-$ of dolomite are registered both in nodules and host rocks, the similar pattern observed for E′ centers of chalcedony. This indicates that the investigated objects were formed together with surrounding rocks, and never were subjected to recrystallization during diagenesis and regional metamorphism in subsequent periods. Furthermore, the signals $R_C$-org with $g = 2.0027$, $\Delta H \approx 0.14$–$0.16$ mT, typical for nonmetamorphized residual organic matter of animal protein (collagen), occur in EPR spectra only after treatment at 600 °C, and there were no any spectra registered which can be related with metamorphized OM, as well as with humus-sapropel or plants matters. It should be noted that measured concentration of paramagnetic centers $R_C$-org in concretion was sometimes considerably higher than in host rock. The well-preserved fossil OM within mineral matrices can be the evidence of high mineralization speeds and microbial activity during concretion growth because no faunal remains were found neither in concretions or in nearby host rocks.

**Table 1** EPR spectral characteristics of studied nodules

| No. | Sample | Mineral composition | The presence of microfossils | The EPR characteristics of the signals in the field of radicals | | | | |
|---|---|---|---|---|---|---|---|---|
| | | | | Radiation centers | | Radicals of organic carbon | | |
| | | | | E′ | $SO_2^-$ $SO_3^-$ | Without heating | 350 °C | 600 °C |
| 1 |   | Dolomite | – | – | + | – | – | + |
| 2 | 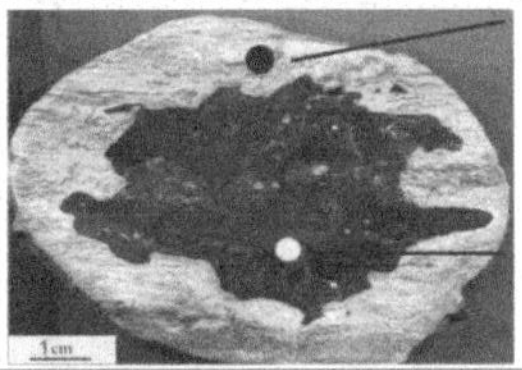  | Dolomite | – | – | + | – | – | + |
| | | Black flint | + | + | – | – | – | + |
| 3 | 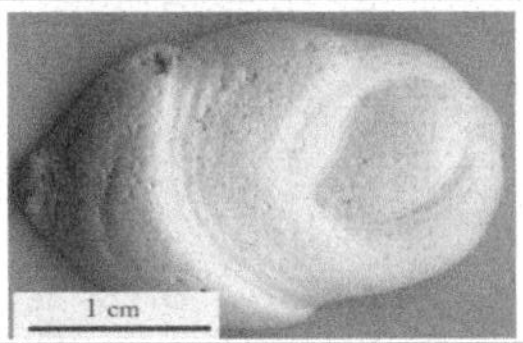  | Dolomite-opal | + | + | – | – | – | + |

(continued)

**Table 1** (continued)

| No. | Sample | Mineral composition | The presence of microfossils | The EPR characteristics of the signals in the field of radicals | | | | |
|---|---|---|---|---|---|---|---|---|
| | | | | Radiation centers | | Radicals of organic carbon | | |
| | | | | E′ | $SO_2^-$ $SO_3^-$ | Without heating | 350 °C | 600 °C |
| 4 | 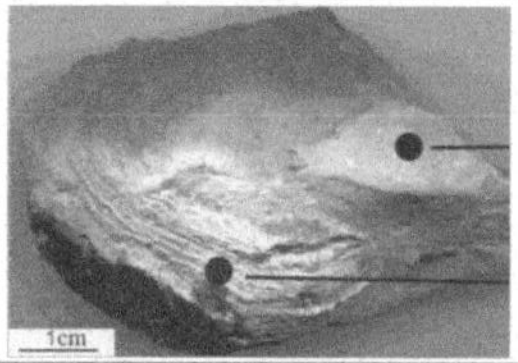 | The host rock dolomite | – | – | + | – | – | + |
| | | Black flint | – | + | – | – | – | + |

# 6 Scanning Electron Microscopy

SEM and EDS analyses show that most nodules are composed by flocculant, drusy aggregations, and carbonate microcrystals (Fig. 5a) with a morphology different from rhombohedral grains of host rocks. Crystals and druses are characterized by similarity in size and by repeatability in shape. However, microbial fossils in the nodules are rare (Fig. 5b). Opal-dolomite nodules are always fine-grained mineral association, sometimes forming hollow spheroids (Fig. 5c), which possibly are mineralized remains of bacterial colonies.

Samples deposited in the laboratory experiment no. 1 contain carbonate crystals (Fig. 5d) similar in morphology to those in fossils or have a specific shape (Fig. 5e, f). There are crystals with biofilm on their surface (Fig. 5g). Microbial-induced precipitates form microscopic nodules (Fig. 5h and i). In experiment no. 2, we established that in pool fingers community, the abundance of microorganisms able to grow in media with organic acids (malic, succinic—Grans,

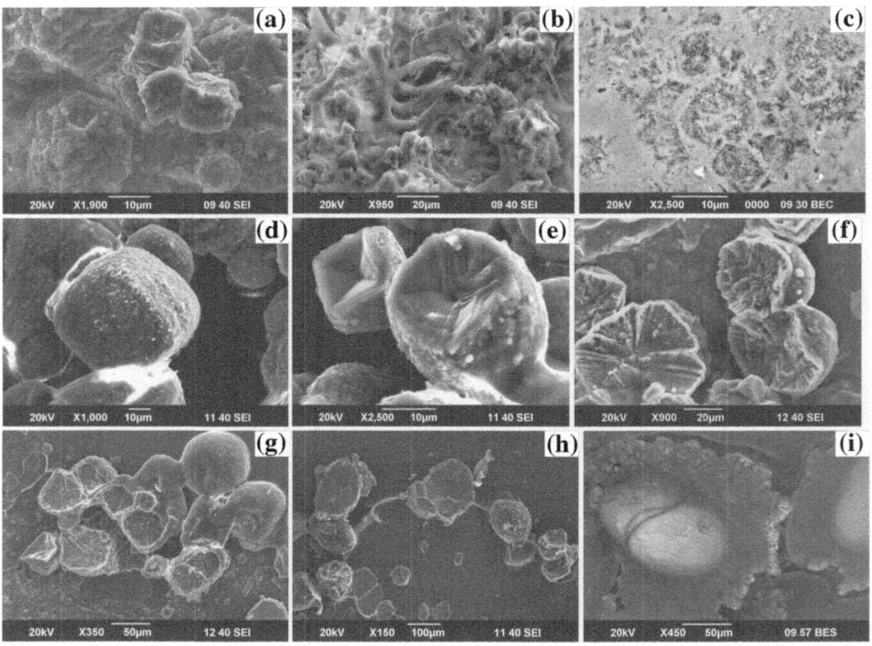


**Fig. 5** **a–i** SEM images of carbonate nodules: **a** specific crystals of carbonates (dolomite and high-Mg-calcite) that constitute nodules in the Upper Permian deposits; **b** bacterial microfossils in the dolomite nodules; **c** microstructure in opal-dolomite nodules from the Upper Permian deposits; **d** bacterial controlled carbonate sediments in experiment No. 1; **e**, **f** bacterial controlled carbonate formation of a specific habitus, experiment No. 1; **g** subsequent stages of carbonate crystals encapsulation by biofilm, which deposits calcium carbonate; **h**, **i** carbonate microconcretions in experiment No. 1

**Table 2** Colony of microorganisms selected to study the mineral sedimentation in bacterial mats

| Strains | Culture media | The shape of the cell | Description colony: color, colony size | Color gram* | The consumption of glucose | Aerobic growth | The habit of the crystals, Plate 1 |
|---|---|---|---|---|---|---|---|
| 1–2a | Variant 3 | Sticks | Colorless with yellow paint small colony | + | + | ± | a |
| 1–2b | Variant 3 | Sticks | Colorless small colony | − | + | ± | b |
| 1–3 | Variant 3 | Cocci | Yellow large colony | − | | + | c |
| 1–3a | Variant 3 | Cocci | Yellow small colony | + | + | + | d |
| 1–4 | Variant 3 | Sticks | Colorless large butyrous colony | − | + | + | e |
| 1–5 | Variant 3 | Cocci | Orange small colony | − | − | + | f |
| 1–6 | Variant 3 + peptone | Sticks | Colorless large butyrous colony | − | + | ± | g |
| 2–2 | Grans medium | Sticks | Colorless large butyrous colony | − | − | ± | h |
| 2–5 | Grans medium | Cocci | Yellow large colony | − | + | ± | i |
| 2–6 | Grans medium | Sticks | Colorless large colony | − | + | + | j |
| 2–7 | Grans medium | Sticks | Colorless small colony | − | + | + | k |

* Test: + Positive, − Negative, ± Optional gone anaerobic

variant 3, and variant 3 + peptone) was about 1.8 ÷ 4.6 × $10^6$ KFU/g. From this broad microbial diversity after several passages and exceptions (loss of ability to precipitate minerals, pollution by fungi at alias), we studied the physiological and biochemical properties of the remaining 11 isolates in order to identify them and to estimate their ability to precipitate minerals. Selected cultures were represented by aerobes and facultative anaerobes, mainly of gram-negative rod-shaped bacteria (6 isolates) and of cocci (3 isolates), and only by two gram-positive isolates (Table 2). The difference in morphology of biominerals precipitated by different isolates was established from SEM analyses. The microscopic images of biocarbonate dominating forms are shown in plate (Plate 1a–l). The influence of nutrient medium on nodules morphology was negligible. As it is seen in Fig. 3, the different isolates 1–5 (Plate 1f), 2–5 (Plate 1i), 2–7 (Plate 1k) precipitate aggregations of similar morphology, even if the isolate 1–5 was cultivated on the medium Variant 3, whereas the isolates 2–5 and 2–7 on Grans medium (Table 2). When observed at high magnification (more than ×4300), certain bacterial cells were found to be buried among crystal particles (Plate 1l).

## 7 X-ray Diffraction

XRD data indicate that calcite is the major crystalline component in sediments for all the analyzed samples. The amount of noncrystalline material (gel) evidenced by the XRD patterns of random mounts which exhibited high background levels and two humps with very broad maxima near 3.5 and 2.2 Å (Fig. 6). Nevertheless, distinct calcite peaks were detected on XRD patterns of such samples. As the gel in the samples dry, the calcite reflections on XRD patterns become more intensive and

**Plate 1 a–l** SEM images of biocarbonate precipitates in experiment no. 2

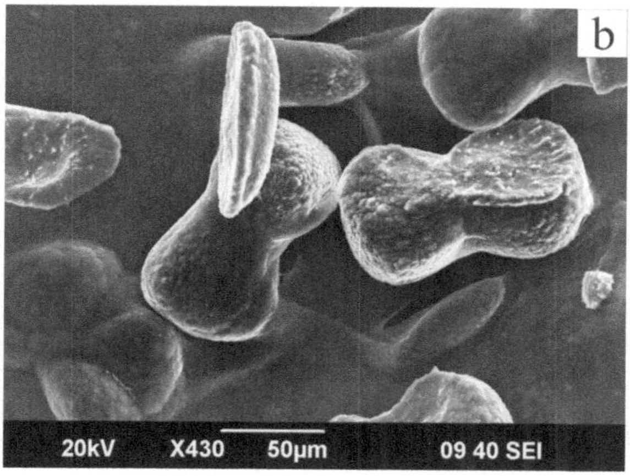


**Plate 1** (continued)

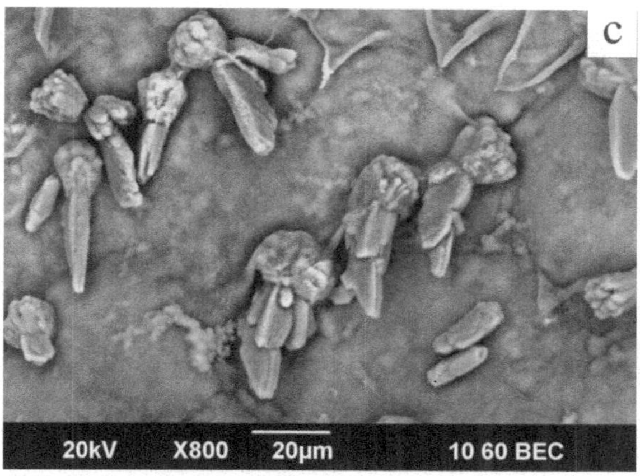


**Plate 1** (continued)

narrow. Cell parameters have been determined for the biogenic calcites from the investigated samples (where possible, with Si as internal standard). Values of these parameters are very close to one another and have minimal deviations (Table 3).

It should be noted that in experimental precipitation induced by bacteria (Wei et al. 2011), the calcite was a dominant mineral phase. Some crystals after experimental interaction with bacteria had bacterial imprints on crystal surfaces. In spite of differences in the tasks of experiments, strains, and mediums chosen, calcite

**Plate 1** (continued)

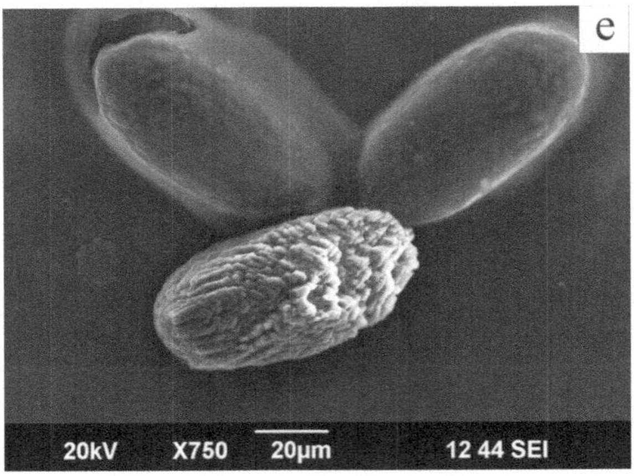


**Plate 1** (continued)

precipitates regularly with increasing the pH of the medium. The presence of Mg salts in the nutrient solution leads to the precipitation of Mg-calcite.

Presence of Mg-calcite was determined in the sample A formed on Peptone nutrient solution ($MgSO_4*7H_2O$—0.5; $CaCl_2$—3 g/l). Cell parameters of this mineral are in good agreement with other data for biogenic Mg-calcite (sample A in Table 3) (Bischoff et al. 1983; Paquette and Reeder 1990).

**Plate 1** (continued)

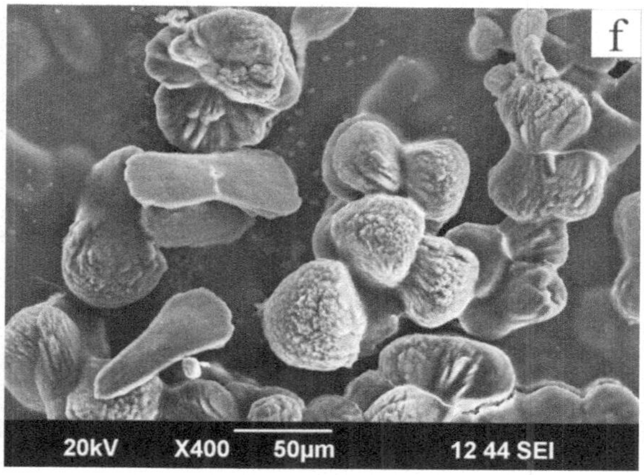


**Plate 1** (continued)

**Plate 1** (continued)

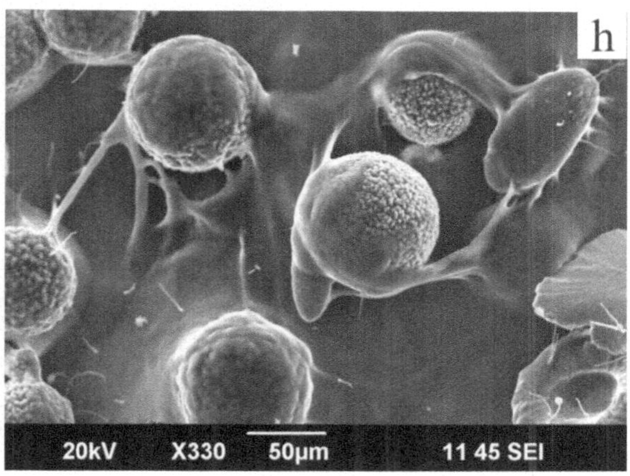

**Plate 1** (continued)

**Plate 1** (continued)

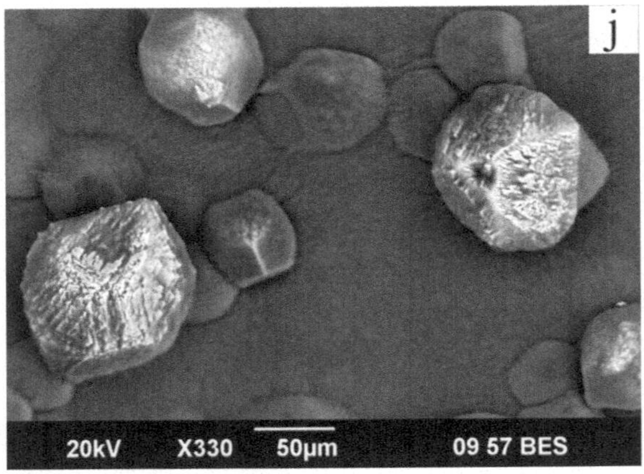


**Plate 1** (continued)

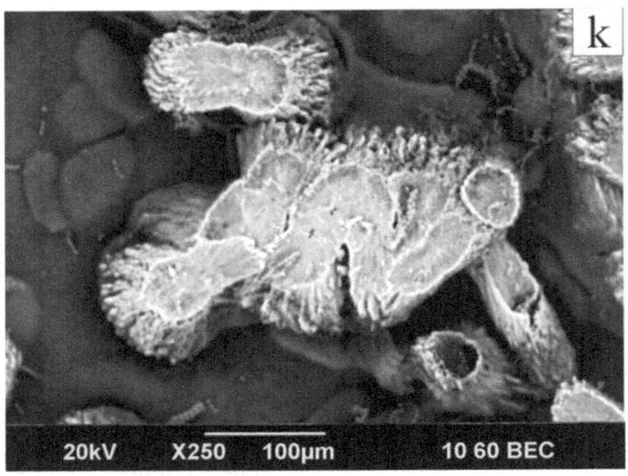

**Plate 1** (continued)

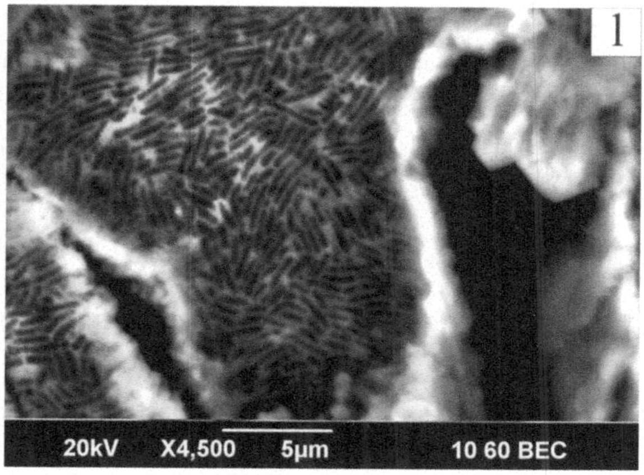


**Fig. 6** X-ray diffraction patterns of sample 2–6: *a* gel state, *b* air-dried. *c* calcite. Spacing in Å

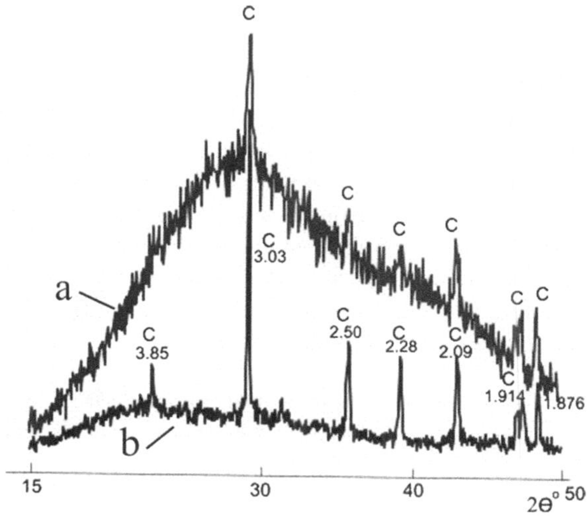


# 8 Conclusions

The complex study of structural and spectral properties of minerals in carbonate nodules evidences on microbial involvement in the genesis of studied natural and laboratory grown samples.

From the results obtained, one can suggest the following conclusions. (1) Studied natural nodule, as well as surrounding rocks, was not subjected to significant regional metamorphization over time after sedimentation. This is evidenced by the presence of stable paramagnetic radiation centers in the crystal lattices of minerals forming nodules and host rocks, as well as by the presence of

**Table 3** The results of XRD analysis of crystalline material precipitated by the bacterial isolates

| No. | No. of sample | Content | Cell parameters | |
|---|---|---|---|---|
| | | | a, Å | c, Å |
| 1 | 1–2a | Calcite | 4.9810 (0.0012) | 17.0367 (0.0081) |
| 2 | 1–2б | Calcite | 4.9784 (0.0010) | 17.0324 (0.0146) |
| 3 | 1–3 | Calcite | 4.9805 (0.0005) | 17.0158 (0.0040) |
| 4 | 1–4 | Calcite | 4.9817 (0.0009) | 16.9963 (0.0082) |
| 5 | 1–5 | Calcite | | |
| 6 | 1–6 | Calcite | 4.9816 (0.0007) | 17.0237 (0.0103) |
| 7 | 2–2 | Calcite | 4.9763 (0.0012) | 17.0529 (0.0201) |
| 8 | 2–5 | Calcite | | |
| 9 | 2–6 | Calcite | 4.9800 (0.0007) | 17.0086 (0.0052) |
| 10 | 2–7 | Calcite | | |
| 11 | A | Calcite | 4.9808 | 17.0295 |
| 11 | 1–6-liquid | Calcite, halite | 4.9816 (0.0009) | 16.9952 (0.0075) |
| 12 | Control | Halite | | |

animal protein (collagen) residues in fossil OM of low metamorphization stage. (2) Similarity in crystal lattice defects for both the dolomite in nodules and host rocks confirms their formation during synsedimentary early diagenesis; (3) Since the rocks nearby sampling areas are depleted in fossil macrofauna, one can assume that the fossilized remains of animal protein occured as a result of bacterial activity, while nodules themselves are the lithified microbial constructions. Findings of fossil bacteria additionally support this suggestion. (4) Experimental laboratory-scale modeling of natural carbonate deposition by microbial communities confirms that bacteria can promote nodules formation.

**Acknowledgements** Some of the study was conducted with the support of a subsidy allocated to the Kazan Federal University for state assignment in the sphere of scientific activities (Project No. 14–69) and Program of UB RAS 15-18-5-49.

## Reference

Antipchuk AF (1979) Microbiological control in ponds.M. Food Industry

Antoshkina AI (2012) Bacterial lithogenesis. In: Obzor konceptual'nykh problem litologii. In: A Review of Conceptual Problems of Lithology, Moscow: GEOS, pp 89–105

Berner RA (1971) Bacterial processes effecting the precipitation of calcium carbonate in sediments. Johns Hopkins Univ Stud Geol 19:247–251

Bischoff WD, Bishop FC, Mackenzie FT (1983) Biogenically produced magnesian calcite: inhomogeneities in chemical and physical properties comparison with synthetic phases. Am Mineral 68:1183–1188

Conard J (1984) EPR in fossil carbonaceous materials. In: Petrakis L, Fraissard JP (eds) Magnetic resonance. Introduction. D. Reidel Publishing Company, Hingham, pp 441–459

Danielli HMC, Edington MA (1983) Bacterial calcification in limestone caves. Geomicrobiol J 3:1–15

De Craen M, Swennen R, Keppens EM, Macaulay CI, Kiriakoulakis K (1999) Bacterially mediated formation of carbonate concretions in the Oligocene Boom Clay of northern Belgium. J Sediment Res 69(5):1098–1106

Geology of Near-Kazan region (2007) Guidebook for field student geological practice. In: Shevelev AI (ed) "Novoie znanie", Kazan

Green AC, Madgwick JC (1991) Microbial formation of manganese oxides. Appl Environ Microbiol 57:1114–1120

Khasanov RR, Galeev AA (2008) Evolution of sin-genetic organic material in Paleozoic deposits of central part of Volga-Ural anteclise. Science. Proceedings of Kazan University. Series of Natural Sciences, vol 150, 3:152–161

Korolev EA, Khuzin IA, Galeev AA, Leonova LV (2014) Epigenetic transformation of dolomite rocks under the influence of hydrocarbonaceous fluids (by the example of Syukeevskoe bituminous deposit). Geol Oil Gas 5:28–32

Mason-Williams MA (1969) Microorganisms in relation to food and energy sources in caves. Proc Br Speleol Assoc 4:69–74

Murav'yev FA (2007) Lithological-mineralogical characterization of Permian marking calcareous horizons in Tatarstan. PhD thesis. Kazan

Paquette J, Reeder RJ (1990) Single-crystal X-ray structure refinements of two biogenic magnesian calcite crystals. Am Mineral 75:1151–1158

Raiswell R (1976) The microbiological formation of carbonate concretions in the Upper Lias of NE England. Chem Geol 18:227–244

Skinner HCW, Fitzpatrick RW (eds) (1992) Biomineralization processes of iron and manganese-modern and ancient environments. CATENA, (Supplement 21)

Soroka EI, Leonova LV, Galeev AA, Gulyaeva TY (2007) EPR properties of organic component of some high-aluminous rocks of Urals. Lithosphere 4:125–128

Stocks-Fischer Sh, Galinat JK, Bang SS (1999) Microbiological precipitation of $CaCO_3$. Soil Biol Biochem 31:1563–1571

Strakhov NM (1962) Foundation of lithogenesis. vol 2

Tazaki K (2000) Formation of banded iron-manganese structures by natural microbial communities. Clays Clay Miner 48(5):511–520

Votyakov SL, Galeev AA, Leonova LV, Galakhova OL, Il'inykh AS (2005) EPR as method of investigation of organic component at biogenic calcareous rocks (Riphean stromatolite-containing rocks from South Urals as an example). Yearbook, Ekaterinburg, pp 39–47

Wei L, Li-Ping L, Peng-Peng Zh, Long C, Long-Jiang Y, Shi-Yun J (2011) Calcite precipitation induced by bacteria and bacterially produced carbonic anhydrase. Res Art Curr Sci 100 (4):502–508

# Structural Features and Composition of Amber from Placers on the East Coast of Sakhalin Island

**Valery V. Kononov, Olga P. Smyshlyaeva and Mihael E. Zelenski**

**Abstract** Information on fossilized resin—a valuable raw material for chemistry, agriculture, and medicine—from the Russian Far East is scarce and insufficient. We studied an extensive collection of samples of amber from beach placers from the mouth of the Najba River to the village of Starodubskoe (South Sakhalin). The sources of amber deposits were erosional local coal deposits from the Paleogene age. The shapes of the wreckage were mostly wrong, squished, flow-like, round and drop-like. The transparency of the samples ranged from perfect to opaque. The textures were solid and flow-sandwich. The color ranged from light yellow to brownish or cherry red. Amber contains numerous microscopic minerals, organic materials, and gaseous inclusions. Based on spectroscopic infrared studies, most amber samples were found to be rumanites, while the rest are an intermediate type between rumanites and retinites. The central part of the sample may have a composition of the rumanite and peripheral-zone retinite. Two main compositional features of these natural resins are the ratio of aromatic to oxygenated hydrocarbons, and the ratio between the ester and the carboxyl groups in the oxygenated compounds. The amber coloration is mainly determined by the distribution of bitumen admixture originating from bitumen crusts of the amber pieces, which also determines the color zoning of the specimens.

**Keywords** Sakhalin Island · Amber · Fossilized resin · Rumanit · IR spectroscopy · Aromatic · Oxygenated hydrocarbons · Mineral inclusions · Bitumen

V.V. Kononov (✉) · O.P. Smyshlyaeva
Russian Academy of Sciences, Far Eastern Geological Institute, Stoletiya Prospect 159, Vladivostok 690022, Russia
e-mail: kononov46@mail.ru

M.E. Zelenski
Institute of Experimental Mineralogy, Chernogolovka, Moscow 142432, Russia



O.V. Frank-Kamenetskaya et al. (eds.), *Biogenic—Abiogenic Interactions in Natural and Anthropogenic Systems*, Lecture Notes in Earth System Sciences,
DOI 10.1007/978-3-319-24987-2_9

# 1 Introduction

Amber is a fossilized resin that is a valuable raw material for jewelry, chemistry, agriculture, and medicine. European amber is well studied, while information on fossilized resins from the Russian Far East is scarce and insufficient. Some information on amber and amber-like fossilized resins from the Russian Far East can be found in review articles, where the results of the study of individual samples from the Far East are compared with extensive data on European amber (Srebrodoljsky 1984; Beck 1986, 1999; Frakey 1990; Gold et al. 1999; Kosmowska-Ceranowiz 1999; Bogdasarov et al. 2008a, b; Golubev et al. 2008; Bogdasarov 2009; Macuy and Granova 2009). Zhighin (1977) presented the results of the survey and assessment work on amber deposits from modern coastal-marine placers located on the east coast of South Sakhalin, but this article has little information about the internal structure and composition of Far East amber. This chapter reveals new data on the physical and chemical characteristics of Far East amber deposits.

We studied an extensive collection of amber samples from beach placers located from the mouth of the Nayba River to Starodubskoe village, South Sakhalin, in the Russian Far East. The placers were confined to modern coastal marine sediments represented by sand-gravel-pebble formations. According to Zhighin (1977), the formation of such placers is associated with the erosion of sand-clay carboniferous amber-containing sediments and local Paleogene coal deposits by longshore sea currents.

The size of studied amber pieces varied from a few millimeters to 5–6 cm. Most of the amber pieces had an irregular or flattened shape; some pieces were pancake shaped or, rarely, round or drop-shaped (Fig. 1a, b). The surfaces of the pieces were mainly irregular, knobby, and wrinkled (Fig. 1a, b). Some smaller grains were drop-shaped with a smooth surface. The peripheral parts of the amber pieces usually had a lower transparency and more intense coloration (Fig. 1c). Most pieces were covered with a dark brown or black bitumen "jacket" (crust) up to 1 mm thick (Fig. 1d, e).

The transparency of the studied samples varied from perfect to opaque. Translucent pieces contained numerous microscopic, mostly colorless, inclusions, thin cracks, and fluidal orange-brown and light brown structures (Fig. 1f, g). The main color of the amber was orange-yellowish and orange-ochre, but it varied within a single sample (Fig. 1h). The color zonality of grains was often observed from light yellow or light ochre in the central part to light brown, orange-brown, or reddish-brown in the periphery (Fig. 2a). Some samples entirely consisted of brownish-red to cherry red substances; the rare yellow grains may have a pale green tint. Under a microscope in reflected light, amber has predominantly brownish-gray or light brown color, but zonal pieces have brown or reddish-brown periphery and some samples show a visible blue tint.

The studied amber had a massive, layered, or conchoidal texture (Fig. 2b) and conchoidal or, rarer, smooth fracture. Many specimens contain thin irregular or dendritic fractures. Rare oval pores are located within the peripheral zone of the

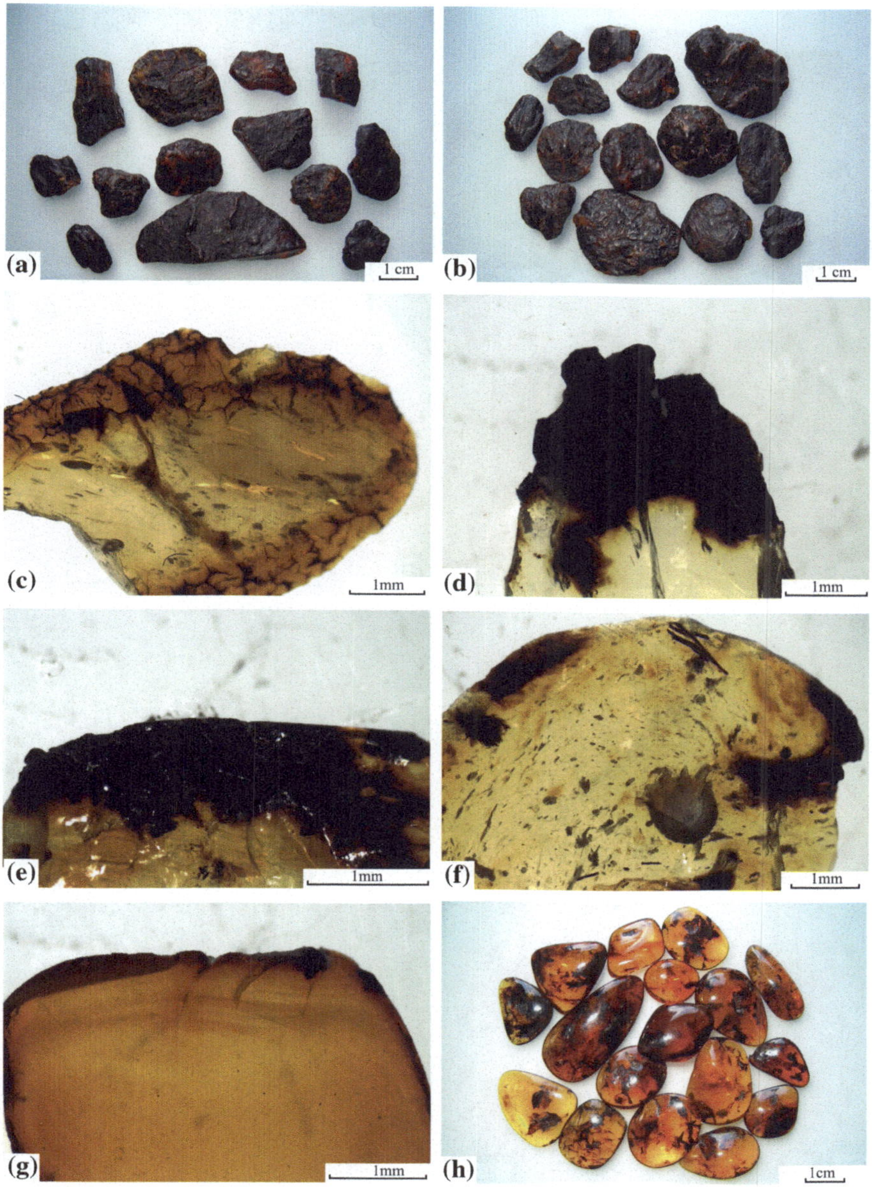


**Fig. 1** A general view and textural features of Sakhalin amber. **a**, **b** Pieces of raw amber. **c** Zonal coloration of a piece of amber with small peripheral cracks filled with *black* bitumen and the central part with indistinct inclusions. **d** A piece with *black* periphery and a semitransparent *dark-brown* zone with inclusions of *gray* clay particles. **e** A piece with *black* peripheral crust and *dark* stains and inclusions under the crust. **f** A semitransparent grain of amber saturated with different inclusions, with a round inclusion of gypsum visible in center of the grain. **g** A transparent piece of amber with indistinct *light-brown* layers in its periphery. **h** Polished amber stones with scattered bitumen inclusions and uneven coloration

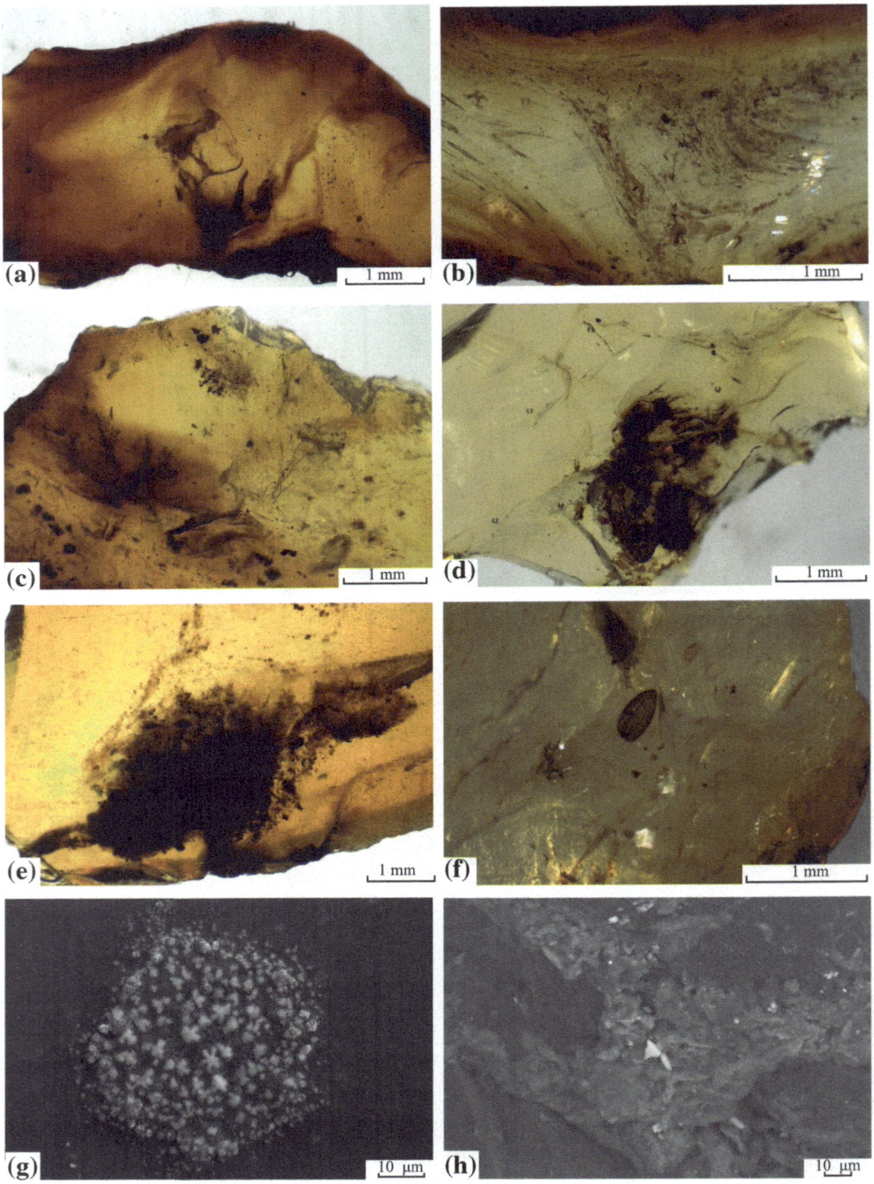


**Fig. 2** Internal structure of amber. **a** A zonal semitransparent piece of amber with *brown* bitumen inclusions. **b** Fluidal structure of amber. **c** Plant inclusions and cracks filled with bitumen. **d** A cluster of *black* and *brown* bitumen inclusions. **e** A cluster of shapeless bitumen inclusions. **f** Fragments of insects. **g** A cluster of star-shaped aggregates of (Na, K)(Cl, Br) crystals on a split surface of an amber grain. **h** The filling of pores with clay minerals (*light gray*) and rare crystals of (Na, K)(Cl, Br) (*white*)

samples. Some amber samples have tubular hollow cavities, oriented parallel to the layers (Fig. 2b). Inclusions in amber are mainly small, dark, and opaque particles or clusters of fine particles (Fig. 2a, c–e). Some samples contained colorless, white, and light gray inclusions (Figs. 1f, 2b). A limited amount of amber pieces possessed fragments of insects and plants (Fig. 2f).

The majority of the samples displayed blue fluorescence in ultraviolet radiation; more transparent specimens showed brighter fluorescence. According to Zhighin (1977), the average composition of amber from the placers near the Starodubskoe village includes 79.30 wt% carbon, 9.90 wt% hydrogen, 10.16 wt% oxygen, 0.23 wt% nitrogen, and 0.35 wt% sulfur.

## 2 Methods

Trace element contents in amber grains of different colors from the same placer were measured using a portable X-ray fluorescence spectrometer (Innov-X, USA; Table 1). Dark-colored amber and bitumen crusts of grains are enriched with Fe, S, Cl, and Ti. Some analyses of amber grains with bitumen crusts show elevated amounts of Cl, S, K, Fe and Ti.

Microscopic studies of split particles and transparent polished wafers of amber in transmitted and reflected polarized light and photomicrography were made using optical microscopes (Leica EZ4D, Zeiss, Germany; ECLIPSE LV100POL, Nikon, Japan).

The study of the internal structure and chemical composition of amber was carried out using a scanning electron microscope (Zeiss EVO 50 XPV, Germany) equipped with an INCA Energy EDS spectrometer. Unpolished splits of amber were chromium-coated and then studied. Infrared (IR) spectroscopic studies were made using Thermo Scientific Nicolet 6700 FT-IR Spectrometer (USA) in the spectral ranges of 4000–350 and 4000–650 $cm^{-1}$. IR spectra were obtained from KBr pellets with a small addition (0.5–1.0 mg) of the analyzed substance.

## 3 Results and Discussion

The matrix of the studied grains of amber contains loosely accreted, intersecting, and often bent fibers that are 0.3–1 μm long and up to 0.1 μm in diameter. Golubev and Kovalyova (2003), Golubev et al. (2005) used atomic-force microscopy to establish that these fibers are aggregates of globular particles from 50 to 140 nm in size. Such intersecting fibers result in the observed mesh structure of the matrix of amber complicated by clusters of clotted aggregates. Chain-like aggregates of such particles prevail in amber with a fluidal texture.

According to energy-dispersive X-ray spectroscopy analyses, the central (lighter) zones of the amber pieces have an average chemical composition (without

**Table 1** Trace element contents in amber from the Starodubskoe beach placer, ppm

| No. | Sample | Ca | K | Cl | S | Ti | Ba | Cr | Mn | Fe | Zn | Pb | Rb | Sr | Zr | Mo |
|---|---|---|---|---|---|---|---|---|---|---|---|---|---|---|---|---|
| 1 | C-1-1 | – | 271 | 707 | 5470 | – | 28 | 16 | 56 | 408 | – | 15 | 13 | 32 | 58 | 68 |
| 2 | C-2-1 | – | 213 | 330 | 3730 | – | – | – | 22 | – | – | 11 | 10 | 18 | 48 | 50 |
| 3 | C-2-2 | – | 236 | 417 | 13300 | – | – | – | 37 | 82 | – | 21 | 13 | 23 | – | 61 |
| 4 | C-3-1 | – | 346 | 726 | 3990 | – | – | – | 20 | 69 | – | – | 11 | 21 | 47 | 52 |
| 5 | C-3-2 | – | 205 | 508 | 4030 | – | – | – | 26 | – | – | – | 11 | 20 | 53 | 62 |
| 6 | C-4-1 | – | 204 | 420 | 5360 | – | – | – | 32 | 38 | – | 10 | 13 | 23 | 51 | 63 |
| 7 | C-13-1 | 427 | 430 | 796 | 7110 | 910 | 56 | 11 | 41 | 960 | 16 | 20 | 15 | 30 | 61 | 57 |
| 8 | C-6-1 | – | 584 | 1540 | 8910 | – | – | – | 58 | 953 | – | 15 | 15 | 29 | 61 | 66 |
| 9 | C-6-2 | – | 323 | 495 | 5000 | – | – | – | 37 | 97 | – | 10 | 13 | 22 | 60 | 67 |
| 10 | C-2-3 | – | 280 | 403 | 2000 | – | – | – | 16 | 68 | – | – | 5 | 30 | 31 | 34 |
| 11 | C-2-4 | – | 508 | 298 | 1600 | 148 | – | – | 27 | 707 | 11 | – | 7 | 32 | 31 | 37 |
| 12 | C-5-1 | – | 233 | 479 | 3500 | – | – | – | 18 | 127 | – | 10 | 10 | 20 | 53 | 52 |
| 13 | C-6-3 | – | 238 | 498 | 1820 | – | – | – | 12 | – | – | – | 9 | 19 | 40 | 52 |
| 14 | C-7-1 | – | 339 | 593 | 4890 | – | – | – | 31 | 131 | – | 21 | 12 | 21 | 49 | 52 |
| 15 | C-10-1 | – | 308 | 862 | 6460 | – | – | – | 36 | 356 | – | 14 | 11 | 24 | 55 | 59 |
| 16 | C-13-2 | 408 | 712 | 947 | 7560 | 134 | 23 | – | 32 | 2230 | – | 10 | – | 24 | 51 | 51 |
| 17 | C-13-3 | – | 358 | 774 | 7290 | 54 | – | – | – | 310 | – | – | 11 | 19 | 48 | 52 |
| 18 | C-11-1 | – | 305 | 495 | 5010 | 38 | – | – | 18 | 125 | 11 | 15 | 7 | 19 | 51 | 53 |
| 19 | C-11-2 | – | 380 | 656 | 12,600 | – | – | – | 32 | 236 | – | – | 11 | 23 | 52 | 58 |
| 20 | C-12-1 | – | 243 | 351 | 2540 | – | – | – | 16 | 96 | 11 | 11 | 11 | 21 | 44 | 50 |
| 21 | C-12-2 | – | 453 | 606 | 6140 | 39 | – | – | 9 | 294 | – | – | 8 | 16 | 43 | 50 |
| 22 | C-10-2 | – | 533 | 1180 | 10,420 | – | – | – | 15 | 138 | – | 10 | 10 | 18 | 44 | 53 |

(continued)

**Table 1** (continued)

| No. | Sample | Ca | K | Cl | S | Ti | Ba | Cr | Mn | Fe | Zn | Pb | Rb | Sr | Zr | Mo |
|---|---|---|---|---|---|---|---|---|---|---|---|---|---|---|---|---|
| 23 | C-9-1 | – | 468 | 1460 | 10,350 | – | – | – | 20 | 640 | – | 13 | 10 | 25 | 53 | 59 |
| 24 | C-9-2 | – | 321 | 482 | 8520 | – | – | – | 18 | 102 | – | 11 | 11 | 180 | 48 | 51 |
| 25 | C-13-4 | – | 701 | 821 | 21,100 | 112 | – | – | 21 | 2260 | – | 11 | 12 | 28 | 52 | 60 |

*Samples* C-1-1 and C-2-1—transparent light yellow; C-2-2—transparent light yellow with green tint; C-3-1—semitransparent greenish yellow; C-3-2—transparent bright yellowish brown; C-4-1—transparent bright brownish yellow; C-13-1—transparent yellow covered by bitumen crust; C-6-1—transparent yellow; C-6-2—transparent brownish yellow; C-2-3 и C-2-4—transparent light-ochre; C-5-1 and C-6-3—transparent yellowish brown; C-7-1—semitransparent light brown; C-10-1—semitransparent brown; C-13-2 and C-13-3—semitransparent brown, covered by bitumen crust; C-11-1 and C-11-2—semitransparent reddish brown, covered by bitumen crust; C-12-1—waxen-brown; C-12-2—semitransparent reddish brown, covered by bitumen crust; C-10-2—transparent reddish brown with fluidal texture; C-9-1 and C-9-2—semitransparent light red with fluidal texture; C-13-4—semitransparent reddish brown, covered by bitumen crust hydrogen) of 87.65 wt% carbon, 12.00 wt% oxygen, and 0.23 wt% sulfur. The average composition of the peripheral zones was 87.34 wt% carbon and 12.06 wt% oxygen. Sulfur in the peripheral zones was measured only at one point (0.37 wt%).

In general, the chemical composition of the grains of amber was more or less constant, with some heterogeneity near the edges. Bitumen crusts around the amber grains contained elevated amounts of oxygen and sulfur. The average composition of bitumen crusts was 80.50 wt% carbon, 18.19 wt% oxygen, and 0.64 wt% sulfur. Some analyzed sites within the crusts contained up to 0.73 wt% calcium, 0.76 wt% sodium, 0.36 wt% potassium, 0.63 wt% silicon, and 6.18 wt% aluminum.

The amber contained microscopic inclusions of NaCl, KCl, NaBr, and KBr. Crystals of these salts formed clusters of star-shaped aggregates on surfaces of split amber grains (Fig. 2g). Such aggregates ranged from 2 to 8.7 μm in size and were composed of crystals of 0.5–1 μm in size. The origin of these structures is possibly connected to the seawater that penetrates through fractures into amber pieces and then evaporates to form chloride and bromide salts. Other mineral inclusions were amphiboles, acid plagioclase, gypsum, iron sulfate, carbonates, quartz, and aluminosilicates of Ca, Fe, Na, and Mg, as well as clay minerals. The size of the inclusions ranged from microns for single crystals to tens of microns for polymineral aggregates (Fig. 2h). An elongated particle of native iron was also observed in one of the healed microcracks. Bitumen crusts contained lens-shaped inclusions of clay-quartz mixture.

In amber from the Starodubskoe placer, IR spectroscopy showed a predominance of saturated hydrocarbons ($CH_2$ and $CH_3$ groups with stretching vibrations, which cause absorption in the range of 2800–3000 $cm^{-1}$). Analysis of the IR spectra of the samples also revealed that the specific features of amber are determined by the ratio of aromatic to oxygenated compounds.

In the case of a high content of aromatic hydrocarbons (if the absorption peak at 1459–1461 $cm^{-1}$ caused by skeletal vibrations of C–H bonds in aromatic ring becomes as high as the absorption peak at 1722–1725 $cm^{-1}$ caused by stretching vibrations of the ester group), IR spectra of the samples also contained the absorption peaks at 818–822 and 600–617 $cm^{-1}$ caused by deformation vibrations of C–H bonds in aromatic rings (Fig. 3a). Some spectra of this type contained a low-intensity absorption band in the range of 1040–1240 $cm^{-1}$ caused by stretching vibrations of C–O ester groups. The samples of this type also contain an admixture of aromatic compounds (the absorption peaks at 1620–1640 $cm^{-1}$ caused by associated C=O group) as well as isopropyl resin, as indicated by the presence of absorption peaks at 1377–1379 $cm^{-1}$. IR spectra of this type of amber contained the absorption peak at 975–978 $cm^{-1}$ connected with non-planar deformation vibrations of C–H.

A group of samples of amber with such a composition can be attributed to rumenite. Some samples have an intermediate composition between rumenite and retinite. Such samples contain both oxygenated and aromatic compounds and can be subdivided into two groups—one group being enriched in ester (absorption peaks at 1722–1725 $cm^{-1}$) and the other group being enriched in carboxyl

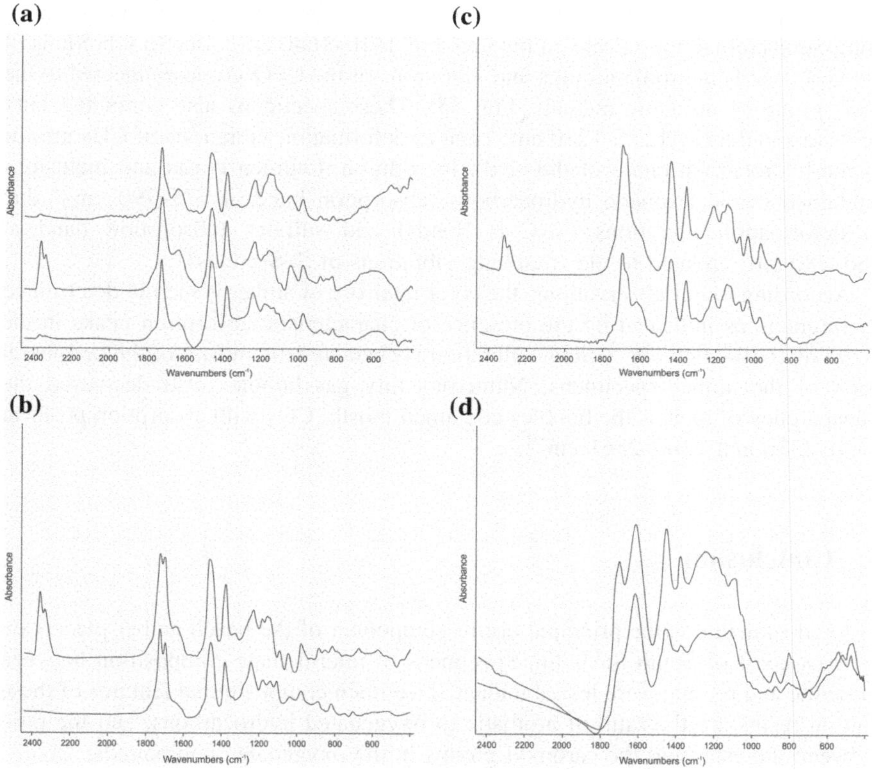


**Fig. 3** Infrared spectra of varieties of amber. **a** IR spectra of rumenite. A strong absorption peak at 1722–1725 $cm^{-1}$ is caused by stretching vibrations of an ester group. The absorption peak at 1698–1700 $cm^{-1}$ is absent. **b** IR spectra of rumenite-retinite. The absorption peaks at 1723–1725 $cm^{-1}$ and 1698–1703 $cm^{-1}$ are caused by vibrations of carboxyl groups and are similar in their intensity. **c** IR spectra of rumenite with some admixture of retinite. The absorption peak at 1701–1703 $cm^{-1}$ is located on a shoulder of the stronger peak at 1722–1724 $cm^{-1}$. **d** IR spectra of bitumen. An intensive absorption band in the range of 1610–1640 $cm^{-1}$ is caused by vibrations of chemical bonds in aromatic compounds. The absorption peak at 1699–1701 $cm^{-1}$ is caused by some retinite admixture

(absorption peaks at 1698–1705 $cm^{-1}$). In a number of cases, the central part of the grain has a composition close to rumenite, whereas peripheral areas are retinite-like.

IR spectra of the samples with approximately equal contents of ester and carboxyl groups (Fig. 3b) display relatively low absorption peaks, in the range of 1040–1150 $cm^{-1}$ (ester compounds), as well as low absorption peaks at 600–617 and 975 $cm^{-1}$, which is characteristic of the spectra of retinites. Such samples contain little free hydroxyl groups. Samples enriched in ester compounds (similar to rumenite, Fig. 3c), in addition to having a strong absorption peak at 1722–1724 $cm^{-1}$, showed other strong peaks in the absorption band of 1040–1170 $cm^{-1}$ due to vibrations of the ester groups and aromatic esters.

Bitumen crusts contained elevated amount of water and some specific aromatic compounds (absorption peaks in the range of 1610–1640 $cm^{-1}$ due to vibrations of the C=C bonds in aromatic rings and vibrations of the C=O group connected to the C=C group of aromatic radicals, Fig. 3d). These specimens also contained fatty acids (a shoulder at 1275–1280 $cm^{-1}$ due to deformation vibrations of $CH_2$ groups in the hydrocarbon chain of the acid). In addition, bitumen crusts and inclusions contained cyclic aromatic hydrocarbons (absorption band at 775–790 $cm^{-1}$ due to deformation vibrations of C–H bonds) and sulfides (absorption band at 460–480 $cm^{-1}$ related to the stretching vibrations of S–S bonds).

According to our observations, the color intensity of amber is mainly determined by bitumen, as indicated by the presence of characteristic absorption peaks in the range of 1610–1640 $cm^{-1}$. Such coloration is characteristic mostly of the peripheral parts of the amber specimens. Numerous tiny gas bubbles also decreased the transparency of amber; the bubbles contained mostly $CO_2$ with absorption peaks at 2330–2336 and 2360–2361 $cm^{-1}$.

## 4 Conclusion

Colored rumenite is the principal useful component of the beach amber placers on the east coast of South Sakhalin. Specimens of intermediate composition between rumenite and retinite were less abundant. Two main compositional features of these natural resins are the ratio of aromatic to oxygenated hydrocarbons, and the ratio between the ester and the carboxyl groups in the oxygenated compounds.

The amber contained numerous microscopic mineral, organic, and gaseous inclusions. The amber's coloration was mainly determined by the distribution of bitumen admixture originating from bitumen crusts of the amber pieces, which also determined the color zoning of the specimens. The oxidation of amber pieces was weak. Some rumenite pieces have peripheral retinite, which possibly can be accounted for by fossilization conditions. The development of retinite might occur due to the fact that some resin was fossilized within wetlands under anaerobic conditions, where the primary resin acids were hydrogenated and decarboxylated to form the retinite phase (Skrigan 1981; Kievlenko and Senkevich 1983).

## References

Beck CW (1986) Spectroscopic investigation of amber. Appl Spectrosc Rev 22(1986):57–110

Beck CW (1999) The chemistry of amber. Estud del Museo de ciencias Nat de Alava Numero especial 14:33–48

Bogdasarov MA (2009) Iskopaemie smoli Severnoy Evrazii. Izd – vo BrGU,. 175 s (in Russian)

Bogdasarov MA, Bushnev DA, Golubev EA, Kovaleva OV, Shanina SN (2008a) Jantar i jantarepodobnie iskopaemie smoli Evrazii. Statya 1. Infrakrasnaya spektroskopiya, differencialny termichesky analiz. Izvestiya vyzov Geologija i razvedka 2:23–30 (in Russian)

Bogdasarov MA, Kovaleva OV, Konovalenko SI (2008b) Novie rezultati issledovany iskopaemih smol Severnoy Evrazii metodom infrakrasnoy spektroskopii. Gemmologiya: Materiali III gemmolog. nauch. konf. Tomsk. S. 16–20 (in Russian)

Frakey E (1990) Jantarj. Mir. 198 s (in Russian)

Gold D, Hazen B, Miller W (1999) Colloidal and polymeric nature of fossil amber. Org Geochem, 30: 971–983

Golubev EA, Kovaleva OV (2003) Vizualizaciya nadmolekulyarnogo stroeniya baltiyskogo jantarya metodom atomno – silovoy mikroskopii. Materiali XIV molodeghnoy konferencii "geologiya, poleznie iskopaemie I geoekologiya severo – zapada Rossii". Petrozavodsk. KarNC RAN. S. 79–80 (in Russian)

Golubev EA, Kovaleva OV (2005) Nadmolekularnaya mikro – i nanostrukturnaya uporyadochennost v tverdih uglerodistih veshestvah. Nanomineralogiya. Ultra – i mikrodispersnoe sostoyanie mineralnogo veshestva, SPb: Nauka. S. 232–246 (in Russian)

Golubev EA, Kovaleva OV, Shanina SN (2008) Sostav, himicheskaya struktura I nadmolekulyarnoe stroenie iskopaemih smol. Minerali i mineraloobrazovanie, struktura, raznoobrazie i evoluciya mineralnogo mira, rolj mineralov v proishoghdenii i razvitii ghizni, biomineralnie vzaimodeystviya. Siktivkar: IG Komi HC UrO RAN. S. 240–258 (in Russian)

Kievlenko EY, Senkevich NN (1983) Geologiya mestoroghdeny podelochnih kamney. M: Nedra 263 s (in Russian)

Kosmowska-Ceranowiz B (1999) Succinite and some other fossil resins in Poland and Europe (deposits, finds, features and differences in JRS). Estudios del Museo de ciencias naturals de Alava, 14. # 2, pp 73–117

Macuy VM, Granova AK (2009) Rossipi jantarya sovremennih morey. Geologiya poleznih iskopaemih Mirovogo okeana 1:69–76 (in Russian)

Skrigan AI (1981) Processi prevrasheniya drevesini i ee himicheskaya pererabotka. Minsk: Nauka i tehnika 206 s (in Russian)

Srebrodoljsky BI (1984) Jantarj. M.: Nauka 112 s (in Russian)

Zhighin AD (1977) Sovremennie pribreghno – morskie rossipi jantarya na vostochnom pobereghje Sahalina. Litologiya i poleznie iskopaemie. Nauka 2:133–137 (in Russian)

# Evidence of Biogenic Activity in Quartz-Hematite Rocks of the Urals VMS Deposits

**Nuriya R. Ayupova, Valeriy V. Maslennikov, Sergei A. Sadykov, Svetlana P. Maslennikova and Leonid V. Danyushevsky**

**Abstract** The textural, mineralogical, and geochemical features of quartz-hematite rocks associated with Urals VMS deposits indicate that the tube microfossils are responsible for immobilization and accumulation of chemical elements during precipitation of authigenic minerals. The crystallization of authigenic minerals is a result of submarine transformation of mixed hyaloclastitic, sulfide, and carbonate sediments and diagenetic processes, which modify the mineralogy and geochemistry of sediments. The tube microfossils about 100 μm across and up to 1 mm long consist of the external rim made up of fine-disperse hematite or hematite-quartz aggregates and of the internal channel filled with hematite and/or transparent quartz, fine-disperse hematite-quartz aggregates, leucoxene, rare sulfides, apatite, Fe-chlorite, and Mn-calcite. The carbon isotopic composition of calcite from quartz-hematite rocks with tube microfossils (up to −26.2 ‰) indicates its biogenic origin. The habitat conditions of the tube microfossils favored the mineral precipitation. The newly formed apatite, rutile, illite, monazite, dolomite, ankerite, siderite, monheimite, REE carbonates, anatase, leucoxene, Mn-oxides, titanomagnetite, and hematitized framboidal pyrite are observed in quartz-hematite matrix with abundant tube organisms in contrast to quartz-hematite rocks free of tube microfossils. Biomorphic hematite contains high contents of Mn (up to 9393 ppm), As (up to 1872 ppm), V (up to 779 ppm), W (up to 1091 ppm), Mo (up to 40 ppm), and U (up to 8.68 ppm), which are indicative of biological mechanisms of accumulation and conservation of these metals in the system.

**Keywords** Tube microfossils · Quartz-hematite rocks · Authigenic minerals · Trace elements · VMS deposits

N.R. Ayupova (✉) · V.V. Maslennikov · S.A. Sadykov · S.P. Maslennikova
Institute of Mineralogy, Ural Branch of Russian Academy of Science, Miass, Russia
e-mail: aupova@mineralogy.ru

N.R. Ayupova · V.V. Maslennikov
South Urals State University, Miass branch, Miass, Russia

L.V. Danyushevsky
Centre for Deposit and Exploration Studies, University of Tasmania, Hobart, Australia



O.V. Frank-Kamenetskaya et al. (eds.), *Biogenic—Abiogenic Interactions in Natural and Anthropogenic Systems*, Lecture Notes in Earth System Sciences,
DOI 10.1007/978-3-319-24987-2_10

# 1 Introduction

Numerous ancient volcanogenic massive sulfide (VMS) deposits are spatially associated with quartz-hematite (siliceous-ferruginous) rocks (Constantinou and Govett 1973; Kalogeropoulos and Scott 1983; Duhig et al. 1992; Peter and Goodfellow 1996; Grenne and Slack 2003; Little et al. 2004; Maslennikov et al. 2012) similarly to the modern hydrothermal sulfide fields associated with silica-rich iron-oxyhydroxide deposits (Adachi et al. 1986; Juniper and Fouquet 1988; Herzig et al. 1991; Hekinian et al. 1993; Halbach et al. 2002; Hein et al. 2008). The quartz-hematite rocks are one of the most important criteria for local prospecting of VMS deposits, however, their formation is still unequivocal and may be due to the low-temperature hydrothermal seafloor growth (Halbach et al. 2002), precipitation from seawater-diluted hydrothermal plume (Grenne and Slack 2003), seafloor oxidation of massive sulfides (Hannington and Scott 1988), or halmyrolysis of sulfides mixed with hyaloclastites and carbonates (Maslennikov et al. 2012). Microbial activity is thought to play an important role in formation of siliceous-ferruginous sediments (Juniper and Fouquet 1988; Little et al. 2004; Ayupova and Maslennikov 2012, 2013), but identification of microfossils in ancient rocks is extremely difficult because of their small sizes and superimposed anadiagenetic or metamorphic processes.

In this paper, we characterize the morphology, mineralogy, and geochemistry of the tube microfossils from quartz-hematite rocks of the Urals VMS deposits, since these organisms are responsible for the immobilization and accumulation of chemical elements during formation of authigenic minerals. The crystallization of authigenic minerals is a result of submarine transformation of mixed hyaloclastitic, sulfide, and carbonate sediments, and diagenetic processes, which modify the mineralogy and geochemistry of sediments. The study of mineralized microfauna may provide a key for understanding the reason of textural-structural and mineralogical-geochemical diversity of quartz-hematite rocks in VMS regions and may help in recognition of their ore-controlling varieties.

# 2 Sampling and Analytical Work

The samples of quartz-hematite rocks with tube microfossils were collected in the open pits and mines of the Talgan, Molodezhnoe, XIX Parts'ezd, Alexandrinka, Uchaly, Sibai, Uzelga, Novyi Shemur, and Yaman-Kasy VMS deposits of the Urals. The polished sections of samples were studied with an Olympus BX51 optical microscope and a REMMA-202 electron microscope equipped with a Link ED-System at the Institute of Mineralogy (IMin UB RAS). SEM analyses were carried out using a 1-micron electron beam, 15 nA beam current, 20 kV accelerating voltage, and a counting time of 120 s for peak.

The carbonates for carbon isotopic analysis were sampled from grains associated with microfossil-bearing quartz-hematite layers, carbonate clasts in quartz-hematite rocks, and hanging wall limestones. The carbon isotopic composition ($\delta^{13}C$, ‰ PDB) was analyzed on a Delta$^+$ Advantage Thermo Finnigan mass-spectrometer at the IMin UB RAS using NBS-18, NBS-19, and IAEA-C-3 standards. The error of the analysis was ±0.1 ‰. The results of isotopic analyses were processed in the ISODAT-2.0 program.

Quantitative LA-ICP-MS analysis of biomorphic hematite and chalcopyrite from quartz-hematite rocks for trace elements (Mn, Zn, As, Se, Mo, Sn, Sb, Te, W, Pb, Bi, Mg, Al, Si, Ca, Ti, V, Cr, Ba, Th, and U) was carried out on a New Wave 213-nm solid-state laser microprobe coupled to an Agilent 4500 quadrupole ICPMS housed at the CODES LA-ICPMS analytical facility, University of Tasmania, following (Danyushevsky et al. 2011). The analyses were performed by ablating spots 40 μm across and 80 μm deep. The detection limit for most trace elements was 1 ppb.

## 3 Geological Setting

The Urals metallogenic belt is subdivided on the giant Tagyl, East and West Magnitogorsk, Sibai, Sakmara, and East Uralian metallogenic zones (Prokin et al. 1985). These zones represent different paleogeodynamic settings and include the fragments of Sakmara marginal sea ), Tagyl, West Magnitogorsk, and East Magnitogorsk island arcs, Sibai inter-arc basin, and Dombarovka, West Mugodzhary, and Rezh back-arc basins (Zaykov 1991) (Fig. 1). The VMS deposits were formed in all paleogeodynamic structures from the Late Silurian to Middle Devonian (Prokin et al. 1985).

The Novyi Shemur deposit located in the Tagyl paleoisland arc is hosted by Silurian andesite-dacite complex. The VMS deposits in Sibai inter-arc paleobasin (Sibai deposit) and West Magnitogorsk paleoisland arc (Uchaly, Talgan, Molodezhnoe, XIX Parts'ezd, and Alexandrinka deposits) occur in the Middle Devonian rhyolite-basaltic complexes of the Karamalytash Formation (Prokin et al. 1985). The Yaman-Kasy deposit is located within the Sakmara paleomarginal sea and is hosted by Silurian basalt-rhyolite association. All these deposits are remarkable for the low degree of metamorphism of massive sulfide ores, which contain fragments of black smoker chimneys, diffusers, and near-vent hydrothermal fauna (Maslennikov 2006).

The horizons of quartz-hematite rocks (former siliceous-ferruginous sediments) are common components of ore-bearing sections of the Urals VMS deposits (Fig. 2). The quartz-hematite layers (1–10 mm up to 50–60 cm) cover clastic sulfide ores or interbed with sulfide turbidites and chlorite hyaloclastites at the thinning out of the sulfide lens. Locally, they form the dispersion halos around degraded sulfide mounds, covering an area about twice to several times that of the ore bodies themselves. In the Urals, these rocks are called "gossanites" (Maslennikov et al. 2012) similarly to

**Fig. 1** The VMS-bearing regions of the Urals after (Prokin et al. 1985)

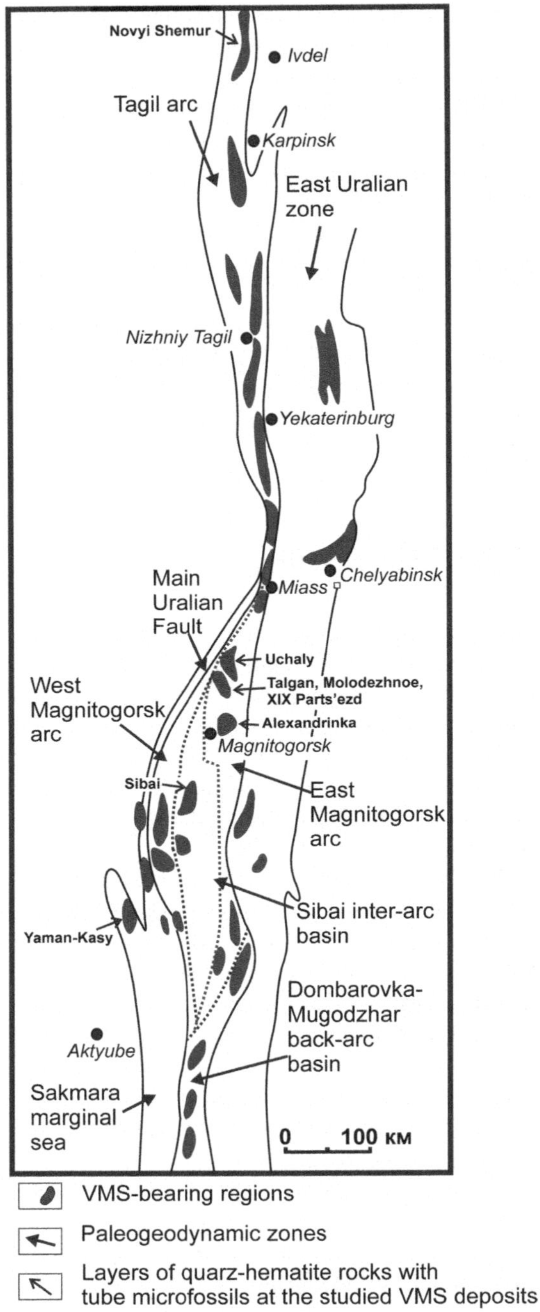

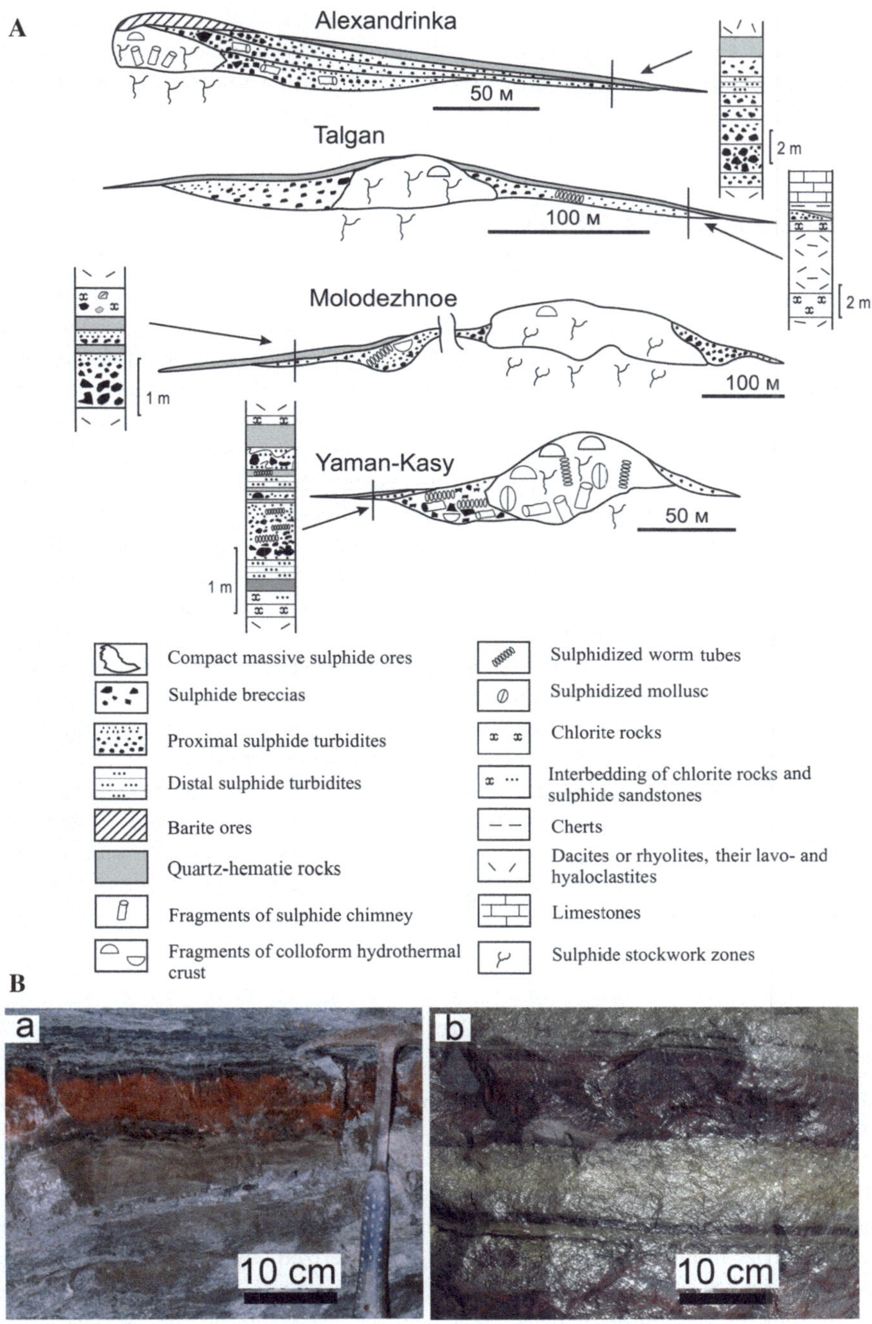


**Fig. 2** Position of quartz-hematite rocks (gossanites) in ore bodies of the VMS deposits of the Urals (*A*) (Maslennikov et al. 2012) and field photo of interbedding hematite-quartz rocks (*red*) with sulfide turbidites (*yellow*) and hyaloclastites (*dark green*) on the flanks of massive sulfide ore bodies: **a** Talgan mine, **b** Molodezhnoe open pit

submarine gossans (Herzig et al. 1991). The replacement textures of sulfide clasts and geochemical similarity of gossanites with sulfide ores indicate the formation of quartz-hematite rocks during seafloor oxidation of clastic sulfides mixed with fragments of hyaloclasts and various carbonates (Maslennikov et al. 2012).

## 4 Biomorphic Signatures

Numerous relics of cylindrical (in cross-section) tube microfossils up to 100 μm across and up to 1 mm long were found in quartz-hematite rocks associated with massive sulfide bodies. The microfossils are separated from the fine-disperse quartz-hematite matrix by a thin rim of various minerals. The areas with tube microfossils are also characterized by the presence of filamentous microbial fossils, which may be interpreted as fossilized bacteria. The tube microfossils consist of the external rim, which is made up of fine-disperse hematite or hematite-quartz aggregates, and of inequigranular, gel-like, and cellular internal channel filled with hematite and/or quartz, fine-disperse hematite-quartz aggregates, rare sulfides, leucoxene, apatite, Fe-chlorite, and Mn-bearing (up to 5 wt% MnO) calcite (Fig. 3). Some of the studied microfossils exhibit well-defined segmentation resulted from precipitation of Ti-minerals (Fig. 3h).

The development of microbial processes provided specific formation conditions favorable for precipitation of authigenic minerals. Newly formed round grains of apatite associated with rutile and illite (locally, with rim made up of manganese oxides), titanomagnetite with leucoxene rim, individual aggregates of leucoxene, rare monazite and REE carbonates, numerous shapeless aggregates of ankerite and siderite, calcite and dolomite crystals, monheimite, and hematitized framboidal pyrite are observed in the quartz-hematite matrix in association with tube microfossils (Fig. 4). As a rule, no these minerals were found in quartz-hematite rocks without tube microfossils.

## 5 Chemical Composition of the Tube Microfossils

The results of LA-ICP-MS analysis show relatively uniform and stable distribution of trace elements in biomorphic hematite (Table 1). At the same time, the high concentration of Mn (up to 9393 ppm), Ca (up to 0.38 wt%), As (up to 1872 ppm), V (779 ppm), W (up to 1091 ppm), Mo (up to 40 ppm), and U (up to 8.68 ppm) are indicative of biological accumulation and conservation of these elements in the system. The biomorphic structures are characterized by higher contents of rock-forming (Si, Al, Mg) and ore (Zn, Pb, Sb) elements borrowed from hyaloclastic material and sulfides, respectively, during fossilization. The elevated contents of ore metals are also typical of the modern tube worms (Lein et al. 1989;

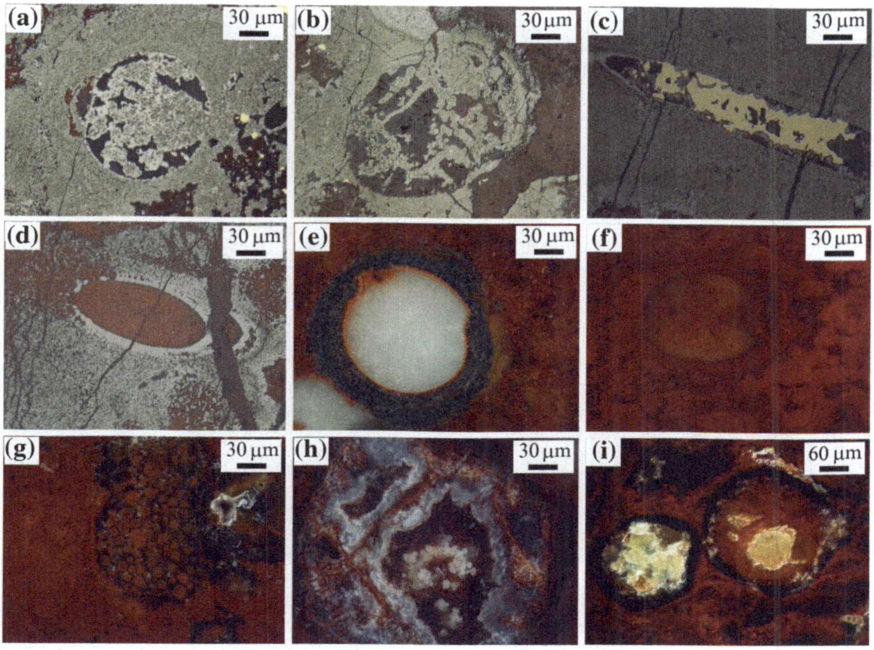


**Fig. 3** Cross-sections of tube microfossils from the Urals quartz-hematite rocks: (**a**–**f**) an internal channel incrusted with hematite (**a**, **b**), **c** chalcopyrite, **d** fine-disperse quartz-hematite aggregates, **e** quartz or **f** Mn-bearing calcite, **g** cellular quartz-hematite aggregate inside the tube organisms , **h** tube microfossil segmented by leucoxene aggregates, **i** microfossil with quartz-hematite channel and monheimite rim (*black*). Reflected light (**e**–**i**, *dark-field* images)

Juniper et al. 1992; Demina et al. 2007), as well as of sulfidized fauna of ancient hydrothermal systems (Maslennikov et al. 2014).

The hematite matrix in quartz-hematite rocks contains lower amounts of most trace elements, including As, W, and V and higher contents of rock-forming elements (Si, Mg, Al, Ca, Ti) and Mn and Mo in contrast to biomorphic hematite. Chalcopyrite, which incrusts the internal channels of the tube microfossils, is characterized by the high contents of rock-forming (Si, Mg, Al, Ca) and chalcophile (Te, Se, Bi) elements and low amounts of As, Mo, W, Ti, Th, and U, whereas V content is comparable with that in biomorphic hematite.

## 6 Carbon Isotopic Composition

The results of carbon isotopic analysis are given in Table 2. The $\delta^{13}C$ values of calcite from the hanging wall organic limestones of the Talgan and XIX Parts'ezd deposits range from −1.09 to +2 ‰ that is typical of marine carbonates (Hoefs

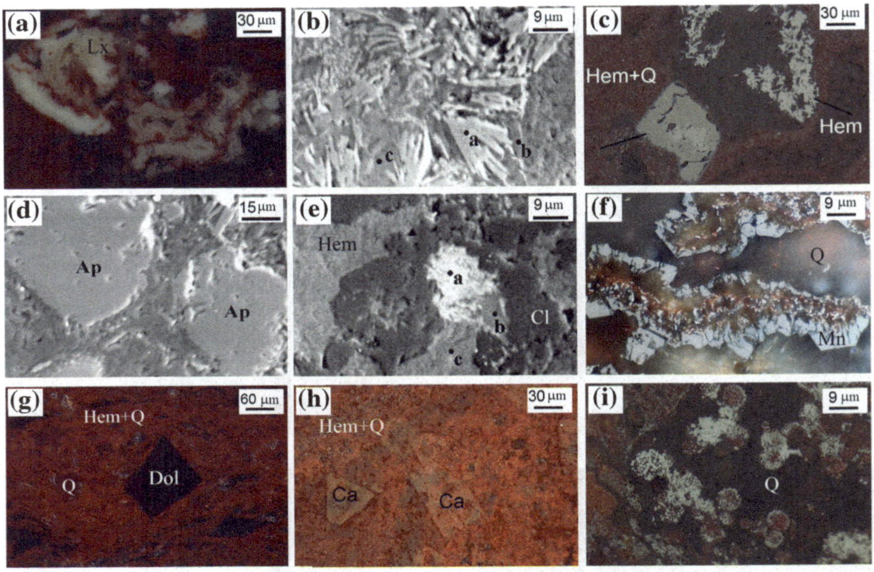


**Fig. 4** Authigenic minerals in quartz-hematite rocks with tube microfossils: **a** leucoxene aggregates (Lx), **b** needle-like rutile crystals (**a**) in quartz-chlorite cement (**b, c**), **c** titanomagnetite with leucoxene rim associated with hematite crystals (Hem), **d** round grains of apatite (Ap), **e** monazite (**a**) in chlorite(Cl)-hematite(Hem) association, (**b, c**—finely dispersed hematite-quartz aggregates), **f** apatite with a rim of manganese oxide in quartz (Q) cement, **g** dolomite crystal (Dol), **h** calcite crystals (Ca), **i** hematitized framboidal pyrite in quarth (Q) cement. *Dark-field* images (**a**, **c**, **f**, **g**, **h**, **i**). SEM-photo (**b**, **d**, **e**)

1997). The quartz-hematite rocks host isometric micritic clasts up to 1–2 cm in size with zigzag boundaries, which are often replaced by the fine-disperse hematite-quartz aggregates. The fragments of brachiopod shells and crinoidea segments may also be found in micritic, partially recrystallized clasts. The micritic and crinoid clasts in quartz-hematite rocks are characterized by $\delta^{13}C$ values varying from −0.09 to +0.97 ‰ and from −0.64 to +1.21 ‰, respectively, which corresponds to that of the hanging wall limestones.

The $\delta^{13}C$ values of the Mn-bearing calcite from the tube microfossils in quartz-hematite rocks vary from −4.2 to −26 ‰. These values span a range of the $\delta^{13}C$ values of the hydrothermal vent tube worms from rift zones of the Pacific Ocean (−11 to −14 ‰), whose trophosome tissue contains high activities of several enzymes associated with chemoautotrophic existence (Felbeck 1981; Gal'chenko et al. 1988). Similar $\delta^{13}C$ values are characteristic of calcite from the axial part of the Vestimentifera tubes of the Sibai deposit (−4.60 ‰) (Maslennikov 1999) and calcite from biomorphic massive sulfide ores of the Yubileinoe deposit in the South Urals (−5.79 ‰) (in-house unpublished data).

**Table 1** Trace element content in biomorphic hematite, inorganic matrix hematite, and chalcopyrite from quartz-hematite rocks of the Talgan VMS deposit (ppm)

| Elements | Mn | Zn | As | Se | Mo | Sn | Sb | Te | W | Pb | Bi | Mg | Al | Si | Ca | Ti | V | Cr | Ba | Th | U |
|---|---|---|---|---|---|---|---|---|---|---|---|---|---|---|---|---|---|---|---|---|---|
| *Biomorphic hematite* | | | | | | | | | | | | | | | | | | | | | |
| | 460 | 168 | 1595 | 1.99 | 20.76 | 2.21 | 209 | 0.15 | 994 | 144 | 0.09 | 134 | 1580 | 3197 | 313 | 84 | 639 | 1.31 | 13.42 | 0.37 | 5.22 |
| | 394 | 113 | 770 | 1.46 | 33.81 | 0.93 | 258 | 0.27 | 828 | 166 | 0.16 | 238 | 2495 | 2921 | 261 | 114 | 780 | 2.53 | 19.92 | 0.43 | 6.55 |
| | 674 | 192 | 1071 | 2.00 | 10.66 | 2.84 | 190 | 0.81 | 589 | 125 | 0.22 | 2481 | 4883 | 6598 | 1454 | 529 | 528 | 4.42 | 16.17 | 1.01 | 8.56 |
| | 502 | 153 | 1873 | 1.22 | 33.57 | 2.01 | 198 | 0.22 | 1091 | 187 | 0.11 | 143 | 1767 | 4731 | 264 | 144 | 647 | 1.82 | 15.86 | 0.51 | 5.31 |
| | 776 | 377 | 1543 | 1.55 | 30.3 | 1.79 | 244 | 0.91 | 800 | 368 | 0.12 | 797 | 2008 | 6665 | 274 | 150 | 557 | 1.37 | 17.76 | 0.44 | 8.63 |
| | 5545 | 675 | 1855 | 1.39 | 26.92 | 1.61 | 277 | 0.69 | 607 | 541 | 0.11 | 168 | 1262 | 6805 | 37,218 | 132 | 425 | 0.33 | 19.92 | 0.43 | 7.98 |
| | 9393 | 146 | 636 | 1.20 | 29.51 | 0.84 | 159 | 0.05 | 669 | 104 | 0.11 | 2292 | 4481 | 5413 | 37,579 | 101 | 754 | 3.08 | 21.47 | 0.42 | 4.79 |
| | 3272 | 220 | 683 | 0.91 | 14.84 | 0.88 | 108 | 0.04 | 601 | 68 | 0.13 | 3710 | 5879 | 7542 | 8571 | 101 | 676 | 2.7 | 15.05 | 0.29 | 3.78 |
| | 1417 | 98 | 833 | 1.44 | 13.82 | 1.25 | 258 | 0.34 | 618 | 1173 | 0.35 | 308 | 2454 | 4103 | 8818 | 66 | 588 | 2.86 | 17.21 | 0.26 | 8.68 |
| | 357 | 45 | 1204 | 1.82 | 22.37 | 1.84 | 124 | 0.26 | 881 | 64 | 0.08 | 189 | 1916 | 36,617 | 271 | 137 | 663 | 1.52 | 11.75 | 0.46 | 4.04 |
| Average | 2143 | 221 | 1211 | 1.54 | 22.75 | 1.64 | 204 | 0.36 | 755 | 292 | 0.14 | 962 | 2762 | 8036 | 8790 | 148 | 615 | 2.21 | 16.75 | 0.44 | 6.53 |
| *Inorganic matrix hematite* | | | | | | | | | | | | | | | | | | | | | |
| | 4341 | 74 | 460 | 0.87 | 63.01 | 1.14 | 310 | 0.15 | 366 | 74 | 0.18 | 322 | 1123 | 3654 | 456 | 1493 | 64 | 0.49 | 8.18 | 0.84 | 1.87 |
| | 3284 | 107 | 424 | 0.53 | 60.31 | 0.66 | 147 | 0.14 | 348 | 52 | 0.1 | 1032 | 2243 | 4567 | 377 | 813 | 75 | 1.37 | 12.7 | 0.41 | 1.29 |
| | 2762 | 95 | 485 | 0.34 | 62.24 | 1 | 129 | 0.16 | 430 | 58 | 0.12 | 2343 | 4387 | 8790 | 44,561 | 1043 | 82 | 0.87 | 8.1 | 0.6 | 1.96 |
| | 2548 | 201 | 446 | 0.97 | 59.08 | 0.77 | 114 | 0.19 | 364 | 48 | 0.08 | 867 | 1323 | 3423 | 32,756 | 875 | 84 | 0.55 | 9.91 | 0.62 | 1.89 |
| | 2501 | 92 | 337 | 1.37 | 52.53 | 0.8 | 125 | 0.02 | 420 | 50 | 0.09 | 3167 | 6398 | 6785 | 9992 | 1112 | 81 | 1.59 | 8.46 | 0.98 | 2.86 |
| | 5367 | 80 | 596 | 0.42 | 72.4 | 0.7 | 291 | 0.14 | 335 | 86 | 0.24 | 564 | 1723 | 13,981 | 7894 | 966 | 66 | 0.57 | 8.49 | 1.29 | 2.89 |
| | 2051 | 100 | 260 | 0.57 | 39.04 | 1.01 | 92 | 0.1 | 290 | 42 | 0.07 | 1876 | 4321 | 43,123 | 504 | 1379 | 79 | 2.96 | 19.42 | 1.13 | 2.45 |
| Average | 3264 | 107 | 429 | 0.72 | 58.37 | 0.87 | 172 | 0.13 | 364 | 58.57 | 0.13 | 1453 | 3074 | 12046 | 13,791 | 1097 | 75 | 1.20 | 10.75 | 0.84 | 2.17 |
| *Chalcopyrite* | | | | | | | | | | | | | | | | | | | | | |
| 18 | 2176 | 399 | 43.78 | 43.66 | 0.41 | 0.29 | 20.88 | 2.11 | 9.02 | 43.17 | 3.52 | 5022 | 12,742 | 17073 | 12,278 | 52 | 642 | 0.43 | 360 | 0.07 | 0.06 |

**Table 2** Carbon isotopic composition of calcite from quartz-hematite rocks of the Urals VMS deposits

| Samples | VMS deposits | $\delta^{13}C$ ‰, PDB |
|---|---|---|
| *Calcite from quartz-hematite rocks with tube microfossils* | | |
| T-5076-139.3 | Talgan | −5.8 |
| T-10 | –"– | −6.1 |
| T-5104 | –"– | −4.2 |
| T-08/4 | –"– | −26.2 |
| T-10/3 | –"– | −7.5 |
| Al-3 | Alexandrinka | −8.5 |
| Al-08/3 | –"– | −21.9 |
| Al-12 | –"– | −8.6 |
| Мол-08/9 | Molodezhnoe | −23.5 |
| XIX-4 | XIX Parts'ezd | −6.4 |
| *Carbonate clasts* | | |
| T-08/3 | Talgan | +0.97 |
| Uz-1 | Uzelga | −0.09 |
| Uz-3 | –"– | −0.40 |
| *Crinoid clasts* | | |
| T-5156-178.9 | Talgan | −0.64 |
| T-08/1 | –"– | +1.21 |
| *Calcite from hanging wall limestones* | | |
| Talg-Limest-1 | Talgan | −1.09 |
| XIX-Limest-1 | XIX Parts'ezd | +1.31 |
| Sib-Limest-1 | Sibai (Maslennikov 1999) | +2.00 |

# 7 Discussion

Submarine transformation of the primary hyaloclasts, sulfides, and carbonates resulted in formation of quartz-hematite rocks is accompanied by accumulation of elements released during the transformation. The accumulation of these elements and formation of authigenic minerals correlate with the presence of tube microfossils in quartz-hematite rocks, which are comparable in size and morphology with fossilized tube worms known in several Mesozoic and Paleozoic VMS deposits (Little et al. 1999; Maslennikov 2006) and modern hydrothermal systems (Juniper et al. 1992; Demina et al. 2007).

The tube worms found in massive sulfide ores of the Urals deposits were intravitally fossilized by bacterial encrustation, leaching, and further infill of the tubes by sulfides (Maslennikov 2006). In contrast to near-vent hydrothermal fauna, the tube organisms in quartz-hematite rocks remote from the hydrothermal vents were fossilized during diagenesis. The chaotic orientation of the tubes indicates their burial inside the unlithified sediments.

Local cellular microtextures of the tube organisms are similar to the typical textures of the trophosomes of the tube worms with intracellular symbiotic chemoautotrophic bacteria (Felbeck 1981). The negative $\delta^{13}C$ values of carbonates from the studied tube microfossils are consistent with biogenic carbon source (Kennedy et al. 2010).

The variations in contents of chemical elements in siliceous-ferruginous sediments depend on proportions of inorganic (hyaloclasts, sulfides, carbonates) and organic matter in sediments. The release of Si, Fe, Al, Mn, Ca, Ti, Na, Mg, K, and chalcophile metals during the transformation of primary detrital sediments and formation of siliceous-ferruginous substrate resulted in crystallization of numerous authigenic minerals and accumulation of some metals during lithification.

The crystallization of Ti-minerals in association with tube microfossils is very important, since little is known about biophile properties of this element. During the formation of quartz-hematite rocks with tube microfossils, titanium remains in the system in contrast to its removal during alteration of primary hyaloclasts without microfossils. The formation of titanium minerals (leucoxene, rutile, titanomagnetite) is, thus, an early relatively low-temperature diagenetic process. It is well known that titanium can passively be accumulated during etching of the oceanic basaltic glass by microbes (Banerjee et al. 2006).

The high Mn contents in biomorphic and inorganic (matrix) hematite correlate with high Ca contents, because the mineralized tissues of tube microfossils are composed of Mn-bearing carbonates. In addition, numerous carbonate minerals were formed in the matrix.

It was found that arsenic is a chemical analogue of phosphorous and can substitute phosphorous under its deficit under the intracellular biochemical reactions (Wolfe-Simon et al. 2010). The accumulation of phosphorous in biomorphoses is supported by finding the authigenic apatite in association with tube microfossils. Arsenic, along with phosphorous, is concentrated in biomorphic hematite and is probably an important component of biomineralization process. Because of its high toxicity, the habitat polluted by arsenic is characterized by low degree of species diversity that was probably reflected on the absence of other species in quartz-hematite rocks with As-bearing tube microfossils in the studied VMS deposits. Correspondingly, no tube microfossils were identified in some quartz-hematite layers with tentaculites, foraminifers, and other species.

The accumulation of potassium and formation of illite occurred most likely in the presence of organic matter, which maintains active energy metabolism (Harder and Dijkhuisen 1983).

The high W, V, and Mo contents in biomorphic hematite are of special interest. Vanadium trapped by organic matter may precipitate along with it and may partially be absorbed by $Fe(OH)_3$ or $SiO_2 \times nH_2O$ gels (Emel'yanov et al. 1979). Tungsten and vanadium in carbon-bearing sediments are considered physiological relics of the Early Archean biosphere involved, along with iron, in biological circulation. It is considered that molybdenum was inaccessible in the oxygen-free biosphere of the early Earth and its functions were performed by other metals (e.g., tungsten, vanadium or even iron) (Kletzin and Adams 1996). This suggestion is supported by distribution of bacteria with corresponding ferments along the geochemical gradients around the modern hydrothermal vents: the hydrotherms inhabitated by hyperthermophiles are rich in W and are depleted in Mo, which precipitates as sulfide (Fedonkin 2003). It is believed that, toward from $H_2S$-saturated hydrotherms, Mo becomes soluble in the presence of oxygen, whereas W becomes inaccessible for metabolic processes and, correspondingly, the composition of microbial communities varies, where molybdenum substitutes tungsten in exchange processes (Fedonkin 2003). This is also typical of the hydrothermal-biogenic ecosystems of the Urals massive sulfide paleohydrothermal fields (Maslennikov 2006).

The presence of authigenic minerals in association with relict tube microfossils in quartz-hematite rocks indicates microbiological contributions to sorption of elements and diverse style of mineral precipitation. In the absence of metabolic activity, passive interactions may occur, in which microbial cells (living or dead) behave as solid phase sorbents of dissolved elements, and heterogeneous nucleation templates for authigenic mineral deposition (McLean et al. 1996). On the other hand, metabolic activities of microorganisms bring about metal precipitation indirectly through production of reactive inorganic ligands, e.g., sulfides, phosphate, dissolved inorganic carbon, or directly through enzyme mediated changes in redox state, e.g., oxidation of reduced iron and manganese (Bachofen 1994).

## 8 Conclusions

Thus, the authigenic minerals and higher contents of some trace elements (e.g., Mn, As, W, V) in biomorphic hematite is a natural process of formation and transformation of siliceous-ferruginous sediments in the presence of tube organisms. The biological activity, along with extreme habitat, most likely resulted in rapid transformation of biogenic organic matter, providing the unique conditions for post-sedimentation transformations of the primary sediments.

**Acknowledgments** The mineralogical work was supported by the State Contract no. 01201350139. The analysis of trace element composition of quartz-hematite rocks was supported by the Russian Science Foundation (project no. 14-17-00691).

## References

Adachi M, Yamamoto K, Sugisaki R (1986) Hydrothermal cherts and associated siliceous rocks from the northern Pacific: their geological significance as indication of ocean ridge activity. Sed Geol 47:125–148

Ayupova NR, Maslennikov VV (2012) Biomineralisation in ferruginous-siliceous sediments of massive sulfide deposits of the Urals). Dokl Earth Sci 442(2):193–195

Ayupova NR, Maslennikov VV (2013) Biomorphic textures in the ferruginous–siliceous rocks of massive sulfide bearing paleohydrothermal fields in the Urals. Lithol Min Res 48:438–455

Bachofen R (1994) Cell structure and metabolism and its relation with the environment. In: Buuffle J, DeVitre RR (eds) Chemical and biological regulation of aquatic systems. Lewis Publishers, Roca Raton

Banerjee NR, Furnes H, Muehlenbachs K, Staudigel H, de Wit M (2006) Preservation of 3.4–3.5 Ga microbial biomarkers in pillow lavas and hyaloclastites from the Barberton greenstone Belt South Africa. Earth Planet Sci Lett 241:707–722

Constantinou G, Govett GJS (1973) Geology, geochemistry, and genesis of Cyprus sulfide deposits. Econ Geol 68:843–858

Danyushevsky L, Robinson Ph, Gilbert S, Norman M, Large R, McGoldrick P, Shelley M (2011) Routine quantitative multi-element analysis of sulphide minerals by laser ablation ICP-MS: Standard development and consideration of matrix effects. Geochem: Explor Environ Anal 11:51–60

Demina LL, Galkin SV, Lein AYu, Lisitsyn AP (2007) First data on microelemental composition of benthic organisms from the 9°50′ N hydrothermal field East Pacific Rise. Dokl Earth Sci 415 (6):905–907

Duhig NC, Stolz J, Davidson GJ, Large RR (1992) Cambrian microbial and silica gel textures in silica iron exhalites from the Mount Windsor volcanic belt, Australia: their petrography, chemistry, and origin. Econ Geol 87:764–784

Emel'yanov EM, Mitropol'skii AYu, Shimkus KM, Mussa AA (1979) Geochemistry of the Mediterranean Sea. Kiev, Naukova dumka (in Russian)

Fedonkin MA (2003) Geochemical impoverishment and eukaryotization of the biosphere: a causal link. Paleontol J 37(6):592–599

Felbeck H (1981) Chemoautotrophic potential of the hydrothermal vent tube worm *Riftia pachyptila* Jones (Vestimentifera). Science 213:336–338

Gal'chenko VF, Galkin SV, Lein AYu, Moskalev LI, Ivanov MV (1988) A role of symbiotic bacteria in nutrition of invertebrates from active submarine hydrotherms. Okeanologiya 28(6):1020–1031

Grenne T, Slack JF (2003) Bedded jaspers of the Ordovician Lökken ophiolite, Norway: seafloor deposition and diagenetic maturation of hydrothermal plume-derived silica-iron gels. Min Dep 38:625–639

Halbach M, Halbach P, Lueders V (2002) Sulfide-impregnated and pure silica precipitates of hydrothermal origin from the central Indian Ocean. Chem Geol 182:357–375

Hannington MD, Scott SD (1988) Mineralogy and geochemistry of a hydrothermal silica-sulfide-sulfate spire in the caldera of Axial Seamount, Juan de Fuca Ridge. Can Min 26:603–625

Harder W, Dijkhuisen L (1983) Physical responses to nutrient limitation. Ann Rev Microbiol 37:1–23

Hein JR, Clague DA, Koski RA, Embley RW, Duhnam RE (2008) Metalliferous sediment and a silica-hematite deposit within the Blanco fracture zone, Northeast Pacific. Mar Geores Geotechnol 26:317–339

Hekinian R, Hoffert M, Larqué P, Cheminée JL, Stoffers P, Bideau D (1993) Hydrothermal Fe and Si oxyhydroxide deposits from South Pacific intraplate volcanoes and East Pacific rise axial and off-axial regions. Econ Geol 88:2099–2121

Herzig PM, Hannington MD, Scott SD, Maliotis G, Rona PA, Thompson G (1991) Gold-rich sea-floor gossans in the Troodos ophiolite and on the Mid-Atlantic ridge. Econ Geol 86:1747–1755

Hoefs J (1997) Stable Isotope Geochemistry. Springer, Berlin, Heidelberg, New York, London, Paris, Tokyo, Hong Kong

Juniper SK, Fouquet Y (1988) Filamentous iron-silica deposits from modern and ancient hydrothermal sites. Can Min 26:859–869

Juniper SK, Jonnasson IR, Tunnicliffe V, Southward AJ (1992) Influence of tube building polychaete on hydrothermal chimney mineralization. Geol 20:895–898

Kalogeropoulos SI, Scott SD (1983) Mineralogy and geochemistry of tuffaceous exhalites (tetsusekiei) of the Fukazawa mine, Hokuroku district, Japan. Econ Geol Monogr 5:412–432

Kennedy CB, Gault AG, Fortin D, Clark ID, Pedersen K, Scott SD, Ferris FG (2010) Carbon isotope fractionation by sircumneutral iron-oxidizing bacterias. Geology 38:1087–1090

Kletzin A, Adams MW (1996) Tungsten in biological systems. FEMS Microbiol Rev 18(1):5–63

Lein AYu, Sedykh EM, Starshinova NP (1989) Distribution of metals in bacteria and animals from submarine hydrothermal fields. Geokhimiya 2:297–303

Little CTS, Maslennikov VV, Morris NJ, Gubanov AP (1999) Two paleozoic hydrothermal vent communities from the southern Ural Mountains, Russia. Paleontology 42:1043–1078

Little CTS, Glynn SEJ, Mills RA (2004) Four-hundred-and-ninety-million year record of bacteriogenic iron oxide precipitation at sea-floor hydrothermal vents. Geomicrobiol J 21:415–429

Maslennikov VV (1999) Sedimentogenesis, halmyrolysis and ecology of the massive sulfide paleohydrothermal fields (on the example of the Southern Urals). Geotur, Miass (in Russian)

Maslennikov VV (2006) Lithogenesis and massive sulfide deposits formation. Institute of Mineralogy UB RAS, Miass (in Russian)

Maslennikov VV, Ayupova NR, Herrington RJ, Danyushevskiy LV, Large RR (2012) Ferruginous and mangniferous haloes around massive sulphide deposits of the Urals. Ore Geol Rev 47:5–41

Maslennikov VV, Ayupova NR, Maslennikova SP, Tret'yakov GA, Melekestzeva IYu et al (2014) Toxic elements in massive sulfide systems. Yekaterinburg, RIO UB RAS (in Russian)

McLean RJC, Fortin D, Brown DA (1996) Microbial metal-binding mechanisms and their relation to nuclear waste disposal. Can J Microbiol 42:392–400

Peter JM, Goodfellow WD (1996) Mineralogy, bulk and rare earth element geochemistry of massive sulphide-associated hydrothermal sediments of the Brunswick horizon, Bathurst mining camp, New Brunswick. Can J Earth Sci 33:252–283

Prokin VA, Necheuhin VM, Sopko PF et al (1985) Copper massive sulfide deposits of the Urals. Geological condition of location. Sverdlovsk: Urals branch of the Academy of Science of USSR; 1985 (in Russian)

Wolfe-Simon F, Blum JS, Kulp TR, Gordon GW Hoeft SE, Pett-Ridge J, Stolz JF, Webb SM, Weber PK, Davies PCW, Anbar AD, Oremland RS (2010) A bacterium that can grow by using arsenic instead of phosphorus. Science 332(6034):1163–1166

Zaykov VV (1991) Volcanism and sulfide mounds of paleoocean margins. Moscow, Nauka (in Russian)

# Part II
# Geochemistry of Biogenic–Abiogenic Systems

# Quantitative Evaluation of Several Geochemical Characteristics of Urban Soils

**Vladimir A. Alekseenko and Alexey V. Alekseenko**

**Abstract** For the first time, the quantitative geochemical data are given for urban soils of several groups of cities which differ in population. The content of chemical elements is considered as well as the specific ecological significance of soil contamination by these elements. The figures were established by authors on the base of average concentrations of chemical elements in the soils of more than 300 cities and settlements. The major part of data (sampling, analyses, and their statistical treatment) was obtained directly by authors as a result of special studies conducted for more than 15 years. The sufficiently numerous published materials of different researchers were also used. The greatest elements accumulation comparing with the Earth's soils (tens of thousands of tons per 1 km$^2$) is associated with an increase in the content of Ca and Mg. Considering the environmental significance of chemical elements accumulation in soils, we note the primary role of Pb and Zn in all groups of cities. Out from the rest pollutants it is necessary, first of all, to note As, Cu, and Cl, which are the main contaminants in four of six cities groups. In two groups of settlements, Cd and Co are important soil pollutants. In three groups, a considerable increase in the Ca content significantly modifies ecological–geochemical state of soils.

**Keywords** Soil contamination · Urban pollution · Urban geochemistry · Environmental geochemistry · Trace elements

V.A. Alekseenko
Southern Federal University, Rostov-on-Don, Russia
e-mail: vl.al.alekseenko@gmail.com

V.A. Alekseenko
Kuban State Agrarian University, Krasnodar, Russia

A.V. Alekseenko (✉)
Department of Geoecology, National Mineral Resources University (Mining University), 21st Line V.O., 2, 199106 Saint Petersburg, Russia
e-mail: al.vl.alekseenko@gmail.com

 
O.V. Frank-Kamenetskaya et al. (eds.), *Biogenic—Abiogenic Interactions in Natural and Anthropogenic Systems*, Lecture Notes in Earth System Sciences,
DOI 10.1007/978-3-319-24987-2_11

Soil is a special biogenic–abiogenic system. Its formation and geochemical characteristics are caused by biogenic influence on rocks, which are abiogenic. Several factors also affecting the results of this influence are geomorphological and climatic features of the area, the duration of interaction, etc. These factors were first discussed in detail in the works of V.V. Dokuchaev, and were widely studied by many other researchers. Technogenic processes have a great (and in some cases even critical) impact on the considered interaction of biogenic and abiogenic substances in recent decades. Their maximum influence on the forming and already formed soil occurs in human settlements. It includes both abiogenic pressure, consisting mainly in the physical and chemical changes in soil, as well as biogenic pressure. The latter is most often associated with the destruction of certain soil organisms and the cultivation of new plant and animal biota. The change of geochemical features of urban soils is a certain result of such interaction. This, in turn, has an influence on many biogenic systems, including humans in settlements.

The overwhelming majority of the Earth's population is concentrated within the boundaries of cities, although settlements occupy less than 5 % of the land area. In this regard, the study of urban soils is of crucial scientific and also practical importance. In this Chapter, we consider the impact of population number in cities on geochemical characteristics of urban soil.

## 1 Introduction

Soils are among the most important natural resources, defining sustainable development and independence of states. Preservation of soil fertility should be a major concern and a priority for economic development. In recent decades, the geochemical characteristics of soils have undergone significant changes under the influence of anthropogenic activities worldwide. The most dangerous changes occur in content and distribution of metals (Motuzova et al. 2014).

Soil as a deposing medium is a summary indicator of ecological–geochemical changes occurring in the study area (Perel'man 1989; Vernadsky 1965). Investigations have shown (Alekseenko and Alekseenko 2013; Kabata-Pendias and Pendias 2001) that in large industrial centres, *urban soils* continue to carry the geochemical information about the pollution happened even after the liquidation of main contamination sources. Let us also remember that human settlements occupying only about 5 % of the land is home to almost the whole population of the Earth. All this allows us to consider the study of soil geochemical characteristics, especially in settlements and agricultural landscapes, as a priority task for ecology and geochemistry (Adriano 2001; Gerasimova et al. 2003; Norra and Stüben 2003; Pashkevich et al. 2015). At the same time, we note that quantitative soil geochemical data is necessary for the subsequent adoption of specific measures to improve environmental situation (Kasimov et al. 2011; Syso et al. 2010; Tsolova et al. 2014; Tume et al. 2011). This paper presents quantitative information about the ecological state of soils in different categories of settlements.

## 2 Materials and Methods

The ongoing studies included sampling, chemical analysis and processing the received information. The geochemical properties of urban soils from more than 300 cities in Europe, Asia, Africa, Australia, and America were evaluated. Methodology of these works with the list of studied cities, methods of analysis of soil samples, their internal and external control and statistical process has been considered in detail (Alekseenko et al. 2013; Alekseenko and Alekseenko 2014b). The number of samples in each locality is ranged from 30 to 1000. Calculation of random and systematic errors showed high analyses repeatability and correctness, and allowed to consider the analytical laboratory work as good. All ordinary analyses were carried out in the single certified and accredited Central Testing Laboratory at "Kavkazgeolsyomka" using emission spectral analysis. The external control was conducted in the laboratories of "CentrKazGeology," "MagadanGeology," Southern Federal University (X-ray fluorescence and gravimetric analyses), and Institute of Ore Geology, Petrography, Mineralogy, and Geochemistry of the Russian Academy of Sciences (neutron activation analyses).

For every city, the average concentrations of elements in soils were determined. To avoid the errors related to unequal number of samples, each city was then represented by only one "averaged" sample. The published data and the materials kindly provided by a number of geochemists were also incorporated into research. As a result for all the elements were written the rows in which each city was presented by an average content of chemical elements in soils. The statistical processing of these data allowed to calculate the average concentrations of chemical elements in urban soils. They can be considered as the abundances (average concentrations) of chemical elements in urban soils (Alekseenko and Alekseenko 2014a).

Several samples selections were prepared for different groups of settlements. Each city was represented only by the average content of chemical elements in soils (as well as in abundances establishing). Criteria for these groups separation are considered further in this work. Statistical processing of the analysis results allowed to establish the average content of elements in soils of all selected groups (Table 1).

In what follows, we compared the average contents with abundances in the Earth's soils (Vinogradov 1959) and with abundances in urban soils (Alekseenko and Alekseenko 2014b). A number of indicators introduced by one of the authors in 1997 were used.

*Indicator of Absolute Accumulation* (IAA) shows the mass of a certain chemical element (or its compound) accumulated in a particular part of the system as a result of geochemical processes occurring per unit area. Soils, all the vegetation of a specific area, plant species, surface or groundwater, etc. could be studied as the parts of landscapes. During the investigations of soils in the study area of 1 km$^2$ the difference is determined between the background content before the start of these processes ($C_1$) and after their completion ($C_2$): IAA = $C_2 - C_1$. When after the end of the processes the content ($C_2$) has decreased (i.e., it has occurred the loss of

**Table 1** The abundances and average concentrations of chemical elements in soils of different cities groups (mg/kg)

| Element | Element number | Abundances | | The average concentrations in soils of | | | | | |
|---|---|---|---|---|---|---|---|---|---|
| | | In the Earth's soil cover | In urban soils | Millionaire cities (>700 thousand people) | Half-millionaire cities (300–700 thousand people) | Cities with a local significance (100–300 thousand people) | Small towns (<100 thousand people) | Small settlements, villages, hamlets | Tourist and recreational areas |
| Ag | 47 | 0.5 | 0.4 | 0.4 | 0.4 | 0.5 | 0.4 | 0.2 | 0.4 |
| Al | 13 | 71,300 | 38,200 | 37,600 | 37,360 | – | 44,840 | – | 42,260 |
| As | 33 | 5.0 | 15.9 | 24.6 | 10.0 | 20.9 | 15.0 | 5.2 | 21.5 |
| B | 5 | 10 | 45 | – | – | – | – | – | – |
| Ba | 56 | 500 | 853 | 881 | 1091 | 603 | 980 | 537 | 984 |
| Be | 4 | 6.0 | 3.3 | 2.9 | 2.8 | – | 1.9 | 8.4 | 2.4 |
| Bi | 83 | – | 1.1 | 1.2 | 1.1 | 2.3 | 1 | 1 | 1.1 |
| C | 6 | – | 45100 | – | – | – | – | – | – |
| Ca | 20 | 13,700 | 53,800 | 17,100 | 77,090 | – | 76,670 | – | 71,180 |
| Cd | 48 | 0.5 | 0.9 | 2.9 | 0.8 | 0.5 | 1.4 | 0.2 | – |
| Cl | 17 | 100 | 285 | 307 | 313 | – | 240 | – | 274 |
| Co | 27 | 8.0 | 14.1 | 15.8 | 14.4 | 12.7 | 14.6 | 10.7 | 18.1 |
| Cr | 24 | 200 | 80.0 | 82.7 | 55.0 | 42.2 | 81.5 | 52.8 | 88.1 |
| Cs | 55 | 5.0 | n·1.0 | – | – | – | – | – | – |
| Cu | 29 | 20.0 | 39.0 | 55.1 | 30.1 | 28.1 | 28.2 | 34.7 | 56.9 |
| Fe | 26 | 38,000 | 22,300 | 19,600 | 20,110 | – | 26,570 | – | 24,900 |
| Ga | 31 | 30.0 | 16.2 | 17.4 | 16.1 | 15.1 | 15.2 | 18.0 | 16.2 |
| Ge | 32 | 5.0 | 1.8 | 1.9 | 1.6 | 1.6 | 2.6 | 1.6 | 1.9 |
| H | 1 | 2300[a] | 15000 | – | – | – | – | – | – |
| Hg | 80 | 0.01 | 0.88 | – | – | – | – | – | – |
| K | 19 | 13,600 | 13,400 | 14,500 | 12,600 | – | 12,770 | – | 12,770 |

(continued)

**Table 1** (continued)

| Element | Element number | Abundances | | The average concentrations in soils of | | | | | |
|---|---|---|---|---|---|---|---|---|---|
| | | In the Earth's soil cover | In urban soils | Millionaire cities (>700 thousand people) | Half-millionaire cities (300–700 thousand people) | Cities with a local significance (100–300 thousand people) | Small towns (<100 thousand people) | Small settlements, villages, hamlets | Tourist and recreational areas |
| La | 57 | 40 | 34 | 15 | 32 | – | – | – | – |
| Li | 3 | 30.0 | 49.5 | 45.4 | 52.5 | 50.3 | 45.9 | 50.4 | 51.2 |
| Mg | 12 | 6300 | 7900 | 6700 | 10,100 | – | 9110 | – | 10,060 |
| Mn | 25 | 850 | 729 | 871 | 716 | 547 | 458 | 675 | 1125 |
| Mo | 42 | 2.0 | 2.4 | 2.3 | 2.2 | 3.6 | 2.6 | 3.7 | 2.0 |
| N | 7 | 1000 | 10,000 | – | – | – | – | – | – |
| Na | 11 | 6300 | 5800 | 6600 | 5270 | – | 5920 | – | 6450 |
| Nb | 41 | – | 15.7 | 16.7 | 15.7 | 24.2 | 17.0 | 15.0 | 16.4 |
| Ni | 28 | 40.0 | 33.0 | 35.4 | 28.0 | 23.7 | 18.4 | 28.4 | 39.8 |
| O | 8 | – | 490,000 | – | – | – | – | – | – |
| P | 15 | 800 | 1201 | 1341 | 1090 | 1327 | 1203 | 1320 | 1198 |
| Pb | 82 | 10.0 | 54.5 | 66.2 | 45.6 | 43.4 | 39.5 | 22.7 | 55.2 |
| Rb | 37 | 100 | 58 | 58 | 51 | – | 63 | – | 63 |
| S | 16 | 850 | 1200 | 1600 | 1140 | – | 1050 | – | 1020 |
| Sb | 51 | – | 1.0 | – | – | – | – | – | – |
| Sc | 21 | 7.0 | 9.4 | 8.6 | 8.4 | 10.1 | 11.8 | 6.0 | 11.8 |
| Se | 34 | – | – | 10.1 | – | – | – | – | – |
| Si | 14 | 330,000 | 289,000 | 329,300 | 261,200 | – | 255,340 | – | 267,010 |
| Sn | 50 | 10.0 | 6.8 | 10.0 | 6.4 | 9.0 | 7.4 | 6.2 | 6.5 |
| Sr | 38 | 300 | 458 | 358 | 359 | 365 | 325 | 232 | 551 |
| Ta | 73 | – | 1.5 | 1.0 | – | – | – | 1 | – |

(continued)

**Table 1** (continued)

| Element | Element number | Abundances | | The average concentrations in soils of | | | | | |
|---|---|---|---|---|---|---|---|---|---|
| | | In the Earth's soil cover | In urban soils | Millionaire cities (>700 thousand people) | Half-millionaire cities (300–700 thousand people) | Cities with a local significance (100–300 thousand people) | Small towns (<100 thousand people) | Small settlements, villages, hamlets | Tourist and recreational areas |
| Ti | 22 | 4600 | 4758 | 4976 | 4378 | 4276 | 4489 | 5583 | 4794 |
| Tl | 81 | – | 1.1 | 1.0 | 1.0 | 10.7 | 2.2 | 1.0 | 1.1 |
| V | 23 | 100 | 105 | 107 | 79 | 99 | 104 | 65 | 115 |
| W | 74 | – | 2.9 | 2.7 | 3.6 | 3.5 | 3.5 | 2.7 | 2.4 |
| Y | 39 | 50.0 | 23.4 | 23.5 | 23.3 | 36.5 | 25.5 | 17.5 | 18.7 |
| Yb | 70 | – | 2.4 | 2.5 | 2.3 | 3.0 | 3.0 | 2.7 | 2.5 |
| Zn | 30 | 50 | 158 | 201 | 116 | 100 | 92 | 98 | 200 |
| Zr | 40 | 300 | 256 | 244 | 331 | 400 | 309 | 132 | 188 |

[a]The abundance of hydrogen in the Earth's soil cover is specified

elements), the IAA value is negative. The study of technogenic changes in the vertical soil profile showed that they usually develop in agricultural and residential landscapes to a depth of 30 cm.

The determination of the average density of soils in Southern Russia allowed to assume after accounts that the increase (decrease) in the concentration of chemical elements in soils by an amount equal to 10 mg/kg, within the upper 30 cm layer corresponds to an increase (decrease) of their weight by 6 tons on an area of 1 $km^2$.

However, such a quantitative measure as IAA required for many decisions related to environmental issues does not provide sufficient information about the most negative impact of certain chemical elements in each case. Let us consider the following example. An increase of hundreds of tons per 1 $km^2$ of *iron* content in soils is less hazardous for organisms than an increase of 10 tons of *mercury* and *arsenic*. It happens due to the different *abundances* of elements (average concentrations) in the Earth's crust. High content of Fe (abundance 4.64 %) is "habitual" for organisms over a long period of development and evolution; otherwise, high concentrations of Hg (83 mg/kg) or As (17 mg/kg) are unaccustomed. The *Indicator of Relative Accumulation* (IRA) was proposed to overcome this geochemical feature.

IRA shows the ratio between the mass of the element accumulated (lost) as a result of the processes occurring in a certain part of the geochemical system (i.e., IAA) and the background content (or the average concentration in the Earth's crust). Thus, we can assume that $\text{IRA} = \text{IAA}/C_1$.

*Absolute Spread* (AS) of chemical elements is a new environmental index to describe the behavior of elements in geochemical systems. It was offered by V.A. Alekseenko in 1997. AS is the ratio of the maximum background element content (abundance) in one part of geochemical system to the element content in the other system part. We have calculated the value of AS for rocks and soils of continents (Alekseenko 2006). All the chemical elements are divided into three groups according to the AS value: (1) AS > 100 (up to 4250); (2) AS from 10 to 100; (3) AS < 10. Elements belonging to each group differ in crystal-chemical parameters: ionisation energy and electronegativity (Fersman 1952–1959).

Some elements are characterised by natural high values of AS. Living organisms become "accustomed" to the large variations of these elements contents during the process of biota development and evolution. In this regard, it can be assumed that fluctuations of the third group elements content (Bi, W, Au, Br, P, Zn, Ag, I, and Be) in environment are the most dangerous for living matter. But it should make a reservation that they are dangerous, if they occur mainly in forms available for organisms.

## 3 Results and Discussion

The analysis of anthropogenic load on urban soil allows to conventionally divide them into six main groups. The key technogenesis indicators were taken as the basis of this division (Alekseenko 2015). Let us briefly consider them.

The volume of food, building materials, and industrial goods imported in cities depends on the number of inhabitants, i.e., the size of settlements. The same factor causes the development of urban transport (both public and private). Transport is a constant source of specific compounds income in the environment: dust from wear of car and bus tires, wheels and rails of trams; exhaust gases of automobiles, etc. The amount of waste, incoming into the landscape as a result of life activity, depends on the number of inhabitants. The population number controls in a great measure the amount of wastes removed from the landscape, their purification rate and methods of disposal (recycling). Thus, *the size of settlement (number of inhabitants) largely determines the balance of technogenic elements migration in landscapes*.

However, the amount and composition of imported raw materials in the settlement, its way of treatment, the amount of waste resulting, as well as the forms of occurrence of elements in raw materials and wastes, are caused by profiles of enterprises operating in this city. Specialisation of main enterprises often has a significant impact on the cleaning system and method of waste disposal from the entire city or from its separate areas. In such a way, *the second most important factor*, largely connected with the number of inhabitants and the defining features of technogenic migration of elements in urban landscapes, *is an area of specialisation of companies operating in city*.

We carry out the classification of residential landscapes considering two most important factors that determine the features of technogenic migration (Fig. 1).

I. In accordance with the scheme above, we first consider separately identifiable *landscapes of millionaire cities*, the industrial centres of national importance. They include cities with population over 700 thousand people. Calculations of many researchers show that only in ordinary life activity inhabitants of the city with one million population every day emit about 0.5 million $m^3$ of carbon dioxide and 1200 $m^3$ of water vapor and sweat gland secretions. The supply of

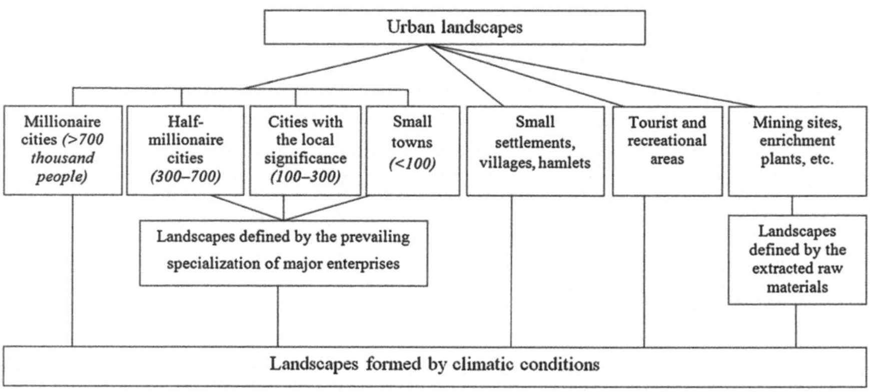


**Fig. 1** The classification scheme of urban landscapes

food and various industrial products in major cities is approximately the same and depends more on the number of inhabitants than on features of natural zone in which the city is situated. Systems of municipal wastewater treatment and disposal methods for solid urban waste are more and more close to typical (standard) in big cities. Roughly the same pollutants in similar amounts are emitted in these residential landscapes by public and private transport. In such a way, the manmade migration of elements in large cities is directly related to the inhabitants life support, and has much more in common than specific differences.

Industry of cities with more than one million inhabitants has some variety, but almost always includes food, chemical, and consumer goods enterprises and industries associated with the processing of metals, and construction companies. These data allow to tentatively assume that in such cities composition and number of polluting elements entering urban landscapes as a result of industrial activity is about the same when calculated per inhabitant. Therefore, the unification of cities with population about 1 million residents in a particular type of landscape has a certain geochemical substantiation.

II. Landscapes of *half-millionaire cities* are settlements with a population of 300,000–700,000 people. Just as in the previous group, for the life support of such number of inhabitants it is necessary to deliver the same amount of food and materials. Approximately, the equal number of companies associated with necessary works for the life support of specified number of people is always located in these cities. People and goods movement through the city with a population of 300,000–700,000 is also provided by the same vehicles.

However, in some cases, development of certain large companies in cities of this group may have a significant impact on the anthropogenic entry and redistribution of elements complexes. Ecological-geochemical impacts of such large enterprises impose specific patterns of prevalence and distribution of chemical elements in air, water, soil, and organisms. In this regard, especially with increasing of studies scale, it is possible to consider such cities separately according to the prevailing companies specialisation. For example cities, where oil refining or cement companies are most common, should be distinguished separately.

III. Large enterprises have even greater impact on the technogenic features of migration in urban landscapes of local importance: *cities with population 100,000–300,000 inhabitants*. Therefore, during the detailed landscape-geochemical studies (especially those related to solution environmental problems), it is advisable to divide landscapes on the same principle as in half-millionaire cities.

IV. In this study, the so-called "*small cities*" *with population less than 100,000 residents* are studied as a separate group. These settlements are quite common, and hence they are home to fairly large number of people in total.

V. *Landscapes of villages, hamlets, and farms* occupy the smallest area among other residential areas. They are like the transition from agricultural landscapes to settlements with quite intense biological cycle of elements. They differ from resorts landscapes by significantly lower number of residents, lower role of transport in technogenic elements addition, and a unique elements' biological cycle, close to ones in surrounding agricultural or biogenic (natural) landscapes.

VI. *Landscapes of recreational and tourist localities* should be considered separately. Among these, a technogenic migration is mainly determined by the vital activity of small number of permanent and a large number of visiting inhabitants, as well as peculiarities of transport and, quite often, the composition of underground saline water used for treatment. A large role of biological cycle in migration of the main elements is often an important characteristic of these landscapes.

Landscapes of *settlements in mines and enrichment plants adjacent areas* are combined into a single group. Within these landscapes, migration and concentration of chemical elements are determined much more by mineral deposit characteristic than the number of inhabitants.

Natural factors influencing the migration of elements (especially those related to climatic conditions), must also be taken into account when dividing the groups of urban areas. For example, the process of various substances migration will be very different in landscapes of Cairo and Novosibirsk, especially in winter season. However, studies and mapping of geochemical landscapes for solving environmental problems are usually carried out in a single climatic zone. In addition, the influence of anthropogenic factors in settlements is incomparably stronger than natural. Given all this, it is suggested to conduct further residential landscapes segregation, if necessary, according to climatic conditions after the separation discussed above.

The data on the average chemical elements content that characterise soils of allocated settlements are shown in Table 1. Earlier studies showed that urban soils differ significantly in the content of chemical elements from the Earth's soil cover, and the association of elements with significantly increased contents has no natural analogues. These differences, in conjunction with the above data on residential landscapes, have forced to identify separately such geochemical system as urban soils (as one of the most important from an environmental viewpoint) with specified average contents, *abundances of chemical elements* (Alekseenko and Alekseenko 2014b).

We pay a particular attention to such elements as Zn, Pb, Ba, Sr, Ca, Hg, B, Cu, and Co in the soils of all settlements. We believe that their high content in this geochemical system is due to mainly technogenic processes, i.e., environmental pollution. This is confirmed by the fact that these elements' content in *the Earth's soil cover* is less than in *the Earth's crust* (Vinogradov 1959). Consequently, the overall processes of soil formation did not result in accumulation of these elements in soils. All of this leads us to identify *Indicators of Absolute and Relative Accumulation* of these metals in geochemical system representing all urban soils in relation to the Earth's soil cover, for making informed decisions to improve the ecological state of human settlements. Somewhat arbitrarily, these figures are characteristic of technogenesis in urban soils, considered as a separate geochemical system. IAA rows of elements accumulation (t/km$^2$) in urban soils are as follows:

Ca(24,060) > Ba(210) > Sr(94.8) > Zn(64.8) > Pb(26.5) > B(21.0) > Cu(11.4) > Co(3.6) > Hg(0.52)

They allow to consider that technogenesis in urban conditions led to the largest (by weight) concentrations of Ca, Ba, and Sr in soils and lower accumulation of Hg, Co, and Cu. In order to establish the highest ecological–geochemical significance of this process, we consider the series of *Indicator of Relative Accumulation* (IRA): Hg > Ba > Pb > B > Ca > Zn > Cu > Co > Sr.

As can be seen from this row, the highest ecological–geochemical changes of urban soils can be associated with increased concentrations of Hg, Ba, and Pb, although they have accumulated in tens and tens of thousands of times smaller quantities (t/km$^2$) than, for example, Ca.

In addition to abundances, the distribution of elements in urban soils is important ecological–geochemical indicator which is established by the values of *Absolute Spread* (AS). The values of AS were set in relation of the maximum average content in soils of one settlements group, allocated by the number of inhabitants, to the minimum average content established in the other group. The value of AS over 2 is marked for 18 chemical elements. They include As, Ba, Be, Bi, Ca, Cd, Cr, Cu, La, Mn, Ni, Pb, Sc, Sr, Tl, Y, Zn, and Zr. For 5 elements, the quantity of AS is greater than 4. They are As, Be, Ca, Cd, and Tl. The main thing is that for Bi and Cd, values of AS in such a geochemical system as urban soils have exceeded values in rocks and soils of continents. The quantity of *Absolute Spread* for Be in urban soils is close to the value in rocks and soils.

If the considered geochemical system (urban soils) was large, it could be possible to talk about the geochemical threat for a living matter. In the meantime, we note that the specified AS definitely affect the change of ecological conditions in allocated cities groups and *clearly points to the significant differences in the processes of income and accumulation of substances in these groups of residential landscapes.*

Changes in the average contents of 18 chemical elements in soils of different groups of settlements in more than 2 times, additionally demonstrates the need for a detailed study of such a geochemical system as urban soils, and correctness of our approach to this division.

Let us now consider some of soil geochemical characteristics in each of these settlements groups. The main of them is the change in prevalence (contents) of chemical elements. The information about the differences between the average concentrations in soils of separate settlements groups and abundances in the Earth's soil cover is shown in Table 2. In each group of cities, only elements which average content was more than 1.5 times higher than in the Earth' soils were considered.

Soil of *settlements with the population over 700,000* (*millionaire cities*) can be considered as a huge Earth's soils lithochemical anomaly, spatially divided into

**Table 2** Indexes of absolute (IAA, t/km$^2$) and relative accumulation (IRA) of chemical elements in soils of different cities groups (in relation to the abundances in the Earth's soils)

| Element | Cities with the number of population | | | | | | | | Small settlements, villages, hamlets | | Tourist and recreational areas | |
|---|---|---|---|---|---|---|---|---|---|---|---|---|
| | Millionaire cities (>700,000 people) | | Half-millionaire cities (300,000–700,000 people) | | Cities with the local significance (100,000–300,000 people) | | Small towns (<100,000 people) | | | | | |
| | IAA | IRA | IAA | IRA | IAA | IRA | IAA | IRA | IAA | IRA | IAA | IRA |
| Ag | – | – | – | – | – | – | – | – | −0.15 | −6 | – | – |
| As | 11.8 | 23.5 | 3 | 6 | 9.54 | 19.1 | – | – | – | – | 9.6 | 19.2 |
| Ba | 228 | 4.6 | 355 | 7.1 | – | – | 288 | 5.8 | – | – | 290.4 | 5.8 |
| Be | −1.9 | −3.1 | −1.9 | −3.1 | – | – | −2.5 | −4.1 | – | – | −2.2 | −3.6 |
| Ca | – | – | 38,036 | 27.8 | – | – | 37784 | 27.6 | – | – | 34,488 | 25.2 |
| Cd | 1.4 | 28 | 0.18 | 3.6 | – | – | 0.6 | 12 | −0.19 | −3.8 | – | – |
| Cl | 124 | 12.4 | 128 | 12.8 | – | – | 83.7 | 8.4 | – | – | 104 | 10.4 |
| Co | 4.7 | 5.85 | 3.8 | 4.8 | 2.8 | 3.5 | 4 | 5 | – | – | 6.1 | 7.6 |
| Cr | −70.4 | −3.5 | −87 | −4.4 | −94.7 | −4.7 | −71.1 | −3.6 | −88.3 | −4.4 | −67.1 | −3.4 |
| Cu | 21 | 10.5 | 6.1 | 3 | – | – | – | – | 8.8 | 4.4 | 22.1 | 11.1 |
| Fe | −11,040 | −2.9 | −10,737 | −2.8 | – | – | – | – | – | – | – | – |
| Li | 9.24 | 3.1 | 13.5 | 4.5 | 12.2 | 4.1 | 9.5 | 3.2 | 12.2 | 4.1 | 12.7 | 4.2 |
| Mg | – | – | 2280 | 3.6 | – | – | – | – | – | – | 2256 | 3.6 |
| Mn | – | – | – | – | −182 | −2.1 | −236 | −2.8 | – | – | – | – |
| Mo | – | – | – | – | 0.96 | 4.8 | – | – | 1 | 5.1 | – | – |
| P | 325 | 4.1 | – | – | 316 | 4 | – | – | 312 | 3.9 | – | – |
| Pb | 33.7 | 33.7 | 21.4 | 21.4 | 20 | 20 | 17.7 | 17.7 | 7.6 | 7.6 | 27.1 | 27.1 |
| Sn | – | – | −2.16 | −2.2 | – | – | – | – | −2.26 | −2.3 | −2.1 | −2.1 |
| Sr | – | – | – | – | – | – | – | – | – | – | 150 | 5 |

(continued)

**Table 2** (continued)

| Element | Cities with the number of population | | | | | | | | Small settlements, villages, hamlets | | Tourist and recreational areas | |
|---|---|---|---|---|---|---|---|---|---|---|---|---|
| | Millionaire cities (>700,000 people) | | Half-millionaire cities (300,000–700,000 people) | | Cities with the local significance (100,000–300,000 people) | | Small towns (<100,000 people) | | | | | |
| | IAA | IRA | IAA | IRA | IAA | IRA | IAA | IRA | IAA | IRA | IAA | IRA |
| V | – | – | – | – | – | – | – | – | −21.3 | −2.1 | – | – |
| Y | −15.9 | −3.2 | −16 | −3.2 | – | – | −14.7 | −2.9 | −19.5 | −3.9 | −18.8 | −3.7 |
| Zn | 90.7 | 18.1 | 39.5 | 7.9 | 29.7 | 5.9 | 25.4 | 5.1 | 28.6 | 5.7 | 90.0 | 18.0 |
| Zr | – | – | – | – | – | – | – | – | −101 | −3.4 | −67.4 | −2.2 |

separate parts. As, Ba, Cd, Cl, Co, Cu, Li, P, Pb, and Zn constitute an anomaly. These elements' mass has increased and makes the following IAA series (in parentheses hereinafter t/km$^2$):

$$\underset{\text{Absolute accumulation (t/km}^2\text{) in soils of millionaire cities}}{\underleftarrow{\text{P}(342.7) > \text{Ba}(228) > \text{Cl}(124\ 2) > \text{Zn}(90.7) > \text{Pb}(33.7) > \text{Cu}(21.0) > \text{As}(11.8) > \text{Li}(9.2) > \text{Co}(4.7) > \text{Cd}(1.4)}}$$

In the same area, as compared with the Earth's soil cover, the following elements content has decreased:

$$\underset{\text{Absolute loss (t/km}^2\text{) in soils of millionaire cities}}{\underleftarrow{\text{Fe}(-11{,}040) > \text{Cr}(-70.4) > \text{Y}(-15.9) > \text{Be}(-1.9)}}$$

These data (t/km$^2$) are necessary to make informed decisions for urban soils rehabilitation. However, these figures do not fully reflect the ecological–geochemical changes that have occurred in soils of cities with population over 700,000 people. For more information, we constructed a series of changes in the value of IRA:

$$\underset{\text{Relative accumulation in soils of millionaire cities}}{\underleftarrow{\text{Pb} > \text{Cd} > \text{As} > \text{Zn} > \text{Cl} > \text{Cu} > \text{Co} > \text{Ba} > \text{P} > \text{Li}}}$$

$$\underset{\text{Relative loss in soils of millionaire cities}}{\underleftarrow{\text{Cr} > \text{Y} > \text{Be} > \text{Fe}}}$$

Thus, the content of P, Ba, and Cl increased the most, but the greatest ecological–geochemical changes are caused by accumulation of Pb, Cd, As, and Zn in soils of these cities. The highest priority should be paid to these elements while developing various ecological measures. Also the special attention should be given to reducing the concentration of Cr, Y, and Be, but not for Fe.

The average contents of As, Ba, Be, Ca, Cd, Cl, Co, Cr, Cu, Fe, Li, Mg, Pb, Sn, Y, and Zn have changed significantly in soils of *cities with population about 500,000 (300,000–700,000) inhabitants*. Their IAA rows have the following forms:

$$\underset{\text{Absolute accumulation (t/km}^2\text{) in soils of half-millionaire cities}}{\underleftarrow{\text{Ca}(38{,}036) > \text{Mg}(2282) > \text{Ba}(354) > \text{Cl}(127) > \text{Zn}(39) > \text{Pb}(21) > \text{Li}(13) > \text{Cu}(6.1) > \text{Co}(3.8) > \text{As} > \text{Cd}(0.2)}}$$

$$\underset{\text{Absolute loss (t/km}^2\text{) in soils of half-millionaire cities}}{\underleftarrow{\text{Fe}(-10{,}737) > \text{Cr}(-87) > \text{Y}(-16) > \text{Sn}(-2.1) > \text{Be}(-1.9)}}$$

Series constructed from the values of the IRA are as follows:

$$\underset{\text{Relative accumulation in soils of half-millionaire cities}}{\underleftarrow{\text{Ca} > \text{Pb} > \text{Cl} > \text{Zn} > \text{Ba} > \text{As} > \text{Co} > \text{Li} > (\text{Cd, Mg}) > \text{Cu}}}$$

$$\xleftarrow[\text{Relative loss in soils of half-millionaire cities}]{\text{Cr} > \text{Y} > \text{Be} > \text{Fe} > \text{Sn}}$$

As seen from the data, Pb, Cl, and Zn are in the last place by accumulated mass, but they provide (along with Ca) a major impact on changes in environmental conditions in soil during the formation of cities with population half a million inhabitants. Cr, Y, and Be have the greatest impact on the ecological–geochemical soil appearance in this cities group among the decreasing elements.

Significant changes in soils of *cities with a population between 100,000 and 300,000 inhabitants* have occurred to the average contents of As, Co, Cr, Li, Mn, Mo, P, Pb, and Zn. Their IAA rows have the following forms:

$$\xleftarrow[\text{Absolute accumulation } \left(\text{t/km}^2\right) \text{ in soils of cities with } 100-300 \text{ thousand inhabitants}]{\text{P}(316) > \text{Zn}(29.7) > \text{Pb}(20.0) > \text{Li}(12.2) > \text{As}(9.5) > \text{Co}(2.8) > \text{Mo}(0.96)}$$

$$\xleftarrow[\text{Absolute loss } \left(\text{t/km}^2\right) \text{ in soils of cities with } 100-300 \text{ thousand inhabitants}]{\text{Mn}(-182) > \text{Cr}(-94.7)}$$

Rows of IRA:

$$\xleftarrow[\text{Relative accumulation in soils of cities with } 100-300 \text{ thousand inhabitants}]{\text{Pb} > \text{As} > \text{Zn} > \text{Mo} > \text{Li} > \text{P} > \text{Co}}$$

$$\xleftarrow[\text{Relative loss in soils of cities with } 100-300 \text{ thousand inhabitants}]{\text{Mn} > \text{Cr}}$$

Pb, As, and Zn have the greatest influence on the change of ecological–geochemical state of soils of this cities group, despite their relatively small accumulated mass.

In soils of group of small towns with population *less than 100 thousand inhabitants* the increased contents are marked for Ba, Ca, Cd, Cl, Co, Li, Pb, and Zn, and low contents are marked for Be, Cr, Mn, and Y. Rows of these elements accumulation (IAA) are as follows:

$$\xleftarrow[\text{Absolute accumulation } \left(\text{t/km}^2\right) \text{ in soils of small cities with } < 100 \text{ thousand inhabitants}]{\text{Ca}(37{,}784) > \text{Ba}(288) > \text{Cl}(83.7) > \text{Zn}(25.4) > \text{Pb}(17.7) > \text{Li}(9.5) > \text{Co}(4.0) > \text{Cd}(0.6)}$$

$$\xleftarrow[\text{Absolute loss } \left(\text{t/km}^2\right) \text{ in soils of small cities with } < 100 \text{ thousand inhabitants}]{\text{Mn}(-235) > \text{Cr}(-71.0) > \text{Y}(-14.7) > \text{Be}(-2.5)}$$

Rows of IRA constructed to establish the relative influence of these elements on ecological–geochemical environmental changes are presented as follows:

$$\overleftarrow{Ca > Pb > Cd > Cl > Ba > Zn > Co > Li}$$

Relative accumulation in soils of small cities with < 100 thousand inhabitants

$$\overleftarrow{Be > Cr > Y > Mn}$$

Relative loss in soils of small cities with < 100 thousand inhabitants

Thus, despite the much smaller increase in masses of such chemical elements as Pb, Cd, and Be, they provide (judging by the soil characteristics) the major ecological–geochemical changes of environment in cities with population less than 100,000.

In soils of *villages and hamlets*, the changes are associated to a greater extent with a significantly reduced (compared with the Earth's soil cover) content of chemical elements: Ag, Cd, Cr, Sn, V, Y, and Zr. The increased contents are marked for Cu, Li, Mo, P, Pb, and Zn. The greatest ecological-geochemical changes noted in soils of this group are associated with the increasing content of Pb, Zn, Mo, and Cu and with decreasing content of Ag and Cr. Rows of these elements accumulation (IAA) are as follows:

$$\overleftarrow{P(311) > Zn(28.6) > Li(12.2) > Cu(8.8) > Pb(7.6) > Mo(1.0)}$$

Absolute accumulation ($t/km^2$) in soils of villages and hamlets

$$\overleftarrow{Zr(-101) > Cr(-88) > V(-21.3) > Y(-19.5) > Sn(-2.26) > Cd(-0.19) > Ag(-0.15)}$$

Absolute loss ($t/km^2$) in soils of villages and hamlets

Their IRA rows are as follows:

$$\overleftarrow{Pb > Zn > Mo > Cu > Li > P}$$

Relative accumulation in soils of villages and hamlets

$$\overleftarrow{Ag > Cr > Y > Cd > Zr > Sn > V}$$

Relative loss in soils of villages and hamlets

*Recreational and tourist centres* have a special place among the settlements. The higher average contents of As, Ba, Ca, Cl, Co, Cu, Li, Mg, Pb, Sr, and Zn and low contents of Be, Cr, Sn, Y, and Zr are marked in their soils. Their accumulation and loss rows (IAA) have the following forms:

$$\overleftarrow{Ca(34{,}488) > Mg(2256) > Ba(290) > Sr(150) > Cl(104) > Zn(90) > Pb(27) > Cu(22) > Li(13) > As(9.6) > Co(6.1)}$$

Absolute accumulation ($t/km^2$) in soils of recreational and tourist centres

$$\overleftarrow{Zr(-67.4) > Cr(-67.1) > Y(-18.8) > Be(-2.2) > Sn(-2.1)}$$

Absolute loss ($t/km^2$) in soils of recreational and tourist centres

Series constructed from the values of the IRA are as follows:

$$\overleftarrow{\text{Pb} > \text{Ca} > \text{As} > \text{Zn} > \text{Cu} > \text{Cl} > \text{Co} > \text{Ba} > \text{Sr} > \text{Li} > \text{Mg}}$$

Relative accumulation in soils of recreational and tourist centres

$$\overleftarrow{\text{Y} > \text{Be} > \text{Cr} > \text{Zr} > \text{Sn}}$$

Relative loss in soils of recreational and tourist centres

The value of *Absolute Spread* (AS), based on the average elements content in urban soils, is also changed quite significantly in different groups of settlements. The distribution of chemical elements with an increased content (compared with the Earth's soils) in the studied *millionaire cities* is sometimes changing in a relatively wide range. For instance, AS is 51 for Cd (from 0.15 in Samara to 7.7 mg/kg in Almaty); is 6.6 for Zn (from 79 in Minsk to 280–520 mg/kg in Hamburg, Shenzhen, and Krasnoyarsk); is 4 for Pb (from 50 in Minsk to 140–200 mg/kg in Cologne, St. Petersburg, Paris, and London).

However, in the majority of cities (excluding single settlements with the minimum and maximum content of certain elements in soils), the fluctuation around the average for the whole group is small. The existence of separate cities with anomalously low and high elements content in soils is explained in most cases by the specificity of human activities. This is also probably related to soil sorption peculiarities (Perel'man 1986); that is mostly typical for Zn distribution.

Let us consider the distribution of the same elements (Cd, Zn, Pb) average contents, but in soils of several cities with a *half-million population*. AS of Cd increased sharply compared with the millionaire cities and reached the value of 1370 (from 0.01 in Irkutsk to 13.7 mg/kg in Pavlodar). For Zn, the AS increased to 40 (from 22 in Brest to 888 mg/kg in Ust-Kamenogorsk). AS for Pb reached the value 32.5 (from 7.8 in Taraz to 254 mg/kg in Ust-Kamenogorsk).

The above data show that the relative uniformity of distribution of chemical elements, common in millionaire cities, violated in half-millionaire cities. This is connected with the fact that some cities of this group have developed manufacture complexes of one branch. For example, in the city of Ust-Kamenogorsk, there are several very large enterprises arising within the scope of nonferrous metallurgy. Such industrial centres are similar to "monotowns" (one-company cities) having one key facility, which is the dominant pollution source.

However, unless taking into account such cities, the distribution of elements in soils of considered settlements is very uniform. For example, the average content of Pb in soils of the vast majority of cities in the group varies from 20 to 67 mg/kg (*AS* is 3.3).

In soils of *cities with the population of 100–300 thousand inhabitants* the AS of these elements content increased and became equal to 32 for Zn (from 10.4 in Uralsk to 320 mg/kg in Bratsk), and 42 for Pb. This is explained by the same factors as the AS increase in the cities group discussed above.

In soils of *villages*, the value of AS for practically all the chemical elements, became the lowest compared with other groups of settlements. The main reason is low average content of elements, close to abundance in the Earth's soils, in some localities.

The value of AS for many elements is quite large in soils of *tourist-recreational centres*. The large AS of the mean values, marked in a number of cases, is primarily explained by the different remote of localities from industrial sites and agricultural land, changing in the transport structure and using medicinal waters and mud with different composition. So the AS for Pb is 46 (from 6.5 on Saipan island to 300 mg/kg in Alupka, Crimea) and for Zn AS is 100 (from 5 on Saipan island to 500 mg/kg in the village on the Black Sea coast).

## 4 Conclusions

1. Quantitative ecological–geochemical data on the composition of soils confirmed the environmental and geographical information on the necessity for separate consideration of urban soils and division of settlements into separate groups according to the number of inhabitants. The greatest elements accumulation comparing with the Earth's soils (tens of thousands of tons per 1 $km^2$) is marked in soils of settlements groups with the half-million and less than 100,000 number of inhabitants, and in recreational and tourist centres. In all these systems, it is associated with an increase in the content of Ca and Mg.
2. Considering the environmental significance of chemical elements accumulation in soils, we note the primary role of Pb and Zn in all groups of cities. The accumulated mass of Pb fluctuates from 17.7 to 33.7 $t/km^2$, decreasing to 7.6 $t/km^2$ in soils of villages comparing with the Earth's soils. The concentration of accumulated Zn usually ranges from 25.0 to 39.0 $t/km^2$ and only in tourist-recreational centres and millionaire cities is up to 90.0 $t/km^2$. Out from the rest pollutants it is necessary, first of all, to note As and then Cu and Cl, which are the main contaminants in four of six cities groups. In two groups of settlements Cd and Co are important soil pollutants. In three groups, a considerable increase in the Ca content significantly modifies ecological-geochemical state of soils.
3. The reduced content of Cr, and often Y and Be noted in soils of all selected settlements groups comparing with the Earth's soils. The greatest mass loss is typical for Fe (over 10,000 $t/km^2$) and Cr (from 67 to 95 $t/km^2$).
4. Highlighted settlements groups also differ from one another in the values of chemical elements Absolute Spread. The greatest variation is typical for groups of settlements, which are characterised by complex of enterprises in one industry, such as nonferrous metallurgy.

**Acknowledgments** The authors would like to acknowledge the researchers from the Institute of Biosphere Geochemistry S.N. Voronets and E.V. Vlasova for their help during the sampling and data earning. We wish to thank S.E. Kalyuga for assistance in analyses performing. We thank organisers of the International Symposium "Biogenic and abiogenic interactions in natural and anthropogenic systems" for invitation to participate in its activity and to take part in the collective monograph. The academician and former Vice President of the Russian Academy of Sciences, N. P. Laverov, is gratefully acknowledged for reviewing an earlier version of the manuscript. This research was partly supported by the joint Russian-German program "DAAD—Mikhail Lomonosov" and Russian Geographical Society.

## References

Adriano DC (2001) Trace elements in terrestrial environments. Springer, New York

Alekseenko VA (2006) Ecological-geochemical changes in the biosphere. Development, estimation. Universitetskaya Kniga, Moscow (in Russian with English Abstract)

Alekseenko VA (2015) Geoecology. Ecological geochemistry. Fenix, Rostov-on-Don (in Russian)

Alekseenko VA, Alekseenko AV (2013) Chemical elements in geochemical systems. The abundances in urban soils. Publishing House of Southern Federal University, Rostov-on-Don (in Russian with English Abstract)

Alekseenko VA, Alekseenko AV (2014a) Chemical elements in urban soils. Logos, Moscow (in Russian)

Alekseenko V, Alekseenko A (2014b) The abundances of chemical elements in urban soils. J Geochem Explor 147(B):245–249

Fersman AE (1952–1959) Selected works, vol 1–5. Publishing House of the USSR Academy of Sciences, Moscow (in Russian)

Gerasimova MI, Stroganova MN, Mozharova NV, Prokof'eva TV (2003) Anthropogenic soils: genesis, geography, and rehabilitation. Oikumena, Smolensk (in Russian with English Abstract)

Kabata-Pendias A, Pendias H (2001) Trace elements in soils and plants. CRC Press LLC, Boca Raton

Kasimov NS, Kosheleva NE, Sorokina OI, Bazha SN, Gunin PD, Enkh-Amgalan S (2011) Ecological-geochemical state of soils in Ulaanbaatar (Mongolia). Eurasian Soil Sci 44 (7):709–721

Motuzova GV, Minkina TM, Karpova EA, Barsova NU, Mandzhieva SS (2014) Soil contamination with heavy metals as a potential and real risk to the environment. J Geochem Explor 144(B):241–246

Norra S, Stüben D (2003) Urban soils. J Soils Sed 3(4):230–233

Pashkevich MA, Alekseenko AV, Vlasova EV (2015) Biogeochemical and geobotanical assessment of marine ecosystems conditions (Novorossiysk city). Water Ecol. 3:67–80

Perel'man AI (1986) Geochemical barriers: theory and practical applications. Appl Geochem 1 (6):669–680

Perel'man AI (1989) Geochemistry. Vysshaya Shkola, Moscow (in Russian)

Syso AI, Smolentsev BA, Yakimenko VN (2010) The soil cover of Novosibirsk Akademgorodok and its eco-agricultural assessment. Contemp Probl Ecol 3(3):253–264

Tsolova VT, Hristova MB, Bech J, Pascual NR, Banov MD (2014) Pb, Cu and Zn geochemistry in reclaimed soils (Technosols) of Bulgaria. J Geochem Explor 144(B):337–344

Tume P, Bech J, Reverter F, Bech J, Longan L, Tume L, Sepúlveda B (2011) Concentration and distribution of twelve metals in Central Catalonia surface soils. J Geochem Explor 109(1–3):92–103

Vernadsky VI (1965) Chemical structure of the Earth's biosphere and its environment. Nauka, Moscow (in Russian)

Vinogradov AP (1959) The geochemistry of rare and dispersed chemical elements in soils, 2nd edn. Consultants Bureau Enterprises, New York

# Thermodynamics of Environmentally Important Natural and Synthetic Phases Containing Selenium

**Marina V. Charykova and Vladimir G. Krivovichev**

**Abstract** Understanding and deciphering the processes proceeding near the surface are the urgent tasks of contemporary mineralogy and geochemistry, which are especially important for resolving ecological challenges and developing principles of rational environmental management. The paper presents our systematized data published about the thermodynamics of selenites, which are formed in the weathering zone of the sulfide ores, and determines approaches to quantitative physicochemical modeling of their formation conditions. The activities of components in natural waters beyond the zones of natural (oxidation zones) and man-made contamination with selenium ($a_{\Sigma Se} = 10^{-9}$, $a_{\Sigma Fe} = 10^{-5}$, $a_{\Sigma Cu} = 10^{-7}$, $a_{\Sigma Zn} = 5 \times 10^{-7}$, $a_{\Sigma Co} = 10^{-8}$, $a_{\Sigma Ni} = 6 \times 10^{-8}$, and $a_{\Sigma Pb} = 10^{-8}$) and in waters formed in the oxidation zone ($a_{\Sigma Se} = 10^{-5}$–$10^{-4}$, $a_{\Sigma Fe} = 10^{-2}$, $a_{\Sigma Cu} = 10^{-2}$, $a_{\Sigma Zn} = 10^{-2}$, $a_{\Sigma Co} = 10^{-3}$, $a_{\Sigma Ni} = 10^{-2}$, $a_{\Sigma Pb} = 10^{-4}$) have been estimated. Eh–pH diagrams were calculated and plotted using the Geochemist's Workbench (GMB 7.0) software package. The database comprises the thermodynamic parameters of 46 elements, 47 main particles, 48 redox pairs, 551 particles in solution, 624 solid phases, and 10 gases. The Eh–pH diagrams of the Me–Se–$H_2O$ systems (Me = Cu, Pb, Co, Ni, Fe, Zn) were plotted for the average contents of these elements in underground water and for their contents in oxidation zones of sulfide deposits. The formation of Co, Ni, Fe, Cu, Zn, and Pb selenites and selenates at the surface is discussed.

**Keywords** Selenites · Near-surface conditions · Environmental mineralogy and geochemistry · Physicochemical modeling · Eh–ph diagrams

In recent years, interest in selenium behavior under supergene conditions has increased owing to the needs of environmental protection. In high concentrations, selenium is toxic and the release of selenium from agricultural and industrial activities had caused diseases and death for wildlife and aquatic animals (Plant et al.

M.V. Charykova (✉) · V.G. Krivovichev
Saint Petersburg University, Saint Petersburg, Russia
e-mail: m.charykova@spbu.ru



O.V. Frank-Kamenetskaya et al. (eds.), *Biogenic—Abiogenic Interactions in Natural and Anthropogenic Systems*, Lecture Notes in Earth System Sciences,
DOI 10.1007/978-3-319-24987-2_12

2014). Due to the toxicity of selenium, it is quite essential to analyze the conditions under which selenites and selenates replace selenides in the oxidation zones of sulfide ore deposits and upon weathering of technological waste. It should be noted that the source of selenium in the environment can be due to the technological activities associated with the nuclear industry, as $^{79}Se$ is a long-lived fission product with a half-life of $1.1 \times 10^6$ years. $^{79}Se$ is chemically and radiologically toxic (Chen et al. 1999). The understanding of mechanisms of the selenium's behavior in the near-surface conditions is one of actual problems of modern mineralogy and geochemistry and it is very important for solving some of the environmental problems.

The toxicity and biological activity of selenium are determined mainly by its speciation in the environment and, therefore, depend on the redox potential (Eh) and the acidity–alkalinity of natural media (surface and near-surface waters). Near the surface, selenium occurs in an oxidized state as selenites and much less frequently as selenates. These minerals are clearly divided into two groups (Krivovichev and Depmeier 2005; Krivovichev and Charykova 2006). Secondary selenites and selenates formed near the surface with actively participating aqueous solutions and seasonal variations of temperature and atmospheric pressure ($\sim$ 1 bar) are categorized as belonging to the first group. Under these conditions, such minerals as chalcomenite, cobaltomenite, and ahlfeldite are found, as a rule, in the oxidation zone of sulfide and selenium-bearing ores. For quantitative thermodynamic modeling of mineral formation in the oxidation zone of sulfide, and selenide-bearing ores, we consider sulfates, selenites, and selenates, the major mineral-forming cations of which are Fe, Cu, Zn, Pb, Co, and Ni. Anhydrous selenites with additional anions (georgbokiite, ilinskite, sophiite) are the products of volcanic exhalations (Tolbachik Eruption, Kamchatka) and belong to the second group. The specific formation conditions of these minerals significantly differ from those of the other selenites and selenates. They precipitate from a gaseous phase at 300–400 °C, atmospheric pressure, and low partial water pressure. Knowledge of the thermal stability of products of volcanic exhalations and hydrous selenites from the oxidation zone of ore deposits is necessary in order to conduct a detailed analysis of their physicochemical formation conditions.

Any progresses in our understanding of the low-temperature mineral assemblages, and especially of those containing toxic elements, strongly depend on our understanding of the thermodynamic stability of the constituting mineral phases. Experimental and thermodynamic modeling is quite essential to analyze the conditions under which selenites and selenates replace selenides, and selen-bearing sulfides in the oxidation zones of sulfide ore deposits or upon weathering of technological waste.

Most selenites are formed by chemical weathering of ores by oxygenated waters, establishing conditions of increased Eh and low or neutral pH (with seasonal fluctuations of temperatures and atmospheric pressure) (Krivovichev and Charykova 2006). These parameters define the migration of selenium, and its precipitation in the form of selenite minerals (chalcomenite $CuSeO_3 \cdot 2H_2O$, cobaltomenite $CoSeO_3 \cdot 2H_2O$, ahlfeldite $NiSeO_3 \cdot 2H_2O$, mandarinoite $Fe_2(SeO_3)_3 \cdot 6H_2O$, molybdomenite $PbSeO_3$). Eh–pH diagrams of the Me–Se–$H_2O$ systems (Me=Co, Ni, Fe,

Cu, Zn, Pb) make it possible to estimate the conditions for the near-surface formation of Se minerals (selenites) (Krivovichev et al. 2011).

The physicochemical modeling is based on the thermodynamic constants of minerals. Reliable solubility product (SP) values of the phases crystallized in the system are necessary for calculation of mineral equilibria. They can be derived from the experimental data (for example, from the data on solubility, if the activity coefficients of the components of a saturated solution are known) or theoretically calculated applying the data on variation of the Gibbs free energy of compound formation. The gained data on Fe, Cu, Zn, Pb, Co, Ni selenites, and selenates (Charykova et al. 2010) indicate that, even when the corresponding parameters are published, they arouse questions and must be specified. It should be kept in mind that in many cases the values of thermodynamic functions of formation taken from distinct sources are based on the same experimental data. For example, for selenites and selanates, the data recommended in the reference book (Olin et al. 2005) should be preferable. This reference book and the article by Seby et al. (2001) present a scrupulous critical reviews of all the published experimental measurements, but the values of logSP submitted there for selenites under our consideration were calculated from the same experimental data on solubility (Chukhlantsev 1956; Chukhlantsev and Tomashevsky 1957), in which the solid phase was not identified reliably and the content of anion in saturated solution was not determined. Both anhydrous salts and crystal hydrates of variable composition could precipitate in this case.

The experimental determination of the thermodynamic constants of rare minerals based on their solubility or calorimetry in natural samples fails due to the insufficient amount of the materials because they mainly occur as fine segregations, fluccans, and illinitions. In addition, natural samples inevitably contain isomorphic admixtures, which affect their properties, in particular, the thermodynamic parameters. Thus, the synthesis of their analogs, the chemical composition, and structure of which strictly correspond to the formulas of natural minerals, is of vital importance for us. Thus $CuSeO_3{\cdot}2H_2O$, $CoSeO_3{\cdot}2H_2O$, $NiSeO_3{\cdot}2H_2O$, $ZnSeO_3{\cdot}2H_2O$, and $ZnSeO_3{\cdot}H_2O$, ahlfeldite $NiSeO_3{\cdot}2H_2O$ have been synthesized. The obtained samples have been identified by X-ray diffraction and IR spectroscopy, and the solubility has been determined by the isothermal saturation method in ampoules at 25 °C (Charykova et al. 2012, 2015). The data obtained are given in Table 1. Using these

**Table 1** Solubility products (logSP) of selenites at 25 °C

| Solid phase (mineral) | Solubility product (logSP) | Reference |
|---|---|---|
| $CuSeO_3{\cdot}2H_2O$ (chalcomenite) | −10.6 | Charykova et al. (2015) |
| $CoSeO_3{\cdot}2H_2O$ (cobaltomenite) | −9.20 | Charykova et al. (2012) |
| $NiSeO_3{\cdot}2H_2O$ (ahlfeldite) | −9.44 | Charykova et al. (2012) |
| $Fe_2(SeO_3)_3{\cdot}6H_2O$ (mandarinoite) | −41.58 | Olin et al. (2005) |
| $PbSeO_3$ (molybdomenite) | −12.5 | Olin et al. (2005) |
| $ZnSeO_3{\cdot}2H_2O$ | −8.35 | Charykova et al. (2015) |
| $ZnSeO_3{\cdot}H_2O$ | −7.96 | Charykova et al. (2015) |

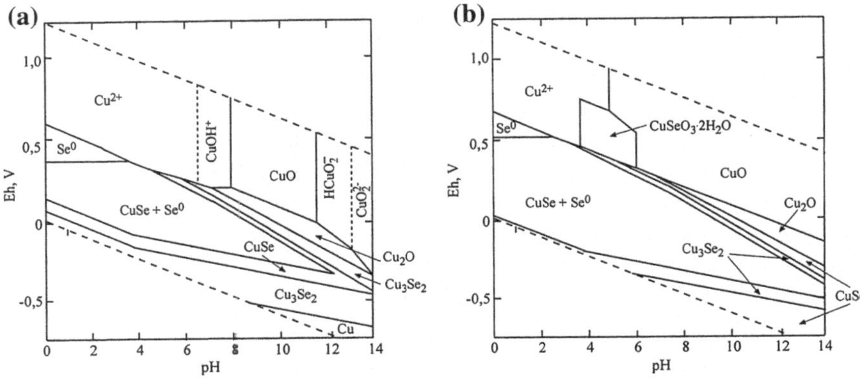


**Fig. 1** Eh–pH diagrams of the Cu–Se–$H_2O$ system at 25 °C and the activities of the components: **a**: $a_{\Sigma Se} = 10^{-9}$, $a_{\Sigma Cu} = 10^{-7}$ and **b** $a_{\Sigma Se} = 10^{-5}$, $a_{\Sigma Cu} = 10^{-2}$. Here and in Figs. 2, 3, 4, 5, 6, 7 and 8, the *solid lines* are the boundaries of the stability fields of the solid phases. The *lines* of equilibria between the particles of metal in the solution are plotted

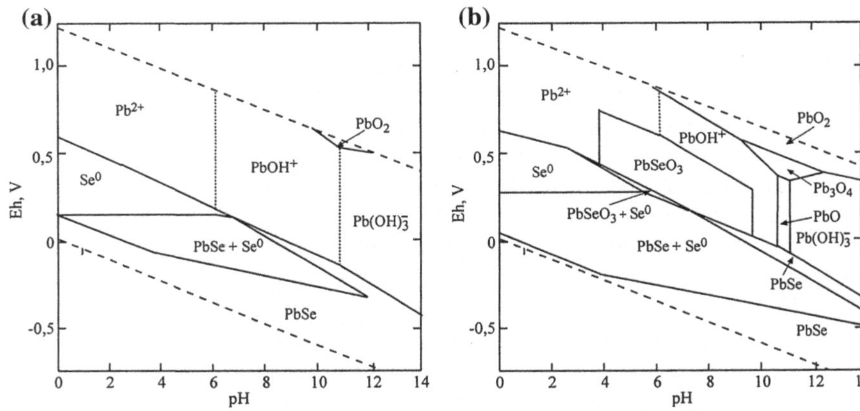


**Fig. 2** Eh–pH diagrams of the Pb–Se–$H_2O$ system at 25 °C and the activities of the components: **a** $a_{\Sigma Se} = 10^{-9}$, $a_{\Sigma Pb} = 10^{-8}$ and $a_{\Sigma Se} = 10^{-5}$, $a_{\Sigma Pb} = 10^{-4}$ (**b**)

values of logSP, the Eh–pH diagrams have been calculated (Figs. 1, 2, 3, 4, 5, 6, 7 and 8).

The Eh–pH diagrams were calculated using the Geochemist's Workbench (GMB 9.0) software package. The database contains the thermodynamic characteristics of 46 elements, 47 main particles, 48 redox pairs, 551 particles in solution, 624 solid phases, and 10 gases recorded as equilibrium constants of dissociation reactions between particles in solutions and reactions of solid phase dissolution. The calculation of the diagrams was predated by the introduction of additional components into the database and the specification of some solubility products

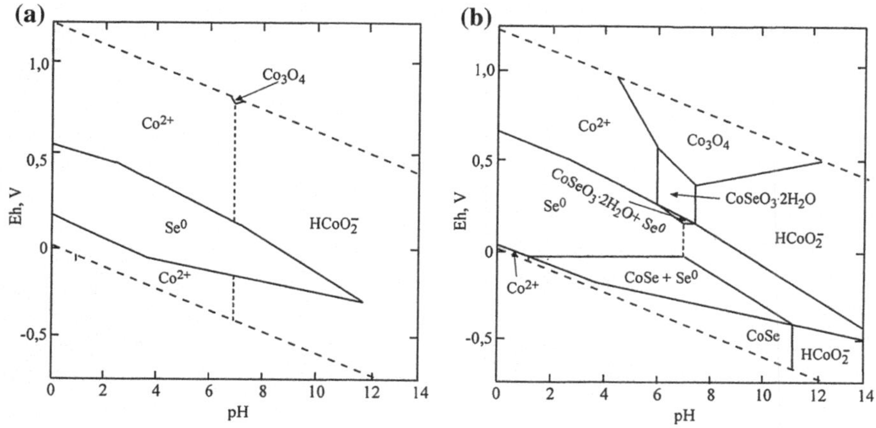


**Fig. 3** Eh–pH diagrams of the system Co–Se–$H_2O$ at 25 °C and the activities of the components: $a_{\Sigma Se} = 10^{-9}$, $a_{\Sigma Co} = 10^{-8}$ (**a**) and $a_{\Sigma Se} = 10^{-5}$, $a_{\Sigma Co} = 10^{-3}$ (**b**)

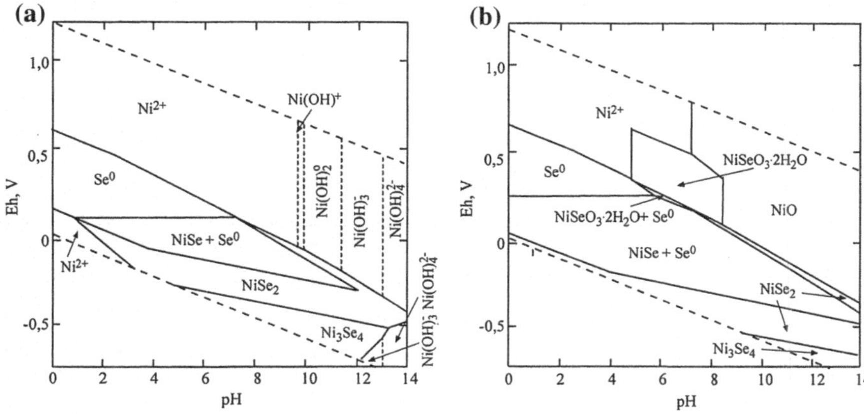


**Fig. 4** Eh–pH diagrams of the system Ni–Se–$H_2O$ at 25 °C and the activities of the components: $a_{\Sigma Se} = 10^{-9}$, $a_{\Sigma Ni} = 6 \times 10^{-8}$ (**a**) and $a_{\Sigma Se} = 10^{-5}$, $a_{\Sigma Ni} = 10^{-2}$ (**b**)

(Krivovichev et al. 2011; Charykova et al. 2012, 2014, 2015). The activity coefficients are calculated from the Debye–Hückel equation.

The stability fields of minerals in the Eh–pH plots are determined (with other things being equal) by the activities of the components in the mineral-forming fluid (Garrels and Christ 1965). Therefore, before plotting and considering diagrams, the activity of the major components in natural crystallization media should be estimated. It is evident that the groundwater and surface waters both natural and resulting from the human impact on the environment are such media, using the literature data (Shvartsev 1998; Krainov et al. 2004), substantiated the choice of

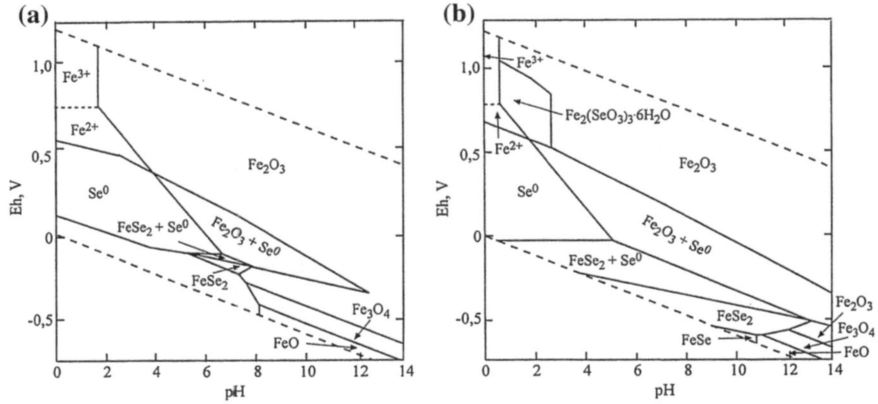


**Fig. 5** Eh–pH diagrams of the Fe–Se–$H_2O$ system at 25 °C and the activities of the components: **a** $a_{\Sigma Se} = 10^{-9}$, $a_{\Sigma Fe} = 10^{-5}$ and **b** $a_{\Sigma Se} = 10^{-4}$, $a_{\Sigma Fe} = 10^{-2}$

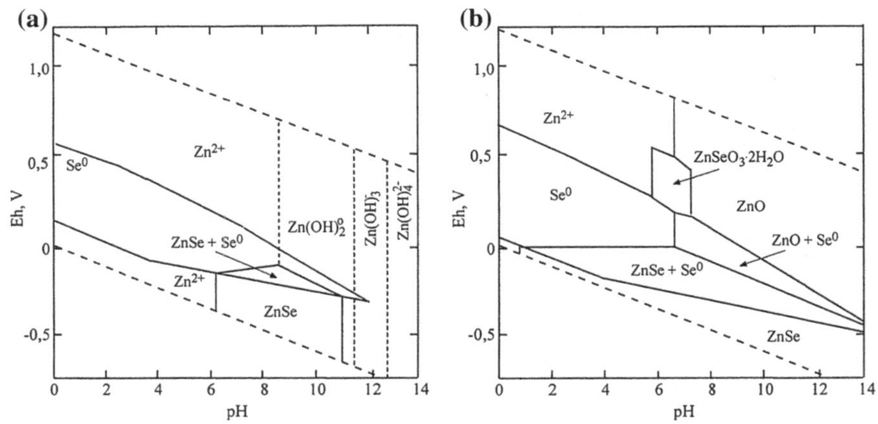


**Fig. 6** Eh–pH diagrams of the system Zn–Se–$H_2O$ at 25 °C and component's activities: $a_{\Sigma Se} = 10^{-9}$, $a_{\Sigma Zn} = 5 \times 10^{-7}$ (**a**) and $a_{\Sigma Se} = 10^{-5}$, $a_{\Sigma Zn} = 10^{-2}$ (**b**)

metal activities to calculate Eh–pH diagrams. The following total activities of the chemical elements in the subsurface and ground ("background") waters of the supergene zone are used in the calculations: $a_{\Sigma Se} = 10^{-9}$, $a_{\Sigma Fe} = 10^{-5}$, $a_{\Sigma Cu} = 10^{-7}$, $a_{\Sigma Zn} = 5 \times 10^{-7}$, $a_{\Sigma Co} = 10^{-8}$, $a_{\Sigma Ni} = 6 \times 10^{-8}$, and $a_{\Sigma Pb} = 10^{-8}$. The following total activities of the ions in solutions from the zones of oxidation and human pollution are used: $a_{\Sigma Se} = 10^{-5}$–$10^{-4}$, $a_{\Sigma Fe} = 10^{-2}$, $a_{\Sigma Cu} = 10^{-2}$, $a_{\Sigma Zn} = 10^{-2}$, $a_{\Sigma Co} = 10^{-3}$, $a_{\Sigma Ni} = 10^{-2}$, $a_{\Sigma Pb} = 10^{-4}$ (Krivovichev et al. 2011).

The Eh–pH diagrams of the Me–Se–$H_2O$ systems (Me = Cu, Pb, Co, Ni, Fe, Zn) corresponding to both the average (background) contents of these elements in the

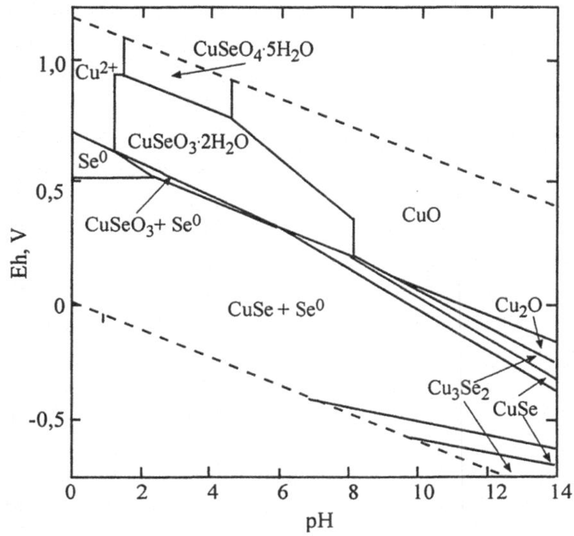


**Fig. 7** Eh–pH diagram of the Cu-Se-$H_2O$ system at 25 °C and the activities of the components: $a_{\Sigma Se} = 10^{-1}$ and $a_{\Sigma Cu} = 10^{-1}$

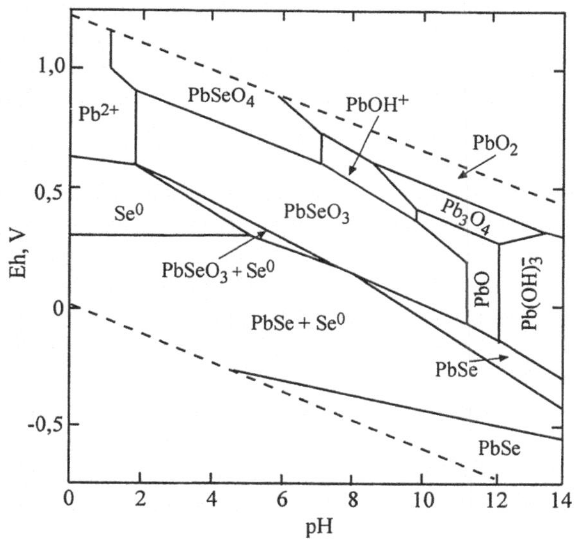


**Fig. 8** Eh–pH diagram of the Pb-Se-$H_2O$ system at 25 °C and the activities of the components: $a_{\Sigma Se} = 10^{-3}$ and $a_{\Sigma Pb} = 10^{-3}$

surface water and their elevated concentrations in the water of the oxidation zone of sulfide deposits are discussed below.

The Eh–pH diagrams of the Cu–Se–$H_2O$ system are shown in Fig. 1. It is seen that, at the background concentrations of Cu and Se (Fig. 1a), selenides (klokmannite, kruta'ite, and umangite), copper oxides (cuprite, tenorite), native selenium, and native copper (in the alkaline region of negative Eh) are stable.

The increase in the activities of these components to the values corresponding to concentration in the oxidation zone of sulfide deposits (Fig. 1b) results in the additional appearance of chalcomenite ($CuSeO_3 \cdot 2H_2O$), which is formed in the weakly acid environment and at a rather high positive Eh. Natural chalcomenite was found for the first time at the Cerro de Cacheuta deposit in Mendoza, Argentina (Des Cloizeaux and Damour 1881), where it occurs as fine (up to 1 mm in length) light blue prismatic crystals associated with cerussite and azurite.

The Eh–pH diagrams of the Pb–Se–$H_2O$ system are shown in Fig. 2. At the background concentration of Pb and Se (Fig. 2a), the diagram is simpler than in the previous case; most of its part does not contain stability fields of solid phases and is characterized only by the fields of the predominant particles in the solution. Only native selenium, PbSe (claustalite), and Pb oxide $PbO_2$ (plattnerite) are crystallized under these conditions. A very small stability field in the alkaline region with an anomalous high positive Eh corresponds to this mineral. The increase in the activities of the components (Fig. 2b) leads to the crystallization of $Pb_3O_4$ (litharge) and PbO (minium) in addition to claustalite and plattnerite. It is noteworthy that a wide stability field of molybdomenite ($PbSeO_3$) appears covering the pH range of 4–9.5. As chalcomenite, molybdomenite was described for the first time from the Cerro de Cacheuta deposit in Mendoza, Argentina (Bertrand 1882), where it occurs as fine (up to 2 mm in length), colorless or white tabular, and acicular crystals.

The Eh–pH diagrams of the Co–Se–$H_2O$ and Ni–Se–$H_2O$ systems are shown in Figs. 3 and 4. The formation conditions of two minerals forming a continuous solid solution $CoSeO_3 \cdot 2H_2O$ (cobaltomenite)–$NiSeO_3 \cdot 2H_2O$ (ahlfeldite) are of interest. In addition to the stability field of native selenium, the Eh–pH plot of the Co–Se–$H_2O$ system with the activities of the components corresponding to their average concentration in the groundwater of the supergene zone (Fig. 3a) contains a very small stability field of $Co_3O_4$, which is unknown in nature. The increase in the activities to the values of the oxidation zone of selenium-bearing sulfide ore results in the expansion of these fields and the appearance of the stability fields of CoSe (freboldite) and cobaltomenite formed in the neutral environment while Eh is not too high (Fig. 3b). The Eh–pH diagrams of the Ni–Se–$H_2O$ system are different in topology. For example, when the activities are low (Fig. 4a), the stability fields of $NiSe_2$ (penroseite) and $Ni_3Se_4$ (wilkmanite) appear in the diagram in addition to native selenium. When the activities of the components increase, a wide field of NiO (bunsenite) and a field of ahlfeldite appear in addition to the stability field of native selenium (Fig. 4b). As molybdomenite and chalcomenite, cobaltomenite was discovered from the Cerro de Cacheuta deposit in Mendoza, Argentina (Bertrand 1882), where it occurs as fine pink-red erythrite-type crystals. Ahlfeldite was found in 1935 (Herzenberg and Ahlfeld 1935) in the Hiaco Mine in Pacajake, which is 30 km north of Colquechaca, Bolivia; its composition and properties were studied in detail much later (Aristarain and Hurlbut 1969).

Figure 5 exhibits the Eh–pH diagrams of the Fe–Se–$H_2O$ system. When the activity of Se and Fe are low ($a_{ZSe} = 10^{-9}$, $a_{ZFe} = 10^{-5}$) (Fig. 5a), this diagram contains wide stability fields of native selenium and hematite and small fields corresponding to the crystallization of oxides (magnetite, wustite) and selenide

$FeSe_2$ (ferroselite). The elevated activity of these components ($a_{\Sigma Se} = 10^{-4}$, $a_{\Sigma Fe} = 10^{-2}$; Fig. 5b) substantially expands the stability field of $FeSe_2$ (ferroselite); FeSe (achavalite), and mandarinoite appear in the alkaline and acid areas, respectively. A very rare natural selenite mandarinoite, $Fe_2(SeO_3)_3{\cdot}6H_2O$, pertaining to this system was found for the first time in the Hiaco Mine in Pacajake, Bolivia (Dunn et al. 1978) as small (up to 3 mm) yellowish-green saber-shaped crystals aggregated into rosettes.

Similar zinc minerals (hydrous selenites) have not yet been found in such settings. Chlorine-bearing zinc selenite sophiite, $Zn_2(SeO_3)Cl_2$, identified as an exhalative product (Tolbachik eruption, Kamchatka) belongs to anhydrous selenites with additional cations, whose formation conditions are very specific: these minerals as noticed above are formed from the gaseous phase at elevated temperatures (300–400 °C), atmospheric pressure, and apparently at a low partial pressure of water. At the same time, in terms of geochemistry, hydrous selenites of zinc were able to be formed in the oxidation zone of selenium-bearing sulfide ores, where zinc is released from sphalerite or stilleite (ZnSe). Rarity, low abundance, and difficulty of their diagnostics are the reasons why zinc selenites have not yet been found in this setting. Diagnostics are partly hampered because Zn (in contrast to Cu, Co, Ni, and Fe) is not a chromophore and therefore its salts are colorless or white. Nevertheless, waterless selenite of zinc (zincomenite, $ZnSeO_3$) was found (Pekov et al. 2014) in the Northern fumarole field, first scoria cone of the Northern Breakthrough of the Great Tolbachik Fissure Eruption, Tolbachik volcano, Kamchatka Peninsula, Russia.

Finally, the Eh–pH plots of the Zn–Se–$H_2O$ system are shown in Fig. 6. This system attracts interest because it allows the estimation of the physicochemical parameters of the formation of Zn selenite, a solid phase that is not yet found in nature. As seen from Fig. 6a, when the activity of the components is low, the diagram contains the stability fields of only native selenium and ZnSe (stilleite). The increase in activity up to the level corresponding to the oxidation zone of sulfide ores (Fig. 6b) results in the addition of the wide stability field of ZnO (zincite) and the stability field of zinc selenite. Different versions of Eh versus pH diagrams for the Zn–Se–$H_2O$ system were constructed to clarify the physicochemical conditions of $ZnSeO_3{\cdot}2H_2O$ and $ZnSeO_3{\cdot}H_2O$ formation. As follows from these diagrams, physicochemical parameters of zinc selenite stability, pH, Eh, and the activity of selenium are close to those of cobalt and nickel selenites (ahlfeldite and cobaltomenite) (Krivovichev et al. 2011), and $ZnSeO_3{\cdot}2H_2O$ is the stable phase at the temperature fluctuations corresponding to the environmental conditions. The crystal monohydrate appears to be a metastable phase, which remains stable beyond its stability field. Thus, the formation of zinc selenite under natural oxidative conditions is quite probable in the form of stable $ZnSeO_3{\cdot}2H_2O$ or metastable $ZnSeO_3{\cdot}H_2O$ species.

In conclusion, we dwell on the probable formation of natural Co, Ni, Fe, Cu, Zn, and Pb selenates. We calculated the Eh–pH diagrams within a wide range of activity of the components and found the values corresponding to the appearance of the stability fields of $CoSeO_4{\cdot}6H_2O$, $CuSeO_4{\cdot}5H_2O$, $NiSeO_4{\cdot}6H_2O$, $PbSeO_4$, and

$ZnSeO_4 \cdot 6H_2O$. The formulas of crystalline hydrates of selenates stable at 25 °C and their thermodynamic constants were taken (or recalculated) from the most comprehensive source concerning selenium and its compounds (Olin et al. 2005). Unfortunately, no data about the thermodynamic functions of the formation of iron selenates were found in the literature.

The calculations indicate that, among all the discussed systems, selenates are formed only in the Pb–Se–$H_2O$ and Cu–Se–$H_2O$ systems at 25 °C with more or less real (although high) activities of metals and Se in solution: $a_{\Sigma Se} = 10^{-3}$ and $a_{\Sigma Pb} = 10^{-3}$; $a_{\Sigma Se} = 10^{-1}$ and $a_{\Sigma Cu} = 10^{-1}$. The activities of Co, Ni, and Zn are extremely high and cannot be realized in nature. The Eh–pH diagrams of the Cu–Se–$H_2O$ and Pb–Se–$H_2O$ systems show crystallization fields of $CuSeO_4 \cdot 5H_2O$ and $PbSeO_4$ (Figs. 7 and 8). Note that, at such high activities of the components, the calculations of the stability fields of the solid phases and the predominant species in the solutions become less accurate, because the program of the Geochemist's Workbench software package applied for the calculation of the activity coefficients uses the Debye–Huckel equation, which gives reliable results only for very diluted solutions of electrolytes.

The results obtained are consistent with the list of the three known to date mineral species containing selenate ions: schmiederite ($Pb_2Cu_2(SeO_3)(SeO_4)(OH)_4$); olsacherite ($Pb_2(SeO_4)(SO_4)$) found at the Dragon and Pacajake deposits in Bolivia and the Baccu Locci deposit in Sardinia, Italy together with selinites; and carlosruizite ($K_6Na_{10}Mg_{10}(IO_3)_{12}(SeO_4)_{12} \cdot 12H_2O$), a very specific mineral differing in its formation conditions found at a niter deposit in association with nitratine, fuenzalidaite, and halite (Krivovichev and Depmeir 2005). Kerstenite $PbSeO_4$ (Mandarino 1994), a single simple selinate, has been discredited to date and turned out to be molybdomenite after a detailed study.

So, the behavior of selenium, the nearest geochemical counterpart of sulfur, in the surface environment can be quantitatively explained by variations of the redox potential and the acidity–basicity of the mineral-forming medium. Precisely, these parameters determine the migration ability of selenium compounds and its precipitation in the form of various solid phases.

**Acknowledgments** This study was supported by St. Petersburg State University (grant number 3.38.286.2015). The equipments of the St. Petersburg State University Resource Centers "X-ray diffraction methods" and "Geomodel" were used.

## References

Aristarain LF, Hurlbut CS (1969) Ahfeldite from Pacajake Bolivia: restudy. Amer Min 54:448–456

Bertrand E (1882) Sur la molybdomenite (selenite de plomb), la cobaltomenite (selenite de cobalt) et l'acide selenieux de Cacheuta (La Plata). Bull Soc Min France 5:90–92

Charykova MV, Krivovichev VG, Depmeier W (2010) Thermodynamics of Arsenates, Selenites, and Sulfates in the Oxidation zone of Sulfide Ores: I. thermodynamic constants at ambient conditions. Geol Ore Deposits 52:689–700

Charykova MV, Krivovichev VG, Yakovenko OS, Semenova VV, Semenov KN, Depmeier W (2012) Thermodynamics of Arsenates, Selenites, and Sulfates in the Oxidation zone of Sulfide Ores: VI. Solubility of synthetic analogs of Ahlfeldite and Cobaltomenite at 25 °C. Geol Ore Deposits 54:638–646

Charykova MV, Krivovichev VG, Lelet MI, Yakovenko OS, Suleimanov EV, Depmeier W, Semenova VV, Zorina ML (2014) A calorimetric and thermodynamic investigation of the synthetic analogues of cobaltomenite, $CoSeO_3{\cdot}2H_2O$, and ahlfeldite, $NiSeO_3{\cdot}2H_2O$. Amer Miner 99:742–748

Charykova MV, Krivovichev VG, Ivanova NM, Semenova VV (2015) Thermodynamics of Arsenates, Selenites, and Sulfates in the Oxidation zone of Sulfide Ores: XI. Solubility of synthetic analogs of chalcomenite and zinc selenites at 25 °C. Zapiski RMO (Proc Russ Mineral Soc) 144:70–80 (In Russian)

Chen F, Burns PC, Ewing RC (1999) $^{79}Se$: geochemical and crystallo-chemical retardation mechanisms. J Nucl Mater 275:81–94

Chukhlantsev VG (1956) Solubility product of selenite of some metals. Zh Neorg Khimii 1:2300–2305 (In Russian)

Chukhlantsev VG, Tomashevsky GP (1957) The solubility of Selenites of certain metals. Zh Anal Khim 12:296–301 (In Russian)

Des Cloizeaux AL, Damour MA (1881) Note sur la chalcomenite, nouvelle espece minerale (selenite de cuivre). Bull Soc Geol France 4:51–55

Dunn PJ, Peacor DR, Sturman BD (1978) Mandarinoite, a new Ferric-Iron Selenite from Bolivia. Can Miner 16:605–609

Garrels R, Christ Ch (1965) Solutions, minerals, and equilibria. Harper and Row, New York

Herzenberg R, Ahlfeld F (1935) Blockit, ein Neues Selenerz aus Bolivien. Zentralbl Miner Geol Palaent Abt A:277–279

Krainov SR, Ryzhenko BN, Shvets VM (2004) Geokhimiya podzemnykh vod. Teoreticheskie, prikladnye i ekologicheskie aspekty (Geochemistry of Subsurface Waters: Theoretical, Applied, and Environmental Aspects). Nauka, Moscow (in Russian)

Krivovichev VG, Charykova MV (2006) Thermodynamic of mineral equilibrium in the system with toxic components. 1. Selenium. St.-Petersburg State University Publishing, St.-Petersburg (in Russian)

Krivovichev VG, Depmeier W (2005) Selenates and selenites: systems $Se–S–H_2O$, $Pb–Se–S–H_2O$, $U–Se–H_2O$, and $U–Se–I–H_2O$—Thermodynamical analysis and geological applications. Zapiski RMO (Proc Russ Mineral Soc) 134:1–15 (In Russian)

Krivovichev VG, Charykova MV, Yakovenko OS, Depmeier W (2011) Thermodynamics of arsenates, selenites and sulphates in oxidising zone of sulphides ore deposits. IV. Eh–pH diagrams of the systems $Me–Se–H_2O$ (Me = Co, Ni, Fe, Cu, Zn, Pb) at 298 K. Geol Ore Deposits 53:514–527

Mandarino JA (1994) Natural and synthetic Selenites and Selenates and their Gladstone-Dale compatibility. Eur J Min 6:337–349

Olin A, Nolang B, Osadchii EG, Ohman L-O, Rosen E (2005) Chemical thermodynamics of Selenium. Elsevier, Amsterdam

Pekov IV, Yapaskurt VO, Britvin SN, Chukanov NV, Sidorov EG (2014) Zincomenite, IMA 2014-014. CNMNC Newsletter. Miner Mag 78:798

Plant JA, Bone J, Voulvoulis N, Kinniburgh DG, Smedley PL, Fordyce FM, Klinck BA (2014) Arsenic and Selenium. In: Lollar BS (ed) Treatise on geochemistry. Environmental Geochemistry, vol 11. Elsevier Pergamon, Amsterdam

Séby F, Potin-Cautier M, Giffaut E, Borge G, Donard OFX (2001) A critical review of thermodynamic data for selenium species at 25 °C. Chem Geol 171:173–194

Shvartsev SL (1998) Gidrogeokhimiya zony gipergeneza. (Hydro-geochemistry of Supergene Zone). Nedra, Moscow (in Russian)

# Migration Models of Cu, Zn, and Cd in Soils Under Irrigation With Urban Wastewater

**Alexandr S. Frid**

**Abstract** A notable migration of Cu, Zn, and Cd has been revealed in the soils used under irrigation for a long period of time in Egypt. To describe the distribution of these elements throughout the soil profiles, mathematical models of migration (diffusion and convective-diffusion models) have been tested, their adequacy has been shown, and their parameters have been estimated. The coefficients of diffusion (D) and convective diffusion ($D_c$) were $10^{-7}$–$10^{-5}$ $cm^2/c$ for Cd and $10^{-7}$–$10^{-6}$ $cm^2/c$ for Cu and Zn. The value for Cd is equal to the diffusion rate in water solutions, which indicates that its fixation in soil layers is almost absent. Changes in the fractional compositions of these elements in the soils caused by irrigation with different-quality water agree with those in the parameters of migration models for Zn and Cd.

**Keywords** Copper · Zinc · Cadmium · Irrigation · Natural waters · Waste waters · Migration models

## 1 Introduction

Wastewater is water released from municipal and industrial production. Urban wastewater containing a great amount of heavy metals (HM) such as Cd, Cu, Fe, Mn, Ni, Pb, and Zn may be utilized for irrigation and fertilization of agricultural fields and may lead to continuous HM accumulation in soils. Heavy metals may accumulate only in the topsoil; however, the irrigated wastewater used for a long period of time contributes to soil contamination at a depth.

The information on mathematical models to show the HM migration in soils is rather limited in the literature (Dube et al. 2001). It is commonly thought that HM accumulated in the topsoil does not migrate downward in the soil or that HM

A.S. Frid (✉)
V.V. Dokuchaev Soil Science Institute, per. Pyzhevskii 7, Moscow 119017, Russia
e-mail: asfrid@mail.ru

 
O.V. Frank-Kamenetskaya et al. (eds.), *Biogenic—Abiogenic Interactions in Natural and Anthropogenic Systems*, Lecture Notes in Earth System Sciences,
DOI 10.1007/978-3-319-24987-2_13

migration is not significant for evaluating the soil contamination. However, it is worth emphasizing that this migration constitutes a risk to contamination of the environment (Labanowski et al. 2008; Dorronsoro et al. 2002; Ruan et al. 2006) and groundwater. For this reason, dynamic mathematical models are an important aspect in scientific research. This chapter aims to estimate the adequate application of migration models for HM transport downward in the soil profile for the case of irrigation with natural and urban wastewater for a long period of time. The heavy metals of interest in this study are Cu, Zn, and Cd.

The data about Zn and especially Cd migration in soils (diffusion coefficients) are scarce in the literature (Frid 1996; Alexeev and Zyrin 1980; Zyrin et al. 1980). There is no information on Cu migration except for our publication (Frid et al. 2014). The available values vary from $1 \times 10^{-11}$ to $3 \times 10^{-7}$ $cm^2/s$ for Zn and $4 \times 10^{-11}$ to $1 \times 10^{-8}$ $cm^2/s$ for Cd (from $1 \times 10^{-7}$ to $12 \times 10^{-7}$ $cm^2/s$ calculated for the pore solution of the same soils). The reference data about diffusion in water solutions (Reference book for a chemist 1964; Zyrin et al. 1965) are as follows:

- Zn: $10–20 \times 10^{-6}$ (according to Okunev 1968 to $4–20 \times 10^{-6}$ in the pH range from 7.2 to 9.2) or $5 \times 10^{-6}$ in the soil solution
- Cu: $7–8 \times 10^{-6}$
- Cd: $3–7 \times 10^{-6}$.

## 2 Materials and Methods

The field experiments were carried out in Egypt (Goma Bothina Saad M.A. near Alexandria). Two test areas were located on alluvial soils (Fluvisol) of lacustrine origin (light-textured sandy loams). One area was irrigated with river water for more than 40 years. The other test area was irrigated with mechanically and biologically treated wastewater for 20 years. The other two test areas were located in an oasis of the Libyan Desert, which was covered by loam sandy yellow-brown desert soil and irrigated with groundwater for 30–40 years and wastewater for 20 years.

In the vegetation period, the average irrigation rate was 7800 $m^3$/ha for Fluvisols and 9400 $m^3$/ha for the desert soil. The Cu concentration was estimated at 0.14–0.16 mg/l in natural water and 0.21 mg/l in the wastewater, whereas the Zn and Cd concentrations varied from 0.02–0.04 and 0.0003 mg/l to 0.18 and 0.0073 mg/l, respectively.

Soil samples were taken from horizons and layers. For modeling the data, almost all of the studied soil layers to a depth of 150 cm found an application. The investigation spanned 3 years. The total HM content in soil was determined by the half-quantitative method of spectral analysis. The fractions of HMs isolated according to the method of Tessier were also studied. The obtained data are presented in Tables 1 and 2.

**Table 1** The total content of heavy metals in soil layers after long-term irrigation with different waters (values averaged for 2008–2010)

| Depth (cm) | Fluvisol | | Yellow-brown desert soil | |
|---|---|---|---|---|
| | River water | Wastewater | Ground water | Wastewater |
| Cu | | | | |
| 0–20 | 30.4/0.0380 | 58.4/0.0712 | 22.8/0.0315 | 48.5/0.0660 |
| 20–40 | 27.3/0.0363 | 49.1/0.0614 | 20.2/0.0281 | 39.3/0.0550 |
| 40–60 | 25.1/0.0339 | 38.2/0.0512 | 19.5/0.0275 | 28.3/0.0402 |
| 60–100 | 24.3/0.0340 | 30.7/0.0427 | 17.2/0.0244 | 17.5/0.0250 |
| 100–150 | 23.4/0.0339 | 29.8/0.0432 | 18.4/0.0265 | 18.1/0.0262 |
| Background | 23/0.0333 | 23/0.0333 | 14/0.0202 | 14/0.0202 |
| Zn | | | | |
| 0–20 | 39.8/0.0485 | 104.7/0.1277 | 34.6/0.0477 | 94.6/0.129 |
| 20–40 | 30.3/0.0379 | 78.2/0.0978 | 16.2/0.0225 | 67.5/0.0945 |
| 40–60 | 20.7/0.0277 | 43.4/0.0582 | 12.1/0.0171 | 29.5/0.0419 |
| 60–100 | 21.2/0.0295 | 23.5/0.0327 | 8.6/0.012 | 10.8/0.01545 |
| 100–150 | 20.4/0.0296 | 22.9/0.0332 | 9.4/0.0135 | 15.3/0.0221 |
| Background | 20/0.0290 | 20/0.0290 | 9.0/0.013 | 9.0/0.013 |
| Cd | | | | |
| 0–20 | 0.90/0.0011 | 2.68/0.00327 | 1.10/0.00152 | 2.04/0.00278 |
| 20–40 | 0.95/0.0013 | 2.57/0.00321 | 0.84/0.0012 | 1.97/0.00276 |
| 40–60 | 0.98/0.0013 | 2.12/0.00284 | 1.24/0.00175 | 1.85/0.00263 |
| 60–100 | 1.35/0.00189 | 2.53/0.00354 | 1.35/0.00192 | 1.44/0.00206 |
| 100–150 | 1.42/0.00206 | 1.25/0.00181 | 1.54/0.00222 | 1.56/0.00226 |
| Background | 0.85/0.0011 | 0.85/0.0012 | 0.80/0.00116 | 0.80/0.00116 |

*Note* Data are given as mg $kg^{-1}$/mg $cm^{-3}$

The shape of the experimental curve of the migration of HMs specifies the parameters of the used models; after they are determined, it is possible to correct the initial (background) contents of HMs in the soils using model calculation.

To simplify the procedure, we assumed that the parameters of the migration models, including the initial HM content, were invariable both in soil profiles and in the period of long-term irrigation. Of course, the use of comparatively simple mathematical migration models needs some simplified assumptions. However, because of the limited parameters, the verification of models becomes simpler and permits one to obtain a correct and qualitative picture. The diffusional model is a constant substance flow at the surface of the half-continuous medium (in this case, at the soil surface) and the substance input to soil occurs during the irrigation season of every year. The convective-diffusional model is a mass exchange of the substance at the surface of the half-continuous medium.

Having subtracted the initial HM content, the interval of values for the total HM content was calculated as ±10 % from the initial one to account for a possible measurement error. Different combinations of model parameters were used to

**Table 2** The fractional composition of heavy metals in soils

| Soil | Irrigation water | Forms of heavy metals (%) | | | | |
|---|---|---|---|---|---|---|
| | | Exchangeable | Sorbed on carbonates | Bound with the organic matter | Bound with Fe and Mn oxides and hydroxides | Residual |
| Cu | | | | | | |
| Fluvisol | River water | $\frac{1.9-2.1}{2.0}$ | $\frac{3.4-3.8}{3.5}$ | $\frac{0.7-1.35}{1.2}$ | $\frac{21.2-28.7}{25}$ | $\frac{64.15-72.9}{67}$ |
| | Wastewater | $\frac{1.8-2.2}{2.0}$ | $\frac{3.3-3.75}{3.4}$ | $\frac{0.7-3.0}{1.75}$ | $\frac{20.0-34.0}{26.6}$ | $\frac{56.1-73.1}{65}$ |
| Desert | Ground water | $\frac{1.6-2.7}{2.2}$ | $\frac{3.6-4.4}{4.2}$ | $\frac{0.7-1.7}{1.3}$ | $\frac{27.6-40.8}{34.4}$ | $\frac{50.4-66.6}{58}$ |
| | Wastewater | $\frac{1.6-3.0}{2.2}$ | $\frac{3.4-6.0}{4.5}$ | $\frac{0.9-3.7}{2.1}$ | $\frac{28.9-34.1}{31.7}$ | $\frac{56.6-62.1}{59.6}$ |
| Zn | | | | | | |
| Fluvisol | River water | $\frac{0.5-1.0}{0.8}$ | $\frac{1.5-1.9}{1.7}$ | $\frac{3.2-4.6}{3.7}$ | $\frac{9.0-11.7}{9.7}$ | $\frac{81.2-84.7}{83.8}$ |
| | Wastewater | $\frac{1.2-2.7}{1.6}$ | $\frac{4.9-9.4}{5.9}$ | $\frac{13.2-15.6}{13.8}$ | $\frac{25.4-34.6}{32.7}$ | $\frac{45.7-46.9}{46.0}$ |
| Desert | Ground water | $\frac{0.6-1.6}{1.0}$ | $\frac{1.1-3.0}{2.4}$ | $\frac{2.3-2.4}{2.4}$ | $\frac{10.8-26.1}{13.9}$ | $\frac{69.3-83.1}{80.2}$ |
| | Wastewater | $\frac{1.0-2.8}{1.7}$ | $\frac{1.8-5.8}{4.4}$ | $\frac{3.7-13.5}{11.3}$ | $\frac{28.3-38.5}{32.7}$ | $\frac{47.1-54.6}{49.9}$ |
| Cd | | | | | | |
| Fluvisol | River water | $\frac{8.8-9.3}{9.1}$ | $\frac{9.1-11.6}{10.5}$ | $\frac{17.5-18.2}{17.9}$ | $\frac{16.1-16.6}{16.3}$ | $\frac{44.6-47.3}{46.2}$ |
| | Wastewater | $\frac{8.8-9.7}{9.2}$ | $\frac{9.0-11.7}{10.5}$ | $\frac{12.2-17.9}{14.7}$ | $\frac{16.1-18.7}{16.8}$ | $\frac{47.0-50.0}{48.8}$ |
| Desert | Ground water | $\frac{5.6-11.0}{8.0}$ | $\frac{6.2-11.0}{9.0}$ | $\frac{14.2-17.4}{15.9}$ | $\frac{20.3-25.8}{22.0}$ | $\frac{40.9-50.3}{45.2}$ |
| | Wastewater | $\frac{7.1-11.0}{9.1}$ | $\frac{8.3-12.0}{9.9}$ | $\frac{13.2-17.8}{15.0}$ | $\frac{19.5-22.5}{20.6}$ | $\frac{38.8-50.0}{45.4}$ |

*Note* Data are given as the variability range in the soil profile over the weighted average along the soil profile

**Table 3** Parameters of copper migration

| Soil | Diffusional model | | Convective-diffusional model | | | Cu accumulation in the soil | Cu input with irrigation water |
|---|---|---|---|---|---|---|---|
| | D $10^6$, $cm^2/s$ | Input to the soil, $mg/cm^2$ | Dc $10^6$, $cm^2/s$ | V $10^9$, cm/s | Input to the soil, $mg/cm^2$ | $mg/cm^2$ | |
| Fluvisol, river water, 50 years | 0.5–1.9 | 0.31 | 0.05–6.0 | 4–8 | 0.70 | 0.46 | 0.62 |
| Fluvisol, wastewater, 20 years | 2.7–3.2 | 1.6 | 1.46–3.5 | 4–9 | 1.9 | 1.7–2.5 | 0.32 |
| Desert soil, groundwater, 40 years | 2.6–4.7 | 0.68 | 0.8–4.7 | −10 to +10 | 0.73 | 0.83–0.96 | 0.53 |
| Desert soil, wastewater, 20 years | 3.0–3.6 | 2.2 | 2–3 | −2 to +10 | 2.3 | 1.8–2.2 | 0.40 |

**Table 4** Parameters of zinc migration

| Soil | Diffusional model | | Convective-diffusional model | | | Zn accumulation in the soil | Zn input with irrigation water |
|---|---|---|---|---|---|---|---|
| | D $10^6$, $cm^2/s$ | Input to the soil, $mg/cm^2$ | Dc $10^6$, $cm^2/s$ | V $10^9$, cm/s | Input to the soil, $mg/cm^2$ | $mg/cm^2$ | |
| Fluvisol, river water, 50 years | 0.43–0.55 | 0.66 | 0.08–0.30 | 3–7 | 0.62 | 0.58–0.89 | 0.078 |
| Fluvisol, wastewater, 20 years | 2.0–2.6 | 4.4 | 0.7–1.0 | 4–6 | 3.2 | 3.6–4.3 | 0.28 |
| Desert soil, groundwater, 40 years | 0.42–0.60 | 1.1 | 0.09–0.23 | 4.0–6.5 | 0.63 | 1.0–1.1 | 0.15 |
| Desert soil, wastewater, 20 years | 2.0–2.4 | 5.1 | 0.8–1.1 | 3.0–5.5 | 3.7 | 4.5–5.0 | 0.34 |

calculate the HM content at a soil depth for the given period of irrigation. The values of parameters seemed applicable if they were within the 10 % interval for all the soil layers. Such an approach allowed evaluation of the interval of values for every parameter. When comparing the data about the HM input to soil together with irrigation water during the whole irrigation period, as well as the HM accumulation in soil determined experimentally and calculated by using the values of model parameters, it seemed reasonable to arrive at a conclusion about the adequate utilization of mathematical models (Tables 3, 4 and 5).

**Table 5** Parameters of cadmium migration

| Soil | Diffusional model | | Convective-diffusional model | | | Cd accumulation in the soil | Cd input with irrigation water |
|---|---|---|---|---|---|---|---|
| | D $10^6$, $cm^2/s$ | Input to the soil, $mg/cm^2$ | Dc $10^6$, $cm^2/s$ | V $10^9$, cm/s | Input to the soil, $mg/cm^2$ | $mg/cm^2$ | |
| Fluvisol, river water, 50 years | – | – | 0.275 | 74 | 0.91–1.6 | 0.04–0.13 | 0.0012 |
| Fluvisol, wastewater, 20 years | 9.7–10.5 | 0.15 | 1.5–7.0 | 40–50 | 0.14 | 0.19–0.24 | 0.011 |
| Desert soil, groundwater, 40 years | – | – | 0.45–0.61 | 100–110 | 1.1–1.35 | 0.08–0.13 | 0.0011 |
| Desert soil, wastewater, 20 years | 50–52 | 0.19–0.24 | 8–40 | 60–100 | 0.28–0.30 | 0.13–0.26 | 0.014 |

# 3 Discussion

It is interesting to compare the obtained parameter values of migration models with those indicated in literature sources. For Cu, the maximum D and Dc values approximate those in the water solution, which indicates slightly expressed Cu sorption in both soils, especially in the desert soil and in the case of irrigation with wastewater. For Cd, the D and Dc values obtained by irrigation with wastewater exceed the values for water solutions, which is explained by the weakly expressed sorption and the significant influence of the convective flow. In all cases, it is possible to emphasize that the salt content is rather high in soils and irrigation water; this fact suggests that the possible HM forms are not absorbed by soil and the rate of their migration should be increased to a considerable extent. Under the given conditions, the parameters of migration models have not been considerably affected by the grown crop.

According to Tessier, the changes in the HM fractional composition of soils irrigated by the wastewater for a long period of time showed that they are found to be in accordance with changes in parameters of migration models. For Zn, a share of the fractions associated with the organic matter and the Fe and Mn oxides-hydroxides becomes higher because the decreasing residual fraction corresponds to the increased values of D and Dc. For Cd, a higher share of the most mobile fractions as compared to Zn and Cu corresponds to a high D level. For Cu, the influence of the fractional composition was weakly expressed.

# 4 Conclusions

During the maize irrigation in the field, both the diffusion and convective-diffusion migration models proved to be successful for assessing the actual distribution of Cu and Zn along the soil profile. The convective-diffusion model was found to be adequate for Cd. In addition, for all heavy metals (and Cd in particular), the parameter values of migration models seemed rather high under the given conditions, which indicates a weakly expressed fixation of metals in soils.

# References

Alexeev AA, Zyrin NG (1980) The Cd diffusion in soils. J Pochvovedenie 3:66–73 In Russian

Dorronsoro C, Martin F, Ortiz I, Garsia I, Simón M, Fernández E, Aguilar J, Fernández J (2002) Migration of trace elements from pyrite tailings in carbonate soils. J Environ Qual 31:829–835

Dube A, Zbytniewski R, Kowalkowski T, Cukrowska E, Buszewski B (2001) Adsorption and migration of heavy metals in soil. Polish J Environ Stud 10(1):1–10

Frid AS (1996) Experience in the experimental assessment of the availability of soil substances to the plant roots as based upon a migration conception. J Agrochem 6:36–46 In Russian

Frid AS, Shuravilin AV, Goma Bothina Saad MA, Borisochkina TI (2014) Copper migration in arid soils irrigated by natural and waste waters in Egypt. J Agrochem 3:60–68 (In Russian)

Labanowski J, Monna F, Bermond A, Cambier Ph, Fernandez Ch, Lamy I, van Oort F (2008) Kinetic extractions to assess mobilization of Zn, Pb, Cu, and Cd in a metal-contaminated soil: EDTA vs. citrate. Environ Pollut 152:693–701

Okunev MS (1968) On colloid-formation in diluted solutions of Zn-bearing salts. Proceed. on chemistry and chemical technology (Gorkiy) 2(20):39–44 (In Russian)

Reference book for a chemist (1964) "Chemistry" Publish. vol. III. House Moscow-Leningrad

Ruan XL, Zhang GL, Zhao YG, Yuan DG, Wu YJ (2006) Distribution and migration of heavy metals in soil profiles by high resolution sampling. Huan Jing Ke Xue. 27:1020–1025 (Article in Chinese)

Zyrin NG, Orlov DS, Vorobieva LA (1965) Reference and calculation tables for physic-chemical methods of soil investigations. MSU Moscow (In Russian)

Zyrin NG, Alexeev AA, Arkhipov NP (1980) The compound forms of Zn in soils, diffusion and input of this element to the plant. J Agrochem 9:95–104 (In Russian)

# Trace Element Composition of Poplar in Mongolian Cities

**Natalia E. Kosheleva, Ivan V. Timofeev, Nikolay S. Kasimov, Tatiana M. Kisselyova, Alexey V. Alekseenko and Olga I. Sorokina**

**Abstract** *Purpose.* The aim of our work was to assess changes in the trace element composition of poplar leaves in large cities and mining centers of Mongolia. The objectives of the study included: (1) to reveal the biogeochemical background features and changes in the trace element composition of poplar leaves in urban and mining landscapes; (2) to determine the degree of technogenic disturbance in the chemical composition of urban vegetation; and (3) to assess the functioning and ecological status of poplars under technogenic impact. *Materials and methods.* Poplar hybrids, which compose about 75 % of the urban woody plantations, were sampled in Ulaanbaatar (77 samples) in the mid-summer of 2008, Erdenet (30 samples) in 2011, Darkhan (19 samples) in 2011, and Sharyngol (21 samples) in 2013. Bulk concentrations of 54 heavy metals in the samples of the dry plant material were measured by inductively coupled plasma mass spectrometry (ICP-MS) on Elan-6100 and Optima-4300 analyzers. *Results and discussion.* The local biogeochemical background of the Mongolian cities under consideration differs from the mean global values in the higher concentrations of Cd, Sr, As, and Zn. The concentrations of Be, V, Pb, Cr, and Ni in plants of the background areas are lower than their global values. The maximum coefficients of the biogeochemical transformation, $Z_v$, were revealed in Ulaanbaatar. In the other cities, the values of $Z_v$ in the industrial zones were higher than those in the residential zones by 1.5–2 times. The trace element ratios characterizing the balance in the provision of metabolic processes confirm the conclusion about the satisfactory state of the urban trees.

**Keywords** Geochemical anomalies · Type of land use · Trace elements · Poplar leaves

N.E. Kosheleva (✉) · I.V. Timofeev · N.S. Kasimov · T.M. Kisselyova · O.I. Sorokina
Department of Landscape Geochemistry and Soil Geography of the Geography Faculty, Lomonosov Moscow State University, Moscow, Russia
e-mail: natalk@mail.ru

A.V. Alekseenko
National Mineral Resources University (Mining University), St. Petersburg, Russia

 
O.V. Frank-Kamenetskaya et al. (eds.), *Biogenic—Abiogenic Interactions in Natural and Anthropogenic Systems*, Lecture Notes in Earth System Sciences, DOI 10.1007/978-3-319-24987-2_14

# 1 Introduction

The urban environment forms specific conditions of interaction between biogenic and abiogenic components of ecosystems, which are discovered primarily in the geochemical changes of biota. The concentration of technogenic contamination sources in cities creates a specific environment, where plants are exposed to the impact of numerous pollutants, among which heavy metals and metalloids (HMs) have a special place (Bashkin 2003). The pollutants are involved in the biogeochemical cycle; they arrive in plants through the soil, hydrosphere, and atmosphere. Woody plants act as an universal natural filter because they extract various elements and concentrate them in the tissues. An excess of certain elements can disturb important physiological and biochemical processes in the plant organism (Bargagli 1998; Stobrawa and Lorenc-Plucińska 2008; Kabata-Pendias 2011). Therefore, intensive studies of the HM accumulation in urban vegetation are carried out in many countries of the world.

Under urban conditions, the general features and mechanisms of HM accumulation by plants undergo changes under the impact of the high concentrations of HMs in the atmospheric air and soils in which the plants grow. The plants contact only with available HM forms, the content of which is determined by the intensity of the atmospheric precipitation and the capacity of the soils to inactivate HMs. Foliar uptake—the input of HMs from gas emissions, smoke, and technogenic dust to plants through the surface of leaves—has gained importance (Ertel et al. 1991; Kopylova 2010; Simon et al. 2011; Serbula et al. 2012, 2013; Pashkevich et al. 2015).

The aim of our work was to assess the changes in the trace element composition of poplar leaves in large cities and mining centers of Mongolia. Ulaanbaatar is a metropolitan city with various industries. Enterprises of ferrous metallurgy, light and chemical industries, and building complexes are concentrated in Darkhan. Erdenet is a big center of mining, concentration, and primary processing of copper and molybdenum ores. One of the largest brown-coal deposits of the northern Mongolia occurs in Sharyngol. The pollution sources in all cities include thermal power plants (TPPs) and private quarters, which release combustion products of brown coal from the Nalaikh, Baganuur, and Sharyngol deposits enriched with Pb, As, and Mo by tens of times and with Cu, Sr, Cd, and Ni by several times compared to their lithosphere content (Kasimov et al. 1995; Kosheleva et al. 2010). Mongolian motor vehicles run on ethylated gasoline with a high content of lead.

The following particular tasks had to be solved:

- Revealing the biogeochemical background features and changes in the trace element composition of poplar leaves in the urban and mining landscapes;
- Determining the degree of technogenic disturbance in the chemical composition of urban vegetation; and
- Assessing the functional and ecological status of poplars under technogenic impact.

## 2 Objects of Study

*Environmental conditions*. All the studied cities are located in intermountain valleys and hollows of the Khangai–Khentii mountain region within the Selenga River basin. This region has a sharply continental climate with cold winters and warm and relatively moist summers with 60–70 % of the annual precipitation (World Meteorological Organization [www.worldweather.wmo.int]). The prevalence of windless anticyclone weather in winter results in the formation of steady temperature inversions, which favor the accumulation of toxicants from the technogenic sources in the air.

The region belongs to the ecotone (transitional) zone of Southern Siberia and Central Asia, where the mountainous forest-steppe prevails, including forests (sometimes typical subtaiga forests) with seasonally frozen soddy mountain-forest soils on the colder and wetter slopes of northern exposure. Warmer slopes of southern aspects are occupied by shrub–dry steppe and meadow-steppe communities that are gradually replaced by dry steppes with thin xerophytic grasses on mountain chernozems and chestnut soils in plain areas (Selenga Basin Ecosystems 2005).

The Ulaanbaatar area is composed of Archean granites, Carboniferous metamorphic clayey slates, and Neogene mottled clays frequently containing soluble salts, gypsum, sands, and conglomerates. The slates and clays are enriched with Fe, Mn, Cr, Co, Pb, Ni, and Ti; granites, sandy sediments, and river alluvium are depleted of them (Batkhishig 1999). In the Darkhan area, there are volcanogenic rocks in combination with marine terrigenous and calcareous rocks and various facies of volcanogenic and tuffogenic sedimentary rocks (Geological map of Mongolia 1998). The ore field in the city of Erdenet is composed of Permian volcanogenic and volcanogenic-sedimentary rocks penetrated by granitoid intrusions of the Selenga complex. It belongs to the molybdenum–copper porphyritic type with industrial concentrations of Re, Ag, and Se and small amounts of Pb, Zn, As, Sr, Bi, Co, Ni, and Cd (Berzina and Sotnikov 2007). Various metamorphosed sedimentary rocks and magmatic rocks (predominantly, granites) predominate in the Sharyngol area.

*Functional zoning*. The industrial and residential zones are the main functional zones in all cities. The residential zone includes multistory and ger blocks. The transport zone is also distinguished in the cities of Ulaanbaatar and Darkhan.

## 3 Methods and Materials

Poplar hybrids, which compose about 75 % of the urban woody plantations, were sampled in Ulaanbaatar (77 samples) in the mid-summer of 2008, Erdenet (30 samples) in 2011, Darkhan (19 samples) in 2011, and Sharyngol (21 samples) in 2013. Leaves were sampled from the trees of similar ages in the phase of vegetation after flowering, after a five-day rainless period. They were washed with running water and rinsed with distilled water to remove the elements deposited on the leaf surface and not participating in the metabolism of the plants. The samples were then dried at 70–80 °C for 24 h.

Bulk concentrations of 54 HMs in the samples of the dry plant material were measured by ICP-MS on Elan-6100 and Optima-4300 analyzers (Perkin Elmer, USA) at the All-Russian Research Institute of Mineral Raw Materials. Fourteen priority pollutants were selected for thorough analysis, including elements of hazard classes I (As, Cd, Pb, Zn), II (Co, Ni, Mo, Cu, Sb, Cr), and III (V, Sr), as well as some others (Be, Bi).

Background concentrations of the elements for each city ($C_b$) were compared with the mean global concentrations ($C_{\text{glob}}$) in the annual production of terrestrial plants (Dobrovol'skii 1998) and with the mean background values $C_{\text{glob}}$ for *Populus nigra* (Djingova et al. 2004) via calculating the clarkes of concentration $\text{CC} = C_b/C_{\text{glob}}$ and dispersion $\text{DC} = C_{\text{glob}}/C_b$. The concentrations of HMs in the urban samples $C_i$ were grouped depending on the functional zone and compared with the local biogeochemical background via calculating the enrichment factor $\text{EF} = C_i/C_b$ and the dispersion factor $\text{DF} = C_b/C_i$. From the obtained EF and DF values, geochemical spectra were obtained for the leaves of poplar trees grown in different cities and functional zones in order to trace geochemical changes in the plants. The specialization of the background (urban) plants was determined from the formula with the values of CC (EF) for concentrating elements in the nominator and the values of DC (DF) for deconcentrating elements in the denominator.

To characterize changes in the trace element composition of the plants, we proposed an integral parameter: the coefficient of biogeochemical transformation $Z_v$ calculated as follows: $Z_v = \sum_1^{n_1} \text{EF} + \sum_1^{n_2} \text{DF} - (n_1 + n_2 - 1)$, where $n_1$ and $n_2$ are the numbers of trace elements for which EF > 1 and DF > 1, respectively (Kasimov et al. 2012). It is known that plants can respond to the deterioration of the environment by either accumulation or depletion of trace elements (Dobrovol'skii 1998; Bargagli 1998; Kabata-Pendias 2011). Coefficient $Z_v$ reflects the disturbance of the normal proportions of trace elements in plant organs typical of their phylo- and ontogenetic specialization. This coefficient quantitatively expresses the imbalance of chemical elements in plants resulting from the anthropogenic load. The degree of technogenic transformation of the trace element composition of plants is characterized by five gradations: minimum ($Z_v$ = 10–20), medium ($Z_v$ = 20–30), high ($Z_v$ = 30–40), very high ($Z_v$ = 40–60), and extremely high ($Z_v$ = 60–80 and higher).

The ecological status of the plants was diagnosed from the Fe/Mn, Pb/Mn, and Cu/Zn ratios. The first ratio is one of the most informative parameters of photosynthetic processes; the second ratio characterizes the relationships between technogenic and biophilic elements, and the third ratio determines the degree of proportionality in the contribution of these biogenic metals to enzyme synthesis (Elpat'evskii and Arzhanova 1990; Novikova 2005; Novikova and Kosheleva 2007; Kasimov et al. 2011b).

## 4 Results and Discussion

***Characterization of the local biogeochemical background***. The background concentrations of HMs in the leaves of poplars growing near the cities differed from the worldwide mean values by the high concentrations of Cd and Sr. Increased concentrations of As and Zn were found in some background areas (Tables 1 and 2). The concentrations of many elements (Be, V, Pb, Cr, Ni) in plants from the background areas were lower than their worldwide values. This can be related to the decrease in the biological uptake of the elements under arid conditions (Lettens et al. 2011), as well as to the predominance of data on plants from humid landscapes with a significantly higher biological uptake of elements in the global database, used in the calculations of the element clarkes (Kasimov et al. 1989).

The poplars growing in Mongolia have some different geochemical features than those of the uncontaminated *Populus nigra* (Djingova et al. 2004). The background element concentrations in the area near Sharyngol are the closest to the mean values for the species; only the Pb concentration in the poplar leaves is 4.9 times lower than the average (clarke) value. In the other cities, deviations from the mean values for *Populus nigra* were observed for several elements. Lower concentrations of many elements were also found for background sites in Darkhan and Erdenet. The elements accumulating in the background poplars include Sr, Cd, and Cu in Darkhan; Cr in Erdenet; and Cr, Be, Co, V, and As in Ulaanbaatar.

***Changes in the trace element composition of poplar leaves in urban and mining landscapes***. The formulas of geochemical specialization for the leaves of urban plants characterize the technogenically induced changes in their composition (Tables 1 and 2). In Ulaanbaatar, poplar leaves accumulate significant amounts of Cd, Zn, and Pb emitted from motor vehicles and many industrial sources, whereas the concentrations of Co, Be, and As are even lower than those in the local geochemical background areas. Darkhan is characterized by the bioaccumulation of Ni, Co, and Mo, which create the most contrasting technogenic anomalies around the enterprises of ferrous metallurgy. The multielemental anomaly developed in the leaves of poplars growing in Erdenet is due to the dust emissions from the mining

**Table 1** Geochemical specialization of poplar under background conditions and the technogenic transformation of its trace element composition in cities and mining centers

| City | CC/DC for the local background against the clarkes | | EF/DF for the whole city against the local background |
|---|---|---|---|
| | Dobrovol'skii (1998) | Djingova et al. (2004) | |
| Ulaanbaatar | $\frac{As_{4.0}Sr_{3.6}Co_{3.3}Cd_{2.2}}{Be_{3.9}Pb_{3.4}Mo_{3.1}Ni_{1.6}V_{1.6}Cu_{1.6}}$ | $\frac{Cr_{5.6}Be_{4.0}Co_{2.4}V_{2.0}As_{1.6}}{Cd_{2.6}Pb_{3.4}}$ | $\frac{Cd_{3.3}Zn_{3.2}Pb_{2.2}}{Co_{2.3}Be_{2.2}As_{1.6}}$ |
| Darkhan | $\frac{Cd_{10.7}Sr_{7.7}Mo_{2.1}Zn_{1.7}As_{1.5}}{Be_{2.2}V_{7.3}Ni_{4.5}Cr_{3.1}Pb_{2.8}}$ | $\frac{Sr_{2.3}Cd_{1.9}Cu_{1.5}}{Ni_{4.0}Pb_{2.8}V_{2.3}As_{1.6}Co_{1.5}Be_{1.5}}$ | $\frac{Ni_{2.7}Co_{2.1}Mo_{2.0}}{Cd_{2.1}V_{2.0}}$ |
| Erdenet | $\frac{Sr_{4.0}Cd_{3.7}Zn_{2.1}}{Be_{21.7}V_{17.4}As_{3.8}Pb_{3.4}Ni_{2.5}Cr_{2.1}}$ | $\frac{Cr_{2.2}}{As_{9.4}V_{5.5}Pb_{3.4}Ni_{2.2}Co_{1.6}Cd_{1.6}}$ | $\frac{As_{3.8}V_{3.3}Ni_{2.1}Mo_{1.9}Cd_{1.7}Be_{1.6}Cu_{1.5}}{-}$ |
| Sharyngol | $\frac{Cd_{5.1}Zn_{4.5}Sr_{4.0}As_{2.4}Co_{1.7}}{Be_{13.8}Pb_{4.9}Cr_{3.2}V_{2.6}}$ | $\frac{-}{Pb_{4.9}}$ | $\frac{Cd_{1.9}}{Co_{1.5}}$ |

**Table 2** Heavy metals in the leaves of *Populus* in the cities of Ulaanbaatar, Darkhan, Erdenet, and Sharyngol

| Parameter, mg/kg dry matter ($n$ is the number of samples) | Hazard class | | | | | | | | | | | | | |
|---|---|---|---|---|---|---|---|---|---|---|---|---|---|---|
| | I | | | | II | | | | | | III | | no | |
| | As | Cd | Pb | Zn | Co | Ni | Mo | Cu | Sb | Cr | V | Sr | Be | Bi |
| *Ulaanbaatar* | | | | | | | | | | | | | | |
| Mean for the city ($n$ = 77) | 0.30 | 0.25 | 0.77 | 70 | 0.71 | 1.1 | 0.44 | 6.6 | 0.04 | 1.8 | 0.85 | 171 | 0.012 | 0.013 |
| Min–Max | 0.01–0.98 | 0.06–0.79 | 0.19–2.6 | 18–270 | 0.18–3.2 | 0.48–4.2 | 0.15–1.5 | 2.7–13 | 0.008–0.32 | 0.92–7.5 | 0.31–1.7 | 75–280 | 0.003–0.035 | 0.008–0.033 |
| Mean for the industrial zone ($n$ = 18) | 0.33 | 0.33 | 0.67 | 64 | 0.72 | 1.1 | 0.51 | 6.7 | 0.02 | 2.2 | 0.96 | 156 | 0.011 | 0.017 |
| Mean for the gers ($n$ = 12) | 0.21 | 0.20 | 0.79 | 67 | 0.86 | 1.3 | 0.45 | 6.9 | 0.05 | 1.7 | 1.04 | 186 | 0.012 | 0.009 |
| Mean for the multi-storey buildings ($n$ = 27) | 0.29 | 0.26 | 0.83 | 81 | 0.73 | 1.1 | 0.43 | 6.5 | 0.05 | 1.7 | 0.79 | 183 | 0.011 | 0.010 |
| Transport ($n$ = 20) | 0.33 | 0.22 | 0.87 | 69 | 0.57 | 1.2 | 0.44 | 6.9 | 0.07 | 1.8 | 0.84 | 170 | 0.012 | 0.014 |
| Background ($n$ = 2) | 0.48 | 0.08 | 0.37 | 22.50 | 1.65 | 1.25 | 0.16 | 5.10 | 0.01 | 2.20 | 0.93 | 125.00 | 0.03 | <0.005 |
| *Darkhan* | | | | | | | | | | | | | | |
| Mean for the city ($n$ = 19) | 0.24 | 0.30 | 0.7 | 63 | 0.52 | 0.64 | 1.45 | 7.55 | 0.01 | 0.94 | 0.11 | 339 | 0.006 | 0.004 |
| Min–Max | 0.04–0.63 | 0.09–0.56 | 0.03–1.4 | 19–130 | 0.27–0.96 | 0.26–1.2 | 0.65–2.9 | 5.5–10 | 0.004–0.021 | 0.47–2.4 | 0.1–0.38 | 210–560 | 0.003–0.011 | 0.002–0.01 |
| Mean for the industrial zone ($n$ = 8) | 0.31 | 0.30 | 0.9 | 87 | 0.47 | 0.60 | 1.49 | 7.16 | 0.01 | 1.25 | 0.14 | 354 | 0.006 | 0.005 |
| Mean for the gers ($n$ = 3) | 0.21 | 0.28 | 0.7 | 45 | 0.54 | 0.53 | 1.66 | 6.73 | 0.01 | 0.82 | 0.10 | 290 | 0.006 | 0.003 |
| Mean for the multi-storey buildings ($n$ = 4) | 0.17 | 0.41 | 0.4 | 51 | 0.49 | 0.45 | 1.03 | 7.78 | 0.01 | 0.62 | 0.10 | 303 | 0.005 | 0.003 |
| Transport ($n$ = 4) | 0.21 | 0.22 | 0.5 | 42 | 0.63 | 0.98 | 1.65 | 8.73 | 0.01 | 0.75 | 0.10 | 385 | 0.006 | 0.003 |
| Background ($n$ = 4) | 0.19 | 0.37 | 0.44 | 50.75 | 0.45 | 0.44 | 1.05 | 7.60 | 0.01 | 0.57 | 0.21 | 270 | 0.004 | 0.003 |

(continued)

**Table 2** (continued)

| Parameter, mg/kg dry matter (*n* is the number of samples) | Hazard class | | | | | | | | | | | | | |
|---|---|---|---|---|---|---|---|---|---|---|---|---|---|---|
| | I | | | | II | | | | | | III | | no | |
| | As | Cd | Pb | Zn | Co | Ni | Mo | Cu | Sb | Cr | V | Sr | Be | Bi |
| *Erdenet* | | | | | | | | | | | | | | |
| Mean for the city ($n = 30$) | 0.12 | 0.22 | 0.53 | 70 | 0.58 | 1.65 | 1.08 | 11 | 0.07 | 1.11 | 0.28 | 177 | 0.008 | 0.01 |
| Min–Max | 0.03 −0.58 | 0.04 −0.44 | 0.12 −5.4 | 26 −140 | 0.25 −1.6 | 0.43 −6.8 | 0.33 −3.4 | 4.5 −32 | 0.01 −0.28 | 0.4 −5.5 | 0.09 −0.78 | 95 −300 | 0.001 −0.03 | 0.003 −0.023 |
| Mean for the industrial zone ($n = 9$) | 0.17 | 0.22 | 0.38 | 66 | 0.66 | 2.81 | 1.50 | 12 | 0.11 | 1.05 | 0.25 | 164 | 0.007 | 0.01 |
| Mean for the gers ($n = 7$) | 0.04 | 0.24 | 1.09 | 73 | 0.55 | 1.12 | 0.97 | 8 | 0.07 | 0.93 | 0.20 | 129 | 0.007 | 0.01 |
| Mean for the multi-storey buildings ($n = 14$) | 0.10 | 0.21 | 0.34 | 70 | 0.55 | 1.16 | 0.87 | 11 | 0.04 | 1.24 | 0.43 | 210 | 0.008 | 0.01 |
| Background ($n = 2$) | 0.03 | 0.13 | 0.37 | 63.50 | 0.44 | 0.81 | 0.56 | 7.15 | 0.02 | 0.85 | 0.09 | 139.50 | 0.005 | 0.01 |
| *Sharyngol* | | | | | | | | | | | | | | |
| Mean for the city ($n = 18$) | 0.28 | 0.33 | 0.31 | 111 | 0.60 | 1.1 | 0.66 | 6 | 0.01 | 0.58 | 0.60 | 144 | 0.009 | 0.004 |
| Min–Max | 0.09 −1.5 | 0.13 −1.2 | 0.12 −1.1 | 21 −360 | 0.24 −2 | 0.51 −2.4 | 0.23 −2.2 | 0.03 −10 | 0.001 −0.017 | 0.32 −1.4 | 0.23 −1.9 | 78 −250 | 0.003 −0.031 | 0.002 −0.02 |
| Mean for the industrial zone ($n = 4$) | 0.69 | 0.68 | 0.50 | 213 | 1.07 | 1.8 | 0.78 | 10 | 0.01 | 0.87 | 1.10 | 133 | 0.016 | 0.008 |
| Mean for the gers ($n = 7$) | 0.16 | 0.32 | 0.26 | 82 | 0.47 | 0.8 | 0.79 | 5 | 0.01 | 0.52 | 0.49 | 120 | 0.007 | 0.004 |
| Mean for the multi-storey buildings ($n = 7$) | 0.17 | 0.20 | 0.28 | 72 | 0.34 | 0.9 | 0.51 | 5 | 0.01 | 0.47 | 0.45 | 177 | 0.007 | 0.003 |
| Background ($n = 3$) | 0.28 | 0.18 | 0.26 | 135.67 | 0.86 | 1.30 | 0.54 | 6.20 | 0.01 | 0.56 | 0.57 | 141.33 | 0.01 | 0.004 |
| *Clarkes* | | | | | | | | | | | | | | |
| Clarke (Dobrovol'skii 1998) | 0.12 | 0.035 | 1.25 | 30 | 0.5 | 2 | 0.5 | 8.00 | – | 1.8 | 1.5 | 35 | 0.1 | – |
| Clarke (Djingova et al. 2004) | 0.3 | 0.2 | 1.25 | – | 0.7 | 1.78 | – | 5.05 | 0.04 | 0.4 | 0.47 | 118 | 0.006 | 0.0103 |

and processing complex (MPC) and the combustion products of brown coals used in the TPPs and for the heating of private houses and gers. Among the pollutants arriving into the environment during the combustion of brown coal, the Sharyngol poplars accumulate the largest amounts of Cd, which is characterized by the highest mobility and availability to plants (Ertel et al. 1991).

Poplar hybrids in the four studied cities of Mongolia are characterized by higher concentrations of Cr and lower concentrations of Pb and Ni as compared with those in the uncontaminated *Populus nigra*. The depletion of Ni can be explained by the hampering effect of Zn and Cu excesses (Temp 1991).

The features of the trace element composition of woody plants in different functional zones of the cities are seen in their geochemical spectra (Fig. 1). In the industrial zone, plants are subjected to the strongest anthropogenic impact; the enrichment factors (*EFs*) against the local background are in the range of 2–3 for most of the elements. In Ulaanbaatar, Cd is a priority pollutant in the industrial zone. The geochemical specialization of industrial enterprises in Darkhan manifests in the accumulation of Zn, Cr, Pb, and Bi by poplar leaves. The identification of the most significant source of pollution in these cities is impossible because their industry includes numerous branches, although the fuel and energy complexes and motor transport are noteworthy.

In the industrial zone of Erdenet, significant concentrations of As, Sb, V, Ni, and Mo are found in poplar leaves; the soils are also contaminated with Mo, Cu, and Sb (Timofeev et al. 2014), which is related to the closeness of the deposit, whose primary and secondary ores are enriched with these elements (Gavrilova et al. 2010). In addition, rock breaking at the Erdenet MPC enhances the aerogenic pollution of the area. The association of the elements accumulated in poplar leaves in the Sharyngol industrial zone includes Cd, As, Sb, Bi, Be, V, and Pb, which is determined by the specific features of brown coals and by the geochemical parameters of the soils (Alekseenko and Kosheleva 2014).

The trace element composition of leaves in the residential blocks does not differ much from the mean values for each city (Table 1). The rate of the biological

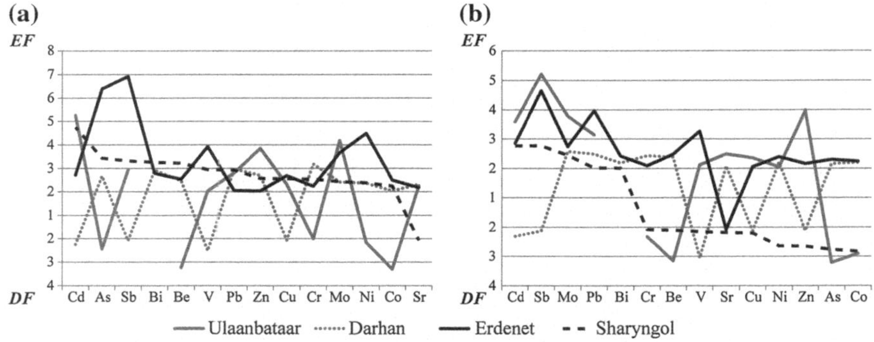


**Fig. 1** Geochemical spectra for poplar leaves in the **a** industrial and **b** ger zones of Mongolian cities

uptake of trace elements is increased in the ger blocks of Ulaanbaatar (Cd, Sb, Mo, Pb, Zn), Erdenet (Sb, Pb, V), Sharyngol (Cd, Sb) (Fig. 1). The geochemical features of the ger zones are determined by the use of brown coals and by their location relative to the industrial enterprises. A more active accumulation of trace elements is observed in the multistory blocks than in the private blocks of Ulaanbaatar (Cd, As, Zn), Darkhan (Zn, Cu), Erdenet (As, Cu, Cr, V, Sr), and Sharyngol (Sr).

***Technogenic transformation of the trace element composition of leaves***. The consequences of technogenic impact on woody plants are summarized by the coefficient of biogeochemical transformation $Z_v$ (Fig. 2). The mean value for the city was maximum in Ulaanbaatar ($Z_v$ = 15.6) and minimum in Sharyngol ($Z_v$ = 4.8). The low degree of the biogeochemical transformation is generally typical for all the cities ($Z_v < 20$). The spatial variation of $Z_v$ in Ulaanbaatar and Erdenet is less significant than that in Darkhan and Sharyngol: the variation coefficient of $Z_v$ is 38–39 % for the former two cities and 58–59 % for the latter two cities. The values of Zv in the industrial zones of mining centers and Darkhan are higher than those in the residential zones by 1.5–2 times due to the proper location of residential blocks with respect to the prevailing wind directions. The situation in Ulaanbaatar is less favorable because of the strong pollution in the center of the city, where a residential–industrial–transport pollution area with the Ni–Co–V, Cu–Cd–Zn, and As–Mo associations was formed (Kasimov et al. 2011a).

Our results indicate that poplar plants are resistant to anthropogenic impact. Among the woody and shrub species, poplar (genus *Populus*) is considered to be one of the best bioconcentrators; it is capable of normally functioning even at high HM concentrations (Sawidis et al. 1995; Stobrawa and Lorenc-Plucińska 2008; Migeon et al. 2009; Wang and Jia 2010). The tolerance of poplar to pollution can be related to the lower sensitivity of high-ash plants to gases and toxicants (Davydova et al. 2013).

The availability of representative sample sets for the areas of Ulaanbaatar and Darkhan allowed composing maps to trace spatial trends in the technogenic transformation of the trace element composition of poplar leaves (Fig. 3). In general, the distribution of $Z_v$ values within the cities is relatively uniform, with most values in the range of 10–20 (i.e., in the range of the minimum degree of

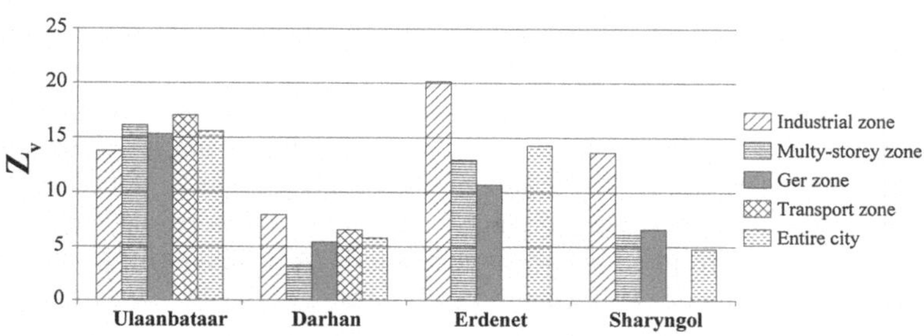


**Fig. 2** Coefficients of the biogeochemical transformation $Z_v$ for poplar leaves in different functional zones and the entire cities

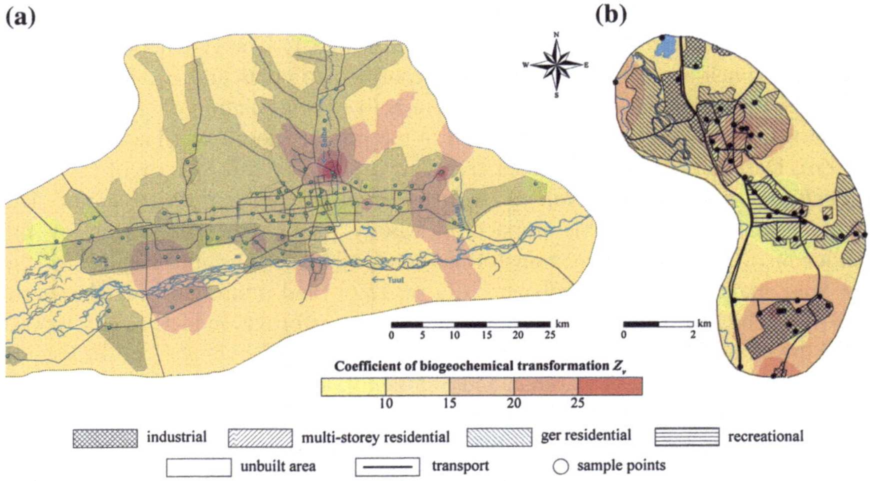


**Fig. 3** Distribution of the Zv values in poplar leaves in **a** Ulaanbaatar and **b** Darkhan

biogeochemical transformation). In Ulaanbaatar, which had mean $Z_v$ values of 10–15, several biogeochemical anomalies can be distinguished. The most striking anomaly is confined to the traffic circles and the main access roads from the Chingeltei district of Ulaanbaatar to the upper road; an extensive contrasting anomaly in the Uliastai quarter of private housing is associated with the aerial transfer of some HMs to the eastern regions of the city; a vast low-contrasting anomaly in the west of the city and small low-contrasting anomalies in the industrial zone are formed under the impact of TPP emissions. A negative biogeochemical anomaly is recorded in the center of the city (on Sukhe Bator Square), where government offices and tourist attractions are located, which can be due to the remediation of soils and the planting of young trees.

In Darkhan, an anomaly with $Z_v$ = 20–30 is located at the boundary between the multistory and single-story residential zones, where the high-rise buildings create a mechanical barrier for the dispersion of pollutants. The second halo is located in the industrial zone, where the major part of HMs arrives with emissions from the metallurgical plants and other enterprises. Larger contributions to the $Z_v$ value are made by Mo, Co, and Pb.

***Assessment of the ecological status of plants under technogenic impact***. The ecological status of poplar leaves was assessed using three parameters: the Fe/Mn, Pb/Mn, and Cu/Zn ratios (Table 2). Among the metals present in plants, iron plays the leading role. Iron-containing organic compounds are necessary for the biochemical processes of respiration and photosynthesis. A deficiency of iron is mainly observed in calcareous or other alkaline soils, where high pH values hamper the metal adsorption (Kopylova 2010). Manganese actively participates in the metabolism and improves physiological processes. It enters into the composition of enzymes; increases their activity; and participates in oxidation processes, nitrate

**Table 3** Parameters of the ecological status of poplar (*Populus*) leaves in Mongolian cities

| City | Fe/Mn | | Pb/Mn | | Cu/Zn | |
|---|---|---|---|---|---|---|
| | Mean | Min–Max | Mean | Min–Max | Mean | Min–Max |
| Ulaanbaatar | 3.85 | 0.7–12.9 | 0.012 | 0.002–0.04 | 0.118 | 0.03–0.39 |
| Darkhan | 3.38 | 1.7–6.6 | 0.009 | 0.0004–0.02 | 0.166 | 0.05–0.43 |
| Erdenet | 2.80 | 0.6–8.9 | 0.005 | 0.002–0.086 | 0.171 | 0.06–0.44 |
| Sharyngol | 3.35 | 1.0–11.8 | 0.004 | 0.001–0.015 | 0.084 | 0.0003–0.23 |

reduction, and photosynthesis. The Fe/Mn ratio of 1.5–2.5 is considered to be optimal for the normal development of plants (Kabata-Pendias 2011). The Fe/Mn ratios averaged for the cities vary from 2.8 in Erdenet to 3.8 in Ulaanbaatar; their variation within the cities was relatively high (Table 3). The most balanced ratio was observed in Darkhan, where variation was minimum (37 %). The maximum Fe/Mn ratios were found in the ger block of Ulaanbaatar (12.9) and the industrial zone of Sharyngol (11.8). Under Mn deficiency, excessive accumulation of active divalent Fe takes place, which results in chlorosis due to iron intoxication (Kopylova 2010).

The high concentration of Mn results in a decrease in the concentration of active divalent Fe. Iron is mobilized in cells as a trivalent organophosphate compound, which provokes chlorosis due to iron deficiency (Kopylova 2010). This situation is noted in the high-rise residential zone and a part of the industrial zone of Ulaanbaatar, where the Fe/Mn ratio in poplar leaves is 0.6–0.8.

The Pb/Mn ratio for the unpolluted terrestrial plants is 0.006, which indicates an extremely small proportion of technogenic elements not involved in physiological processes. Poplars are subject to the least technogenic load in the mining centers, where the mean ratios are lower, and the maximum ratios values exceed the values in the unpolluted plants by 2.6 times in Erdenet and by 3.6 times in Sharyngol. In Erdenet, the maximum ratio (0.086) was found in the local anomaly within the residential zone of gers. A different situation is observed in the cities, where the mean Pb/Mn ratio exceeds the global background value by 1.5–2 times. In Ulaanbaatar, the most significant technogenic changes in the composition of leaves are revealed along motor roads, where the Pb/Mn ratio exceeds its value for the unpolluted terrestrial plants by more than 6.5 times.

The optimum Cu/Zn ratio for the unpolluted terrestrial plants is 0.27. The woody plants growing in Sharyngol undergo the most significant imbalance in the provision of enzyme synthesis because of Zn excess and Cu deficiency.

A comparison with the known ranges of the deficient, normal, and toxic concentrations of Zn and Cu in mature leaf tissues, which were established from the generalized data for many species (Kabata-Pendias 2011), revealed excessive concentrations of Zn and Cr and a deficient concentration of Cu in some of the tested plants in all cities. Analogous results were obtained for woody plants in the affected zone of the TPP in Serbia, where a low general level of contamination and deficits of some important trace elements (Cu and Mn) were found (Pavlović et al. 2004).

## 5 Conclusions

1. The local biogeochemical background of the Mongolian cities under consideration differs from the mean global values in the higher concentrations of Cd, Sr, As, and Zn. The concentrations of Be, V, Pb, Cr, and Ni in plants of the background areas are lower than their global clarkes. The technogenic impact in the cities is manifested in the accumulation and dispersion of HMs by woody plants. In Ulaanbaatar, poplar leaves are enriched with Cd, Zn, and Pb and depleted of Co, Be, and As compared to the local background. Bioaccumulation of Ni, Co, and Mo arriving with the emissions from enterprises of ferrous metallurgy is typical of Darkhan. The multielement anomaly in the leaves of poplar growing in Erdenet is due to the dust emissions from the MPC and combustion products of brown coals. In Sharyngol, the highly mobile Cd is most actively accumulated by poplars. The poplar hybrids in all the four Mongolian cities are characterized by higher concentrations of Cr and lower concentrations of Pb and Ni than those typical of the uncontaminated trees of *Populus nigra*.
2. The changes in the trace element composition of urban poplar leaves are generally insignificant, which indicates their high tolerance toward the anthropogenic impact. The maximum coefficients of the biogeochemical transformation, $Z_v$, were revealed in Ulaanbaatar. In the capital of Mongolia, the differences between the functional zones are small because of the formation of a common residential–industrial–transport pollution area in the center of the city. In the other cities, the values of $Z_v$ in the industrial zones are higher than those in the residential zones by 1.5–2 times.
3. The trace element ratios characterizing the balance in the provision of metabolic processes confirm the conclusion about the satisfactory state of the urban trees. The Fe/Mn ratio is best balanced in Darkhan and maximally differs from the optimum value in the ger zone of Ulaanbaatar and the industrial zone of Sharyngol. Poplars undergo the most significant technogenic load in the transport zone of Ulaanbaatar, where the Pb/Mn ratio exceeds its value for the unpolluted terrestrial plants by more than 6.5 times. The woody plants growing in Sharyngol undergo imbalance in the provision of enzyme synthesis because of Zn excess and Cu deficiency.

**Acknowledgments** This study was carried out in the framework of the research supported by Russian Scientific Foundation (project No. 14-27-00083). The authors thank Prof. P.D. Gunin and the staff members of the Russian-Mongolian joint biological expedition who help in the field investigations.

# References

(1998) Geological map of Mongolia, scale 1:1 000 000. Ulaanbaatar: 44

(2005) Selenga Basin Ecosystems. In: Biological resources and natural conditions of Mongolia: Proceedings of the Joint Russian-Mongolian integral biological expedition. In: Vostokova EA, Gunin PD (eds) Nauka, Moscow: 395 (in Russian)

Alekseenko AV, Kosheleva NE (2014) Ecological and geochemical consequences of the open mining of brown coal (Sharyngol, Mongolia). In: Proceedings of the Buryat republican branch of the Russian Geographical Society, Ulan-Ude: 7–13 (in Russian)

Bargagli R (1998) Trace elements in terrestrial plants. An ecophysiological approach to biomonitoring and biorecovery. Springer, Berlin, Germany: 344

Bashkin VN (2003) Environmental chemistry: Asian lessons. Springer; Kluwer Academic Publishers: 488

Batkhishig O (1999) Soil-geochemical features of the Tuul river valley (Extended Abstract of Cand. Sci. (Geogr.) Dissertation), Ulaanbaatar: 23

Berzina AP, Sotnikov VI (2007) Character of formation of the Erdenet-Ovoo porphyry Cu-Mo magmatic center (northern Mongolia) in the zone of influence of a Permo-Triassic plume. Russ Geol Geophys 48:141–156. doi:10.1016/j.rgg.2007.01.001

Davydova ND, Znamenskaya TI, Lopatkin DA (2013) Identification of chemical elements as pollutants and their primary distribution in steppes of the southern Minusinsk depression. Contemp Prob Ecol 6(2):228–235. doi:10.1134/S1995425513020029

Djingova R, Kuleff I, Markert B (2004) Chemical fingerprinting of plants. Ecol Res 19:3–11. doi:10.1111/j.1440-1703.2003.00602.x

Dobrovol'skii VV (1998) Fundamentals of biogeochemistry, Vysshaya Shkola, Moscow: 413 (in Russian)

Elpat'evskii PV, Arzhanova VS (1990) Geochemistry of landscapes and Technogenesis, Nauka, Moscow: 196 (in Russian)

Ertel J, Prohl G, Paratzke HG (1991) The contamination of plants by cadmium due to root uptake and aerial deposition. Water Air Soil Pollut 57–58:861–874

Gavrilova SP, Maksimyuk IE, Orolmaa D (2010) Erdenet molybdenum–copper–porphyritic deposit (Mongolia), IMGRE, Moscow: 270 (in Russian)

Kabata-Pendias A (2011) Trace elements in soils and plants, Fourth Edition. CRC Press:439

Kasimov NS, Evdokimova AK, Rotshild EV, Urtnasan Zh (1989) Biogeochemical specialization of plants in Central Mongolia. Geogr Prir Res 2:112–119 (in Russian)

Kasimov NS, Lychagin MYu, Evdokimova AK, Golovanov DL, Pikovskii YuI (1995) Ulaanbaatar, Mongolia (heat power industry). Intermountain depression. In: Ecogeochemistry of urban landscapes, MGU, Moscow: 231–248. (in Russian)

Kasimov NS, Kosheleva NE, Sorokina OI, Gunin PD, Bazha SN, Enkh-Amgalan S (2011a) Ecological-geochemical state of soils in Ulaanbaatar (Mongolia). Eur Soil Sci 44(7):709–721. doi:10.1134/S106422931107009X

Kasimov NS, Kosheleva NE, Sorokina OI, Gunin PD, Bazha SN, Enkh-Amgalan S (2011b) An ecological-geochemical assessment of the state of woody vegetation in Ulaanbatar city (Mongolia). Arid Ecosyst 1(4):201–213. doi:10.1134/S2079096111040081

Kasimov NS, Bityukova BP, Kislov AV, Kosheleva NE, Nikiforova EM, Malkhazova SM, Shartova NV (2012) Ecogeochemical problems of large cities. Razv Okhr Nedr 7:8–13 (in Russian)

Kopylova LV (2010) Accumulation of iron and manganese in leaves of woody plants in technogenic regions of the Transbaikalia. Izv Sam Nauchn Tsentra RAN 12(1):709–712 in Russian

Kosheleva NE, Kasimov NS, Dorjgotov D, Baja SN, Golovanov DL, Sorokina OI, Enkh-Amgalan S (2010) Assessment of heavy metal pollution of soils in industrial cities of Mongolia. Geogr Environ Sustain 3(2):51–65

Lettens S, Vandecasteele B, De Vos B, Vansteenkiste D, Verschelde P (2011) Intra- and inter-annual variation of Cd, Zn, Mn and Cu in foliage of poplars on contaminated soil. Sci Total Environ 409:2306–2316. doi:10.1016/j.scitotenv

Migeon A, Richaud P, Guinet F, Chalot M, Blaudez D (2009) Metal accumulation by woody species on contaminated sites in the North of France. Water Air Soil Pollut 204(1–4):89–101. doi:10.1007/s11270-009-0029-5

Novikova OV (2005) Ecological-geochemical assessment of the state of woody plants in urban landscapes (with Moscow and Quito as examples) (Cand. (Geogr.) Dissertation), Moscow:164 (in Russian)

Novikova OV, Kosheleva NE (2007) Ecological-geochemical assessment of the state of woody plants in Quito (Ecuador). Vestn. MGU, Ser. 5. Geogr., 6:43–48. (in Russian)

Pashkevich MA, Alekseenko AV, Vlasova EV (2015) Biogeochemical and geobotanical assessment of marine ecosystems conditions (Novorossiysk city). Water Ecol 3:67–80. (in Russian)

Pavlović P, Mitrović M, Djurdjević L (2004) An ecophysiological study of plants growing on the fly ash deposits from the "Nikola Tesla–A" thermal power station in Serbia. Environ Manage 33(5):654–663

Sawidis T, Marnasidis A, Zachariadis G, Stratis J (1995) A study of air pollution with heavy metals in Thessaloniki city (Greece) using trees as biological indicators. Environ Contam Toxicol 28(1):118–124

Serbula SM, Miljkovic DDj, Kovacevic RM, Ilic AA (2012) Assessment of airborne heavy metal pollution using plant parts and topsoil. Ecotoxicol Environ Saf 76:209–214. doi:10.1016/j.ecoenv.2011.10.009

Serbula SM, Kalinovic TS, Ilic AA, Kalinovic JV, Steharnik MM (2013) Assessment of airborne heavy metal pollution using Pinus spp. and Tilia spp. Aerosol Air Qual Res 13:563–573. doi:10.4209/aaqr.2012.06.0153

Simon E, Braun M, Vidic A, Bogyó D, Fábián I, Tóthmérész B (2011) Air pollution assessment based on elemental concentration of leaves tissue and foliage dust along an urbanization gradient in Vienna. Environ Pollut 159:1229–1233. doi:10.1016/j.envpol.2011.01.034

Stobrawa K, Lorenc-Plucińska G (2008) Thresholds of heavy-metal toxicity in cuttings of European black poplar (*Populus nigra L.*) determined according to antioxidant status of fine roots and morphometrical disorders. Sci Total Environ 390:86–96

Temp GA (1991) Nickel in plants and its toxicity. In: Resistance of wild species to heavy metals. Leningrad:139–146 (in Russian)

Timofeev IV, Kosheleva NE, Bazha SN, Enkh-Amlagan S (2014) Geochemical transformation of soil cover in the copper-molybdenum ore mining region (Erdenet, Mongolia). Inzhenernye Izyskaniya 12:26–35 (in Russian)

Wang X, Jia Y (2010) Study on adsorption and remediation of heavy metals by poplar and larch in contaminated soil. Environ Sci Pollut Res 17(7):1331–1338. doi:10.1007/s11356-010-0313-3

## Internet Reference

World Meteorological Organization. Global web site presents official weather observations, weather forecasts and climatological information for Mongolia by National Meteorological & Hydrological Services. https://www.wmo.int/pages/index_en.html

# Regional Peculiarities of Micro-element Accumulation in Objects in the Transural Region of the Republic of Bashkortostan

**Irina N. Semenova, Yuliya S. Rafikova, Yalil T. Suyundukov and Gulnaz Ya. Biktimerova**

**Abstract** *Objective* The objective was to study the peculiarities of micro-element accumulation in environmental objects (soil, vegetation, human body) in the region of the Southern (Bashkir) Transurals, which lacks large industrial plants. *Method* The research was conducted throughout 2014 in the area of the Southern (Bashkir) Transurals, which includes the Burzyansky, Zilairsky, and Zianchurinsky administrative districts of the republic of Bashkortostan. Samples were collected from 4 to 5 sample areas in each district. The obtained data were averaged. *Results* The soils were contaminated by copper and zinc (Zilairsky and Zianchurinsky District), iron (Zianchurinsky, Zilairsky, Burzyansky District), manganese (almost all studied soils), cadmium (most of the studied soils, especially Zilairsky District). Despite the differences in the levels of heavy metals, all the studied soils had an acceptable level of contamination. The heavy metal concentration levels in the dry matter of plants growing in the soils of the Southern Transurals exceeded the acceptable limits in some cases, in particular lead (more than 10 times), cadmium (more than 6 times), and zinc (2 times). The research shows that the level of chemical elements in the soil can affect their accumulation in the bodies of people living in the area. *Conclusions* The results show the perspectives of related scientific fields (between medicine, geology, and ecology), with an aim to study the elemental "portrait" of the population of biogeochemical provinces to which the Southern Urals belong, with the object of scientific research and the implementation of programs designed for elimination of elemental imbalance in the human body.

**Keywords** Geochemical provinces · Southern (Bashkir) transurals · Heavy metals · Morbidity

---

This publication was prepared within the framework supported by the Russian Humanitarian Science Foundation research project №. 15-16-02003.1.

---

I.N. Semenova (✉) · Y.S. Rafikova · Y.T. Suyundukov · G.Ya. Biktimerova
State Autonomous Scientific Institution, "Institute of Regional Research of the Republic of Bashkortostan", Sibay, Russia
e-mail: ifalab@rambler.ru



O.V. Frank-Kamenetskaya et al. (eds.), *Biogenic—Abiogenic Interactions in Natural and Anthropogenic Systems*, Lecture Notes in Earth System Sciences,
DOI 10.1007/978-3-319-24987-2_15

## 1 Introduction

The environment of the Southern (Bashkir) Transurals has been evolving due to the influence of several planetary and some unique regional processes. First is the complexity of the Urals and Transurals peneplain's geological structure. These mountain chains were formed with rocks, aged from the Archean up to the modern age. The transition from landscapes to plains defines the characteristics of the rocks' weathering, and, thus, the peculiarities of the horizontal transfer of the soils and subsoil rocks.

Other factors were the presence of several geochemical provinces, rich in a wide variety of elements and differentiated by their composition and concentration; climate, depending largely on the Urals being "the barrier"; and the specifics of flora and the vegetation's historical development. In most cases, these processes influenced the territory's elements quite independently. A variety of uniquely composed and balanced habitats have formed (Usmanov et al. 2014).

The habitats' composition and macro/microelement content are closely related to human life sphere because the elements take part in the human's most important adaptive mechanisms, including functioning of major systems such as the cardiovascular, respiratory, digestive, immune, endocrine, and others (Skalny 2004). High concentrations of some elements in the numerous polymetallic deposits of the Southern Urals have resulted in their accumulation in soils, water, vegetation, and other ecosystem parts, including the human body. Long and intensive mining industry development has tremendously contributed to environmental contamination with heavy metals, which may undermine people's health. According to research in Bashkortostan, the concentrations of certain microelements in the environment are closely associated with certain diseases, such as Graves' disease, diabetes, multiple sclerosis, and others. Numerous works prove the correlation between the health of people living in mining regions and certain levels of heavy metals and other pollutants in the environment. For example, some profound research has been carried out on the correlation between work environments and the level of health of concentrating mill workers regarding health hazards for the mining and concentration plant workers and the population living in the area (Tchuyenkova et al. 2012). Additional research incudes a toxicological-hygienic assessment of environmental contamination and its effect on the health status of the population of Bashkortostan's mining region (Semenova et al. 2011), examination of biological microelement composition and health of people living in the region, and comprehensive studies on Bashkir Transurals mining and concentration plant landscape contamination (Rafikova et al. 2014). The results obtained indicate the significance of the element status for different organs and system performance, as well as the necessity of studying the correlation between somatic and endocrine diseases and the environment element composition, which to a great extent depends on geologic-geomorphic conditions of the region. This research has been conducted in areas suffering from major technogenic pollution. We found the utmost interest in the element composition of the people who live in areas with minor technogenic pollution. Such research is aimed at determining the region's normal element levels in the people living in a certain area.

This chapter examines the peculiarities of microelement accumulation in environmental objects (soils, vegetation, human body) in the region of the Southern (Bashkir) Transurals, which lack large industrial plants.

## 2 Objects and Methods

The research was carried out in 2014 in the area of the Southern (Bashkir) Transurals, including the Burzyansky, Zilairsky, and Zianchurinsky administrative districts of the Republic of Bashkortostan. Samples were collected from 4 to 5 areas in each district. The obtained data were averaged.

The Burzyansky District is an administrative district in the Republic of Bashkortostan, Russia, with coordinates of 53°06′N 57°26′E. It is located in the south of the republic and shares borders with the Beloretsky District to the north, the Abzelilovsky District to the east, the Baymaksky District to the southeast, the Zilairsky District to the south, the Kugarchinsky District to the southwest, the Meleuzovsky District to the west, and the Ishimbaysky District to the west and northwest. The area of the district is 4444 $km^2$ (1716 $mi^2$). Its administrative center is the rural locality of Starosubkhangulovo.

The territory is limited by the dividing ridge Uraltau to the east and the ridge Kalu to the west. Between them lie the ridges Kraka and Yurmatau (850–1040 m above the sea level). Almost in the middle of the territory from the north to the southwest flows the river Belaya. On the western outskirts of the region flows the river Nugush. This area occupies a mountain relief. It is rich in barium sulphate, building stone, slate, brick earth, sand, marl, dolomite, and magnesite deposits, which have not been profoundly studied in a geological sense.

The Zilairsky District (coordinates: 52°13′N 57°26′E) is located in the south of the republic and borders the Burzyansky District to the north, the Baymaksky District to the east, the Khaybullinsky District to the southeast and south, the Zianchurinsky District into the southwest and west, and the Kugarchinsky District to the west and northwest. The area of the district is 5773.99 $km^2$ (2229.35 $mi^2$). Its administrative center is the rural locality of Zilair. The district's area is located within the Zilair plateau, dissected by canyon-making valleys of rivers such as the Barakal, the Zilair, the Bolshoy Suren, and the upstreams of the Bolshoy Ik: the Avashla, the Berdyash, and the Tekal. The soils here are mountain black soils and mountain-forest gray and black soils, with ordinary black soils in the east.

The Zianchurinsky District (coordinates: 52°11′N 56°34′E) is an administrative district in the Republic of Bashkortostan, Russia. It is located in the south of the republic and borders the Kugarchinsky District to the north, the Zilairsky District to the northeast and east, the Khaybullinsky District to the east, and the Orenburg Oblast to the south and west. The area of the district is 3342.35 $km^2$ (1290.49 $mi^2$). Its administrative center is the rural locality of Isyangulovo. The district is located in the western part of the Zilair plateau and is characterised by a high level of dissection and prevalence of ridges and slopes striking from the north to the south. It is rich in oil, sand, limestone, and dolomite deposits. The climate is continental with low humidity. The following rivers are flowing in the district: the Bolshoy Ik, the Bolshaya, the Malaya Suren, the Assel, the Kasmarka, the Kazanbulak, the Kuruil, and the Uskalyk. Soils vary depending on the landform, from leached and podzolic black soils to dark gray forest and low power rocky soil.

Sampling was carried out according to standard techniques.

To evaluate the content of heavy metals in soil and plants, the following indicators were used: bulk earth content and heavy metal concentration limits for soils and plants (Opekunova 2004). Atomic absorption determination of metals in soils, water, and plants was performed according to standard procedures.

The bioelement status of people living in the region was determined on the basis of hair spectral analysis. Analytical work was carried out at the Centre for Biotic Medicine (Moscow). The levels of Cu, Zn, Fe, Mn, Pb, and Cd were determined in hair using atomic emission spectrometry and mass spectrometry with inductively-coupled argon plasma. Features of the element status of the inhabitants had been estimated checking the levels of chemical elements found in the hair against a biologically acceptable level (Skalny 2004). Statistical data processing was performed using Microsoft Excel 2003 and Statistica 6.0.

# 3 Results and Discussion

## *3.1 Heavy Metal Levels in the Soils of the Southern Transurals*

Lead, copper, zinc, and cadmium are regarded as the priority pollutants and are referred to as the first (Cd, Pb, Zn) and second (Cu) classes of hazard.

The levels of mobile forms of copper and zinc in the examined soils exceeded the concentration limits in the Zilairsky and Zianchurinsky Districts.

The maximum concentration level of iron, exceeding the regional rates, was detected in soils of the Zianchurinsky District. Elevated concentration levels of iron in mobile form were also revealed in Zilairsky and Burzyansky Districts.

The concentration level of total and mobile manganese, exceeding the concentration limits, was detected in almost all of the studied soils.

The concentration level of total and mobile lead did not exceed the concentration limits.

Levels of cadmium exceeded the concentration limits in the soils of the Zilairsky and Zianchurinsky Districts. The levels of mobile cadmium exceeded the concentration limits almost everywhere, especially in the soils of the Zilairsky District (Table 1).

**Table 1** The concentration levels of heavy metals in the soils of the Southern Transurals (mg/kg)

| № | The study area | Cu | Zn | Fe | Mn | Pb | Cd |
|---|---|---|---|---|---|---|---|
| 1 | Burzyansky district | $\frac{14.55}{1.50}$ | $\frac{44.60}{22.75}$ | $\frac{11,097.50}{137.20}$ | $\frac{5845^{a}}{109.9}$ | $\frac{13.63}{0.08}$ | $\frac{2.98^{a}}{0.23^{a}}$ |
| 2 | Zilairsky district | $\frac{90.56 \pm 98.01^{a}}{13.92 \pm 19.48^{a}}$ | $\frac{100.71 \pm 35.83^{a}}{23.08 \pm 14.19^{a}}$ | $\frac{18.804 \pm 5273.68^{a}}{90.81 \pm 41.45}$ | $\frac{4729 \pm 1362^{a}}{216.05 \pm 98.94^{a}}$ | $\frac{8.83 \pm 5.12}{0.89 \pm 1.05}$ | $\frac{3.10 \pm 0.52}{0.84 \pm 0.92^{a}}$ |
| 3 | Zianchurinsky district | $\frac{26.33 \pm 2.81}{4.05 \pm 2.17^{a}}$ | $\frac{98.22 \pm 2.60}{23.01 \pm 2.29^{a}}$ | $\frac{46.125 \pm 39.516^{a}}{73.66 \pm 17.69}$ | $\frac{13,444.17 \pm 11,759.19^{a}}{188.37 \pm 109.27^{a}}$ | $\frac{4.29 \pm 3.67}{1.84 \pm 1.24}$ | $\frac{3.31 \pm 0.29^{a}}{0.37 \pm 0.06^{a}}$ |
| | Maximum permissible concentration | $\frac{55}{3}$ | $\frac{100}{23}$ | $\frac{25,000}{-}$ | $\frac{1500}{140}$ | $\frac{32}{6}$ | $\frac{2.00}{0.20}$ |

[a]Exceeds the concentration limit

Thus, the examined soils were contaminated by copper and zinc (Zilairsky and Zianchurinsky Districts), iron (Zianchurinsky, Zilairsky, and Burzyansky Districts), manganese (almost all studied soils), and cadmium (most of the studied soils, especially the Zilairsky District). Despite the differences in the levels of heavy metals, all of the studied soils had acceptable levels of contamination.

## *3.2 Levels of Heavy Metals in Plants of the Southern Transurals*

The major part of heavy metals reaches the soil, which is a powerful accumulator that does not eliminate them over time. In particular, heavy metals become fixed in the upper layer of humus, which is the most fertile layer. Therefore, contaminated soil becomes a secondary source of pollution, leading to decreased drinking water quality, with substantial amounts of these elements entering the bodies of animals and humans. The metal translocation from soils into plants is inevitable, which leads to negative consequences, as well as the degradation of the quality of medicinal plants. One of the ways for heavy metals to get into the human body is by using medication based on vegetable raw materials. However, the concentration levels of heavy metals in medicinal plants, including wild ones, are still not standardized. Therefore, many researchers use indicators for plant-based biologically active food additives in the hygienic evaluation of medicinal vegetable raw materials.

The levels of heavy metals in plants found in the studied areas are presented in Table 2. According to the results of the research, the concentration of heavy metals in plants forms the following decreasing series: Fe > Mn > Zn > Cu > Pb > Cd.

The highest levels in all studied plants belonged to iron: its level varied from 315.17 mg/kg *Artemísia absínthium* L. (Zilairsky District) to 692.63 mg/kg in *Achilléa millefólium* L. (Burzyansky District). Manganese had the next highest

**Table 2** The levels of heavy metals in plants of the Southern Transurals (mg/kg)

| Cu | Zn | Fe | Mn | Pb | Cd |
|---|---|---|---|---|---|
| *A. millefólium*, Burzyansky District | | | | | |
| 24.54 ± 15.25 | 32.00 ± 4.58 | 692.63 ± 123.55 | 110.08 ± 21.69 | 2.34[a] ± 1.12 | 0.30 ± 0.05 |
| *A. millefólium*, Zilairsky District | | | | | |
| 22.90 ± 17.21 | 108.58[a] ± 103.31 | 657.85 ± 409.14 | 116.37 ± 91.39 | 2.79[a] ± 1.81 | 1.96[a] ± 1.88 |
| *A. absínthium*, Zilairsky District | | | | | |
| 12.21 ± 6.10 | 133.13[a] = 52.02 | 315.17 ± 265.32 | 41.73 ± 21.75 | 2.94[a] ± 3.05 | 1.89[a] ± 0.36 |
| *A. nóbilis*, Zianchurinsky District | | | | | |
| 12.60 ± 3.25 | 20.73 ± 3.90 | 331.13 ± 126.38 | 40.07 ± 8.89 | 5.95[a] ± 1.32 | 1.68[a] ± 0.50 |
| Concentration limits | | | | | |
| 30 | 50 | – | – | 0.5 | 0.3 |

[a]Exceeds maximum permissible concentration

concentration levels: it ranged from 40.07 mg/kg (*Achilléa nóbilis* L., Zianchurinsky District) to 116.37 (*A. millefólium* L., Zilairsky District). Zinc ranked third (from 20.73 mg/kg *A. nóbilis* L. in Zianchurinsky District to 133.13 mg/kg *A. absínthium* L. in Zilairsky District). The copper levels ranged from 12.21 mg/kg (*A. absínthium* L., Zilairsky District) to 24.54 (*A. millefólium* L., Burzyansky District). Lead content ranged from 2.34 mg/kg (*A. millefólium* L., Burzyansky District) to 5.95 (*A. nóbilis* L., Zianchurinsky District), and cadmium ranged from 0.30 mg/kg (*A. millefólium* L., Burzyansky District) to 1.96 mg/kg (*A. millefólium* L., Zilairsky District).

Thus, the heavy metal concentration levels in the dry matter of plants growing in the soils of Southern Transurals exceeded the acceptable limits in some cases, particularly for lead (more than 10 times), cadmium (more than 6 times), and zinc (2 times).

## 3.3 Elemental Composition of the Hair of Southern Transurals Population

The Southern Transurals, like the majority of mining territories, is environmentally unfriendly for people. The elemental imbalance of soils, water, and plants adds to the problem. The elemental imbalance of soils due to biochemical food chains forms a similar imbalance of mineral substances in people's diets. We studied the elemental composition of the hair of the population residing in the area in question. A total of 92 representatives of the Bashkir Transurals population were under our supervision, including 26 women and 39 men. The average age was 19.8 ± 10.7 years.

The bioelemental status of the inhabitants was determined on the basis of the spectral analysis of hair. Hair is an active metabolic tissue, and the levels of micro- and macro-elements in hair correlate with those in the external and internal environments of the body. The concentration of heavy metals in hair (toxic elements) is directly proportional to their content in the environment and the intensity of its elimination from the body (Skalny 2004). Essential element accumulation is interpreted aselimination from the body and the formation of endogenous deficit.

The conducted study of the micro-element composition of the Bashkir Transurals population's hair showed that the levels of a number of essential chemical elements in the body of the examined people did not match the appropriate figures: elevated levels of iron in the Zilairsky District, manganese (all studied regions), and low levels of copper (Burzyansky District) (Table 3). The inhabitants of the Burzyansky area had lower levels of copper, zinc, and iron in their hair compared with other individuals examined. Residents of the Zilairsky area were characterized by high levels of iron, manganese, and lead in their hair.

The concentration of the studied chemical elements in the hair of people living in the Southern Transurals correlated with their mobile form concentrations in soils to varying degrees. A close correlation was shown for copper concentration ($r = 0.94$),

**Table 3** The levels of heavy metals in the hair of the population of the Southern Transurals (mg/kg)

| The study area | Cu | Zn | Fe | Mn | Pb | Cd |
|---|---|---|---|---|---|---|
| Burzyansky district | 8.78 ± 2.61[b] | 186.65 ± 71.10 | 13.89 ± 4.60 | 2.39 ± 1.08[a] | 1.63 ± 1.94 | 0.07 ± 0.05 |
| Zilairsky district | 9.43 ± 3.07 | 199.51 ± 131.40 | 34.92 ± 31.12[a] | 3.43 ± 4.57[a] | 2.05 ± 3.30 | 0.09 ± 0.09 |
| Zianchurinsky district | 9.12 ± 1.51 | 207.67 ± 45.55 | 23.98 ± 20.71 | 1.95 ± [a]1.87 | 0.37 ± 0.23 | 0.05 ± 0.04 |
| Interval biologically acceptable values | 9.00–14.00 | 125.00–250.00 | 11.00–24.00 | 0.32–1.13 | 0–5.00 | 0–0.50 |

[a]Increased level
[b]Lower level

zinc ($r = 0.83$), and cadmium ($r = 0.74$), whereas an average degree of correlation was observed between levels of manganese in hair and soils ($r = 0.47$). At the same time, a strongly negative correlation was found between the levels of lead and iron in hair and soils ($r = -0.75$ and $r = -0.69$, respectively). Thus, the research shows that the levels of chemical elements in the soil can affect their accumulation in the bodies of people living in the area.

The results show the perspectives of related scientific fields (medicine, geology, and ecology), with an aim to determine an elemental "portrait" of the population of biogeochemical provinces in the Southern Urals. The scientific research and the implementation of programs were designed for the elimination of uncovered elemental imbalances in the human body.

## References

Opekunova M (2004) Bioindication of pollution. Saint Petersburg State University Press, Saint Petersburg

Rafikova YS, Semenova IN, Drovosekova IV (2014) Detectability of some microelements in Republic Bashkortostan's population living in the technogeneous biogeochemical areas (Sibay city). Adv Curr Nat Sci 2:36–39 (in Russian)

Semenova IN, Abdullina LA, Rafikova YS (2011) Pollution of object of environment in the zone of influence of Buribaevsky mountain-concentrating industrial complex and indicators of disease of the population. Fundam Res 10:558–560 (in Russian)

Skalny A (2004) Chemical elements in human physiology and ecology. Publishing House Onyx 21, Moscow

Tchuyenkova GA, Askarova ZF, Askarov RA, Denisov EI (2012) The impact of ecologic risk factors in mining cities of South Ural. Prob Social Hyg Health Med Hist 2:13–17 (in Russian)

Usmanov I, Semenova IN, Scherbakov AV, Suyundukov Y (2014) Endemic ecological niches of South (Bashkir) Trans-Urals: multidimensional and fluctuating modes. Vestnik BGAU 1 (29):16–22 (in Russian)

# Assessment of Possible Ecologo–Demographic Effects of Air Emissions by the Example of Karelia

**Dmitrii S. Rybakov**

**Abstract** A close linear correlation between reducing air pollutant emissions from stationary sources (except emissions from iron ore mining in Kostomuksha) and mortality coefficients (annual number of deaths per 1000 people) was revealed: including mortality coefficients from diseases of the circulatory system ($r = 0.98$), from external causes ($r = 0.97$) and the total mortality coefficient ($r = 0.97$) for 2004–2013. The possible total number of deaths caused by these emissions over a 10-year period was estimated in the interval 12.2–31.6 % using linear statistical models. The spasmodic growth of the mortality coefficient from neoplasms in 2008–2011 could have been the result of a sharp rise in motor transport emissions in 2006–2009.

**Keywords** Emissions of pollutants · Types of economic activities · Mortality coefficients · Main causes of death

## 1 Introduction

From the point of view of ecological geology, sources of environmental impact (technical objects, etc.) and biota (humans, etc.) are the components of natural–technical ecologo–geological systems (side by side with the lithosphere). The atmosphere, in turn, is understood as a constituent of the environment, which borders on the lithosphere and interacts closely with ecologo–geological systems. It is through the air in the zone affected by pollution emissions from industrial, energy, and other technical objects that the most rapid harmful effect on living organisms is exerted. This fact fully applies to humans.

World Health Organization (WHO) estimates show that air pollution is the greatest environmental risk for human health. They also indicate that about

D.S. Rybakov (✉)
Institute of Geology, Karelian Research Centre, RAS, Petrozavodsk, Russia
e-mail: rybakovd@krc.karelia.ru

 
O.V. Frank-Kamenetskaya et al. (eds.), *Biogenic—Abiogenic Interactions in Natural and Anthropogenic Systems*, Lecture Notes in Earth System Sciences,
DOI 10.1007/978-3-319-24987-2_16

3.7 million extra deaths in 2012 were caused by atmospheric air pollution and 4.3 million by indoor air pollution. As many people are affected by both the household and ambient polluted air, the number of deaths caused by these factors cannot be merely summed up. Therefore, according to WHO, the final number of pollution victims in 2012 is estimated at about 7 million (WHO 2014).

The pollution of the atmospheric air by harmful chemicals has been studied in the last few decades as a problem closely related to changes in human health and mortality rates in many countries. There are thousands of publications on this subject, many of which have been referred to in the relevant literature, guidelines, recommendations, and reports of international experts (ATSDR 1998; Gichev 2002; Onishchenko et al. 2002; Revich et al. 2004; Air Quality Guidelines 2006; Sulfur dioxide 2008; Pietrodangelo et al. 2010; Chambliss et al. 2013). Special studies, in which mathematical statistics is used, are conducted in some cities (Boev and Bystrykh 1999; Sinitsyn 2011; Hoi et al. 2014).

Health problems that arise from direct air pollution are most closely related to diseases of the circulatory system, respiration and oncological diseases, psychoneurological disturbances, etc. (Gichev 2002).

Special attention should be paid to the possible indirect effect of the inhalation of harmful substances. As a result, unfavorable and dangerous factors, which seem to result (e.g., via ecologically dependent psychoneurological disturbances) in more deaths caused by accidental alcohol poisoning, murders, suicides, etc., are activated.

The consumption of alcoholic drinks by the population of the industrial cities of North European Russia was observed to rise as early as the 1960s–1980s. The per capita alcohol consumption rates in these cities were higher than average regional and average republican estimates (Kolokolchikova 2011). Abdaliev (1991) refers to many researchers who have proved that industrial workers display the highest level of alcoholism. These workers represent a high-risk group of people who suffer from severe forms of alcoholism with a bad outcome. At the same time, medicolegal experts have shown that small quantities of ethyl alcohol in the blood have a favorable effect on a person who suffers from carbon monoxide (CO) poisoning, and high concentrations enhance the toxic effect of carboxyhemoglobin, thus increasing the percentage of death risk (Abdukarimov and Iskandarov 2010).

Psychoneurological disturbances are known to be related to their possible severe consequences such as murders, suicides, fatal occupational injuries, and road accidents. Gichev (2002) offers numerous arguments and refers to various publications, including the papers published in 1955–2001, which indicate that ecological factors are related to the psychological and neurological disturbances of the human organism. The correlation required can be obtained by summing up relevant evidence and conducting special studies. For example, South Korean researchers have correlated the number of committed suicides with concentrations of particulate matters in the atmospheric air estimated 2 days before the suicides (Kim et al. 2010). Bakian et al. (2015) observed maximum heightened odds of suicide associated with interquartile range increases in nitrogen dioxide during cumulative lag 3 (average of 3 days preceding suicide) and fine particulate matter (diameter ≤2.5 μm) on lag day 2 (day 2 before suicide) in Salt Lake County, Utah from 2000 to 2010.

As the above problems still exist and the ecological and demographic rates joint estimated for the study region are not yet convincing enough, the goal of the present paper is to identify man-induced factors which, together with other factors, may contribute to increased mortality in areas with advanced industry and transport and a distinctive weather and climate pattern.

It should be noted that weather and climatic factors, including meteorological ones (air temperature, humidity, and wind regime), exert a positive or negative effect on human health under polluted or unpolluted air conditions (Danilov-Danilyan 2003). In the latter case, existing meteorological conditions influence the distribution of pollutants emitted into the atmosphere, thus reducing or increasing the exposure of the population.

An attempt is made, therefore, to assess a statistical correlation between mortality rates and atmospheric air pollution in the study region, to identify types of economic activities (TEA) responsible for an additional statistically significant mortality risk brought about by air pollution and to roughly estimate the number of additional deaths caused by anthropogenic emissions. In fact, a conceptual framework for assessment of further environmental risk for the population is thus formed to estimate the effect of harmful chemicals, present in polluted air, on the human organism, to identify risk groups of population with regard to the distribution of pollution sources and to carefully develop recommendations for decision-making in management.

## 2 Methods

To construct statistical models and to make correlations, we used data on air pollution from stationary and mobile (motor transport) sources and medical statistical data, published in 1998–2013 in annual state reports on the environmental status in the Republic of Karelia in 2013 (2014), and 2012–2013 Rosstat (Department of Statistics) data on motor transport (General information 2012; Generalized data 2013). Mortality coefficients (annual number of deaths per 1000 people) were corrected with regard to the information available at the Kareliastat web site (Mortality by main causes of death 2014).

Long-term trends were constructed and statistical parameters (e.g., correlation coefficients) were calculated chiefly to assess changes in the health status of the population affected by harmful environmental pollution (e.g., Gichev 2002). Annual data on total pollutant emissions in the region were used.

The efficient use of total emissions data depends on the specialized nature of Karelia's regional economy dominated by some TEA: (1) mining, (2) pulp, paper, and cardboard production, (3) metallurgy (chiefly aluminium production in the Segezha municipal district), and (4) power, gas, vapor, and hot water production. This economic structure was formed when many small companies, institutions and military units, which could make it hard to understand the correlations, were closed down. In 1998, for example, emission reports were submitted by 413 companies of the region, in 2002 by 402 and in 2008 by 367. In 2012, there were 225 air-polluting companies in Karelia.

The minimum number of deaths caused by air pollution was calculated with regard to the adequacy of the statistical models constructed (adequacy was estimated using determination coefficients), using current and minimum mortality coefficients as well as the size and total mortality of the population.

The maximum number of deaths was calculated, assuming that the termination of the effect of air pollution on the population is theoretically possible and that the linear models constructed will remain. The mean annual size of Karelia's population, the annual total number of deaths and mortality coefficients as well as the linear regression equations obtained were also taken into account.

All statistical calculations were made and plots were constructed using the Data Analysis package of the Microsoft Excel software.

## 3 Results and Discussion

The amount of air pollutants from all stationary sources in the Republic of Karelia decreased from 147.8–154.9 thousand tons in 1998–2000 to a minimum of 96.0 thousand tons in 2011, but in 2012 and 2013 it increased again to 106.6 and 118.5 thousand tons, respectively (Fig. 1). The decrease is due to the closing down of

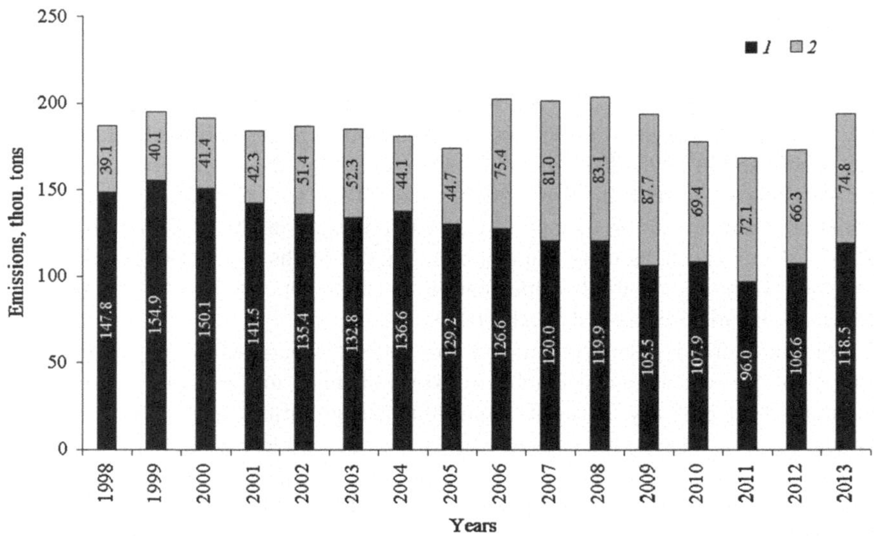


**Fig. 1** Emissions of air pollutants in the Republic of Karelia in 1998–2013: *1* from stationary sources, *2* from motor transport

some small enterprises and boiler plants, the decline in production rate at some big enterprises, the use of natural gas and wood fuel instead of fuel oil and coal and the increased efficiency of new gas purifying facilities.

Pollutants are dominantly emitted by the Karelsky Okatysh Iron Mining Company and processing plants (total values varied from 74.9 % in 2000 to 88.7 % in 2013). Emissions from the processing plants decreased gradually from 72.8 thousand tons in 2000 to 30.1–31.3 thousand tons in 2012 and 2013 respectively, while emissions from Karelsky Okatysh increased from 39.5 to 73.9 thousand tons.

The structure of emissions from the processing companies is dominated by the pulp and paper industry which is concentrated in the Segezha, Kondopoga, and Pitkäranta municipal districts (83.7–88.9 % in 2000–2004, and 92.0–95.3 % in 2005–2013). These TEA, together with wood processing, are responsible for the emission of 50.7–53.2 thousand tons of pollutants in 2000–2004 and 20.3–25.8 thousand tons in 2011–2013.

The amounts of emissions from motor transport from 1998 varied from 39.1 to 87.7 thousand tons and reached 74.8 thousand tons in 2013 (see Fig. 1). One should take into consideration a possible "artificial" decrease in emissions from motor transport, caused by changes in the methods for estimation of emissions from mobile sources in 2010. As well as the structure update and strength of the available motor transport of Euro-3 and Euro-4 ecological classes had been taken into account from the same year [State reports on the environmental status in the Republic of Karelia in 2010 (2011)].

Deaths in the Republic of Karelia in 1998–2013 were caused dominantly by diseases of the circulatory system (53.4–56.1 % of cases), external causes (18.9–9.6 %), and oncological diseases (10.5–16.1 %). From 2008, malignant tumors as the cause of mortality persisted as a second main cause of death (Fig. 2), shifting external causes to position 3 (in 2013 they accounted for 16.1 % and 9.6 %, respectively).

The course of variations in the total mortality coefficient and the coefficients of mortality from diseases of the circulatory system and external causes, shown in Fig. 2, can be divided into two periods. For example, in 1998–2003 the rates increased (cases per 1000 people): from 13.28 to 19.86, from 7.44 to 10.85, and from 2.24 to 3.36, respectively. From 2003 onwards they generally decreased: to 14.61 in 2013, to 7.92–8.28 in 2011–2013, and to 1.40 in 2013, respectively.

Variations in the mortality coefficient from neoplasms display a different pattern. In 1998–2007, mortality varied from 2.00 to 2.17 cases per 1000 people, while in 2008 and later it ranged from 2.25 to 2.41 (see Fig. 2).

Table 1 shows the results of linear correlation analysis attempted to establish a statistical relation between air emissions and mortality rates in the Republic of Karelia. Three periods: 1998–2013, 1998–2003 and 2004–2013 were selected for calculations. The latter two periods display inverse trends in the mortality coefficient (see Fig. 2).

The structural pattern of correlations is due to multidirectional trends in emissions from various sources. Emissions from stationary sources generally decreased, while those from motor transport increased (see Fig. 1). Obviously, increase of the

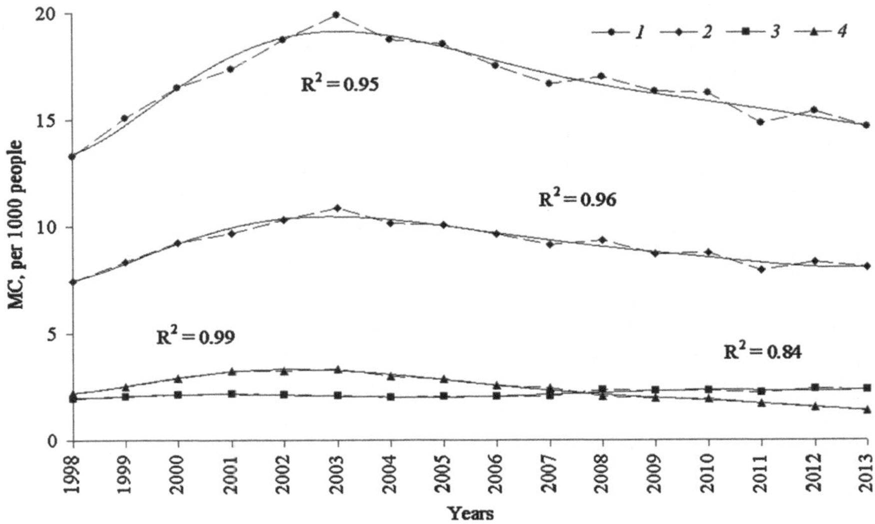


**Fig. 2** Population mortality in the Republic of Karelia in 1998–2013: mortality coefficients (MC): *1* total, *2* from diseases of the circulatory system, *3* from neoplasms, *4* from external causes; trends 6-degree polynomial; $R^2$ coefficients of determination

amounts of air pollutants from the Karelsky Okatysh structures located 13 km from the residential area of Kostomuksha, a third largest city in the Republic of Karelia, not affected to the trends of decreasing of the rates of total mortality and mortality caused by diseases of the circulatory system in the region. This can be explained by the relatively young age of the permanent population of Kostomuksha, a city founded in 1977: according to the Kareliastat data as of January 1, 2012, men older than 59 and women older than 54 made up 16.3 % in Kostomuksha against 23.5 % in the Republic of Karelia (Population 2014). Furthermore, the "dropping out" of Karelsky Okatysh from the general regional model discussed could have been affected by dominant northwestern, western, and southwestern winds that contribute to the transport of pollutants from the city toward a less populated area.

A general decline in harmful emissions from stationary sources does not lead to a consistent decrease in mortality from neoplasms. On the contrary, mortality, caused by these factors, increases. As a result, the relevant parameters are inversely correlated (see Table 1). Attention should be paid to an emerging positive correlation between emissions from motor transport and mortality from neoplasms.

A considerable reduction in emissions from the pulp and paper and wood processing industries and, consequently, a better environmental status in the Segezha and Kondopoga districts are accompanied by the decrease of total mortality, mortality caused by diseases of the circulatory system and external causes that resulted in a close correlation between corresponding rates (see Table 1). The linear correlation between the above parameters in the last decade is especially strong.

**Table 1** Pairwise correlation coefficients between the emissions of air pollutants and mortality coefficients

| Emission sources | TMC | | | $MC_{DCS}$ | | | $MC_{Neopl.}$ | | | $MC_{Ext.}$ | | |
|---|---|---|---|---|---|---|---|---|---|---|---|---|
| | 1998–2013 | 1998–2003 | 2004–2013 | 1998–2013 | 1998–2003 | 2004–2013 | 1998–2013 | 1998–2003 | 2004–2013 | 1998–2013 | 1998–2003 | 2004–2013 |
| Motor transport + stationary | 0.05 | −0.45 | 0.13 | 0.12 | −0.45 | 0.21 | −0.09 | −0.16 | −0.06 | 0.15 | −0.52 | 0.11 |
| Motor transport | −0.09 | **0.90** | −0.57 | −0.16 | **0.89** | −0.53 | **0.62** | 0.36 | 0.53 | **−0.61** | 0.77 | −0.56 |
| Stationary | 0.12 | **−0.83** | **0.81** | 0.24 | **−0.82** | **0.86** | **−0.68** | −0.32 | **−0.70** | **0.70** | −0.78 | **0.78** |
| Karelsky Okatish, JSC (log) | −0.41 | −0.53 | **−0.75** | −0.49 | −0.57 | **−0.70** | **0.59** | **−0.93** | 0.61 | **−0.85** | −0.72 | **−0.78** |
| Stationary, except Karelsky Okatish, JSC | 0.20 | −0.66 | **0.97** | 0.31 | −0.64 | **0.98** | **−0.73** | −0.02 | **−0.82** | **0.77** | −0.55 | **0.97** |
| Pulp and paper + woodworking | **0.57** | 0.79 | **0.95** | **0.66** | 0.81 | **0.96** | **−0.75** | **0.86** | **−0.84** | **0.94** | **0.84** | **0.96** |

*Note* TMC—total mortality coefficient; mortality coefficients: $MC_{DCS}$—from diseases of the circulatory system, $MC_{Neopl.}$—from neoplasms, $MC_{Ext.}$—from external causes (accidents, poisoning, and injuries); statistical samples: for 1998–2013 ($n$ = 16), for 1998–2003 ($n$ = 6), for 2004–2013 ($n$ = 10); significant coefficients at 95 % reliability level are highlighted in bold font

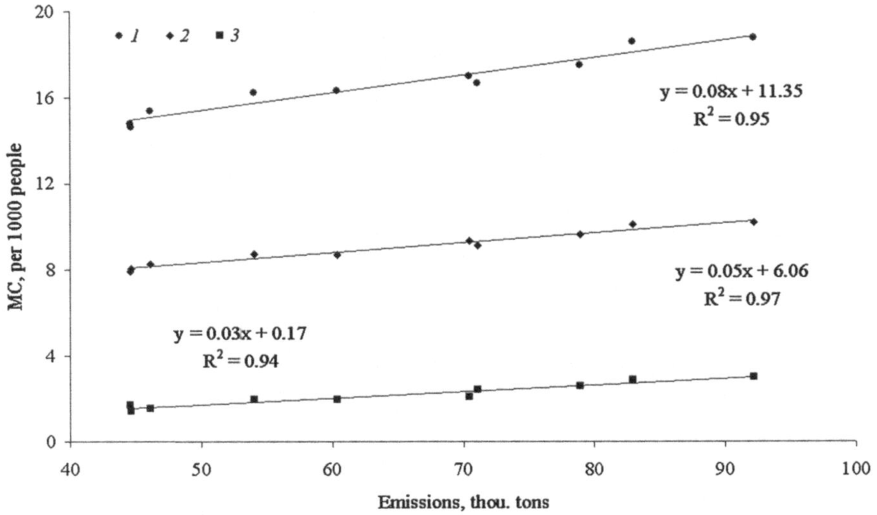


**Fig. 3** Emissions of air pollutants from stationary sources (except emissions from Karelsky Okatysh, JSC) and population mortality in 2004–2013: MC: *1* total, *2* from diseases of the circulatory system, *3* from external causes; trends—linear with the corresponding equations

A statistical correlation between harmful emissions and mortality rates is reliably described by linear trends (Fig. 3).

The possible number of extra deaths caused by anthropogenic emissions was estimated (Table 2) using the linear statistical model equations shown in Fig. 3. For example, the minimum number of such cases in 2013 made up 0.8 % of the total number of deaths in the Republic of Karelia that year, while the maximum number is 22.3 %, provided the emissions considered in our calculations are completely terminated, which is theoretically possible, and the linear model persists. In this case, the number of deaths caused by air pollution from stationary sources (except Karelsky Okatysh) in 2004–2013 accounts for 12.2–31.6 % of the total number of deaths over that period of time. The estimates of possible deaths caused in 2013 by diseases of the circulatory system, brought about by the above emissions, make up 1.5–24.6 % of the total number of deaths due to same causes of death (see Table 2).

A rapid increase in emissions from motor transport in 2006–2009 (see Fig. 1) could be responsible for a spasmodic transition to a higher mortality from neoplasms in 2008–2011 (Fig. 4). When assessing ecological risks upon quantity-to-quality transitions, a lag between the moment of exposure and the emergence of the effect, notably the lack of survivors with malignant tumors over a certain period of time, should be considered. It is known that the leading cancer pathologies in the regional structure of oncopathologies of the Republic of Karelia includes cancers of the lung, trachea, bronchi, breast, and stomach (State Reports on the Environmental Status in the Republic of Karelia in 2006 2007). At the same time, the impact of harmful emissions, such as nitrogen dioxide and sulfur dioxide (Primary prevention

**Table 2** Assessment of the likely number of the deaths due to illnesses related to emissions of pollutants in 2004–2013

| Years | MC, per 1000 people | | $DCS_i$–$DCS_{min}$ | | | | $TMC_i$–$TMC_{min}$ | | | | Mid-year population, 1000 people | Total deaths, people | Deaths due to emissions, people | |
|---|---|---|---|---|---|---|---|---|---|---|---|---|---|---|
| | TMC | DCS | Min | | Max | | Min | | Max | | | | Min | Max |
| | | | Per 1000 people | % | Per 1000 people | % | Per 1000 people | % | Per 1000 people | % | | | | |
| 2004 | 18.73 | 10.13 | 2.21 | 21.9 | 4.07 | 40.2 | 4.12 | 22.0 | 7.38 | 39.4 | 699.148 | 13,092 | 2880 | 5157 |
| 2005 | 18.53 | 10.04 | 2.12 | 21.2 | 3.98 | 39.6 | 3.92 | 21.2 | 7.18 | 38.7 | 682.800 | 12,649 | 2676 | 4899 |
| 2006 | 17.47 | 9.61 | 1.69 | 17.6 | 3.55 | 36.9 | 2.88 | 16.5 | 6.12 | 35.0 | 670.500 | 11,716 | 1929 | 4106 |
| 2007 | 16.62 | 9.12 | 1.20 | 13.1 | 3.06 | 33.5 | 2.23 | 13.4 | 5.27 | 31.7 | 662.200 | 11,007 | 1473 | 3491 |
| 2008 | 16.96 | 9.32 | 1.41 | 15.1 | 3.26 | 35.0 | 2.35 | 13.9 | 5.61 | 33.1 | 656.583 | 11,134 | 1544 | 3682 |
| 2009 | 16.28 | 8.69 | 0.77 | 8.8 | 2.63 | 30.2 | 1.67 | 10.3 | 4.93 | 30.3 | 651.233 | 10,599 | 1087 | 3208 |
| 2010 | 16.19 | 8.72 | 0.80 | 9.2 | 2.66 | 30.5 | 1.58 | 9.8 | 4.84 | 29.9 | 645.641 | 10,451 | 1020 | 3123 |
| 2011 | 14.78 | **7.92** | 0.00 | 0.0 | 1.86 | 23.5 | 0.32 | 2.1 | 3.43 | 23.2 | 641.132 | 9474 | 203 | 2197 |
| 2012 | 15.36 | 8.28 | 0.37 | 4.4 | 2.22 | 26.8 | 0.75 | 4.9 | 4.01 | 26.1 | 638.307 | 9804 | 480 | 2559 |
| 2013 | **14.61** | 8.04 | 0.12 | **1.5** | 1.98 | **24.6** | 0.12 | **0.8** | 3.26 | **22.3** | 635.668 | 9285 | 76 | 2070 |
| | | | | | | | | | | | **Сумма**: | **109,211** | **13,367** | **34,492** |
| | | | | | | | | | | | % | **100.0** | **12.2** | **31.6** |

*Note* Mid-year population—arithmetic mean of the abundances of the beginning and end of the year (according to government statistics); min—assessment of the minimum number of the deaths using the minimum values $TMC_{min}$ и $DCS_{min}$ 14.61 и 7.92 respectively; max—assessment of the maximum number of the deaths using the equations of linear models (Fig. 3), where 11.35 и 6.06—coefficients, coinciding with the predicted values $TMC_{min}$ и $DCS_{min}$, respectively, in the hypothetical absence of emissions ($x = 0$)

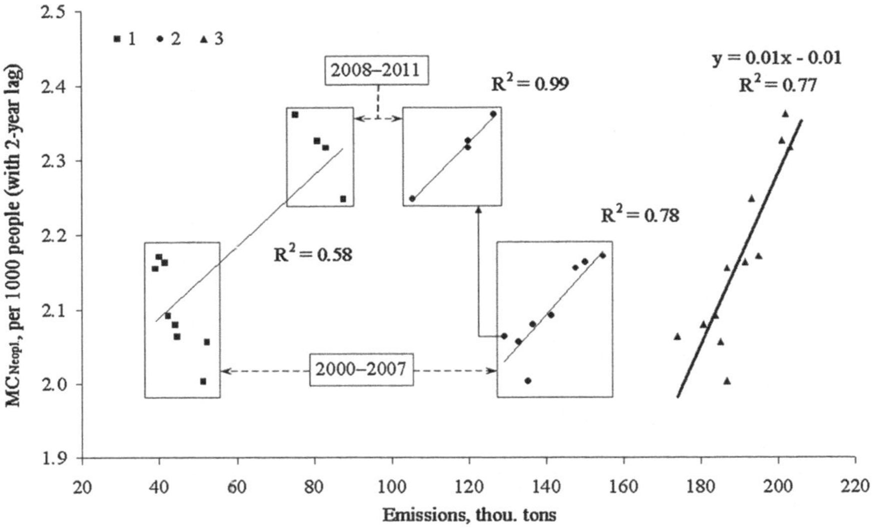


**Fig. 4** Emissions of air pollutants in 1998–2009: *1* from motor transport, *2* from all stationary sources, *3* total emissions (*1* + *2*), and mortality coefficients from neoplasms ($MC_{Neopl.}$) in 2000–2011

2011), on the human organism may be superimposed on other factors (smoking, alcohol, bad working conditions, etc.) and to accelerate the development of cancer in general (since the period of precancer and ending with a lethal outcome). In addition, it should be taken into account, such as fulminant lung cancer, which can lead to death after inhaling pollutants in a very short period of time. A linear correlation between total emissions and the mortality coefficient from neoplasms, estimated with regard for a two-year lag, is shown in Fig. 4.

Table 3 shows correlation between emission of pollutants and mortality from neoplasms rate. Paired correlation coefficients were calculated using the moving averages of mortality coefficients from neoplasms consecutively with smoothing intervals of 2–5. Here, a moving average value with a lag of 1–4 years, respectively, was consistent with each emission value. It is clear from the Table 3 that as the smoothing interval increases in width, a positive statistical correlation between emissions from motor transport and mortality rate increases. Negative correlation for stationary sources increases in similar manner. A maximum correlation coefficient of 0.77 between total emissions from stationary sources and motor transport and the rate of mortality from neoplasms was obtained for a three-year period (arithmetic mean of the mortality coefficients over a three-year period, including the year of the emissions and two subsequent years).

A strong statistical correlation between the amounts of emissions and mortality from external causes, which gradually decreased in the period discussed (see Fig. 2), was revealed (see Fig. 3).

**Table 3** Pairwise correlation coefficients between air pollutant emissions in 1998–2009 and moving average of mortality coefficients from neoplasms

| | Stationary sources | Motor transport | Stationary sources + motor transport |
|---|---|---|---|
| Stationary sources | 1 | | |
| Motor transport | **−0.89** | 1 | |
| Stationary sources + motor transport | −0.30 | **0.70** | 1 |
| $MC_{Neopl.}$ (1) | −0.54 | **0.60** | 0.41 |
| $MC_{Neopl.}$ (2) | **−0.59** | **0.71** | 0.57 |
| $MC_{Neopl.}$ (3) | **−0.61** | **0.83** | **0.77** |
| $MC_{Neopl.}$ (4) | **−0.72** | **0.87** | **0.71** |
| $MC_{Neopl.}$ (5) | **−0.78** | **0.89** | **0.63** |

*Note* Numbers in parentheses indicate that average mortality coefficient values were calculated for: *1* emission year (without a lag), *2–5*: for the emission year and: *2* for the following year, *3* for the two, *4* three, *5* four following years; statistically significant correlation coefficients at critical $r_{P\,=\,0,95;\,f\,=\,10} = 0.58$ are shown in bold type

Mortality from external causes in the Republic of Karelia is formed of the following factors: accidental alcohol poisoning, murders, suicides, etc. (State Reports on the Environmental Status in the Republic of Karelia in 2006 2007; Mortality by main causes of death 2014). It should be noted that, according to statistical data for the Republic of Karelia, the percentage of accidental alcohol poisoning in the total number of deaths due to external factors has decreased: 21.1–21.2 % (2004–2005), 11.1–18.5 % (2006–2011) and 8.2–9.0 % (2012–2013). At the same time, available evidence indicates that the amounts of carbon monoxide emissions from stationary sources, which may increase the mortality risk of poisoning provided ethyl alcohol concentration in the blood is high (Abdukarimov and Iskandarov 2010), have decreased in the region by 27.0 % from 20.7 thousand tons in 2005 to 15.1 thousand tons in 2009 (State Reports on the Environmental Status in the Republic of Karelia in 2009 2010). Thus, single excessive drinking with a lethal outcome could be one of the reasons for the “paradoxical” dependence of mortality from external factors on atmospheric air pollution level (see Fig. 3).

## 4 Conclusions

Mathematical methods of statistics were used to correlate the amounts of pollutants emitted into the atmosphere from stationary and mobile (motor transport) sources with the mortality parameters of Karelia’s population. For example, the reduction of the total mortality coefficient and mortality coefficients from diseases of the circulatory system and external causes in 2004–2013 is closely correlated with the reduction of emissions from stationary sources, except for emissions from the iron ore complex operated by Karelsky Okatysh Company ($r = 0.97{-}0.98$).

A considerable rise in the mortality coefficient from neoplasms in 2008–2012, following a rapid growth of motor transport emissions in 2006–2009, is noteworthy. Assessing environmental risks, arising upon such quantity-to-quality transitions, requires taking into account the presence of a lag between the time of population exposure to an air pollutant after emissions and the emergence of a tangible effect on measurable mortality rates from neoplasms.

The minimum total mortality level for the Republic of Karelia, related to diseases caused by regional emissions of pollutants from stationary sources, was estimated at the end of 2013 at 0.8 % of the total number of deaths that year, and the maximum level—at 22.3 % and 12.2–31.6 % of the deaths attributable to emissions from stationary sources for 2004–2013.

## References

Abdaliev AM (1991) Therapeutic remissions in male alcoholic patients, workers employed in industrial production. Abstr. Ph.D. thesis (Medicine), Vsesoyuz. nauch. tsentr med.-biol. problem narkologii, Moscow. URL: http://medical-diss.com/medicina/terapevticheskie-remissii-u-bolnyh-alkogolizmom-muzhchin-rabochih-zanyatyh-v-promyshlennom-proizvod-stve. Accessed 15 Jan 2015 (in Russian)

Abdukarimov BA, Iskandarov AI (2010) Forensic medical toxicometry of acute carbon monoxide poisoning during alcoholic intoxication. Sudebno-meditsinskaya expertiza 1:30–33 (in Russian). URL: http://www.mediasphera.ru/journals/smekc/651/eng/10033/ (abstract). Accessed 15 Jan 2015 (in English)

Air Quality Guidelines. Global Update 2005 (2006) Particulate matter, ozone, nitrogen dioxide and sulfur dioxide. WHO Regional Office for Europe, Copenhagen. URL: http://www.euro.who.int/__data/assets/pdf_file/0005/78638/E90038.pdf. Accessed 15 Jan 2015

ATSDR (Agency for Toxic Substances and Disease Registry) (1998) Toxicological profile for sulfur dioxide. US Public Health Service. URL: www.atsdr.cdc.gov/toxprofiles/tp116.pdf. Accessed 15 Jan 2015

Bakian AV, Huber RS, Coon H, Gray D, Wilson P, McMahon WM, Renshaw PF (2015) Acute air pollution exposure and risk of suicide completion. Am J Epidemiol 181(5):295–303. doi:10.1093/aje/kwu341

Boev BM, Bystrykh VV (1999) Anthropogenic atmospheric air pollution and human health. In: Integrated estimation of air quality in the industrial cities of the Orenburg region. Published by OGU, Orenburg. URL: http://zshluz.com/articles/air.htm. Accessed 15 Jan 2015 (in Russian)

Chambliss S, Miller J, Façanha C, Minjares R, Blumberg K (2013) The impact of stringent fuel and vehicle standards on premature mortality and emissions. Int Council Clean Transport, Washington DC. URL: http://theicct.org/global-health-roadmap. Accessed 15 Jan 2015

Danilov-Danilyan VI (ed) (2003) Climatic changes: the view from Russia. TEIS, Moscow (in Russian)

General information on atmospheric air pollution by mobile sources (automobiles and railway transport) in the cities, regions and districts of the Russian Federation in 2012 (2013) URL: solidwaste.ru/i/ndocs/765/za_2012krasnoyarsk.doc. Accessed 15 Jan 2015 (in Russian)

Generalized data on atmospheric air pollution from mobile sources (automobiles and railway transport) in the cities, regions and districts of the Russian Federation in 2013 (2014) URL: http://rpn.gov.ru/sites/all/files/users/navdoctor/attachedfiles/prilozhenie_dannye_vybrosov_za_2013_rosstat.doc. Accessed 15 Jan 2015 (in Russian)

Gichev YuP (2002) Environmental pollution and human health. (The sad experience of Russia). SB RAMS, Novosibirsk (in Russian)

Hoi K-I, Zhang D-B, Mok K-M, Yuen K-V (2014) Association of human mortality with air pollution of Hong Kong. Toxics 2(2):158–164

Kim C, Jung SH, Kang DR, Kim HC, Moon KT, Hur NW, Shin DC, Suh I (2010) Ambient particulate matter as a risk factor for suicide. Am J Psychiatry 167:1100–1107

Kolokolchikova RS (2011) Social problems of the population: drunkenness and alcoholism in the industrial cities of North European Russia (mid-1960s–mid-1980s). Izvestiya Rossiiskogo gosudarstvennogo pedagogicheskogo universiteta im. AI Hertzen, Saint-Petersburg 130:9–18. URL: http://cyberleninka.ru/article/n/sotsialnye-problemy-naseleniya-pyanstvo-i-alkogolizm-v-industrialnyh-gorodah-evropeyskogo-severa-rossii-ser-1960-h-ser-1980-h-gg. Accessed 15 Jan 2015 (in Russian)

Mortality by main causes of death (2014) Territorial body of the Federal Statistical Service for the Republic of Karelia. URL: http://krl.gks.ru/wps/wcm/connect/rosstat_ts/krl/resources/dfb25c8043ae0164a172b5d06954faf7/71021.pdf. Accessed 15 Jan 2015 (in Russian)

Onishchenko GG, Novikov SM, Rakhmanin YuA, Avaliani SL, Bushtueva KA (2002) A framework for health risk assessment when exposed to chemicals that pollute the environment. AN Sysin GU NII Ech i Gos, RAMN, Moscow (in Russian)

Pietrodangelo A, Bencardino M, Cecinato A, Decesari S, Perrino C, Sprovieri F, Pirrone N, Facchini MC (2010) Role of atmospheric pollution on harmful health effects. In: Bianchi F, Cori L, Moretti PF (eds) CNR environment and health inter-departmental project: present knowledge and prospects for future research. Consiglio Nazionale delle Ricerche, Rome. URL: http://issuu.com/cnr-dta/docs/pias_2010. Accessed 15 Jan 2015

Population subdivision into main age groups in the context of urban and municipal districts as of 1 January (2014) Territorial body of the federal state statistical service for the Republic of Karelia. URL: http://krl.gks.ru/wps/wcm/connect/rosstat_ts/krl/resources/a42d608041f7e2989-35bdf2d59c15b71/74541.htm. Accessed 15 Jan 2015 (in Russian)

Primary prevention of cancer in modern Russia: collection of information-methodological letters (2011) URL: http://standartgost.ru/g/pkey-14293799626. Accessed 15 Jan 2015 (in Russian)

Revich BA, Avaliani SL, Tikhonova GI (2004) Ecological epidemiology. Akademia, Moscow (in Russian)

Sinitsyn IS (2011) Assessing the impact of air pollution of Yaroslavl on the incidence respiratory organs. Yaroslavsky pedagogical vestnik V III (Natural sciences) 1:190–194. URL: http://vestnik.yspu.org/releases/2011_1e/36.pdf. Accessed 15 Jan 2015 (in Russian)

State Report on the Environmental Status in the Republic of Karelia in 2006 (2007) Ministry of agriculture, fishery and ecology of the Republic of Karelia, Petrozavodsk (in Russian)

State Report on the Environmental Status in the Republic of Karelia in 2009 (2010) Ministry of agriculture, fishery and ecology of the Republic of Karelia, Petrozavodsk. URL: http://www.gov.karelia.ru/Power/Committee/Forest/Docum/gd2009.pdf. Accessed 15 Jan 2015 (in Russian)

State Report on the Environmental Status in the Republic of Karelia in 2010 (2011) Ministry of natural resources and ecology of the Republic of Karelia, Petrozavodsk. URL: http://www.gov.karelia.ru/Power/Committee/Forest/Docum/gd2010.pdf. Accessed 15 Jan 2015 (in Russian)

State Report on the Environmental Status in the Republic of Karelia in 2013 (2014) Ministry of natural resources and ecology for the Republic of Karelia, Petrozavodsk (in Russian)

Sulfur dioxide (CAS Reg. No. 7446–09-5) (2008) Final acute exposure guideline levels (AEGLs). URL: www.epa.gov/oppt/aegl/pubs/sulfur_dioxide_interim_may_2008_v1.pdf. Accessed 15 Jan 2015

WHO (2014) Burden of disease from ambient and household air pollution. URL: http://www.who.int/phe/health_topics/outdoorair/databases/en/. Accessed 15 Jan 2015

# Part III
# Biomineral Interactions in Soil

# Soil-like Patterns Inside the Rocks: Structure, Genesis, and Research Techniques

**Nikita S. Mergelov, Ilya G. Shorkunov, Victor O. Targulian, Andrey V. Dolgikh, Konstantin N. Abrosimov, Elya P. Zazovskaya and Sergey V. Goryachkin**

**Abstract** Microprofiles established due to the activity of endolithic communities inside the solid rocks of East Antarctica were studied with the approaches of soil science. Major products of endolithic rock transformation in situ are the silty-sandy fine earth and abundant organo-mineral films that are formed within the porous space of endolithic system. Such films are the result of interaction between biofilms and mineral surfaces and reflect elemental composition of both components, mainly comprising C, O, Si, Al, Fe, K, Ca, Na, and Mg. Morphology observed on different hierarchical levels and microtomography data indicated that different layers of endolithic system are connected with the fracture network serving for the elements transfer in the subsurface part of solid rocks. Examined profiles in granites with high quartz content had clear eluvial–illuvial differentiation patterns similar to macro-profile of a common Podzol (Spodsol) on loose substrates. It is shown, that subaerial segment of hard rocks is not sealed and is potentially permeable for dissolved products of endolithic weathering and pedogenesis. As a unique result—the soil-like pattern is established inside the massive, crystalline rock. Understanding modern processes in endolithic systems is of fundamental importance to decrypt paleosol record, as such systems may be the closest modern analogues of protosoils that existed on our planet before the higher vascular plants with root systems established.

**Keywords** Extreme environment · Endoliths · Soil-like bodies · Exfoliation · Eluvial–illuvial differentiation · Podzols

N.S. Mergelov (✉) · I.G. Shorkunov · V.O. Targulian · A.V. Dolgikh · E.P. Zazovskaya
S.V. Goryachkin
Institute of Geography, Russian Academy of Sciences, Moscow, Russia
e-mail: nikvox@yandex.ru

K.N. Abrosimov
V.V. Dokuchaev Soil Science Institute, Russian Academy of Sciences, Moscow, Russia



O.V. Frank-Kamenetskaya et al. (eds.), *Biogenic—Abiogenic Interactions in Natural and Anthropogenic Systems*, Lecture Notes in Earth System Sciences,
DOI 10.1007/978-3-319-24987-2_17

# 1 Introduction

Communities of endolithic organisms and the microenvironment they create inside the rocks are among the least-explored terrestrial ecosystems with pronounced photoautotrophic component. Their potential ranges are enormous: rock outcrops occupy up to 20 % of the Earth's land surface. However, actual habitats are largely attributed to various areas of external stress where development of biota at the surface is inhibited by the lack of moisture, UV radiation, wind corrasion, etc. Under such conditions, organisms find their ecological niche inside the rocks which interact with minerals and establish specific endolithic systems. These biotic–abiotic bodies possess many of soil attributes. However, they are never perceived as soils or at least soil-like bodies. Dominant autotrophic components of endolithic systems are cyanobacteria and green algae existing primarily in the form of biofilms. The most suitable rocks for colonization by endolithic photoautotrophs are sandstones, various granitoids, marbles and any others with a significant content of translucent and/or transparent mineral grains, which make primary production of organic matter possible inside the rock even under limited levels of available light.

Organisms inhabiting subsurface layer inside solid rocks captured attention of K. Ehrenberg, the founder of micropaleontology. Their possible role in biochemical weathering and exfoliation was indicated by Glazovskaya (1958) in her pioneering study of initial pedogenesis in Antarctica, but conceptually this phenomenon was described by geobiologist Friedmann et al. (1967), Friedmann and Ocampo (1976), Friedmann (1982), etc. who applied the term "endoliths" to such organisms. Later, they were divided into crypto-, chasmo-, and euendoliths according to the differences in mode of development and penetration into cracks and structural cavities of rocks (Golubic et al. 1981). Since 1967, when microscopic endolithic algae and cyanobacteria were discovered by I. Friedmann in Sinai and Negev deserts they continue to be systematically studied in various extreme environments of the world by geo(micro)biologists and ecologists focusing on diversity and growth conditions. In recent years, the emphasis has shifted to their metagenome decoding (Sigler et al. 2003; De los Rios et al. 2007; Walker and Pace 2007; Horath and Bachofen 2009 and many others). Studies on biochemical weathering and biomineralization in endolithic systems (De los Rios et al. 2003, 2014; Wierzchos et al. 2005 and others) are not so numerous and we still experience a lack of comprehensive understanding of their (bio)geochemistry.

Except for trace fossils and biosignature issues (Golubic and Schneider 2003) endolithic bio–abiotic systems are not fully recognized as paleo protosoils and modern soil-like bodies and are not investigated using the methods and techniques of pedology. However, they are actual precursors to more advanced soil formations, and often the only equilibrium soil-like bodies with external factors in many areas of cold and hot deserts and high mountains of the planet. Endolithic bio–abiotic systems are relatively simple formations and are among best suitable objects to study organo-mineral interactions at a detailed level. This is extremely important for the modern soil science, where so called "organo-mineral complexes" are considered to be one of the major mechanisms of organic matter stabilization (Kögel-Knabner et al. 2008; Schmidt et al. 2011). There is also a knowledge gap in biogeography concerning diversity and spatial patterns of endolithic biotic complexes. Understanding modern processes in endolithic systems is of fundamental importance to decrypt paleosol record, as such systems may be the closest modern analogues of protosoils that existed on our planet before the higher vascular plants with root systems established.

In recent decades, soil scientists realized that the challenges they face are more than just the study of loose subaerial bio-abiotic formations on the surface of Earth (Targulian and Goryachkin 2011). Previously, soil-like bodies (Dmitriev 1996) or semi-soils (Sokolov 1996) soil scientists began to study under water (Ivlev and Nesterova 2004; Roslikova 2006), and later in caves (Semikolennykh and Targulian 2010). This work is also devoted to nontraditional objects of soil science that, however, occur on the Earth surface and are influenced by the "usual" factors of pedogenesis but under highly specific extreme conditions.

It was suggested earlier (Gilichinsky et al. 2010; Mergelov et al. 2012) that endolithic systems could represent the most widely spread soil-like bodies or even soils in Antarctica. However, it is still disputable whether superficial rock layers biochemically transformed by endoliths could be qualified as soils. The main objective of this study was to evaluate properties and processes occurring under the influence of organisms living inside the rocks in East Antarctica from the perspective of soil science.

## 2 Materials and Methods

The samples of endolithic systems were collected from the exposed bedrock surfaces in coastal oases of East Antarctica: the Larsemann Hills (S69° 20′, E76° 20′) and the Thala Hills (S67° 40′, E45° 20′). The bedrock in both cases was represented by granitoid and gneiss formations with granites and granite gneisses (consisting of feldspars, quartz, garnet, and biotite) in the Larsemann Hills and with orthogneiss and enderbites (feldspars, quartz, hypersthene, diopside, amphiboles (hornblende), and biotite in various combinations) in Thala Hills. According to temperature parameters that we measured in situ (Mergelov et al. 2012), the parts of hard bedrock with endolithic communities occur under relatively favorable (for Antarctica) conditions due to insolation of dark surfaces with rusty to reddish varnish/silica glaze. The most developed endolithic communities are usually found on the warm (in the Southern hemisphere) north-facing slopes where duration of positive temperatures period reaches 4.5 months, including 2 months with day temperatures ranging between 20 and 30 °C. On the rocky slopes of southern aspect, the temperature conditions are colder and are approximately the same as in loose substrates of the local valleys: the period with $T > 0$ °C lasts up to 2.5 months, and the surface soil (rock) day temperature reaches 10 °C during 1.5 months. Thus, the duration of the period with positive temperatures on the rock surfaces with well-developed endolithic communities is 1.8–2.0 times longer, and the absolute summer day temperatures are 2–3 times higher than those on other surfaces in oases.

We sampled multicomponent formations at the bedrock surfaces, including (1) exfoliation plates with endolithic communities on the bottom sides and rock varnish/silica glaze on the upper sides, (2) mineral fine earth and biomass of the endolithic organisms from the system of fractures immediately under such plates, and (3) nondisturbed rock interior under the plates without endolithic organisms. Totally, 10 target sites were sampled on granitoids from Larsemann Hills and Thala Hills oases. Among them, three were specially selected for excavation of micromonoliths (0.7–1.5 $cm^3$ in volume) for microtomography survey.

To set the framework of research, we suggest that these three components together can be designated as an endolithic system. The structure and morphology of this system resemble a soil microprofile, and a preliminary term—endolithic soils—was suggested earlier (Mergelov et al. 2012). This proposal was more thoroughly examined in the course of our study.

The morphology of samples was studied on different hierarchical levels—under Leica MZ6 binocular, Nikon Eclipse E200 microscope both with digital cameras, and under a scanning electron microscope JSM-6610LV with an X-ray microanalyzer by Oxford Instruments making it possible to determine elemental composition of the substrate. The samples were observed both in native state and in thin sections (under incident light in parallel nicols). Internal structure and fracture network of rock micromonoliths were studied using X-ray high-resolution computer microtomography on SkyScan 1172. The carbon and nitrogen contents in fine earth were determined by the dry combustion method using a Vario ELIII analyzer; the radiocarbon age of the organic matter—by accelerator mass spectrometry (AMS) using a 1.5SDH-1 Pelletron AMS device (we included data obtained by this method from our previous publication—Mergelov et al. 2012).

## 3 Results

***Macrostructure of endolithic system*** in its native state and its schematic representation are shown in Fig. 1. The removal of an exfoliation plate (0.5–1.0 cm) covered with rock varnish and/or silica glaze reveals a community of endolithic organisms including green algae and cyanobacteria as primary components, their mort biomass, and fine earth mainly of coarse silty and fine sandy fractions. Fine earth material partly penetrates into the deeper rock zone along vertical fissures. We argue for mostly in situ origin of the fine earth since the attachment of exfoliation plates was very strong and the transverse size of superficial fractures was less than the size of observed fine earth fractions. Mineral grains are covered by biofilms, which are thick enough to be visible even at macro level. Endolithic biota and fine earth material compose together a specific layer, which we called endolithic organo-mineral horizon. Its thickness varies between 0.1 and 1.0 cm.

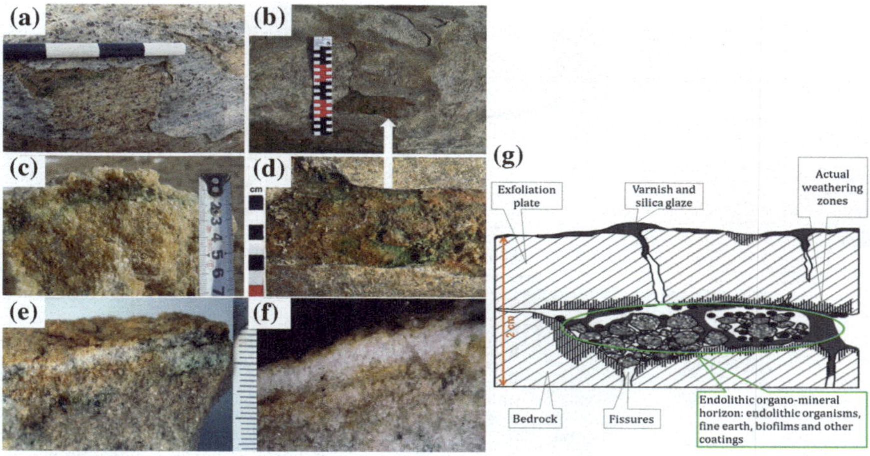


**Fig. 1** Macromorphology of endolithic system on granitoids: **a** exfoliation patterns on granites in Larsemann Hills; **b** exfoliation patterns on orthogneiss in Thala Hills; **c** vertical section with distinct organo-mineral horizon but without bleached zone on gneiss in Larsemann Hills; **d** horizontal section through endolithic organo-mineral horizon on orthogneiss in Thala Hills; **e**, **f** eluvial–illuvial differentiation and distinct bleached horizon on granites with high quartz content in Thala Hills; **g** schematic structure of endolithic system

***Mesomorphological features of exfoliation plates and endolithic organo-mineral horizon***. Mesomorphological study suggests the association of Fe-loci to microdepressions at the surface of exfoliation plate. Fe-film does not have a continuous distribution and only partially cover mineral grains (10–60 % of the surface) (Fig. 2a). On the contrary, the lower part of the exfoliation plate is depleted in iron compounds (Fig. 2b): Fe-locus are absent, some mineral grains are covered by cyanobacterial biofilm and fossilized organo-mineral coatings. Endolithic organo-mineral layer reveals several features usually attributed to organogenic horizons of a common soil: (1) it is a hot spot of biota-to-rock interactions with most rapid mineral transformations (Fig. 2c); (2) it is a locus of fine earth formation mainly fine sand and coarse silt; (3) it has blocky subangular and up to granular structure, where major aggregation agents are cyanobacterial and green algae biofilms rich in EPS (Fig. 2d).

***Micromorphology of endolithic system*** is exemplary in terms of visualization how cyanobacteria colonize mineral surfaces at more intimate level and even produce pseudomorphic features (Fig. 3a). Sites of intensive weathering and accumulation of Fe-(hydr)oxides are closely associated with cyanobacteria loci (Fig. 3b, d). These degraded zones are confined to the upper limit of bleached eluvial layer in profile of endolithic system. In a horizontal cross-section, an extensive network of fractures was revealed (Fig. 3c) indicating possible migration of Fe-(hydr)oxides along fissures between quartz and feldspar grains taking source in biotite degradation loci.

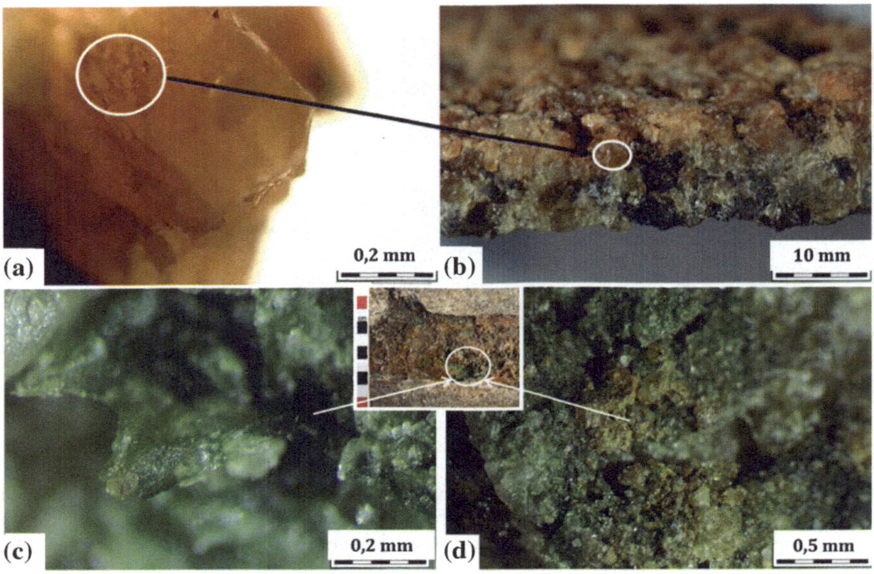


**Fig. 2** Mesomorphology of endolithic system. **a** Fe film on mineral grains in oxidized part of exfoliation plate; **b** two zones in exfoliation plate: oxidized top with varnish and silica gaze and depleted in iron bottom with minerals coated by cyanobacterial biofilms and fossilized organo-mineral films; **c** cyanobacterial biofilm on highly degraded quartz grain; **d** sandy/silty fine earth aggregated by cyanobacterial biofilm

***Submicromorphology***. Figure 4a show biofilms with cyanobacterial cells embedded in extracellular matrix, which covers most mineral surfaces in the photic part of endolithic organo-mineral horizon. The loci of melanin and hyaline pigmented fungal hyphae (Fig. 4b) are attributed to the top of the bleached eluvial zone with strong weathering patterns, most pronounced along the whole system (Fig. 4c). Hyaline hyphae also spread deeper and penetrate into green cyanobacterial loci. Some hyphae are closely associated with single algae or cyanobacterial cells suggesting that observed part of community is an endolithic protolichen. Particular coatings are fossilized to a great extent (Fig. 4d in blue color).

The most common case is when relatively fresh biofilms (green color on Fig. 4e) alternate in space with mineralized/fossilized coatings (blue color). The extracellular matrix is also subjected to mineralization (Fig. 4f, g). Silicon predominates in the composition of these bodies. In general, organo-mineral films are mainly composed of C, O, Si, Al, Fe, K, Ca, Na, and Mg. Their elemental set reflects the chemical composition of biofilms and primary minerals interacting with each other. From our point of view, the newly formed organo-mineral films represent in situ microproducts of pedogenesis. Figure 4h shows how cells and clay-sized particles penetrate through fractures between mineral grains. Elemental map of the thin section (Fig. 4i) vividly demonstrates that quartz (shown in red color) and feldspar

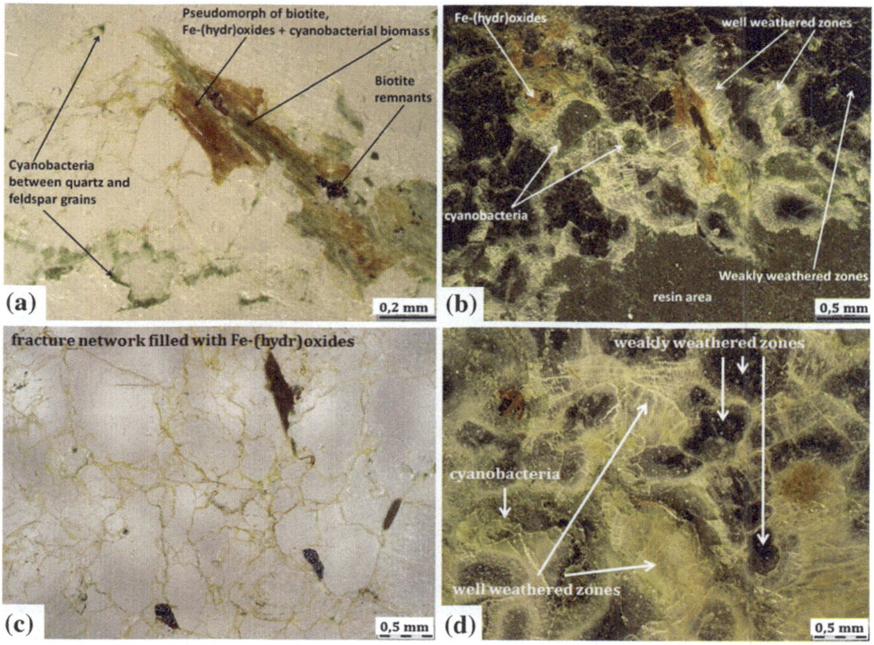


**Fig. 3** Micromorphology of endolithic system with eluvial–illuvial differentiation and bleached horizon in Thala Hills. *Left part* (reflected light)—thin sections are illuminated by incident light from above, *bright white background* is used. *Light areas* indicate grains of quartz and feldspar. *Rusty-brown zones* are non-silicate iron compounds and *green-olive patterns*—areas of cyanobacteria colonization. *Right part* (reflected light)—thin sections are illuminated by incident light from above, black background is used—*dark areas* indicate that light passes well through the mineral grains and is absorbed by the black background of the substrate. *Bright areas* indicate that the light is well reflected due to strongly weathered surfaces, developed fracture network, and high specific surface area. Such bright sites correspond to the bleached eluvial zones. *Rusty-brown patterns* indicate non-silicate iron compounds. *Green-olive patterns*—areas of cyanobacteria colonization. Further explanations are given in the text

(brown) particles are ubiquitously interlayered with biofilms and organo-mineral coatings (blue).

Thus, one of the major products of endoliths-rock interaction are the numerous coatings and films covering the inner rock surfaces under exfoliation plate. In all the films, the major role is played by Si and Al compounds and by C. As follows from the morphological analysis of the films, Si and Al compounds are present in them in the predominantly amorphous forms, including silica.

***Some chemical and isotopic properties of endolithic horizon***. Reaction of the medium in bulk samples is close to neutral (Table 1). The carbon content varies within 0.2–3.3 %, and nitrogen—0.02–0.47 %. These values are usually higher on stable horizontal surfaces than on the vertical ones, where exfoliation is enhanced by the gravitational force. The same is true for warmer slopes of the northern aspect

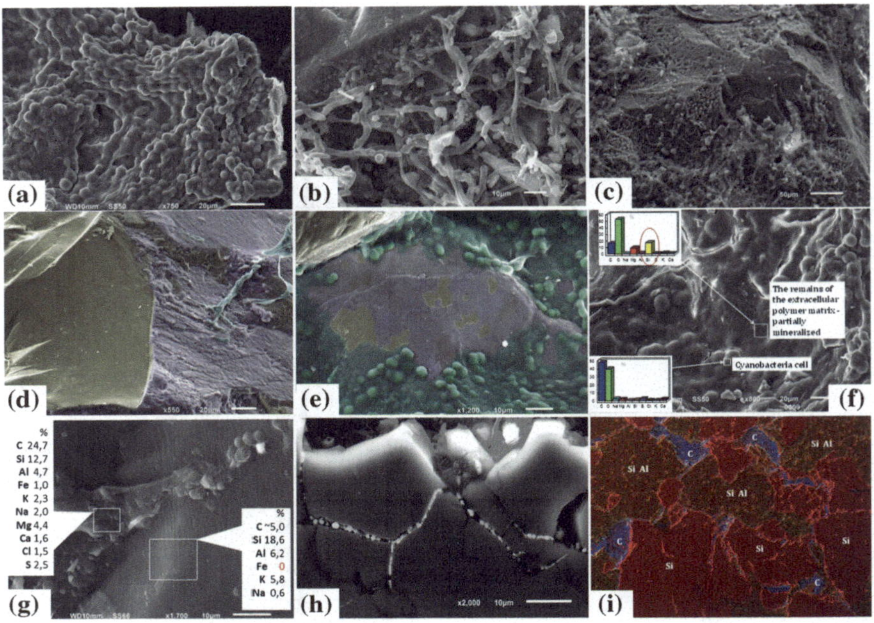


**Fig. 4** Endolithic biofilms and fossilized organo-mineral films inside the rocks—products of endolithic pedogenesis. Explanations are given in the text

**Table 1** Some properties of endolithic organo-mineral horizon (Larsemann Hills oasis)

| No. | Sample | pH | C (%) | N (%) | C/N | δ13C (‰) |
|---|---|---|---|---|---|---|
| 1 | 10-45 B | 6.4 | 3.33 | 0.47 | 7.1 | −23.7 |
| 2 | 10-47 2B1 | 6.8 | 1.54 | 0.21 | 7.3 | −21.0 |
| 3 | 10-45 B1 | 6.5 | 0.21 | 0.02 | 10.5 | – |
| 4 | 10-47 B2 | 6.7 | 1.71 | 0.09 | 19.0 | – |
| 5 | 10-54 B1 | 7.8 | 1.04 | 0.12 | 8.7 | −22.0 |
| 6 | 10-58 B1 | 6.8 | 2.76 | 0.25 | 11.0 | – |
| 7 | 10-64 B1 | 7.8 | 3.01 | 0.39 | 7.7 | −23.4 |
| 8 | 10-64 B2 | 7.6 | 0.86 | 0.08 | 10.8 | −22.0 |
| 9 | 10-61 B1 | 6.3 | 2.48 | 0.27 | 9.2 | −19.6 |
| 10 | 10-65 B2 | 6.8 | 1.69 | 0.14 | 12.1 | – |
|  | μ | 6.9 | 1.86 | 0.20 | 10.3 | – |

in comparison to colder ones of the southern aspect. Under stable and thermally favorable conditions the C/N ratio decreases to 7.1, which probably indicates more “mature” state of endolithic systems in such environment. Variability of measured parameters in bulk samples of organo-mineral horizon always depends on the ratio between organic and mineral components. We are dealing with very small quantities of a substrate and very thin organic coatings, so that just several silt-size mineral grains without biofilms could significantly affect the total estimate. More than that the integral chemical parameters measured in bulk samples by standard methods may be misleading for describing interactions of the biofilms with minerals which occur mostly at the microlevel. For example, neutral reaction observed in endolithic horizon could not explain morphology of quartz as shown earlier in Fig. 2c, as well as various forms of secondary organo-mineral coatings produced via transformation of minerals by biofilms (Fig. 4). The presence of biofilms on the mineral surface supports a specific geochemical microenvironment that may enhance or retard weathering processes. The capacity of biofilms to alter the mineral surface is enormous. Some works attest to the possibility of the local decrease of pH values to 3.0–4.0 directly in the endolithic cyanobacterial biofilms covering mineral grains (De los Rios et al. 2003). It was also shown that cyanobacterial biofilms are capable to initiate quartz dissolution via shifting pH in small loci from 3.4 to 9.0 during photosynthesis (Brehm et al. 2005).

The radiocarbon "age" of organic matter was determined in two samples from: (a) vertical surface of rock cliff of the warm northern aspect and (b) horizontal surface of the same rock (Table 2). Though the volumes of the analyzed biomass and fine earth were approximately the same, the first sample had very young radiocarbon "age" (less than 60–80 years), whereas, in the second one the mean residence time (MRT) of organic matter reached 480 ± 25 years.

This result is quite reasonable considering that the lifetime of endolithic system at the north facing vertical cliff is reduced due to more intensive exfoliation driven by gravitation. Besides that, favorable thermal conditions promote here biochemical weathering and increase rate of organic matter turnover. In case of stable horizontal surface, the MRT of organic matter suggests endolithic organo-mineral horizon is not an ephemeral body. And even more, some part of organic matter in this endolithic system is much older than obtained MRT value, however, it is mixed with younger components (e.g., biomass of the living organisms) that rejuvenate the average 14C "age." Therefore, we may conclude that the absolute age of endolithic systems under stable conditions, i.e., the time that has passed since endolithic organisms inhabited the rock, is more than 500 years. However, this assumption needs additional verification. The literature data indicate that some endolithic systems formed on stable surfaces (without active exfoliation) in Antarctic oases may be very old; their age is estimated as several thousand years (Johnston and

**Table 2** 14C "age" of organic matter in endolithic organo-mineral horizon (Mergelov et al. 2012)

| Material | Sample | % of modern C | 14C "age"/ MRT, years |
|---|---|---|---|
| Organic matter from endolithic horizon | 10-45 B<br>warm northern aspect, vertical surface | 102.86 ± 0.28 | Modern |
| | 10-47 2B1<br>horizontal surface | 94.19 ± 0.26 | 480 ± 25 |

Vestal 1991; Sun and Friedmann 1999). In this case, we operate almost at geological timescale, and this allowed Johnson and Vestal to suppose that endolithic microecosystems might be the slowest growing communities on Earth. However, taking into account the observed intensity of exfoliation and the "ages" estimated by us, including the date attesting to the modern radiocarbon age of organic matter, such old endolithic systems may represent an exception rather than the rule. In Antarctica extreme environment, they may only be developed in relatively rare "shelters." The values of δ13C obtained for the samples being dated (Table 1) attest to a somewhat heavier isotopic composition of the organic matter in comparison with the values typical of C3 plants (usually, −24 to −30 ‰), as well as to the values obtained for endolithic material from the Dry Valleys of Antarctica (Hopkins et al. 2009). The heavier isotopic composition of the samples studied by us may be specified by the significant contribution of cyanobacteria and green algae to the organic matter at the expense of heterotrophic organisms (micromycetes and bacteria). For a more reliable interpretation of the "isotopic memory" of the system, a larger number of measurements are required. It is also necessary to obtain data on the isotopic composition of nitrogen.

***Potential carbon stocks and pools in endolithic systems***. We have obtained the preliminary estimate of average endolithically generated organic C stock in granites of Larsemann Hills which is 0.037 ± 0.019 g C/cm$^2$ (3,7 Mg C/ha). We also calculated the potential total carbon pool of endolithic systems at a key site in the wet valley of Larsemann Hills (S69° 20′, E76° 20′, 27 ha area), which is 36 Mg C (Fig. 5). It is comparable with the carbon pool of the same site created by bryophyte communities in loose soils (40 Mg C). More than that, endolithic carbon pool occurred to be significantly higher than in soils on lacustrine sediments (9 Mg C) and ornithogenic soils (2 Mg C), and in comparison to epilithic (<1 Mg C) and hypolithic (8 Mg C) pools.

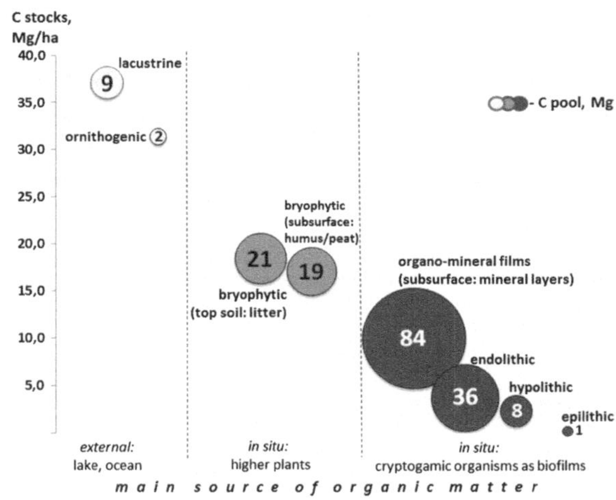


**Fig. 5** Carbon stocks and pools at the key site in Larsemann Hills (S69° 20′, E76° 20′, wet valley, 27 ha area)

***Specific* in situ *products of weathering and pedogenesis*** that contain carbon in organic or inorganic form are frequently encountered in the interior of endolithic system. Among them are carbonates which accumulate on an external surface of cyanobacterial biofilms (Fig. 6a); crystals of carbonates could also be found embedded in the cyanobacterial EPS (Fig. 6b). Recent studies widely confirm possibility of carbonates precipitation by cyanobacteria. Two mechanisms are involved: $CaCO_3$ crystals precipitation on the cell surface layer and extracellular sheath calcification in situ promoted by pH rise due to $CO_2$ depletion and $OH^-$ increase during photosynthesis (Dupraz et al. 2009; Fundamentals of Geobiology 2012). Common features of a bleached eluvial zone and its vicinities are the oxalates crystals produced when exudates of endolithic fungi including oxalic acid and its derivates react with silicates and/or nonsilicate Fe compounds (Fig. 6c, d). The main sources of cations stabilizing oxalate and carbonate ions could be the feldspars. New born oxalates and carbonates are the products of pure in situ bio-mineral interactions, thus may also be considered as a proper pedogenic attribute.

***Transportation network of endolithic system***. Microtomography study helped to reveal the fractures, weathering foci and areas colonized by biofilms inside sampled monoliths, which were invisible from the surface. It is important that diagnostics

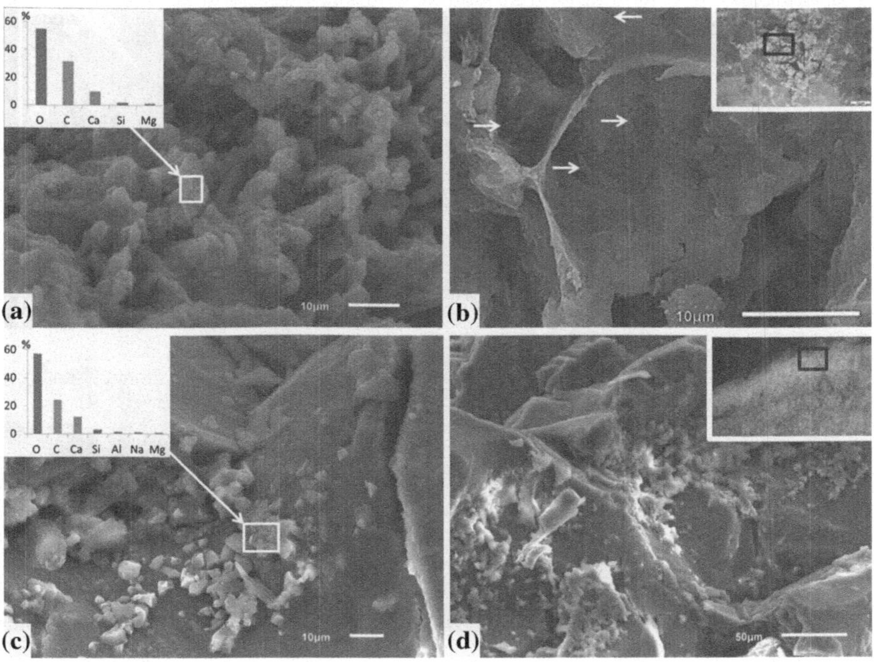


**Fig. 6** Specific in situ products of endolithic weathering and pedogenesis. **a** Carbonates on an exterior of cyanobacterial biofilm; **b** carbonates formed within cyanobacterial EPS (pointed by *arrows*); **c**, **d** oxalates in bleached eluvial horizon

was held by noninvasive way without exfoliation plate being detached or destroyed. Target areas for further qualitative and quantitative study on SEM were identified.

The fracture network (Fig. 7) can clearly be divided into (a) perforating slightly branched subhorizontal fractures (20–200 μm) and (b) filamentous highly branched subvertical fractures (2–20 μm). Exfoliation of 10 mm thick plate occurs along major subhorizontal fissures which are populated by endolithic organisms to a maximum extent in comparison to the rest of the rock. The weathering zones of 0.1–1.0 mm size are formed here and can be clearly distinguished on tomography image due to the well-established grid of microfractures and numerous roundish pores creating a "perforated" microcellular pattern (Fig. 7a). The most advanced stage would have endolithic organomineral horizon formed along subhorizontal fractures which consists of individual mineral grains, dead and living biomass of endoliths, and sometimes even aggregates (according to mesomorphology study—Fig. 2d).

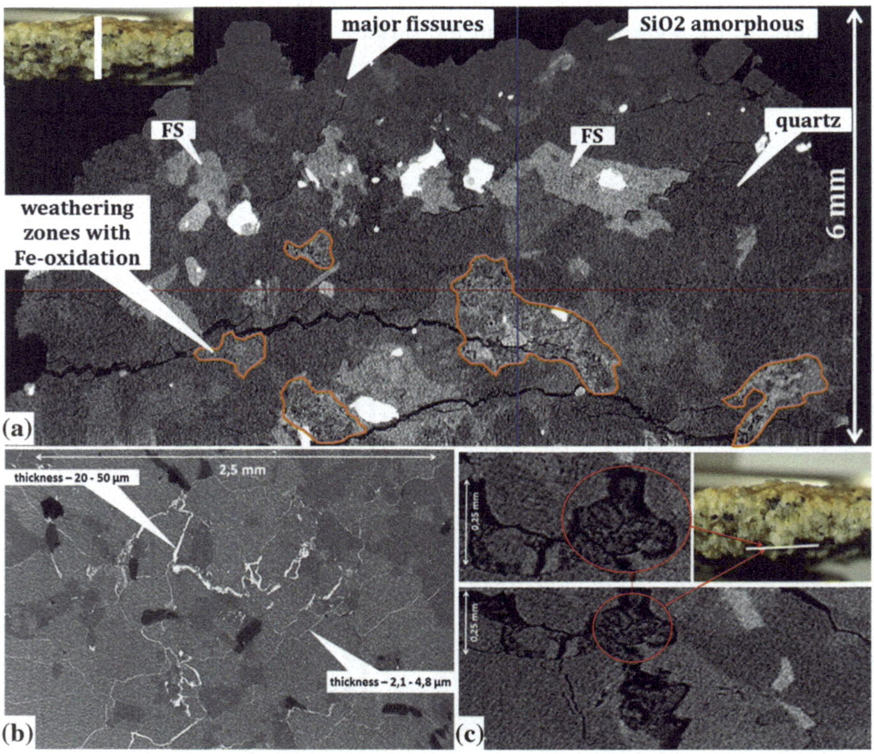


**Fig. 7** X-ray microtomography (max resolution 1 μm) of endolithic system. **a** Vertical profile of exfoliation plate; **b** horizontal cross section—fissures network raising to the day surface of rock (image is in inverted color and fractures are shown in *white*); **c** pocket-like microzones with initial stage of fine earth formation on the lower surface of exfoliation plate

The most significant finding was the thin subvertical fractures in exfoliation plate rising to its current day surface (Fig. 7a, b) and connected to each other with thin subhorizontal ones. The fractures size may decrease to 1–2 μm, i.e., right to the limit resolution of the tomograph. The broken fractures that penetrate between the individual grains of minerals are widespread. 5–20 % of fractures have smoother pattern, they dissect weathered grains, mainly of feldspars. Weathering zones with microcellular patterns are formed in points of intersection between subvertical and subhorizontal fractures right in the body of exfoliation plate (Fig. 7a). According to X-ray microanalyzer (SEM), such areas have a higher content of Fe and C. Morphologically, either true biofilm or their derivatives—organomineral coatings containing, as impurities, Si, Al, C, K, Ca, Mg (thickness—1 or more microns) are present here. Initial stages of in situ fine earth formation were detected in pocket-like microzones (Fig. 7c).

The network of fine fractures connects zones depleted in Fe at the bottom of exfoliation plate with the oxidized Fe-enriched loci at its surface. Perhaps such a network acts as a transport system when an upward migration of iron in dissolved form occurs toward the surface of the plate right to the oxidative geochemical barrier. Mobilization/deposition of Fe can occur during rare wetting events and subsequent long desiccation of the granitoids surface in Antarctica. Thermogradient forces can also be involved.

***Soil-like eluvial–illuvial differentiation on granitoids***. Even the first works by Friedmann and Ocampo (1976), Friedmann (1982) examining functioning of endolithic communities in Beacon sandstones in Dry Valleys of Antarctica indicated bright deprived of Fe-(hydr)oxides layers less than several millimeters thick. It was assumed that they are linked with the vital activity of fungi, cyanobacteria and lichens that live in the interior of sandstone. However, exact mechanisms of this phenomenon have not been investigated in details and no analogies with the formation of bleached eluvial horizons of classical soils have been drawn.

In 2000s, series of studies primarily relating to techniques adjustment for exploring organo-mineral substrates by the means of Raman spectroscopy used Beacon sandstone as a research material. The approach was more about methodology and did not address the genesis of material, its environmental or soil interpretations. A change in composition of non-silicate minerals including transformation of hematite to goethite received particular attention in these studies. Mechanism of Fe mobilization from hematite, its migration and subsequent immobilization as goethite was proposed. This process formed microzones depleted and enriched in Fe. Besides that, oxalates were detected as specific products of endolithic lichen interaction with the minerals of sandstone (Villar et al. 2005; Russell et al. 1998; Edwards et al. 2004, 2005). Some bleached eluvial layers described in these studies were very well developed and their thickness reached 1 cm (Edwards et al. 2005).

We propose that the observed eluvial–illuvial soil-like differentiation inside granitoids of Thala Hills (Fig. 8a) is driven by endolithic biogenic pump, possibly endolithic protolichen (as described in Wierzchos and Ascaso 2001) and to a less extent by classical gravitational mechanism. Establishment of perpendicular

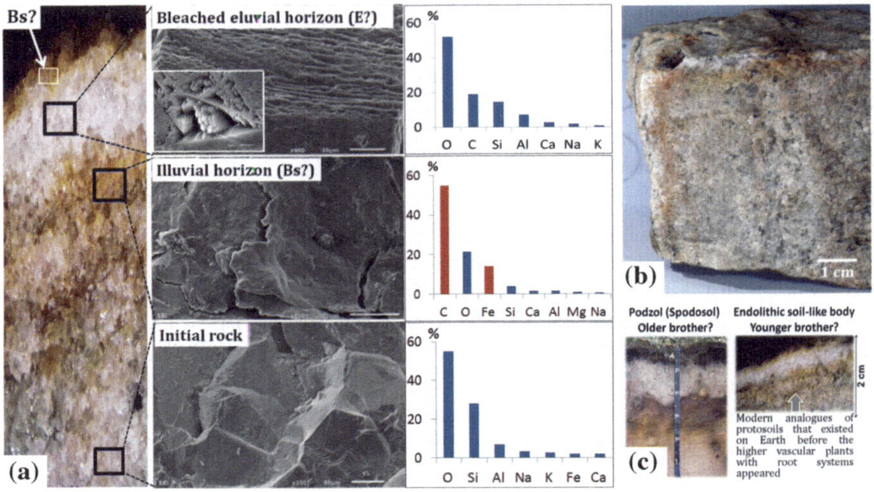


**Fig. 8** Soil-like eluvial–illuvial differentiation on granitoids resembles "classical' podzol. **a** Examples of morphology and elemental composition of bleached eluvial horizon, illuvial horizon and initial rock (*left* photo is out of scale and stretched vertically for display convenience); **b** monolith with endolithic system indicating that eluvial and illuvial horizons are established along perpendicular axes; **c** podzol and endolithic soil-like body with distinct eluvial–illuvial differentiation—older and younger "brothers"?

bleached horizons as well as oxidized ones in natural conditions is the best illustration of primarily nongravitational origin of this phenomenon (Fig. 8b).

Prior to this study, bleached layers of endolithic origin were described only for the sandstones (in Antarctica, Canada and South Africa), but never on granitoids. Could it be that common Podzol and endolithic soil-like body with distinct eluvial–illuvial differentiation—are the older and younger "brothers" (Fig. 8c)? It is disputable.

## 4 Discussion

Studies in East Antarctica oases propose that endolithic system has major features attributed to soils: (a) rock layer exposed to external abiogenic factors, (b) lithomatrix inhabited by living organisms which are synthesizing and decomposing organic matter, and (c) as a result initial parent rock (lithomatrix) is transformed in situ by biogenic and abiogenic factors, the products of transformation are retained and/or removed, the vertical heterogeneity is established in a form of microhorizons composing microprofile. Crucial feature also related to pedogenesis is the presence of endolithic organo-mineral horizon—the hotspot of biota-to-rock interactions, which is similar in its basic functions to organogenic and organo-mineral horizon of a common soil. Major products of endolithic rock transformation are the silty-sandy fine earth and abundant organo-mineral films that are formed within the porous space of endolithic system. Evidently, such films are the result of interaction between biofilms and mineral surfaces and reflect elemental composition of both components, mainly comprising C, O, Si, Al, Fe, K, Ca, Na, and Mg. Biofilms as complex formations create continuous organic layers, whose properties radically differ from the properties of the mineral matrix owing to the presence of polymeric substance. Biofilms on the mineral surface support a specific geochemical microenvironment that enhances substrate transformation. From our point of view, the newly formed organo-mineral films can be considered as in situ microproducts of pedogenesis.

Microtomography data shows that different layers of endolithic system are connected with the fracture network. Subtle branched subvertical (2–20 μm) and larger perforating subhorizontal fractures serve as a uniform network in the subsurface part of studied granitoids in East Antarctica. Thus, subaerial segment of explored rocks has mineral matrix, which is not sealed and is potentially permeable for dissolved products of endolithic weathering and pedogenesis. The fractures network is the possible transport system for the elements transfer in the interior of the upper 1–2 cm of granitoid that makes eluvial–illuvial differentiation possible. Differentiation occurs as a result of an upward and downward migration of solutions due to biogenic transfer of elements between components of endolithic systems (e.g., transfer between photobiont and mycobiont in endolithic lichen, siderophores carrier functions, etc.), hydro and thermal gradients, and to a less under the influence of downward gravity flows as common to "classic" Podzols. The real mechanisms should be confirmed by additional studies, including computer microtomography, SEM, Mössbauer spectroscopy, and chemical analysis of Fe compounds. But, in any case, the presence of an integrated transport system in hard rock, connecting the individual parts of endolithic system (e.g., organo-mineral horizons, exfoliation plates, bedrock, etc.) is a fact and it is crucial for relating such bodies to soils and soil-like bodies.

The more developed endolithic system becomes, the greater are the chances for its destruction and lower probability that initial endolithic soil-like body will turn into a common full-scale soil. Macrohorizons are never formed due to periodic rejuvenation of the rock surface by exfoliation. Besides that, the fine earth is removed by wind and allocated in accumulative positions where it enriches elemental and isotopic record of loose soils (Hopkins et al. 2009). Such kind of initial pedogenesis was called previously as "self-destructive" (Mergelov et al. 2012). Bleached eluvial microhorizons are the borderlines and weakest links that readily promote exfoliation and restrict full-scale pedogenesis in situ. This mode is common to other soils of extreme environment whose thin primitive profiles are often found in the quasi-equilibrium state with external conditions for hundreds and thousands of years. They are not transformed into the mature full-scale soil bodies because of the very low biogeochemical transformation rate and the high intensity of catastrophic processes and disturbances (exfoliation in our case). Such bodies can be referred as "infantile" soils: though their absolute age may be considerable

($n \times 10^2$–$10^4$ years), their morphology corresponds to the very initial stages of pedogenesis that are relatively quickly (in about $n \times 10^1$ years) replaced in more favorable environment by the more advanced stages. Such soils can be considered as modern analogues of protosoil bodies that existed on our planet before higher vascular plants with root systems appeared and contributed as major sources of humus (Kudinova et al. 2015).

## 5 Conclusion

The study of endolithic weathering front with the approaches of soil science showed that microprofiles established in granitoids morphologically and functionally are very similar to a common soil. Different horizons of this body are connected with the fracture network, which serves as a transport system for elements transfer. It leads to a unique result—the soil-like pattern is established inside the massive-crystalline rock. The profiles being examined have clear eluvial–illuvial differentiation patterns similar to macroprofile of a common Podzol (Spodsol) on loose substrates. It is unusual that this phenomenon is found in formally arid conditions and in the absence of higher plants. Due to the nature of the substrate (massive crystalline rock), position in the landscape (rock outcrops), and bioclimatic extremes (endolithic ecological niche), the thickness of profiles and horizons are one to three orders of magnitude less than in common Podzols. We propose quite a disputable statement that common Podzol and endolithic soil-like body with distinct eluvial–illuvial differentiation—are the older and younger "brothers." Besides that, endolithic soil-like bodies could be the modern closest analogues of protosoils that existed on Earth before the higher vascular plants with root systems established. More systematization and understanding of the place these objects occupy in the soils world is needed since endolithic weathering patterns are among the most spatially abundant soil-like bodies in extreme environment of Antarctica; besides that, they secure primarily-produced organic matter pool in these barren rocky landscapes.

**Acknowledgments** This work has been supported by the Russian Science Foundation; project no. 14-27-00133. Authors are grateful to Prof. Sofia Lessovaia and Prof. Dmitry Vlasov from Saint Petersburg State University for their assistance in micromorphology studies and inspiration.

## References

Brehm U, Gorbushina A, Mottershead D (2005) The role of microorganisms and biofilms in the breakdown and dissolution of quartz and glass. Palaeogeogr Palaeoclimatol Palaeoecol 219 (1):117–129

De Los Ríos A, Wierzchos J, Sancho LG, Ascaso C (2003) Acid microenvironments in microbial biofilms of Antarctic endolithic microecosystems. Environ Microbiol 5(4):231–237

De Los Ríos A, Grube M, Sancho LG, Ascaso C (2007) Ultrastructural and genetic characteristics of endolithic cyanobacterial biofilms colonizing Antarctic granite rocks. FEMS Microbiol Ecol 59(2):386–395

De Los Ríos A, Wierzchos J, Ascaso C (2014) The lithic microbial ecosystems of Antarctica's McMurdo Dry Valleys. Antarct Sci 26(05):459–477

Dmitriev EA (1996) Soils and soil-like bodies. Eurasian Soil Sci 29(3):275–282

Dupraz C, Reid RP, Braissant O, Decho AW, Norman RS, Visscher PT (2009) Processes of carbonate precipitation in modern microbial mats. Earth Sci Rev 96(3):141–162

Edwards HG, Wynn-Williams DD, Villar SEJ (2004) Biological modification of haematite in Antarctic cryptoendolithic communities. J Raman Spectrosc 35(6):470–474

Edwards HG, Moody CD, Villar SEJ, Wynn-Williams DD (2005) Raman spectroscopic detection of key biomarkers of cyanobacteria and lichen symbiosis in extreme Antarctic habitats: evaluation for Mars Lander missions. Icarus 174(2):560–571

Friedmann EI (1982) Endolithic microorganisms in the Antarctic cold desert. Science 215 (4536):1045–1053

Friedmann EI, Ocampo R (1976) Endolithic blue-green algae in the dry valleys: primary producers in the Antarctic desert ecosystem. Science 193(4259):1247–1249

Friedman I, Lipkin Y, Roseli OP (1967) Desert algae of the Negev. Phycologia 6:185–200

Gilichinsky D, Abakumov E, Abramov A, Fyodorov-Davydov D, Goryachkin S, Lupachev A, Mergelov N, Zazovskaya E (2010) Soils of mid and low Antarctic: diversity, geography, temperature regime. In: Proceedings of the 19th world congress of soil science. Published on DVD. www.iuss.org. Symposium WG 1.4. Cold soils in a changing world. Brisbane, Australia, pp 32–35

Glazovskaya MA (1958) Weathering and initial soil formation in Antarctica. Nauch Dokl Vyssh Shkoly Geol Geogr-Nauki 1:63–76 (in Russian)

Golubic S, Schneider J (2003) Microbial endoliths as internal biofilms. In: Fossil and recent biofilms. Springer, Netherlands

Golubic S, Friedmann I, Schneider J (1981) The lithobiontic ecological niche, with special reference to microorganisms. J Sediment Res 51(2):475–478

Hopkins DW, Sparrow AD, Gregorich EG, Elberling B, Novis P, Fraser F, … Greenfield LG (2009) Isotopic evidence for the provenance and turnover of organic carbon by soil microorganisms in the Antarctic dry valleys. Environ Microbiol 11(3):597–608

Horath T, Bachofen R (2009) Molecular characterization of an endolithic microbial community in dolomite rock in the Central Alps (Switzerland). Microb Ecol 58(2):290–306

Ivlev AM, Nesterova OV (2004) On the study of aquasols. Vestn DVO RAN 4:47–52 (in Russian)

Johnston CG, Vestal JR (1991) Photosynthetic carbon incorporation and turnover in Antarctic cryptoendolithic microbial communities: are they the slowest-growing communities on earth? Appl Environ Microbiol 57(8):2308–2311

Knoll AH, Canfield DE, Konhauser KO (eds) (2012) Fundamentals of geobiology. Wiley, New York

Kögel-Knabner I, Guggenberger G, Kleber M, Kandeler E, Kalbitz K, Scheu S, … Leinweber P (2008) Organo-mineral associations in temperate soils: integrating biology, mineralogy, and organic matter chemistry. J Plant Nutr Soil Sci 171(1):61–82

Kudinova AG, Lysak LV, Soina VS, Mergelov NS, Dolgikh AV, Shorkunov IG (2015) Bacterial communities in the soils of cryptogamic barrens of East Antarctica (the Larsemann Hills and Thala Hills oases). Eurasian Soil Sci 48(3):276–287

Mergelov NS, Goryachkin SV, Shorkunov IG, Zazovskaya EP, Cherkinsky AE (2012) Endolithic pedogenesis and rock varnish on massive crystalline rocks in East Antarctica. Eurasian Soil Sci 45(10):901–917

Roslikova VA (2006) Modern notions on the subaqual pedogenesis. Tikhookean Geol 25(4):97–103

Russell NC, Edwards HGM, Wynn-Williams DD (1998) FT-Raman spectroscopic analysis of endolithic microbial communities from Beacon sandstone in Victoria Land, Antarctica. Antarct Sci 10(01):63–74

Schmidt MW, Torn MS, Abiven S, Dittmar T, Guggenberger G, Janssens IA, … Trumbore SE (2011) Persistence of soil organic matter as an ecosystem property. Nature 478(7367):49–56
Semikolennykh AA, Targulian VO (2010) Soil-like bodies of autochemolithotrophic ecosystems in the caves of the Kugitangtau Ridge, eastern Turkmenistan. Eurasian Soil Sci 43(6):614–627
Sigler WV, Bachofen R, Zeyer J (2003) Molecular characterization of endolithic cyanobacteria inhabiting exposed dolomite in central Switzerland. Environ Microbiol 5(7):618–627
Sokolov IA (1996) The paradigm of pedology from Dokuchaev to the present day. Eurasian Soil Sci 29(3):222–232
Sun HJ, Friedmann EI (1999) Growth on geological time scales in the Antarctic cryptoendolithic microbial community. Geomicrobiol J 16(2):193–202
Targulian VO, Goryachkin SV (2011) The 19th world congress of soil science. Eurasian Soil Sci 44(9):1041–1047
Villar SEJ, Edwards HG, Cockell CS (2005) Raman spectroscopy of endoliths from Antarctic cold desert environments. Analyst 130(2):156–162
Walker JJ, Pace NR (2007) Phylogenetic composition of Rocky Mountain endolithic microbial ecosystems. Appl Environ Microbiol 73(11):3497–3504
Wierzchos J, Ascaso C (2001) Life, decay and fossilisation of endolithic microorganisms from the Ross Desert, Antarctica. Polar Biol 24(11):863–868
Wierzchos J, Sancho LG, Ascaso C (2005) Biomineralization of endolithic microbes in rocks from the McMurdo Dry Valleys of Antarctica: implications for microbial fossil formation and their detection. Environ Microbiol 7(4):566–575

# Abiotic and Biotic Processes of Mineral Weathering in Tundra Soils on Ultramafic and Mafic Rocks of the Polar Urals, Russia

**Sofia N. Lessovaia, Sergey Goryachkin, Yury Polekhovsky, Viktoria Ershova and Alexey Filimonov**

**Abstract** The weathering of mafic and ultramafic rocks in soil environment was investigated in weakly developed soil profiles in order to determine the origin of phyllosilicate association in the soils formed in humid cold climate of the mountainous tundra of the Polar Urals. The objects of the study are represented by soils formed (i) on and underlain by the ultramafic rock and (ii) on the moraine composed of the mafic rock with an admixture of the ultramafic rock fragments. The minerals found in the clay fraction (<1 μm) of the profiles are the same, characterized by the presence of smectite (saponite), which is absent in both mafic and ultramafic rocks; serpentine and talc identified in ultramafic rock; and chlorite. Chlorite was found in both types of rocks. It was shown that the appearance of smectite (saponite) in the weakly developed soil is not related to pedogenesis. But these soil profiles illustrate the possibility of soil formation on "mature" fine earth formed from a high-sensitive ultramafic rock due to chemical weathering. In cold soil environment the more weatherable ultramafic material plays the more important role as a prerequisite for the weathering trends and soil formation than a mafic rock. The admixture of ultramafic materials mitigates the development of Entic Podzols which were earlier found in the Polar Urals on the pure mafic materials. So, the presence of ultramafic materials either predominating or even in admixture results in the "extreme lithological environment" for a pedogenesis and in the formation of weakly developed soils—Regosols and Leptosols.

**Keywords** Clay minerals · Extreme environment for pedogenesis · Metagabbro–amphibolite · Serpentinous dunite · Weakly developed soils

---

S.N. Lessovaia (✉) · Y. Polekhovsky · V. Ershova
Institute of Earth Sciences, St. Petersburg State University, St. Petersburg, Russia
e-mail: s.lesovaya@spbu.ru

S. Goryachkin
Russian Academy of Science, Institute of Geography, Moscow, Russia

S.N. Lessovaia · A. Filimonov
St. Petersburg State Polytechnical University, St. Petersburg, Russia

 
O.V. Frank-Kamenetskaya et al. (eds.), *Biogenic—Abiogenic Interactions in Natural and Anthropogenic Systems*, Lecture Notes in Earth System Sciences,
DOI 10.1007/978-3-319-24987-2_18

# 1 Introduction

The challenge of the twenty-first century for both clay mineralogy and pedology includes the processes of weathering in extreme environment on the Earth that brings the bridge for understanding the relevant processes on the Mars and other planets. It concerns the desert environment as well as the polar one. The development of abiotic physical processes, including the disintegration initiated by freezing–thawing cycles is the most acceptable scenario of rock weathering in the cold environments. Nevertheless, the progress of chemical weathering limited by moisture availability in cold conditions has been shown (Hall et al. 2002). The processes (physical and chemical) are coexisted in cold environments. That was confirmed by findings that disaggregation of a rock was stressed in the chemical weathering (Hoch et al. 1999) due to an increase of a surface area (Arnaud and Whiteside 1963; Allen 2002). Besides that, the microstructure properties of rocks influence the element release from them (Meunier et al. 2007). Intensive chemical weathering affected by pedogenesis, which was led to the smectite (saponite) appearance in the soils underlain by serpentinous dunite from the Polar Urals, has been also shown previously (Lessovaia and Polekhovsky 2009; Lesovaya et al. 2012). The saponite–trioctahedral, Mg reach smectite is rare described as a pedogenic mineral. For example, in serpentinite soils in northeast Scotland the saponite, the content of which decreased from the bottom towards the more acidic upper soil horizons was identified (Wilson and Berrow 1978). Based on these findings, the pedogenic origin of saponite in the soils underlain by serpentinous dunite from the Polar Urals was supposed (Lessovaia et al. 2012). The weakly developed soils—Haplic Regosol (Eutric) and Stagnic Leptosol (Eutric) according to the World Reference Base for Soil Resources (2006) were described here together with more developed soils such as Haplic Cryosols (Reductaquic). The mineral association of the fine size fractions of weakly developed soils characterized by the slightly pronounced pedogenic features was the same as in mature profiles. But the pedogenic origin of saponite supposing the intensive chemical weathering is difficult to explain in weakly developed soils. So, the aim of present research is to investigate the weathering trends of mafic and ultramafic rocks, the origin of phyllosilicate association and the role of abiotic and biotic pedogenic processes in weakly developed soils formed in humid cold climate of the mountainous tundra of the Polar Urals.

# 2 Objects

The study sites are located at the same altitude of ~300 m in the mountainous tundra from the Polar Urals on the southern slope of the Rai-Iz massif and outwith the Rai-Iz massif (Fig. 1). The Rai-Iz massif is mostly consists of the Early Paleozoic dunite–harzburgite ultramafic igneous rocks (harzburgite is peridotite with orthopyroxene) complex, whereas the mafic rocks are adjacent to the Rai-Iz

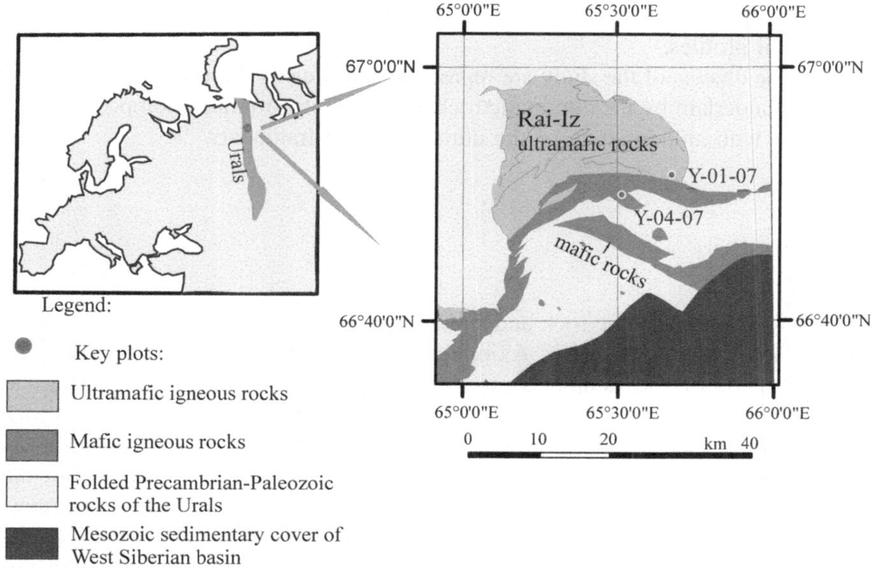


**Fig. 1** Location of the key plot in the Urals

massif (Major Ore-Bearing Geological–Geochemical Systems of the Urals 1990). The landscapes of study area are mainly controlled by glacial landforms formed during Middle Pleistocene glaciation with several distinct moraine lobes around the Rai-Iz massif (Svendsen et al. 2014).

The first study site on the southern slope angled at ~20° of the Rai-Iz massif covered by colluvial blocks of ultramafic rock determines the formation of well-drained soils in the accumulation of fine earth. The fine earth is a result of redistributed weathering products of ultramafic rock. The surface is covered by the sedge–grass–lichen vegetation community with rare shrubs of blueberry and ledum. Rare trees are represented by larches with the diameter of 8–10 cm, which are not higher than 5 m. Haplic Regosol (Eutric) (WRB 2006) (Pit Y-01-07) was described here.

The second key plot located outwith the Rai-Iz massif, to the north of Lake Yareity, on a moraine ridge. The stoniness of the area is ~60–80 % with pronounced cryogenic patterned ground. The moraine is greenish-gray and consisted mostly of mafic material with admixture of ultramafic one pointed to erosion and redeposition of Rai-Iz massive igneous rocks. The flat surface determines the development of stagnic features in the bottom of the studied soil profile classified as Stagnic Leptosol (Eutric) (Pit Y-04-07). The surface is covered (70–80 %) by moss–dwarf shrub tundra vegetation community with sedge and rare suppressed larch, dwarf Arctic birch tree (*Betula nana*), creeping form of willow, blueberry, and ledum. Permafrost was not detected at the pits of the studied area even at the end of June—beginning of July of the year with normal level of precipitation, when the field research was performed. It is because of the close lithic contact and shallow soil profiles.

Thus, the objects of the study are represented by weakly developed soils formed (i) on and underlain by the ultramafic rock and (ii) on the moraine composed of the mafic rock with an admixture of the ultramafic rock fragments.

## 3 Methods

Mineral associations of the rock and soil horizons were studied in thin sections by optical microscopy using Zeiss Axioplan 2 and Polam P-312 microscopes. The <1 mm fraction of soil samples was obtained by gently grinding in a mortar and subsequent dry sieving. Bulk chemical composition of the <1 mm fraction and rock was determined by X-ray fluorescence analysis (Tefa-611, EG&G Instruments Ortec). The soluble Fe forms in the fractions <1 mm were obtained from the dithionite citrate bicarbonate method of Mehra and Jackson (1958) and extractions with oxalate (Jackson et al. 1986), and pyrophosphate (Bascomb 1968). In the oxalate extract Al and Si were also analyzed. pH values were measured potentiometrically in $H_2O$ with a soil: water ratio of 1:2.5. C-content of the <1 mm fraction was determined by wet combustion with a mixture of 0.14 M $K_2Cr_2O_7$ and concentrated $H_2SO_4$ 1:1 at 150 °C for 20 min and titration with ferrous sulfate solution according to Tyurin (1931). In the A and Ah horizons rich in organic matter, loss on ignition was determined.

The content of the <1 μm fraction in the fine soil fraction (<1 mm) was determined by sedimentation with the pipette method. Ammonium hydrate was used as a peptizing agent. The XRD patterns were obtained from oriented specimens using DRON-2 X-ray diffractometer, with CoK$\alpha$ radiation and a monochromator in the diffracted beam. Pretreatment of samples included saturation with Mg, ethylene glycol solvation, and heating 550 °C. Additionally, FTIR spectra in the 400–4000 $cm^{-1}$ range of the <1 μm fraction were obtained using a Tensor 27 spectrometer (Bruker) with a resolution of 4 $cm^{-1}$ in ambient air and at room temperature. Spectra were recorded in the transmission mode using the KBr pellet method. Here 1 mg of sample and 300 mg of KBr were thoroughly mixed and pressed to a transparent pellet.

## 4 Results

*Rocks characteristic.* The rock fragments from soil horizons including bottom one of Haplic Regosol (Eutric), Pit Y-01-07 from the first key plot were identified as serpentinous dunite with predominance of olivine in the mineral association (Fig. 2). Some alteration of initial rock affected by metamorphic processes resulted in appearance of serpentine, talc, and chlorite as well as traces of amphibole and mica.

In the microcracks in olivine, serpentine, and chlorite the accumulation of Fe oxides due to rock weathering leads to a reddening.

The rare boulders of ultramafic rock from the moraine ridge of the second key plot located outside the Rai-Iz massif were also identified as serpentinous dunite. Based on the data of bulk chemical composition the content of $SiO_2$ as well as $Al_2O_3$, and $K_2O$ in the rock fragments from the bottom R horizon of Pit Y-01-07 is higher than it should be in ultramafic rock (Table 1). Supposedly, that could be explained by decrease of proportion of Mg-rich minerals first of all olivine and residual accumulation of possible source of $Al_2O_3$ and $K_2O$.

The rock fragments from the soil profile of Stagnic Leptosol (Eutric), Pit Y-04-07 from the second key study is enriched by total iron and characterized by mafic composition based on the percentage of $SiO_2$ (Table 1). The rock samples from the profile as well as mafic samples from the moraine ridge are identified as metagabbro–amphibolite due to mineral association (Fig. 2). Hornblende predominated in the rock is a result of replacement of pyroxenes all over the rock during high temperature regional metamorphism, which is also caused to recrystallization of some melanocratic minerals in initial rock and appearance of high proportion of quartz. The initial rock was most probably represented by the magmatic type of mafic composition. Later amphibole was partially substituted by epidote and chlorite. Grains of plagioclase are characterized by the development of the saussuritization and sericitization.

In the thin sections from soil horizons of this profile (Pit Y-04-07) the fragments of mafic and ultramafic rocks, which size is up to 10–15 mm illustrate the influence of pyroxenite, serpentinite, metagabbro–amphibolite and epidosite, the latter is composed by epidote (Fig. 2). In addition, individual minerals: hornblende, plagioclase and quartz, which size is up to 1.5 mm were also identified. Several rock samples as well as plagioclase and especially pyroxene are characterized by the development of Fe oxides along the cleavages reflecting iron removal from silicates caused to cement soil substrate. Thus, these findings confirm that soil profile of Stagnic Leptosol (Eutric), Pit Y-04-07 is enriched by mafic as well as ultramafic material. As opposed to ultramafic rocks, fragments of mafic one are less weathered based on reddening rims.

*Soils' characteristics.* The upper horizons (A and Ah) of studied profiles are enriched by total carbon mostly due to the decomposed soil organic matter, not linked with mineral matter that is confirmed by high value of loss on ignition. However, the depth of A horizons is very shallow—2 cm only, in Haplic Regosol (Eutric) the gradient of C content decrease is much higher than in Stagnic Leptosol (Eutric). Haplic Regosol (Eutric), Pit Y-01-07 is alkaline with the exception of the uppermost (A) horizon, whereas Stagnic Leptosol (Eutric), Pit Y-04-07 is neutral in the upper part and slightly alkaline in the bottom horizons despite the absence of calcareous material (Table 1).

Higher proportions of $Al_2O_3$ and especially $SiO_2$ in the soil horizon of Pit Y-01-07 in comparison with rock samples (R horizon) suppose the influence of allochthonous material. The percentage of MgO in the bulk chemical composition,

**Fig. 2** Micromorphology of soil (a, d, e) and rocks (b, c) based on thin sections data: **a** general view of Bwh horizon, Pit Y-04-07 with fragments of mafic and ultramafic rocks—metagabbro–amphibolite (*1*), pyroxenite (*2*), serpentinite (*3*), and epidosite (*4*); **b** sample of metagabbro–amphibolite from Pit Y-04-07; **c** rock fragment of serpentinous dunite from Pit Y-01-07; **d** pronounced development of Fe oxides along the cleavages of pyroxene from Bwh horizon, Pit Y-04-07; **e** weakly (*upper*) and well developed (*down*) Fe oxides in grains of pyroxene and fresh grains of plagioclase and serpentine from BC horizon, Pit Y-04-07. Determinations on thin sections at (*I*) plain polars and (*II*) crossed nicols. Indexes are due to Whitney and Evans (2010) in modification: *Chl*—chlorite, *Ep*—epidote, *Ol*—olivine, *Pl*—plagioclase, *Px*—pyroxene, *Srp*—serpentine; *Fe*—Fe oxides

**Table 1** Some characteristics of studied soils

| Horizon, depth, cm | pH $H_2O$ | C/LI | <1 µm | Chemical composition of the rock and the <1 mm fraction of soil, % in ignited sample | | | | | | | | | $Fe_2O_{3d}$ | $Fe_2O_{3P}$ | $Fe_2O_{3O}$ | $Al_2O_{3O}$ | $SiO_{2O}$ | $Fe_2O_{3o}$/ $Fe_2O_{3d}$ | $Fe_2O_{3p}$/ $Fe_2O_{3o}$ |
|---|---|---|---|---|---|---|---|---|---|---|---|---|---|---|---|---|---|---|---|
| | | (%) | | $SiO_2$ | $Al_2O_3$ | $Fe_2O_3$ | CaO | MgO | $K_2O$ | $Na_2O$ | MnO | $TiO_2$ | (%) | | | | | | |
| *Haplic Regosols (Eutric), Pit Y-01-07* | | | | | | | | | | | | | | | | | | | |
| A 0-2 | 6.6 | 7.5*/ 28.2 | – | – | – | – | – | – | – | – | – | – | – | – | – | – | – | – | – |
| C1 2-10 | 8.1 | 0.6 | 16.6 | 56.7 | 6.53 | 8.43 | 1.43 | 23.93 | 0.93 | 1.11 | 0.12 | 0.49 | 1.48 | 0.02 | 0.60 | 0.11 | 0.12 | 40.54 | 3.33 |
| C2 10-25 | 8.3 | 0.5 | 15.7 | 55.76 | 6.24 | 8.71 | 1.50 | 25.13 | 0.84 | 0.77 | 0.16 | 0.46 | 1.48 | 0.02 | 0.58 | 0.09 | 0.06 | 39.19 | 3.45 |
| C2 25-52 | 8.4 | 0.6 | 15.4 | 57.23 | 6.67 | 8.50 | 1.43 | 23.54 | 0.93 | 0.74 | 0.12 | 0.50 | 1.28 | 0.02 | 0.53 | 0.10 | 0.11 | 41.40 | 3.77 |
| C3 52-70 | 8.6 | 0.5 | 17.2 | 55.12 | 6.16 | 8.85 | 1.36 | 26.04 | 0.82 | 0.71 | 0.15 | 0.43 | 1.82 | 0.01 | 0.56 | 0.10 | 0.11 | 30.76 | 1.78 |
| R | – | – | – | 49.20 | 3.83 | 9.46 | 0.96 | 32.43 | 3.01 | 0.37 | 0.13 | 0.25 | – | – | – | – | – | – | – |
| *Stagnic Leptosol (Eutric), Pit Y-04-07* | | | | | | | | | | | | | | | | | | | |
| Ah 0-2 | 7.1 | 3.5*/ 14.5 | – | 53.22 | 6.82 | 10.99 | 2.77 | 23.56 | 0.70 | 0.71 | 0.22 | 0.44 | 1.90 | 0.10 | 1.10 | 0.21 | 0.20 | 57.89 | 9.09 |
| Bwh 2-10 | 7.1 | 2.4 | 8.2 | 53.57 | 7.01 | 11.22 | 2.76 | 22.77 | 0.68 | 0.74 | 0.23 | 0.45 | 2.37 | 0.14 | 1.17 | 0.26 | 0.17 | 49.37 | 11.96 |
| BC 10-40 | 7.6 | 0.9 | 19.4 | 59.06 | 8.49 | 10.03 | 2.04 | 17.39 | 1.00 | 0.87 | 0.17 | 0.56 | 1.66 | 0.08 | 0.78 | 0.22 | 0.12 | 46.99 | 10.26 |
| BCg 40-60 | 7.4 | 0.6 | 21.2 | 62.89 | 9.52 | 9.37 | 2.24 | 12.62 | 1.19 | 1.08 | 0.14 | 0.65 | 1.86 | 0.09 | 1.01 | 0.17 | 0.24 | 54.30 | 8.91 |
| R | – | – | – | 52.11 | 17.85 | 13.68 | 5.15 | 7.65 | 0.79 | 1.64 | 0.40 | 0.63 | – | – | – | – | – | – | – |

Notes: *C* total carbon; *LI* loss on ignition: <1 µm content of particle size <1 µm in the <1 mm fraction; '–' no data available; *d* dithionite, *o* oxalate, and *p* pyrophosphate extractable forms; *R* rock samples from the bottom of soil profile especially in the upper horizons of Pit Y-04-07, reflects the influence of ultramafic material; however, the correlation between pH value and the content of MgO is absent. The considerable proportions of the <1 μm fraction in Haplic Regosol (Eutric), Pit Y-01-07, and especially in the bottom horizons of Stagnic Leptosol (Eutric), Pit Y-04-07 demonstrate a sensitivity to the weathering and alteration of parent material of hard rock(s).

Weathering intensity of rock(s) in soil environment is illustrated by oxalate-soluble Si as well as pronounced content of soluble Fe. That is affected by dissolving of Fe-rich minerals and formation of pedogenic Fe oxides. Almost 50 % of soluble Fe, especially in the more developed profile of Stagnic Leptosol (Eutric), Pit Y-04-07 is represented by well-crystallized oxides, whereas the proportion of pyrophosphate extractable Fe, which is related to iron released from the complexes with organic matter, is relatively low. Eluvial–illuvial distribution of dithionite extractable Fe in Pit Y-04-07 also reflects weakly pronounced development of pedogenic processes of podzolization in spite of neutral pH values. In Pit Y-04-07 the soluble Al confirms that the mafic rock is a source of the fine earth taking into account that the portion of Al is extremely low in the ultramafic rock.

*Soils' clay mineralogy.* The minerals identified in the clay fraction (<1 μm) of both profiles are as follows: (i) smectite(s), (ii) serpentine, (iii) chlorite, (iv) talc, (v) mineral of mica group, and (vi) quartz based on XRD data (Figs. 3 and 4). Smectite(s) is confirmed by a peak at 14.2 Å in the Mg-saturated air-dry state shifts to a fundamental 001 spacing of 17.1 Å after ethylene glycol solvation. As the 002 reflection at 8.6 Å, which is characteristic of a pure smectite, is detected indicating that the expandable phase is represented by individual smectite(s) (Moore and Reynolds 1997). In the studied samples the identification of tri- or dioctahedral nature of smectite(s) due to the 060 (data is not showing) reflection is complicated because of presence of talc, serpentine, chlorite, and a mineral of the mica group. Additional information was due to FTIR data. The absence of OH bending bands in the 950–800 $cm^{-1}$ region in the FTIR spectra of clay size fractions points to the absence of dioctahedral coordinated OH groups of smectite (Komadel et al. 2000). The absorption band near 3680 $cm^{-1}$ assigned to stretching vibrations of $Mg_3OH$ (Farmer 1974), the absorption band near 3624 $cm^{-1}$, ascribed to the $Mg_2Fe^{3+}OH$ vibrations in the spectrum of ferruginous saponite (Farmer 1974), and the absorption band near 533 $cm^{-1}$ (Si–O–Al bending vibration), documented in the spectrum of saponite (Madejová et al. 1992) indicate presence of trioctahedral smectite–saponite in the both profiles. This smectite was also identified previously in the mature profile of Haplic Cryosols (Reductaquic) on serpentinous dunite from mountainous tundra, the Polar Urals (Lessovaia et al. 2012).

Chlorite identification in the presence of smectite(s) is based on the basal reflections at 7.1, 4.72, and 3.57 Å that are stable after ethylene glycol solvation and showing a slight contraction of the 001 peak to 13.8 Å after 550 °C treatment.

Serpentine is recognized in the presence of chlorite by splitting of ~7 Å peak at 7.24 (serpentine) and 7.1 Å (chlorite) and presence of 3.64 Å peak.

Talc identification is based on the peaks at 9.3 and 3.12 Å, stable after ethylene glycol solvation and 550 °C treatment. Mineral of mica group is confirmed at the

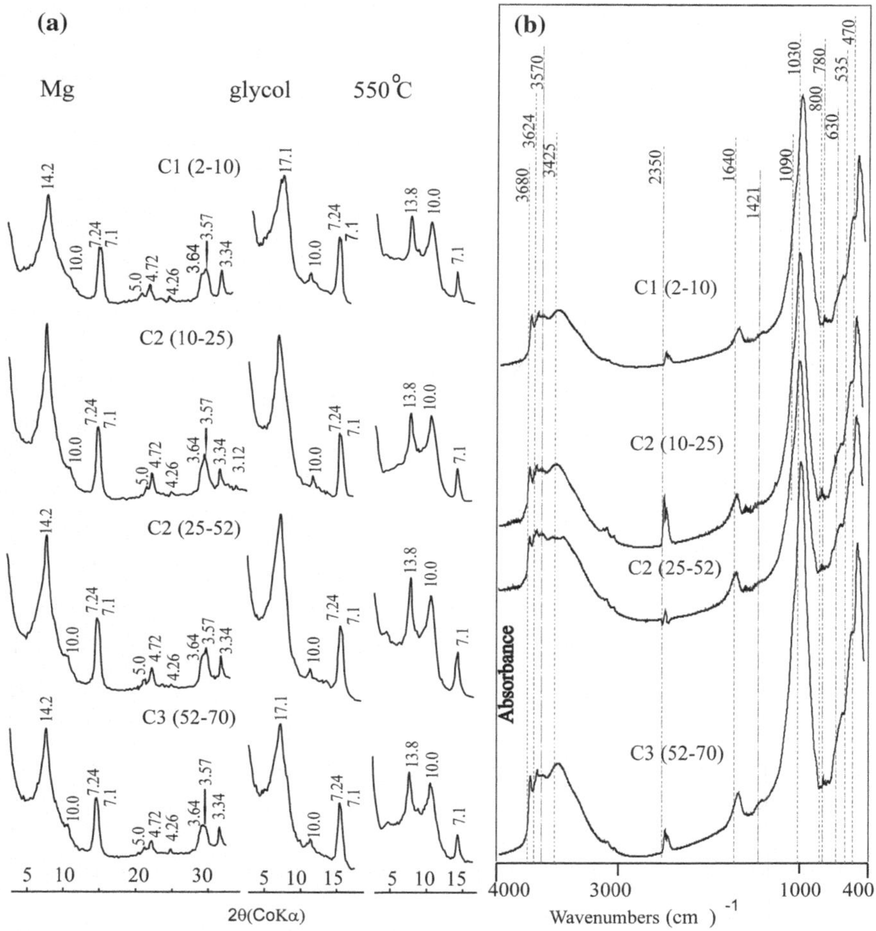


**Fig. 3** XRD patterns (**a**) and FTIR spectra (**b**) of the <1 μm fraction of the soil horizons, Pit Y-01-07

001 and 002 basal reflections at 10.0 and 5.0 Å, respectively, that remain stable after ethylene glycol and heat treatment.

Quartz is identified by its characteristic XRD peaks at 3.34 and 4.26 Å and by its most typical IR absorption band doublet at 800–780 $cm^{-1}$ (Hlavay et al. 1978; Madejová and Komadel 2005).

Absence of absorption bands near 3700 $cm^{-1}$ (the vibration of inner surface OH groups), as well as 914 $cm^{-1}$ (bending vibration of inner OH groups) indicates that kaolinite is not identified in the soil profiles (Madejová and Komadel 2001; Pentrak et al. 2009). So, kaolinite does not affect the ~7 and 3.5 Å peaks confirmed by

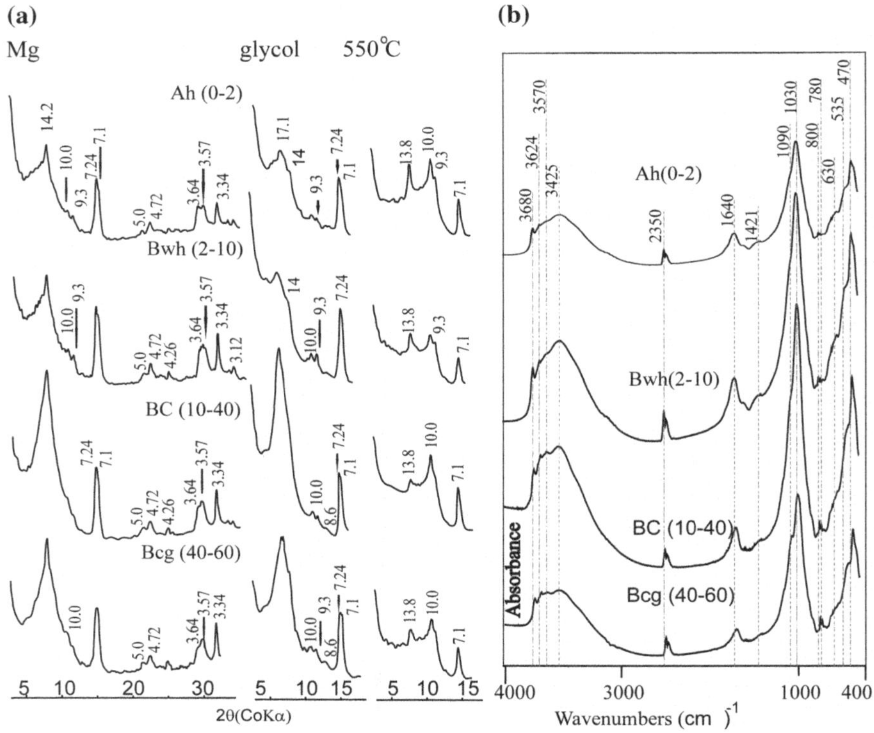


**Fig. 4** XRD patterns (**a**) and FTIR spectra (**b**) of the <1 μm fraction of the soil horizons, Pit Y-04-07

FTIR spectroscopy data, which is more sensitive method for the detection of kaolinite (Delvaux et al. 1989).

The distribution of the clay minerals in the profile based on mentioned above criteria is characterized by slightly decrease of smectite portion in the upper horizons, which is more pronounced in Pit Y-04-07. Proportion of chlorite based on peak intensity is higher in Pit Y-01-07.

## 5 Discussion

The origin of the allochthonous material supposed by proportions of $SiO_2$ and $Al_2O_3$ in Haplic Regosol (Eutric), Pit Y-01-07 could be related to eolian or water transfer. Nevertheless, the strong affect of ultramafic rock is pronounced based on MgO percentage. In Stagnic Leptosol (Eutric), Pit Y-04-07 the influence of ultramafic rock is concluded based on MgO proportion in the bulk chemical composition, especially in the upper horizons. The increase of $SiO_2$ percentage in the soil fine earth (the <1 mm fraction) in comparison with rock samples (R horizon) is most probably a result of the residual accumulation of inherited quartz, which content is high in metagabbro–amphibolite. Based on these findings BC and BCg are mostly affected by mafic rock.

High pH values in Haplic Regosol (Eutric), Pit Y-01-07 is mostly assigned by ultramafic material. Ultramafic rock resulted in soil alkalinity in spite of the absence of calcite keeping by high amounts of exchangeable Mg, which was shown previously (Lesovaya et al. 2012). pH values in Stagnic Leptosol (Eutric), Pit Y-04-07, which are higher than in the mature profile of Epileptic Entic Podzols on metagabbro–amphibolite (Lessovaia et al. 2014 in press) is also affected by ultramafic substrate.

Both profiles are characterized by poorly developed organic horizons (2 cm in depth) of the decayed organic material mixed with mineral mass. Stagnic Leptosol (Eutric), Pit Y-04-07 on the moraine has more developed organic profile with higher content of organic C in Bwh horizon. That is related to more toxic conditions in case of ultramafic material. The problem is also in the absolutely different roles of ultramafic and mafic rocks in the ecosystems and soil development—strongly deteriorative one in first case (Alexander et al. 2007) but usually enriching one in case of basic material (Lal 2006). However, the outcrops of ultramafic rocks are used to be adjacent in space to the mafic rocks (Kusky 2005) but study of soils and weathering trends of this transition in cold environment is still poorly known.

A sensitivity to weathering and alteration of the parent materials based on the proportion of <1 μm fraction is confirmed by pronounced reddening of ultramafic rock fragments and proportions of soluble forms of Si, Al, and especially Fe. Iron removal from silicates matrix, especially the fine-grained parent rocks is reflected by increase of dithionite—and, oxalate extractable iron, that is more pronounced in the upper part of Pit Y-04-07. Here in Bwh horizon thin films of Fe–Al-organo complexes on stones were found. Previously, the development of weathering rinds, in which the weathering is isolated from other soil processes was detected (Colman 1982). These findings indicate the translocation of poorly crystalline iron and aluminum oxides, although the studied soil is characterized by neutral pH values most probably due to podzolization. However, the admixture of ultramafic materials mitigates the development of Entic Podzols which were found in the Polar Urals on the pure mafic materials (Lessovaia and Polekhovsky 2009). Based on the findings of thin section from soil horizons of this profile, we can conclude that the source of iron coursed the reddening rims of rock fragments is ultramafic substrate. Generally, the sensitivity of ultramafic rocks to weathering can strongly influence not only soil macrofeatures but also landscape stability (Alexander 2009; Alexander and DuShey 2011). Thus, the features affected by chemical weathering are obvious in the weakly developed soils.

The identity of mineral associations of clay size fraction from both profiles confirms that the soils are characterized by the same sources of them. Talc and serpentine are definitely affected by ultramafic substrate. Chlorite is also inherited from ultramafic substrate based on its presence in serpentinous dunite. Additionally, the presence of chlorite as well as mica and quartz in Stagnic Leptosols (Eutric),

Pit Y-04-07 can be affected by metagabbro–amphibolite. Whereas mica and quartz in Haplic Regosol (Eutric), Pit Y-01-07 are allochthonous. The presence of quartz here can explain the high proportion of $SiO_2$ in the profile.

The smectite identified as saponite from both studied profiles is absent in the rocks of serpentinous dunite as well as in metagabbro–amphibolite. Generally, the smectite appeared in the soils could be a result of the crystallization caused by weathering of (i) amphibole, which was detected in metagabbro–amphibolite (Proust et al. 2006; Wilson and Farmer 1970) or (ii) olivine (Bulmer 1992; Eggleton 1984; Wilson 2004) and/or transformation of serpentine from serpentinous dunite. But previously the smectite was identified only in Haplic Cryosols (Reductaquic) on serpentinous dunite but not in Entic Podzols on metagabbro–amphibolite (Lessovaia and Polekhovsky 2009). Thus, the appearance of smectite in the weakly developed profiles can be affected by the weathering only of ultramafic rock. In Stagnic Leptosols (Eutric), Pit Y-04-07 influenced by mafic and ultramafic rocks the proportion of smectite in clay size fraction does not correlate to the percentages of MgO or $SiO_2$, which can be explained that smectite is located in fine size fractions, whereas proportions of oxides characterize the bulk chemical composition (<1 mm). Besides that, the main source of Mg in the ultramafic rock is olivine concentrated in the course fractions.

Generally, the processes affected by only pedogenesis could not be split from ones influenced by chemical processes. Nevertheless as opposed to the studied weakly developed profiles, in the more mature profile of the Haplic Cryosol (Reductaquic) on serpentinous dunite the pedogenesis resulted in significant decrease of proportion of smectite and the appearance of vermiculite due to chlorite transformation in the upper horizons (Lessovaia et al. 2012). Thus, it can be concluded that the appearance of smectite (saponite) in the weakly developed soils is not related to pedogenesis. These soil profiles illustrate the soil formation from "mature" fine earth from high-sensitive ultramafic rock due to chemical weathering.

## 6 Conclusion

In cold soil environment the more weatherable ultramafic material plays the more important role as a prerequisite for the weathering trends and soil formation than a mafic rock.

Appearance of smectite in the weakly developed soil profiles illustrates the scenario of soil formation slightly affected by pedogenic biotic processes, but from the "mature" fine earth due to the active abiotic chemical weathering in the cold region.

The admixture of ultramafic materials mitigates the development of Entic Podzols which were earlier found in the Polar Urals on the pure mafic materials. So, the presence of ultramafic materials either predominating or even in admixture results in the "extreme lithological environment" for a pedogenesis and in the formation of weakly developed soils—Regosols and Leptosols.

**Acknowledgments** This study was supported by the Russian Foundation for Basic Research (14-04-00327), the St. Petersburg State University (18.38.418.2015). The general concept of the paper and the intellective analysis and synthesis of the data was supported by the Russian Science Foundation (14-27-00133). XRD study of soils was carried out in the X-ray Diffraction Centre of St. Petersburg State University.

## References

Alexander EB (2009) Soil and vegetation differences from peridotite to serpentinite. Northeastern Nat 16(sp 5):178–192

Alexander EB, Coleman RG, Keeler-Wolfe T, Harrison SP (2007) Serpentine geoecology of Western North America: geology, soils and vegetation. Oxford University Press, New York

Alexander EB, DuShey J (2011) Topographic and soil differences from peridotite to serpentinite. Geomorphology 135:271–276

Allen ChE (2002) The influence of schistocity on soil weathering on large boulder tops, Kärkevagge, Sweden. Catena 49:157–169

Arnaud RJST, Whiteside EP (1963) Physical breakdown in relation to soil development. J Soil Sci 14:267–281

Bascomb CL (1968) Distribution of pyrophosphate-extractable iron and organic carbon in soils of various groups. Eur J Soil Sci 19:251–268

Bulmer CE (1992) Pedogenesis of soils derived from ultramafic rocks and tephra in southwestern British Columbia. PhD thesis, University of British Columbia, Vancouver

Colman SM (1982) Clay mineralogy of weathering rinds and possible implications concerning the sources of clay minerals in soils. Geology 10:370–375

Delvaux B, Mestdagh MM, Vielvoye L, Herbillon AJ (1989) XRD, IR and ESR study of experimental alteration of Al-nontronite into mixed-layer kaolinite/smectite. Clay Miner 24:617–630

Eggleton RA (1984) Formation of iddingsite rims on olivine: a transmission electron microscope study. Clays Clay Miner 32:1–11

Farmer VC (1974) The layer silicates. In: Famer VC (ed) The infrared spectra of minerals. Monograph 4, Mineralogical Society, London

Hall K, Thorn CE, Matsuoka N, Prick A (2002) Weathering in cold regions: some thoughts and perspectives. Prog Phys Geogr 26:577–603

Hlavay J, Jonas K, Elek S, Inczedy J (1978) Characterization of the particle size and the crystallinity of certain minerals by IR spectrophotometry and other instrumental methods-II. Investigations on quartz and feldspar. Clays Clay Miner 26:139–143

Hoch AR, Reddy MM, Drever JI (1999) Importance of mechanical disaggregation in chemical weathering in a cold alpine environment, San Juan Mountains, Colorado. Geol Soc Am Bull 111(2):304–314

Jackson ML, Lim CH, Zelazny LW (1986) Oxides, hydroxides, and aluminosilicates. In: Klute A (ed) Methods of soil analysis. Part 1. 2nd edn. SSSA, Book Series 5, Madison, Wiscontin, USA

Komadel P, Madejová J, Laird DA, Xia Y, Stucki JW (2000) Reduction of Fe(III) in griffithite. Clay Miner 35:625–634

Kusky TM (2005) Encyclopedia of earth science facts on file inc. New York

Lal R (ed) (2006) Encyclopedia of soil science, 2nd edn, vol I, II. Taylor & Francis, Boca Raton, FL

Lessovaia SN, Polekhovsky YS (2009) Mineralogical composition of shallow soils on basic and ultrabasic rocks of East Fennoscandia and of the Ural Mountains, Russia. Clays Clay Miner 57:476–485

Lessovaia S, Dultz S, Polekhovsky Yu, Krupskaya V, Vigasina M, Melchakova L (2012) Rock control of pedogenic clay mineral formation in a shallow soil from serpentinous dunite in the Polar Urals, Russia. Appl Clay Sci 64:4–11

Lessovaia S, Dultz S, Plötze M, Andreeva N, Polekhovsky Yu, Filimonov A, Momotova O (2014 in press) Soil development on basic and ultrabasic rocks in cold environments of Russia traced by mineralogical composition and pore space characteristics. Catena. doi:10.1016/j.catena.2014.11.020

Lesovaya SN, Goryachkin SV, Polekhovskii YS (2012) Soil formation and weathering on ultramafic rocks in themountainous tundra of the Rai-Iz Massif, Polar Urals. Eurasian Soil Sci 45:33–44

Madejová J, Putyera K, Cicel B (1992) Proportion of central atoms in octahedra of smectites calculated from infrared spectra. Geologica Carpathica Clays 1:117–120

Madejová J, Komadel P (2001) Baseline studies of the clay minerals society source clays: infrared methods. Clays Clay Miner 49:410–432

Madejová J, Komadel P (2005) Information available from infrared spectra of the fine fraction of bentonites. In: Kloprogge JT (ed) The application of vibrational spectroscopy of clay minerals and layered double hydroxides. CMS Workshop Lectures, vol 13. The Clay Mineral Society, Aurora, CO

Major Ore-Bearing Geological-Geochemical Systems of the Urals. (Glavnye rudnye geologo-geokhimicheskie sistemy Urala) (1990) Nauka, Moscow (in Russian)

Mehra OP, Jackson ML (1958) Iron oxide removal from soils and clays by a dithionite-citrate system buffered with sodium bicarbonate. Clays Clay Miner 7:317–327

Meunier A, Sardini P, Robinet JC, Prêt D (2007) The petrography of weathering processes: facts and outlooks. Clay Miner 42:415–435

Moore DM, Reynolds RC (1997) X-ray diffraction and the identification and analysis of clay minerals. Oxford University Press, New York

Pentrak M, Madejová J, Komadel P (2009) Acid and alkali treatment of kaolins. Clay Miner 44:511–523

Proust D, Caillaud J, Fontaine C (2006) Clay minerals in early amphibole weathering: tri- to dioctahedral sequence as a function of crystallization sites in the amphibole. Clays Clay Miner 54:351–362

Svendsen JI, Krüger LC, Mangerud J, Astakhov VI, Paus A, Nazarov D, Murray A (2014) Glacial and vegetation history of the Polar Ural Mountains in northern Russia during the last ice age, marine isotope stages 5–2. Quatern Sci Rev 92:409–428

Tyurin EV (1931) New modification of volume method for determination of humus by chromic acid (Novoe vidoizmenenie ob'emnogo metoda opredeleniya gumusa s pomosh'u khromovoi kisloty). Pochvovedenie 37:515–524 (in Russian)

Whitney DL, Evans BW (2010) Abbreviations for names of rock-forming minerals. Am Miner 95:185–187

Wilson MJ (2004) Weathering of the primary rock-forming minerals: processes, products and rates. Clay Miner 39:233–266

Wilson MJ, Farmer VC (1970) A study of weathering in a soil derived from a biotite-hornblende rock II. The weathering of hornblende. Clay Miner 8:435–444

Wilson MJ, Berrow ML (1978) The mineralogy and heavy metal content of some serpentinite soils in north-east Scotland. Chem Erde 37:181–205

World Reference Base for Soil Resources (2006) World soil resources Reports 103, 2nd edn. FAO, Rome

# Biogenic–Abiogenic Interaction in Antarctic Ornithogenic Soils

**Evgeniy V. Abakumov, Ivan Yu. Parnikoza, Dmitry Yu. Vlasov and Alexey V. Lupachev**

**Abstract** In severe climatic and specific landscape conditions of Antarctica birds play an important role in transportation of organic matter of guano to the coastal landscapes. It has been shown that redistribution of guano components affects the speed of soil cover spatial development and formation of new polypedons of soils in environments, surrounding rookeries. Soils development is also affected by flying birds' transportation activity, while they transport the viable diasporas of plants, material of limpet shells, etc. This affects an initial or additional colonization of rocks being in distance from the coasts, sources of seed, and organic matter. Soils of Antarctica formed under effect of bird activity are the following: the most known typical ornithosols of the current penguin rockeries, post-Ornithosols, developed during post-Ornithogenic succession and organic lithosols formed in the areas of flying sea birds nesting and feeding areas due to limited nitrification and viable plant material and diasporas transportation.

**Keywords** Antarctic soils · Biogenic–abiogenic interactions · Birds affect of soils · Organic material transportation · Ornithogenic material

E.V. Abakumov (✉) · D.Yu.Vlasov
Saint-Petersburg State University, 199178 Universitetskaya emb. 7/9, Saint-Petersburg, Russia
e-mail: e_abakumov@mail.ru

I.Yu.Parnikoza
Institute of Molecular Biology and Genetics, National Academy of Sciences of Ukraine, 150 Zabolotnogo Str., Kyiev 03680, Ukraine

A.V. Lupachev
Institute of Biological and Physico-Chemical Problems of Soil Science, Puschino, Russia

 
O.V. Frank-Kamenetskaya et al. (eds.), *Biogenic—Abiogenic Interactions in Natural and Anthropogenic Systems*, Lecture Notes in Earth System Sciences,
DOI 10.1007/978-3-319-24987-2_19

# 1 Introduction

Soil formation is known as a set of processes of biogenic–abiogenic interactions. Usually, the pedogenic processes are presented mainly by transformation of organic remnants within the mineral matrix with essential fine-earth content. Nevertheless, there are many soils and soil-like bodies in extreme environments that are formed under the accumulation of zoogenic organic matter on relatively unweathered parent rocks with low fine-earth portion. Soil formation in Antarctica has been investigated by many researchers (Kubiena 1970; Campbell and Claridge 1987; Bockheim and Ugolini 1990; Zhao 2000; Korsun et al. 2008; Parnikoza et al. 2007, 2011; Mergelov et al. 2012; Lupachev and Abakumov 2013). It was shown that areas, occupied by soils are concentrated in ice- and snow-free areas (Bockheim 2015), which are estimated as 0.35 % of total Antarctic area or 44,000 $km^2$. Soils are mainly form under the mono-species populations of fungi, lichens, mosses, and vascular plants. Multispecies ecosystems are also described as significant source of soil organic matter. Meanwhile, the low intensity of photosynthesis and low productivity of ecosystems result in relatively low accumulation of humus and raw organic matter (Campbell and Claridge 1987; Abakumov 2010). The average content of organic carbon in soils is 1–2 % in continental Antarctic and 3–4 % in the region of Antarctic Peninsula (Abakumov 2010). The C/N ratio changes in limits of 10–60 units which are strongly affected by the organic material quality, intensity of its transformation, and period of biological activity in each given region. So, the classical soils, formed under plants are presented by Lithosols, Leptosols, Gleysols, Histosols, Cryosols, Gelisols, and others. Beside them, the major representative of soil in Antarctica is ornithogenic soils. There is a verificator "Ornithic" in WRB classification system (WRB 2006), but there is no special type of ornithogenic soils in this classification. Ornithogenic material is a material with strong morphological and chemical influence of birds on soils. Meanwhile, the term ornithogenic soils became more and more usable for those soils, which have morphological and chemical features, related to ornithic definition. Thus, it is possible to classify soil as Ornithic Gelisols, Ornithic Lithosols, Ornithic Cryosols, etc.

Ornithic soils have been revealed in all the coastal parts of Antarctica form the northernmost (South Shetland Islands) to the deed gulfs, penetrated to the coastal parts (Bockheim et al. 2015). These territories have been studied as habitats of soil fungal communities which can be indicator of soil biological activity (Vlasov et al. 2012). In general, penguins play the most important role in organic matter accumulation, rock transformation, chemical weathering, and ornithogenic soil formation (Simas et al. 2007; Shaefer et al. 2008). Penguin guano is enriched by organic carbon, nitrogen, and phosphorous. Places of stable guano accumulation in Antarctica are known as penguin, (or other birds), rookeries and, single nest places. They occupy different levels of elevation, and therefore, the redistribution of organic matter in landscape becomes possible. That is why the places of landscape which are not directly affected by birds are exposed to enrichment by biogenic compounds due to the redistribution with water streams. This phenomenon has been described as ornithogenic impact on landscape geochemistry spatial pattern. The guano accumulations affect soils not only in vertical scale, but play important role in initial accumulation of biogenic elements in geochemically subordinated positions. As the result, the soils of the coastal zone of Antarctica are subjected to the direct or indirect guano impact during their history. This is the most characteristic feature of soils, formed in the maritime Antarctic and rookeries in coastal zone of Eastern Antarctica. The phenomenon of post-ornithogenic succession has been described by many scientists (Ramsay 1983; Simas 2007; Abakumov and Mukhametova 2014). On the exposed rocks with unfavorable conditions for plant growth an accumulation of ornithogenic organic matter leads to development of so-called "ornithogenic" lichens (e.g., *Caloplaca*) and nitrophilous algae *Prasiola crispa* (Lightfoot) Kützing.

In areas with more favorable conditions for the development of the various formations of the Antarctic terrestrial vegetation, fairly simple ecosystems can form, with the participation of mosses, lichens, and in particularly favorable conditions—vascular plants, such as *Deshampsia antarctica* Desv (Kozeretska et al. 2010). So, the bird's biogeochemical activity is known as the reason of intensive soil development in Antarctic environment (Simas et al. 2007). In oases with dissected relief there is a gradient of microenvironments, with different levels of plant cover development (Krywult et al. 2003; Smykla et al. 2006; Kozeretska et al. 2010). This gradient is presented by localities with fresh, undecomposed guano, then places with initial colonization and finally covers by 2–5 plant species. The second aspect is the direct transportation of the vegetative parts and diasporas of plants form on new places or from the coastal part to the central parts of oasis's, previously unaffected by this species. This is an outstanding example of bird's effect on local biodiversity, expansion of some plants on the territories of the glaciers retreat, and bird's effect on soil processes intensification in Antarctica (Parnikoza et al. 2008, 2012a, b, 2014). This chapter is aimed:

- *to describe phenomenon of guano organic matter accumulation and its effect of solid parent rocks and their fine-earth containing debris*
- *to analyze the diversity of ornithogenic soils in different climatic and lithological conditions of Antarctic region and their influence on the soil fungal communities formation*
- *to assess the role of bird transportation of plants as nest material on the spatial distribution of organic weak developed soils, formed due to food and breeding activity mainly of kelp gull and other species of Antarctic sea birds.*

## 2 Materials and Methods

Soil biogenic–abiogenic interactions have been investigated on examples of different latitude plots form Western and Eastern Antarctica (Table 1). The northernmost key sites are situated on the King George Island (62°S), while the

**Table 1** General characteristics of the studied Antarctic areas with bird's impact: penguin rookeries and flying sea birds nesting sites

| No. | Plot name or station name | Coordinates | Location | Bedrock type | Mean annual air temperature (°C) | Mean annual precipitation (mm) |
|---|---|---|---|---|---|---|
| *Eastern Antarctic* | | | | | | |
| 1 | Station Progress | 69° 23.04′S 76° 22.57′E | Princess Elizabeth, Land, Prydz Bay | Granites, pegmatites | −8.5 | 250 |
| 2 | Seasonal station Druzhnaya 4 | 69° 74.79′S 73° 70.88′E | | Granites | −8.5 | 250 |
| 3 | Station Mirny | 66° 33.09′S 93° 00.58′E | Davis Sea, Pravda Coast | Granites, gneises, plagioclases | −11.5 | 585 |
| 4 | Haswell archipelago, close to the Mirny station | 66° 31.34′S 92° 59.55′E | Davis Sea, Pravda Coast | Charnokites | −11.5 | 585 |
| *Western Antarctic* | | | | | | |
| 5 | Station Russkaya | 74° 45.90′S 136° 48.38′ W | Marie Byrd Land, Hobbs Coast | Gneisses | −12.4 | 1190 |
| 6 | Lindsey Island | 73° 36.80′S 103° 02.00′ W | Amundsen sea, Canisteo peninsula | Gneisses, Andesitic basalts | No data | No data |
| 7 | King George Island, Fildes Peninsula and Ardley Island | 62° 12.00′S 58° 58.00′ W | Maritime Antarctica, South-Shetlands Islands | Debris of the massive rocks, marine sediments, moraines, fluvioglacial rocks | −2.8 | 729 |
| 8 | Galindez Island, Argentine archipelago | 65° 14.44′S 64° 15.28′ W | Maritime Antarctica, Argentine Islands | Debris of andesites, fresh moraines | −2.8 | 450 |

southernmost ones are situated on the Russkaya station (74°S). The highest diversity of penguins revealed in the maritime Antarctica, while on the southern islands and coasts mainly one penguin species dominates—*Pygoscelis adeliae*. Other birds species are the South Polar skua (*Catharacta maccormicki)* and Kelp gull (*Larus dominicanus*). These birds are very important in transportation of plants diasporas in the ecosystem. Soil nitrification is also caused by imperial shag (*Phalacrocorax atriceps*) colonies (Peklo 2007).

As described in Table 1, geographical locations of the investigated penguin rockeries are quite different and situated in various latitudes and climates. They can be situated in the coastal zones of relatively warm and humid Antarctic Peninsula or relatively dry and cold East Antarctica.

Soil samples with ornithogenic influence have been investigated in all the listed in Table 1 locations. Soil pits were made with sampling of each soil horizon and bedrock material for analyses of chemical, biological characteristics and particle-size distribution, and micro-morphological analyses (thin sections). Soils were analyzed in fine-earth fraction (<1 mm) for carbon and nitrogen contents (Walkley 1935), organic carbon mineralization rate (Methods…1992), pH values. Investigation of the role of birds in spatial distribution of soil was conducted mainly in field by analyzing of transported material nature, thickness of new-formed soil profile. The diversity and structure of microscopical fungi (micromycetes) communities in ornithogenic soils and zoogenic substrates were analyzed by complex of mycological methods (Vlasov et al. 2012).

## 3 Results and Discussion

The most intensive affect on soil formation is expresses on the penguin rookeries (Barcikowski et al. 2005). Typical penguin rockeries on Lindsey Island are presented on Fig. 1. This is a one of the southernmost key site of intensive penguin effect of landscape and soil formation. There are no evidences of soil formation here

**Fig. 1** Penguin rookeries in Lindsey Island

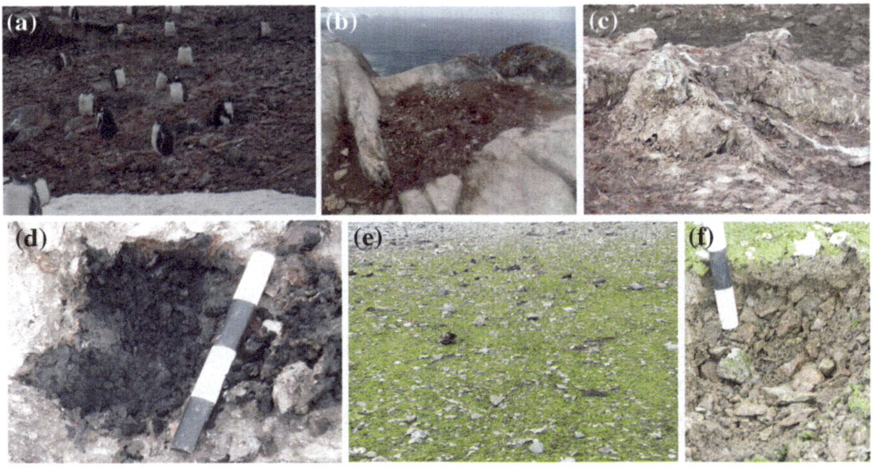


**Fig. 2** Examples of soils, formed on the penguin rookeries: **a**, **b** penguin rookeries on the Galindez Island, Argentine Islands, **c** penguin guano on the Fildes Peninsula, King George Island, **d** profile of fresh guano layer, on the Fildes Peninsula, **e** first stage post-ornithogenic succession—colonization of soil by *Prasiola crispa*, Fildes Peninsula, **f** typical post-ornithogenic soil with undercomposed guano, Fildes Peninsula

without guano effect. Plant cover distribution is also strongly connected with distribution of the former locations of penguin rockeries.

Examples of soils of modern and former penguin rookeries are presented on the Fig. 2. Typical ornithic soils contain huge amounts of undecomposed guano. Definitely, they differ from regular soils with expressed organo–mineral interactions and developed soil horizons system. Many of those soils are not colonized by any plants due to toxicity of guano and current activity of penguins. Main soil chemical processes expressed here is mineralization of guano. Intensity of mineralization is closely connected with the duration of warm period. Namely, measured speed of guano mineralization in maritime Antarctica is 5–6 time higher than in coastal high-latitudinal Antarctic. That is why, cold Antarctic soils accumulate thick and well-expressed layers of guano (up to 3 m), while there are only 0.3–0.5 m in maritime Antarctica.

The second important geochemical process in the surroundings of penguin rookeries is the redistribution of water-soluble organic compounds in the ecosystem. Data obtained shows that the water-soluble part of total guano matter of different stage of decomposition change from 15 to 35 %. Not only organic forms of nutrients can migrate but also mineral forms of nitrogen and phosphorous. These substances play a key role in initiation of post-ornithogenic succession.

Such significant organic input, that penguin colony provided, causes the additional strong environmental gradient from colony edge to the periphery, where guano input gradually attenuated. In general, the impact of the penguin rockeries provided local, although quite strong perturbations within the general ecological gradient that formed from the edge of the ocean to the edge of the retreating glacier (Parnikoza et al. 2008, 2009). The common colonist of nutrient enriched microenvironments on the active penguin rockery margins or first colonist of former bird's colonies is alga *Prasiola crispa*. Their biogenic crusts covers the soils, friable grounds, stones, and rocks while they are enriched by nitrogen, derivated from guano. After *Prasiola* margin zone usually zone of well-developed other vegetation types present. Due to gradual decomposition of guano the common vegetation can spread on area of former colony. In maritime Antarctica it may be Antarctic herb tundra formation *Deshampsia antarctica* or different lichens and mosses communities in all vegetated regions of the continent. All these communities start from *Prasiola* one provide guano decomposition and post-ornithogenic soils formation. Post-ornithogenic soil is a soil with the gray-humus horizon, features of organo-mineral interactions, formation of soil structure and aggregates, penetration of guano compounds into cracks and even with the formation of organic or organo-mineral cutans. Typical ornithogenic soils contain about 20–30 % of total carbon in the top horizons, while the post-ornithogenic soils show only 4–5 % and less content of organic carbon. This is definitely caused by mineralization and humification of guano. The humification intensity usually assessed by degree of humification, i.e., ratio of humic and fulvic acids content. The lowest values of this ratio in investigated soils is about 0.12 (modern penguin rookeries), while the highest one is 0.55 (post-ornithogenic soils). Moreover, the intensity of humification is higher in maritime Antarctic conditions than in soils of high latitudes. The pH values of typical ornithogenic soil shows the variance from 4.5 to 7.7, decreasing in soils of post-ornithogenic succession to 1.0 unit due to fine-earth acidification by the products of organic matter decomposition.

The results of mycological studies of ornithogenic soils show an increase of the fungal diversity and complicating of the microbial communities structure in comparison with another soils types. The reasons of this increase are the accumulation of the organic matter (source of nutrition for fungi), concentration of the zoogenic substrates (feathers, bones, etc.) as well as the development of cyanobacterial mats and green algae *Prasiola crispa.* The number of fungi in some soil samples containing guano was more than ten thousand of colony forming units per 1 g of ornithogenic soil. Anamorphic fungus *Geomyces pannorum* (Link) Sigler & J.W. Carmich was dominant species for ornithogenic habitats in different regions of Antarctica. It is increased diversity of potential pathogenic micromycetes (for humans and animals) of ornithogenic soils especially near Antarctic stations. Some opportunistic species of the genus Aspergillus which are known as the causative agents of birds disease (particularly penguins) as well as capable of causing human disease were found in ornithogenic habitats near polar stations. We cannot exclude the possible role of birds in the circulation of pathogen organisms in Antarctica, particularly opportunistic micromycetes, which can be accumulated in ornithogenic soils.

The last stage of post-Ornithogenic soil succession is the transformation to Cryosol, Lithosol etc. However, while the direct geochemical effect of guano is not evident in soils, colonized by plants, there is an essential residual effect of those substances, which amended the soil on earlier stages of succession. Some soils of coastal territories were exposed to cycles of ornithogenic and post-ornithogenic successions many times.

Thus, a fundamental influence on the modern penguin colonies on richness of vegetation indicated in particular oasis for instance Point Thomas Oasis, King George Island (Barcikowski et al. 2005). At the same time, there some areas without significant bird influence are described this may be caused by the decomposition of organic matter of colonies that existed here before (Korsun et al. 2008). That is why we can conclude that, ornithogenic soil formation by direct or indirect way activities the soil biological process regularly and seriously affect the soil diversity of the continent.

An activity of flying sea birds is the second leading factor in parent rock colonization by plants and initial soil formation. It is expressed primarily at first in enhancing soil organic matter due to the influx of relatively—guano and wastes power. This effect is similar to the effects of a local disturbance of penguins, though of course on a much smaller scale. However, sometimes it has a pronounced effect, such as in the case of stains ornithophilous Antarctic lichen *Caloplaca*. The second aspect is the transfer of the various components of the Antarctic vegetation as breeding material by mainly *Larus dominicanus*, more seldom by skua species and *Macronectus giganteus* (Peklo 2007). Thus, transferred diasporas can generate new plants or indwelling new components into existing ecosystems. As a consequence soil formation significantly moving forward on base of its initial types. Initial soil formation under and surround the bird's nests and activity of birds in organic remnants transportation have been investigated in Argentine Islands region. Here, the kelp gulls occupy the costal parts for nesting in the early spring. In same place this birds had its feeding places. They use the limpets (*Nacella concinna*) as a food source. The products of shells transformation is an additional source of organic matter for soils. In same time mainly on these coastal higher areas communities with *Deschampsia antartica* spread. Kelp gull can transfer *Deshampsia antarctica* also on high points of small rocky islands, where this species previously absent. In conscience of this it is possible that most of the existing population can have ornithogenic origin. At the same time, during the transfer *Deshampsia antarctica* or other species material can be lost, thereby may form this species population located inland of islands. The secondary plants redistribution is affected also by *Catharacta maccormicki* activity (Parnikoza et al. 2014). Examples of bird activity in soil formation are given on the Fig. 3.

Transportation activity of organic materials is quite diverse in coastal landscapes of Antarctica, especially in zone of maritime Antarctica. First of all, birds bring the viable diasporas of plants to the uplands and rockeries from the coastal zones. Colonization of weathered rocks by lower or higher plants initiates soil formation. Initially soils form in microenvironment, created by one or several plant organisms, then the density of plant population increases and soils became covered by regular organic layer. The building of nests also plays an important role in accumulation of organic matter on the surface of existing soils or uncolonized rocks. Transportation of the limpet shells is typical for *Larus dominicanus* nest places. Its partially

**Fig. 3** Transportation activity of birds affect the soil formation: **a** The evidence of *Deshampsia antarctica* collections under snow by kelp gull *Larus dominicanus*, Galindez Islands, Argentine Islands, **b** Cushion of *Deshampsia antarctica* lost during the transportation by kelp gull, Galindez Island, **c** accumulation of plant material in *Larus dominicanus* nest, Galindes Island, **d** *Larus dominicanus* typical nesting-feeding place, Galindes Island, **e** limpets shells layers, colonized by *Deshampsia antarctica*, Galindes Island, **f** transportation of the Saniona by *Catharacta maccormicki*, Fildes Peninsula, King George Island, **g**, **h** examples of organic lithosol profiles, formed under Saniona-Deschampsia community, probably initiated by birds nest material transfer, Galindez Island

decomposed condition plays an important role in creation of coarse fraction in new-formed soil. An example of soils formed under transported vegetative plant fragments with seeds and shells are presented on the Fig. 3. They are identified as organic lithosols with well-expressed organic horizons, containing raw humus with low degree of humification. The average thickness of this horizon is about 10–15 cm and organic carbon in the fine-earth content is about 15–35 %. An admixture of shell remnants to the soil results in formation of aggregated mass of the fine-earth. Lithosols of this type, usually occupy microcaves and cracks, previously not colonized by plants, only the birds' transportation activity give the possibility for development of biogenic–abiogenic interactions in this type of Antarctic environment.

Microprocesses of different forms of soil formation are presented on the Fig. 4. This data illustrates the interaction of guano organic material with mineral material can be deep and intensive, when organic matter is decomposed, mixed with mineral

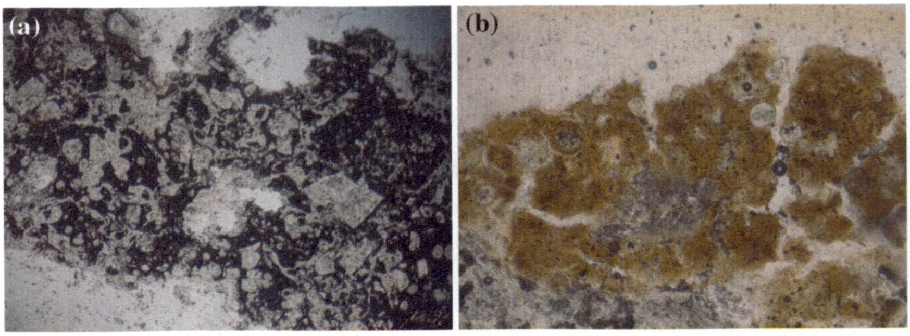


**Fig. 4** Ornithic soils in thin sections: **a** well decomposed guano, aggregated with mineral particles, **b** fresh guano on the surface of stone, Fildes Peninsula (King George Island)

particles and play the role of the glue, which aggregates soil particles together. In other case, guano just covers the stone surface and this is only initial stage of organo–mineral interactions.

## 4 Conclusions

Soil cover of Antarctica is known as isolated from other continents and biomes. Severe climatic conditions, short duration of austral summer, limited number of plant species, and sporadic distribution of plant cover made the soil formation weakly expressed. In these conditions, birds play an important role in supply of organic matter of guano to the coastal landscapes. Redistribution of guano components affects the speed of soil cover spatial development and formation of new polypedons of soils in environments, surrounding rookeries. Not only direct geochemical effect of guano on soil formation is important for development of soil in Antarctic environment. The ornithogenic influence on the soil fungal community's diversity and structure is also essential. Soils development is also affected by flying birds' transportation activity, while they transport the viable diasporas of plants, material of limpet shells, etc. This affects an additional or initial colonization of rocks being in distance from the coasts, sources of seed, and organic matter. Finally, we can summarize that there are at least three types of soils in Antarctica, which have been formed under effect of birds activity: the most known typical ornithosols of the current penguin rockeries, post-ornithosols, developed during post-Ornithogenic succession, and organic Lithosols formed in the areas of flying sea birds nesting and feeding areas due to limited nitrification and viable plant material and diasporas transportation.

**Acknowledgments** Great acknowledgments are made to L. Kurbatova and I. Dyyky for kindly granted photos, to National Scientific Antarctic Center MS of Ukraine and Russian Antarctic

Expedition, NSF and V. Papitashvili, Dr. I. Kozeretska and prof. V. Kunakh for logistic assistance and scientific support. This material realized partly in frame of NAS of Ukraine and PAS join project 2015–17 "Adaptive strategy of mutual survival of organisms in extreme environmental conditions" and Russian Foundation for Basic Reasearch, project 15-04-06118.

# References

Abakumov EV (2010) The sources and composition of humus in some soils of West Antarctica Eur. Soil Sci 43:499–508

Abakumov E, Mukhametova N (2014) Microbial biomass and basal respiration of selected Sub-Antarctic and Antarctic soils in the areas of some Russian polar stations. Solid Earth 5:705–712

Barcikowski A, Lyszkiewic A, Loro P et al (2005) Keystone species and ecosystems functioning: the role of penguin colonies in differentiation of the terrestrial vegetation in the maritime Antarctic. Ecol Questions 6:117–128

Bockheim, J (ed) (2015) The soils of Antarctica, 290 p

Bockheim JG, Ugolini FC (1990) A review of pedogenic zonation in well-drained soils of the southern circumpolar region. Quaternary Res 34:47–66

Campbell IB, Claridge GGC (1987) Antarctica: soils, weathering processes and environment. Elsvier, Amsterdam, p 368

Korsun S, Kozeretska I, Parnikoza I, Skarivska L, Lugovska K, Klimenko I (2008) Effect of natural and anthropogenic factors on the chemical composition of soils of the King George in littoral Antarctic. Agroecological J 4:45–52

Kozeretska IA, Parnikoza IY, Mustafa O, Tyschenko OV, Korsun SG, Convey P (2010) Development of Antarctic herb tundra vegetation near Arctowski station, King George Island. Polar Sci 3:254–261

Krywult M, Smykla J, Wincenciak A (2003) Influence of ornithogenic fertilization on nitrogen metabolism of the Antarctic vegetation. The functioning of polar ecosystems as viewed against global environmental changes. The Institute of Botany of the Jagiellonian University, Krakow, pp 123–127

Kubiena WL (1970) Micromorpohologic investigation of Antarctic soils. Antarct J 5(4):105–106

Lupachev AV, Abakumov EV (2013) Soils of Marie Byrd Land, West Antarctica. Eur Soil Sci 46:994–1006

Mergelov NS, Goryachkin SV, Shorkunov IG, Zazovskaya EP, Cherkinskii AE (2012) Endolithic pedogenesis and rock varnish on massive crystalline rocks in East Antarctica. Eur Soil Sci 45:901–918

Parnikoza IY, Abakumov EV, Dykyy IV, Pilipenko DV, Shvydun PP, Kozeretska IA, Kunakh VA (2014) The influence of birds for spatial distribution of Deschampsia antarctica Desv. on Galindez Island (Argentine Island, maritime Antarctica). Russian Ornithological J 23 (1056):3095–3107

Parnikoza I, Miryuta NY, Maidanyuk DN, Loparev SA, Korsun SG, Budzanivska IG, Shevchenko TP, Polischuk VP, Kunakh VA, Kozeretska IA (2007) Habitat and leaf cytogenetic characteristics of *Deschampsia antarctica* Desv. in maritime Antarctic. Polar Sci 1:121–127

Parnikoza I, Kozeretska O, Kozeretska I (2008a) Is a translocation of indigenous plant material successful in the maritime Antarctic? Polarforshung 78(1–2):25–27

Parnikoza I, Inozemtseva DM, Tyshenko OV, Mustafa O, Kozeretska IA (2008b) Antarctic herb tundra colonization zones in the context of ecological gradient of glacial retreat. Ukrainian Botan J 65:504–511

Parnikoza I, Smykla J, Kozeretska I, Kunakh V (2009) Details of Antarctic tundra in two ecological gradients conditions. Visn ukr tov genet sel 7(2):218–226

Parnikoza I, Korsun S, Kozeretska I, Kunakh V (2011) A discussion note on soil development under the influence of terrestrial vegetation at two distant regions of the maritime Antarctic. Polarforschung 80(3):181–185

Parnikoza I, Dykyy I, Ivanets V, Kozeretska I, Kunakh V, Rozhok Ochyra R, Convey P (2012a) Use of Deschampsia antarctica for nest building by the kelp gull in the Argentine Island area (maritime Antarctica). Polar Biol. doi:10.1007/s00300-012-1212-5

Parnikoza I, Dykyy I, Ivanets V, Kozeretska I, Rozok A, Kunakh V (2012b) Transfer of Antarctic herb tundra formation components by kelp gull in the Argentine Islands area (maritime Antarctica). Ukrainian Antarct J 10–11:272–281

Peklo AM (2007) Birds of Argentine Islands and Piterman Island. Kriviy Rig, Mineral 264 p

Ramsay AJ (1983) Bacterial biomass in ornithogenic soils of Antarctica. J Polar Biol 1:221–225

Shaefer CEGR, Simas FNB, Gilkes RJ, Mathison C, da Costa LM, Albuquerque A (2008) Micromorphology and microchemistry of selected cryosoils from maritime Antarctica. Geoderma 144:104–115

Simas FNB, Schaefer CEGR, Melo VF, Albuquerque-Filho MR, Michel RFM, Pereira VV, Gomes MRM, da Costa LM (2007) Ornithogenic cryosols from Maritime Antarctica: phosphatization as a soil forming process. Geoderma 138:191–203

Smykla J, Wołek J, Barcikowski A et al (2006) Vegetation patterns around penguin rookeries at Admiralty Bay, King George Island, maritime Antarctic: preliminary results. Polish Botanical Stud 22:449–459

Vlasov DY, Zelenskaya MS, Kirtsideli IY, Abakumov EV, Krylenkov VA, Lukin VV (2012) Fungi on natural and anthropogenic substrates in Western Antarctica. Mycol Phytopathol 46 (1):20–26

Walkley A (1935) An examination of methods for determining organic carbon and nitrogen in soils. J Agr Sci 25:598–609

World Reference Base for Soil Resources (2006) A framework for international classification, correlation and communication. World soil resources reports 103

Zhao Ye (2000) The soil and environment in the Fildes Peninsula of Kind George Island, Antarctica, China, Bejing

# Clay Minerals in the Loose Substrate of Quarries Affected by Vegetation in the Cold Environment (Siberia, Russia)

**Olga I. Sumina and Sofia N. Lessovaia**

**Abstract** Pioneer plant communities play an important role in the process of parent substrate colonization by biota and a successful restoration of ecosystem as well, especially in the first stages of recovery successions. The aim of the present research is to study the influence of plant communities of the initial stages of primary succession on the mineral composition of substrates from sandy quarries situated in the forest-tundra zone to understand the specificity of substrate transformation initiated by vegetation, Western Siberia close to the town of Labytnangi. Sandy substrate was quarried here in former open woodlands, in communities with spruce, larch, and birch in the overstory and dwarf shrubs, mosses, and lichens in the ground layer. The time of vegetation development in quarries varies from 15 to 40 years. In substrates, pH values decrease simultaneously with the rise of moistening as well as plant canopy closure, which also influences the moistening. Clay size fraction of all samples is characterized by the same mineral association, as follows: highly smectitic clay, minerals of the mica group, chlorite, and kaolinite. In addition, traces of quartz were also identified. According to our findings, the changes in substrate mineralogy affected by the plant community decrease from mosses (reduced proportion of highly smectitic clay and transformation of chlorite into random mixed-layer chlorite-smectite), lichens (reduced proportion of highly smectitic clay), and vascular plants (absence of changes).

**Keywords** Clay minerals · Sandy quarries · Primary succession · Plant community development

O.I. Sumina (✉)
Faculty of Biology, St. Petersburg State University, St. Petersburg, Russia
e-mail: o.sumina@spbu.ru

S.N. Lessovaia
Institute of Earth Sciences, St. Petersburg State University, St. Petersburg, Russia
e-mail: s.lesovaya@spbul.ru

 
O.V. Frank-Kamenetskaya et al. (eds.), *Biogenic—Abiogenic Interactions in Natural and Anthropogenic Systems*, Lecture Notes in Earth System Sciences,
DOI 10.1007/978-3-319-24987-2_20

# 1 Introduction

Primary succession is a model that can be used to study the colonization processes of starting off from scratch of bare substrates by plants, the initial stages of plant community development, and trends of the process of terrestrial ecosystem formation. Primary successions "demonstrate the creative role of nature in building up complete and complex ecosystems from simple beginnings" (Bradshaw 1993). In addition to the above, studying primary successions provides new knowledge concerning vegetation recovery on areas significantly affected by anthropogenic influence (Moral del et al. 2007; Walker and Moral del 2008). Quarries can be considered as a perfect object for studying primary successions, in which total destruction of soil-plant cover is the cause of distribution on the surface of unweathered substrate on huge areas. Pioneer plant communities play an important role in the process of parent substrate colonization by biota and a successful restoration of ecosystem as well, especially in the first stages of recovery successions.

Summarizing data, the most significant role of substrate colonization by biota has been with lichens (Torbjørg and Thorseth 2002). The biological impact on weathering was recognized by Elenkin (1901) and further developed by Polynov (1945), who in the 1940s further described the role of lichens in the weathering of hard rocks. Morphologically simple and microbially dominated ecosystems termed "biofilms" existed later. Such systems form at interfaces of solid materials with gas or liquid phases (Burford et al. 2003; Gorbushina 2007).

The influence of vascular plants has been mostly studied in experiments. More pronounced dissolution of minerals under conifer than deciduous trees has been shown for the example of feldspars (Augusto et al. 2000). Differences of granite and dolomite pebble dissolution in the soil environment of forest and meadow ecosystems were explained by moisture conditions (Dixon et al. 2001). It was shown that the properties of substrate surfaces determine the final distribution of seeds (Chambers et al. 1990). In addition, the speed of colonization and quantity of pioneer species strongly depends on the proportion of the clay size fraction in the substrate (Borgegård 1990; Densmore 1994; Harper and Kershaw 1997; Forbes 1997; Cannone and Gerdol 2003).

The aim of present research was to study the influence of plant communities in the initial stages of primary succession on the mineral composition of substrates from sandy quarries situated in the forest-tundra zone to understand the specificity of substrate transformation initiated by vegetation.

## 2 Objects

The studied quarries were located in the forest–tundra zone of Western Siberia close to the town of Labytnangi (66° 39′N, 66° 25′E) (Fig. 1). Sandy substrate was quarried here on the place of former open woodlands, with communities of spruce, larch, and birch in the overstory and dwarf shrubs, mosses, and lichens in the ground layer. The time of vegetation development on the quarries varied from 15 to 40 years (Sumina 2013). The time of quarry overgrowth was not enough to form a mature soil profile, even in the oldest quarry. This fact determined our strategy of soil sampling. The selected key plots characterize the main ecotopes in the quarries and the stages of primary succession. The depths of soil sampling were 1–2 and 5–7 cm in the key plots, with domination of vascular plants in the plant community. In the case of a significant proportion of mosses and lichens, the sampling took place directly under the moss or lichen cover and at a depth of 3 cm.

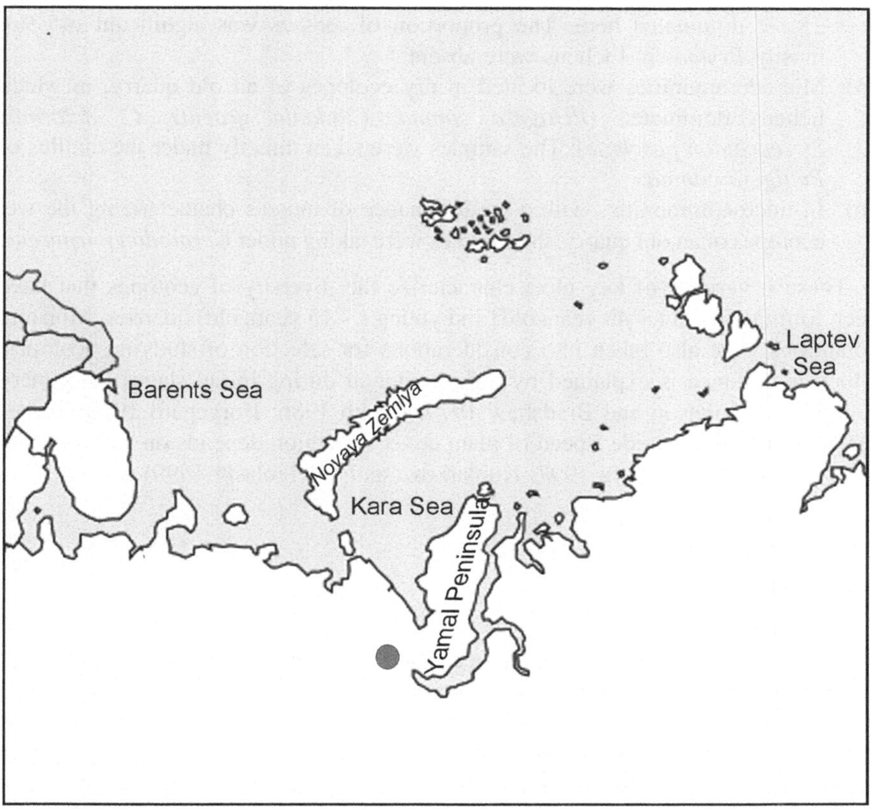


**Fig. 1** Location of studied quarries in Western Siberia

The key plots characterize the following plant communities:

(1) An open pioneer community with a total plant cover of 5–10 % formed in dry ecotopes of young quarry, with an age of ~15 years (young quarry). *Chamaenerion angustifolium* and *Salix viminalis* dominated here. Mosses and lichens were absent.
(2) Wet ecotopes of a young quarry, plant community developed with a total cover of 30–40 %, in which the coverage of *Salix viminalis* was 20 % and *Alopecurus aequalis* was 10 %. Mosses and lichens were also absent here.
(3) A plant community with total plant cover reaching close to 70–75 % formed in wet ecotopes of an old quarry (40 years old), which is significantly higher than in key plots (1) and (2). *Equisetum arvense* dominated here (40–50 %), with an admixture of *Festuca ovina* (5–10 %) and *Salix phylicifolia* and *S. viminalis*. The plant cover of two species of willows was 10–15 % in total. The soil surface was covered by mosses (20 %), mostly *Ceratodon purpureus*. Lichens were very rare.
(4) A plant community with total plant cover close to 95 % was located in wet ecotopes of an old quarry. *Festuca ovina* (30 %) and *Salix phylicifolia* (20–25 %) dominated here. The proportion of mosses was significant (45 %), mostly *Bryum* sp. Lichens were absent.
(5) Microcommunities were located in dry ecotopes of an old quarry, in which lichens dominated (*Peltigera canina, Cladonia gracilis, C. deformis, Stereocaulon paschale*). The samples were taken directly under the thalllus of *Peltigera canina*.
(6) In microcommunities with a predominance of mosses characterizing the wet ecotopes of an old quarry, the samples were taking under *Ceratodon purpureus*.

These 6 variants of key plots characterize the diversity of ecotopes that have been forming in old (~40 years old) and young (~15 years old) quarries. Moisture conditions were also taken into considerations for selection of studying ecotopes, which importance is explained by nutrient deficit during initial stages of primary successions (Johnson and Bradshaw 1979; Grubb 1986; Borgegård 1990; Forbes 1997). From another side, speed of plant cover formation depends on soil moisture (Ebersole 1985; Stranberg 1997; Komarkova and Wielgolaski 1999).

## 3 Methods

The studied samples were carbonate free. Therefore, a <1 mm fraction was obtained by gently grinding in a mortar and subsequently dry sieving. The pH values were measured potentiometrically in $H_2O$ with a soil-to-water ratio of 1:2.5. After dispersion of the sample by the addition of some drops of ammonia as a peptizing agent, the <1 μm fraction from soil horizons was separated by sedimentation and decantation. The mineralogical composition of the <1 μm fractions were studied by

X-ray diffraction (XRD; DRON-2) with CoKα radiation and a monochromator in the diffracted beam. Pretreatment of the <1 μm fractions before XRD included saturation with Mg; heating at 550 °C for 3 h, and ethylene glycol solvation.

The samples studied by X-ray diffraction were also investigated by FTIR (Fourier transform infrared) spectroscopy. FTIR spectra in the 400–4000 $cm^{-1}$ range were obtained using a Tensor 27 spectrometer (Bruker) with a resolution of 4 $cm^{-1}$ in ambient air and at room temperature. Spectra were recorded in the transmission mode using KBr pellets. The KBr and clay samples were heated in a furnace overnight at 105 °C to minimize adsorbed water. Then, 1 mg of the sample and 300 mg of KBr were thoroughly mixed and pressed to a transparent pellet. Their infrared spectra were recorded.

## 4 Results

Based on the findings of pH value, most acidic substrates are in the wet ecotopes of an old quarry in key plots (3), (4), and (6). The pH values in key plot (4) were 5.2 on the surface and 5.0 at depth. The tendency for an insignificant increase of pH values at depth was fixed in the other ecotopes: 4.8 at the surface and 5.2 at depth in key plot (3), and 5.0 at the surface and 5.2 at depth in key plot (6). In the dry ecotopes of an old quarry, as in key plot (5), the substrate is slightly less acidic, even when lichen were present: 5.3 on the surface and 5.6 at depth. As opposed to those in young quarries, these substrates are characterized by a decrease in acidity even in wet ecotopes [5.6 on the surface and 5.7 at depth in key plot (2)], whereas the substrate is neutral in dry ecotopes [6.3 without any changes at a depth in key plot (1)]. Thus, in substrates, pH values decrease simultaneously with the rise of moisture as well as plant cover density, which also influences the moisture level.

The clay size fraction of all samples is characterized by the same mineral association, which is the following: highly smectitic clay, minerals of the mica group, chlorite, and kaolinite. In addition, traces of quartz were also identified.

Highly smectitic clay is identified by a response to ethylene glycol solvation, whereby a peak at ~14 Å in the Mg-saturated air-dry state shifts to a fundamental 001 spacing of ~17 Å. As the 002 reflection at 8.6 Å (an indistinct peak in XRD patterns of several samples) is characteristic of a pure smectite mineral, it is probable that the expandable phase partially or overall represents a pure mineral of smectite group. In addition, the high background towards the low angle side of the 17 Å peak is typical of a random mixed-layer illite-smectite structure where the smectite layers exceed 50 % (Reynolds 1980).

Chlorite identification is based on the 001 and 003 basal reflections at ~14.0 and 4.75 Å, respectively, which are stable after ethylene glycol solvation and show a slight contraction of the 001 peak to 13.8 Å after 550 °C treatment. In the XRD patterns of several samples, the 001 peak shifted to ~12 Å after 550 °C treatment, hinting at the presence of random mixed layer structures as a result of chlorite transformation.

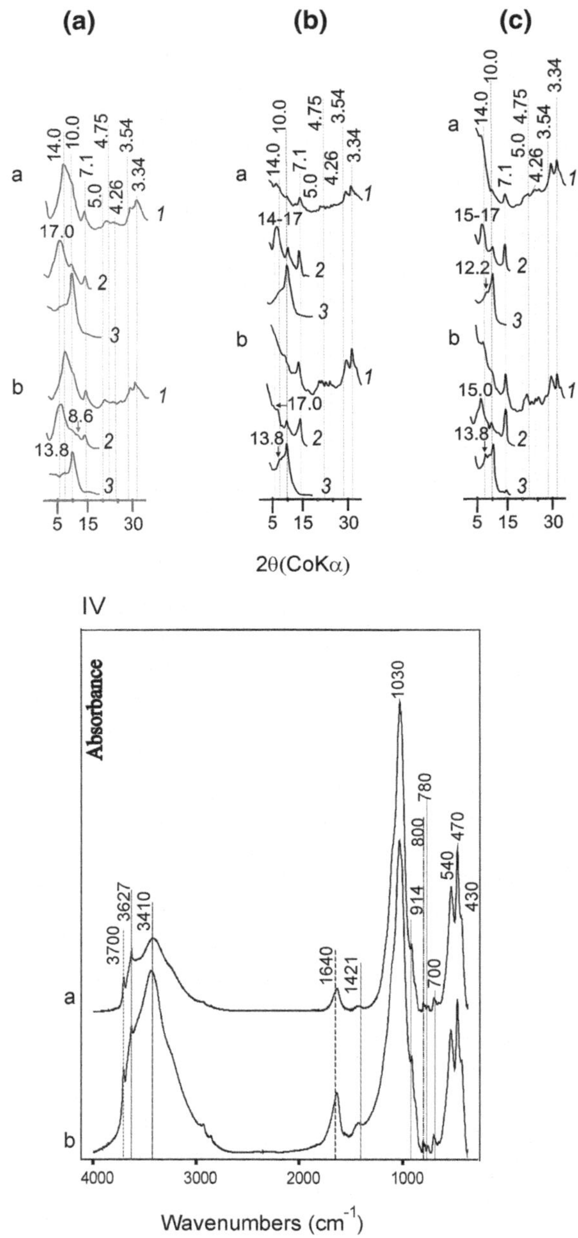


**Fig. 2** XRD patterns (**a–c**) of the <1 μm fraction of substrate samples under the following: **a** vascular plants, key plot (*2*); **b** lichens, key plot (*5*); and **c** mosses, key plot (*6*); *1*—Mg-saturated, *2*—ethylene glycol solvated, *3*—550 °C treatment, *d*-values are given in Angstrom; *a*—sample from the surface and *b*—sample from depth. **d** FTIR spectra of samples from key plot (*2*)

Minerals of the mica group are recognized by peaks at the 001 and 002 basal reflections at 10.0 and 5.0 Å, respectively, which remain stable after ethylene glycol and heat treatments.

Kaolinite is recognized by its 001 reflection at 7.1 Å, which disappears after heating at 550 °C. The identification is confirmed by FTIR spectroscopy data, which enable kaolinite to be identified in the presence of chlorite. IR spectroscopy is a sensitive method for the detection of kaolinite (Delvaux et al. 1989). The absorption bands near 3700 $cm^{-1}$ (the vibration of inner surface OH groups), 3627 $cm^{-1}$ (the stretching vibration of inner OH groups), 914 $cm^{-1}$ (bending vibration of inner OH groups), 540 $cm^{-1}$ (Si–O–Al bending vibrations), and 470 $cm^{-1}$ (Si–O–Si bending vibrations) all indicate the presence of kaolinite (Madejová and Komadel 2001; Pentrak et al. 2009).

Quartz is identified by its characteristic XRD peaks at 3.34 and 4.26 Å and by its most typical infrared absorption band doublet at 800 and 780 $cm^{-1}$ (Hlavay et al. 1978; Madejová and Komadel 2005) (Fig. 2).

Using the above criteria, the distribution of the various clay minerals can be described as follows. The changes of mineral composition affected by vascular plants or vascular plants with admixture of mosses were not fixed despite the differences in moisture and even low pH values (4.8). The influence of the plant community on clay minerals in substrates also was not fixed in the ecotopes with relatively high proportions of mosses in the total cover: 20–45 % in key plots (3) and (4). On the contrary, significant changes were found in the substrates under micro-communities of mosses and lichens [key plots (5) and (6)]. Under thallus *Peltigera canina,* the proportion of highly smectitic clay considerably decreased based on the intensity of the $d_{001}$ peak in XRD patterns of ethylene glycol solvated samples. An interesting fact is that the decrease of highly smectitic clay is more pronounced at a depth of 2–3 cm than under thallus. Under mosses, especially in the sample from the surface, the decrease of highly smectitic clay proportion was also found. Furthermore, in the surface sample, the transformation of chlorite into random mixed layer structures was identified based on a 001 peak shift to ~12 Å after 550 °C treatment.

## 5 Discussion

The differences in sensitivity to vegetation influence based on the mineralogy of clay size fraction do not depend on ecotope moisture in quarries of different ages and with different varieties of total plant cover. Based on our findings, changes in substrate mineralogy were affected by vegetation decrease: (i) mosses (reduced proportion of highly smectitic clay and transformation of chlorite into random mixed-layer chlorite-smectite); (ii) lichens (reduced proportion of highly smectitic clay); and (iii) vascular plants (absence of changes).

The high sensitivity of highly smectitic clay in the studied samples is in good agreement with hypothesis of selective destruction of clay minerals in acidic soils developed in the example of soils from the northern taiga zone (Tonkonogov et al. 1987). In accordance with the hypothesis, smectite(s) is the most instable mineral. More pronounced destruction of smectite(s) was identified in soil profiles developed on ultrabasic rock (the Polar Urals, Russia), where the surface is covered by lichens and moss, and less intensive in profiles where vegetation cover is absent; this may be due to the decomposing acidic effects of moss and lichens (Lessovaia and Polekhovsky 2009). Markedly reduced inherited smectitic clay in the upper acidic part of soils formed on the calcareous substrate in the ultra-continental climate of Central Yakutia plain (East Siberia, Russia) has also been observed (Lessovaia et al. 2013). Similar transformations caused a loss of expandability of smectitic clays in the acidic upper horizons, which are very common for soils such as Podzols and Brown forest soils (Wilson et al. 1984; Wilson 1999).

The situation with regard to chlorite can also be explained by the acidic effect of mosses, taking into account that pH value could be significantly lower in the rhizosphere than in the bulk sample. Complete decomposition of the clay mineral occurs in some soils (Bain 1977), which is not surprising considering its susceptibility to acid dissolution in the laboratory; in other soils, it has been proposed that the mineral forms interlayered and mixed-layer clays (De Coninck et al. 1983; Righi et al. 1993). However, there is not enough evidence to decide upon these two possibilities, only transformation into random mixed-layer structures fixed by the appearance of 12 Å after 550 °C treatment or also decomposition. The transformation of chlorite into randomly interstratified chlorite-smectite in circumneutral to weakly acidic conditions has been documented for permafrost-affected soils Kriozems (Cryosols) formed in the cold, continental, climate of the Kolyma Lowland, North-East Siberia, as has the extremely weak intensity of chemical weathering of the Yedoma deposits (Alekseev et al. 2003).

Thus, the effect of vegetation on the mineral composition of substrates increases simultaneously with the presence of mosses and lichens. Mosses can appear as pioneer species, such as with *Ceratodon purpureus* and species of genus *Bryum*, *Polytrichum*, as well as representatives of advanced stages (*Aulacomnium turgidum*, *A. palustre*, *Tomentypnum nitens*). Lichens (excluding one-two pioneer species, mostly *Peltigera didactyla*, *Dibaeis baeomyces*) appear in the advanced stages of successions (Sumina and Koptzeva 2004), whereas vascular plants dominate in the plant communities in the initial stages of succession of quarry overgrowth (Sumina 1995, 2013; Koptzeva and Sumina 2001).

As opposed to the traditional viewpoint that the speed of primary succession becomes slower, our results support the hypothesis that acceleration of succession takes place during the "middle" stages, simultaneously with increasing proportions of mosses and lichens. Mosses and lichens can significantly change the substrate properties, creating new environments for vascular plants. Formation of moss-lichen cover concurrently with the development of overstory composed by shrubs and later by young growth of trees increases the effects of a plant community on the abiotic environment (Sumina 2013). The acceleration of ecosystem formation may occur due to the appearance of new heterotrophs, whereas the species composition of plant–producers will be stable from the "middle" stages.

# 6 Conclusions

Mineral associations of clay size fraction from studied substrates—in which smectitic components, minerals of the mica group, and chlorite were identified—allowed us to characterize them as "rich" based on the presence of nutrients easily accessible for the plant community. Based on our findings, the changes in substrate mineralogy affected by the plant community decrease mosses (reduced proportion of highly smectitic clay and transformation of chlorite into random mixed-layer chlorite-smectite), lichens (reduced proportion of highly smectitic clay), and vascular plants (absence of changes).

**Acknowledgements** This study was supported by the Russian Foundation for Basic Research (14-04-00327). The XRD study was carried out in the X-ray Diffraction Centre of St. Petersburg State University.

# References

Alekseev A, Alekseeva T, Ostroumov V, Siegert C, Gradusov B (2003) Mineral transformation in permafrost-affected soils, North Kolyma Lowland, Russia. Soil Sci Soc Am J 67:596–605

Augusto L, Turpault M-P, Ranger J (2000) Impact of forest tree species on feldspar weathering rates. Geoderma 96(3):215–237

Bain DC (1977) The weathering of some chlorite minerals in Scottish soils. J Soil Sci 28:144–164

Borgegård SO (1990) Vegetation development in abandoned gravel pits: effects for surrounding vegetation, substrate, and region. J Veg Sci 1:675–682

Bradshaw AD (1993) Introduction: understanding the fundamentals of succession. Primary succession on land. In: Miles J, Walson DWH (eds) Special publication number 12 of the British Ecological Society. Blackwell Scientific Publications, Oxford, London

Burford EP, Fomina M, Gadd GM (2003) Fungal involvement in bioweathering and biotransformation of rocks and minerals. Mineral Mag 67:1127–1155

Cannone N, Gerdol R (2003) Vegetation as an ecological indicator of surface instability in Rock Glaciers. Arct Antarct Alp Res 35(3):384–390

Chambers JC, MacMahon JA, Brown RW (1990) Alpine seedling establishment: the influence of disturbance type. Ecology 71(4):1323–1341

De Coninck F, van Ranst E, Jensen W (1983) Trioctahedral and dioctahedral chlorite in soils: examples of a Dystrochrept (Corsica), a Cryorthod (Norway) and a Hapludalf (France). In: Nahon D, Noack Y (eds) Pétrologie des altérations et des sols, Sciences géologiques. Mémoire, Strasbourg, 73:75–84

del Moral R, Walker LR, Bakker JP (2007) Insights gained from succession for the restoration of landscape structure and function. In: Walker LR, Walker J, Hobbs RJ (eds) Linking restoration and ecological succession. Springer, New York

Delvaux B, Mestdagh MM, Vielvoye L, Herbillon AJ (1989) XRD, IR and ESR study of experimental alteration of Al-nontronite into mixed-layer kaolinite/smectite. Clay Minerals 24:617–630

Densmore R (1994) Succession on degraded placer mine spoil in Alaska, U.S.A., in relation to initial site characteristics. Arct Antarct Alp Res 26(4):354–363

Dixon JC, Thorn CE, Darmody RG, Schlyter P (2001) Weathering rates of fine pebbles at the soil surface in Kärkevagge, Swedish Lapland. Catena 45:273–286

Ebersole JJ (1985) Vegetation disturbance and recovery at the Oumalik Oil Well, Arctic Coastal Plain, Alaska. PhD thesis, University of Colorado, Boulder

Elenkin AA (1901) Lichens and soils (Lishainiki i pochva). Pochvovedenie 4:319–324 (in Russian)

Forbes BC (1997) Tundra disturbance studies. IV. Species establishment on anthropogenic primary surfaces, Yamal Peninsula, Northwest Siberia, Russia. Polar Geogr 21(2):79–100

Gorbushina AA (2007) Life on the rocks. Environ Microbiol 9:1613–1631

Grubb PJ (1986) The ecology of establishment. In: Bradshaw AD, Goode DA, Thorp EHS (eds) Ecology and design in landscape. Blackwell Scientific Publications, Oxford

Harper KA, Kershaw GP (1997) Soil characteristics of 48-year-old borrow pits and vehicle tracks in shrub tundra along the CANOL No. 1 Pipeline corridor, Northwest Territories, Canada. Arct Antarct Alp Res 29:105–111

Hlavay J, Jonas K, Elek S, Inczedy J (1978) Characterization of the particle size and the crystallinity of certain minerals by IR spectrophotometry and other instrumental methods—II. Investigations on quartz and feldspar. Clays Clay Miner 26:139–143

Johnson MS, Bradshaw AD (1979) Ecological principles for the restoration of disturbed and degradated land. Appl Biol 4:141–200

Komarkova V, Wielgolaski FE (1999) Stress and disturbance in cold region ecosystems. In: Walker LR (ed) Ecosystems of disturbed ground. Ecosystems of the World, vol 16. Elsevier, Amsterdam, New York, Oxford

Koptzeva EM, Sumina OI (2001) Pants of anthropogenic and natural habitats on route of constructing railway (South Yamal) (Rasteniya tekhnogennykh i estestvennykh mestoobitanii na trasse stroyasheisya zheleznoi dorogi). Botanicheskii Zhurnal 86(9):95–108 (in Russian)

Lessovaia SN, Polekhovsky YS (2009) Mineralogical composition of shallow soils on basic and ultrabasic rocks of East Fennoscandia and of the Ural Mountains, Russia. Clays Clay Miner 57:476–485

Lessovaia SN, Goryachkin SV, Desyatkin RV, Okoneshnikova MV (2013) Pedoweathering and mineralogical change in Cryosols in an ultracontinental climate (Central Yakutia, Russia). Acta Geodyn. Geomater 10(4)(172):465–473

Madejová J, Komadel P (2001) Baseline studies of the clay minerals society source clays: infrared methods. Clays Clay Miner 49:410–432

Madejová J, Komadel P (2005) Information available from infrared spectra of the fine fraction of bentonites. In: Kloprogge JT (ed) The application of vibrational spectroscopy of clay minerals and layered double hydroxides, CMS workshop lectures, vol 13. The Clay Mineral Society, Aurora CO

Pentrak M, Madejová J, Komadel P (2009) Acid and alkali treatment of kaolins. Clay Miner 44:511–523

Polynov BB (1945) The first stages of pedogenesis on massive crystalline rocks. (Pervye stadii pochvoobrazovaniya na massivno-kristallicheskekh porodakh). Pochvovedenie 7:327–339 (in Russian)

Reynolds RC (1980) Interstratified clay minerals. In: Brindley GW, Brown G (eds) Crystal structure of clay minerals and their X-ray identification. Mineralogical Society, Monograph. No. 5, London

Righi D, Petit S, Bouchet A (1993) Characterization of hydroxy-interlayered vermiculite and illite/smectite interstratified minerals from the weathering of chlorite in a Cryorthod. Clays Clay Miner 41:484–495

Stranberg B (1997) Vegetation recovery following anthropogenic disturbances in Greenland. In: Crawford RMM (ed) Disturbance and recovery in arctic lands: an ecological perspective. Academic Press, Dordrecht

Sumina OI (1995) About the classification of vegetation of anthropogenic habitats of the Arctic (isthmus of Chukchi Peninsula) (O klassifikazii rastitel'nosti tekhnogennykh mestoobitanii Arktiki (peresheek Chukotskogo poluostrova). Botanicheskii zhurnal 80(10):79–90 (in Russian)

Sumina OI (2013) Development of vegetation on human-made habitats of the Russian Far North (Formirovanie rastitel'nosti na technogennykh mestoobitaniyakh Krainego Severa Rossii). Inform-Navigator, Saint-Petersburg. (in Russian)
Sumina OI, Koptzeva EM (2004) Diversity and dynamic of quarries vegetation in forest-tundra zone of Western Siberia (vicinity of town Labytnangi, Yamal-Nenets Autonomous District) (Raznoobrazie i dinamika rastitel'nosti kar'erov v lesotundre Zapadnoi Sibiri (okrestnosti g. Labytnangi, Yamal-Nenets AO)). Zhurnal Rastitel'nost' Rossii 6:83–103 (in Russian)
Tonkonogov VD, Gradusov BP, Rubilina NE, Targulian VO, Chizhikova NP (1987) On the differentiation of the mineralogical and chemical compositions of soddy podzolic and podzolic soils (K differenziazii mineralogicheskogo i khimicheskogo sostava dernovo-podzolistykh i podzolistykh pochv). Pochvovedenie 3:68–81 (in Russian)
Torbjørg B, Thorseth IH (2002) Comparative studies of lichen—rock interface of four lichens in Vingen, western Norway. Chem Geol 192:81–98
Walker LR, del Moral R (2008) Lessons from primary succession for restoration of severely damaged habitats. Appl Veg Sci 12:55–67
Wilson MJ (1999) The origin and formation of clay minerals in soils: past, present and future perspectives. Clay Miner 34:7–25
Wilson MJ, Bain DC, Duthie DML (1984) The soil clays of Great Britain. II Scotland. Clay Miner 19:709–735

# Rare Earth Elements and Yttrium in the Soil Forming Materials and Ploughing Horizons of North-West Russia

**Natalia N. Matinian, Kseniia A. Bakhmatova and Anastasiia A. Sheshukova**

**Abstract** This study presents the rare earth elements (REE) and Y concentrations in the main soil forming materials and ploughing horizons of North-West Russia agricultural soils. The study area covers Leningrad region, west part of Novgorod and Pskov regions and southern Karelia. Investigated area (200,000 km$^2$) was subdivided into a net of square cells 50 km on 50 km in size. Eighty-one samples of topsoil (0–25 cm, ploughing layer) and subsoil (C-horizon) were collected. REE and Y concentrations were measured using ICP-MS (Perkin Elmer Sciex Elan 5000). The median concentrations of the investigated elements in the soil forming materials were as follows (mg/kg): Y—15.3, La—21.7, Ce—45.0, Pr—5.5, Nd—20.9, Sm—3.9, Eu—0.8, Gd—3.5, Tb—0.5, Dy—2.7, Ho—0.5, Er—1.5, Tm—0.2, Yb—1.5, Lu—0.2. It was found that the content of REE in northwestern Russia soil forming materials is less than in Earth crust. Varve clays and carbonate moraines are enriched in REE content relatively another soil forming materials of northwestern Russia (non-carbonate moraines, lacustrine-glacial loams, sandy loams and sands, fluvioglacial sands and alluvial sands). Content of REE in soils depends on their concentrations in soil forming materials, therefore, spatial distribution of REE in ploughing horizons of agricultural soils reflects regional abundance of different soil forming materials.

**Keywords** Rare earth elements · Lanthanides · Yttrium · North-West Russia · Parent materials · Arable soils

## 1 Introduction

Properties and chemical composition of the deposits and the soils of the North-West region of Russia are fairly well investigated; however, the data on the content of rare earth elements (REE) are extremely scarce and the role of REE in soil

N.N. Matinian (✉) · K.A. Bakhmatova · A.A. Sheshukova
Saint Petersburg State University, Saint Petersburg, Russia
e-mail: geosoil@mail.ru



O.V. Frank-Kamenetskaya et al. (eds.), *Biogenic—Abiogenic Interactions in Natural and Anthropogenic Systems*, Lecture Notes in Earth System Sciences,
DOI 10.1007/978-3-319-24987-2_21

formation is still poorly understood. The term "REE" refers to lanthanides (Taylor and McLennan 2003), but in many studies Y is considered in that group due to its similar ionic radius. Until recently, the quantitative determination of REE in soils has been limited for a long time because of the lack of sensitive analytical technique. For the last two decades, the geochemistry of REE has been extensively investigated because of their scientific and economical applications (e.g., medicine, agriculture, high-tech industry). The significant impact on the REE studies has had the development of ICP-MS analytical technique.

REE have rather similar chemical and physical properties. Most of REE are stable in a trivalent state, but Ce may occur in +4 valence and Eu—in +2 valence. Generally, REE are divided in light REE (LREE, from La to Eu) and heavy REE (HREE, from Gd to Lu). The REE are found in several accessory minerals (phosphates, carbonates, fluorides, silicates). They occur particularly in such igneous rocks as pegmatites and granites (Tyler 2004). Many silicate minerals may contain REE, in particular LREEs (cerite, allanite). Zr mineral zircon may be enriched by HREEs (Pan et al. 1994). REE are concentrated in phosphorites, acid igneous rocks and argillaceous sediments (Kabata-Pendias and Pendias 2000). A major amount of REE in clastic sedimentary deposits is found in fraction <2 m km. The minerals in the heavy fraction of sand deposits are also enriched by REE, especially HREE (Taylor and McLennan 1985). In the recent years, many studies dealing with the geochemistry and biogeochemistry of REE in soils and soil-plant systems have been carried out (Hu et al. 2006; Liang et al. 2008; Davranche et al. 2011; Miao et al. 2011; Vodyanitskii et al. 2011; Perelomov et al. 2012; Vodyanitskii 2012; Zhu et al. 2012; Gonsalez et al. 2014; Pedrot et al. 2015). Laveuf and Cornu (2009) considered the use of REE to trace pedogenic processes, particularly argilluviation and redox processes, Chen et al. (2014) used REE for sediment provenances, parent material uniformity and soil evolution investigation. It was shown that REE concentrations in soils depend on the mineralogy of soil forming material (Brioschi et al. 2013). The absorption of REE onto clays and amorphous Fe–Mn oxides, formation of phosphate or organic complexes with REE are the factors that affect the geochemical fractionation and mobility of REE; the mobility tends to increase with decreasing pH (Mihajlovic et al. 2014).

Literature data analysis indicated significant variety in concentration of REE in sedimentary rocks, especially in argillic material. In common, sedimentary cover of Russian plate contains less quantity of REE than earth's crust (Table 1).

The aim of the present study is to investigate the concentrations of REE and Y in different soil forming materials and ploughing horizons of arable soils of North-West Russia. The significance of investigation of baseline REE concentration in soils is determined by the growing technogenic input of these elements into the environment. The more that, both positive and negative effects of REE on plants were found (Gonzalez et al. 2014), so the concentration of REE should be monitored and taken into account in the environmental assessment.

**Table 1** Total REE and Y concentration (mean) in the earth crust, and soil forming materials (mg/kg)

| Y | La | Ce | Pr | Nd | Sm | Eu | Gd | Tb | Dy | Ho | Er | Tm | Yb | Lu |
|---|---|---|---|---|---|---|---|---|---|---|---|---|---|---|
| *Earth's crust*[a] | | | | | | | | | | | | | | |
| 31 | 35 | 66.0 | 9.1 | 40.0 | 7.0 | 2.1 | 6.1 | 1.2 | 4.5 | 1.3 | 3.5 | 0.5 | 3.1 | 0.8 |
| *Sedimentary cover of Russian plate*[b] | | | | | | | | | | | | | | |
| 16.5 | 24.4 | 43.9 | 5.8 | 20.8 | 4.1 | 1.0 | 3.6 | 0.6 | 3.0 | 0.7 | 1.9 | 0.3 | 1.7 | 0.3 |
| *Sedimentary rocks:*[c] | | | | | | | | | | | | | | |
| Argillaceous | | | | | | | | | | | | | | |
| 25–40 | 30–90 | 3–90 | 6.0–10.0 | 18.0–35.0 | 5.0–7.0 | 1.0–2.0 | 5.0–7.5 | 0.9–1.1 | 4.0–6.0 | 1.0–1.8 | 2.5–4.0 | 0.2–0.6 | 2.2–4.0 | 0.2–0.7 |
| Sandstones | | | | | | | | | | | | | | |
| 15–250 | 17–40 | 25–80 | 4.0–9.0 | 16.0–48.0 | 4.0–10.0 | 0.7–2.0 | 3.0–10.0 | 1.6–2.0 | 2.6–7.2 | 0.05–2.0 | 1.5–6.0 | 0.3–0.5 | 1.2–4.4 | 0.8–1.2 |
| Calcareous | | | | | | | | | | | | | | |
| 4.0–30.0 | 4–10 | 7–20 | 1.0–2.5 | 5.0–9.0 | 1.0–2.0 | 0.2–0.4 | 1.3–2.7 | 0.2–0.4 | 0.8–2.0 | 0.2–0.3 | 0.4–1.7 | 0.03–0.2 | 0.3–1.6 | 0.03–0.2 |

[a]Greenwood and Ernshaw (1984)
[b]Ronov and Migdisov (1995)
[c]Kabata-Pendias and Mukherjee (2007)

## 2 Objects and Methods

Our study is a part of international ecological project «Baltic soil survey (BSS-project)» (Reimann et al. 2003). During 1996 and 1997, soil forming materials and ploughing horizons of agricultural soils from 10 northern European countries (western Belarus, Estonia, Finland, northern Germany, Latvia, Lithuania, Norway, Poland, northwestern Russia and Sweden) were collected.

This study presents the results of REE content investigation in the main soil forming materials of agricultural soils of North-West Russia. The study area covers Leningrad region, west part of Novgorod and Pskov regions and southern Karelia. The geology varies significantly over this area. The investigated soil parent materials are represented by several different types of quaternary sediments: moraine loams (both carbonate and non-carbonate), varved clays, lacustrine-glacial sands and sandy loams, fluvioglacial and alluvial sands.

Investigated area (part of northwestern Russia, square 200,000 km$^2$) was subdivided into a net of square cells 50 km on 50 km in size. The aim was to collect representative samples of soil forming materials and agricultural soils, forming in it from each of these net cells. It was given 81 samples of topsoil and 81 samples of subsoil (Fig. 1). Topsoil was collected from a depth of 0–25 cm (ploughing layer,

**Fig. 1** The plan of sampling

Ap-horizon) and subsoil (C-horizon) was collected at an approximate depth more than 75–80 cm. The samples were 8 l in volume and a composite of 5 to 13 sub-samples. After sampling, all samples were air-dried and then randomized. Rare earth elements concentrations were measured in HF-extraction with ICP-MS at GTK's Espoo Geolaboratory. The decomposition of the soil sample was done on two steps. First, the organic matter was removed by ashing. Afterwards an acid dissolution with hydrofluoric and perchloric acid was performed. A 0.5 g sub-sample was weighed into a porcelain dish. The crucible was placed into a muffle furnace and the temperature was slowly raised from temperature to 450 °C. The final temperature was reached after 1 h. The sample was kept at 450 °C for 3 h. After cooling, the sample was moved to a Teflon dish. As the second step of the decomposition procedure, 10 ml of 40 % hydrofluoric acid and 4 ml of 70 % perchloric acid were added, and evaporated on a hot plate, followed by dissolution in 20 ml of 8 mol/l nitric acid. 1 ml of 30 % hydrogen peroxide was added and the solution was made up to 100 ml (1.8 mol/l nitric acid). Concentrations of REE were measured using ICP-MS (Perkin Elmer Sciex Elan 5000). Lower limits of detection were (in mg/kg): for Nd—0.25, Sm—0.2, Ce, Gd, Er—0.15, La, Dy, Yb, Y—0.1, Tb, Tm, Pr, Eu, Ho, Lu—0.05.

After mathematical treatment by classical statistical methods analytical results were interpreted. Geochemical colour maps were drawn for showing regional specifics on REE distribution in agricultural soils and soil forming materials of investigation area. The ESRI ArcView™ software package was employed for mapping and interpolation. The inverse distance weighted (IDW) interpolator was used for gridding. In colour maps blue colours represent the low local regional values of REE. Green colour indicates normal values. Yellow and orange represent the increased values of REE. And red colour was used for the higher values. For class selection were chosen percentiles (5th, 25th, 50th, 75th, 90th and 98th) of the smoothed data distribution.

## 3 Results and Discussion

The investigation area is divided into three geomorphological zones: (1) Aqueoglacial lake low plain area (include Lovatsko-Volhovskiy, Velikorechkiy and Chudskoy districts); (2) Terminal moraine area (Velikolukskiy district); (3) Morainic lowland (contain Pskovskiy, Izhorskiy, Predvaldaiskiy and Lugsko-Shelonskiy districts) (Fig. 2). Sands (about 50 % in Chudskoy district) and varve clays—(40 % in Lovatsko-Volhovskiy district and almost 40 % in Velikorechkiy district)—are prevail in aqueoglacial lake low plain zone. At the territory of terminal moraine zone along similar present three type of unconsolidated quaternary sediments: sands, two-layered deposits and lateral moraines (about 30 % each of them). There is maximal variety of soil forming materials in morainic lowland zone: carbonate moraine dominate in Izorskiy (98 %) and Lugsko-Shelonskiy districts (70 %), present in Predvaldaiskiy district (37 %); non-carbonate moraines occur

**Fig. 2** Plan of geomorphological division of the North-West of Russia. *I*—Lovatsko-Volhovskiy district, *II*—Velikorechkiy district, *III*—Chudskoy district, *IV*—Velikolukskiy district, *V*—Pskovskiy district, *VI*—Izhorskiy district, *VII*—Predvaldaiskiy district, *VIII*—Lugsko-Shelonskiy district, *IX*—Vepsovsko-Andomskiy district

in Pskovskiy (38 %), Predvaldaiskiy (40 %) and Lugsko-Shelonskiy (21 %) districts; sands are located in Pskovskiy (35 %) and Predvaldaiskiy (25 %) districts, varve clays are spread only in Predvaldaiskiy (12 %) district (Gagarina et al. 1995).

At the beginning consider the obtained results for the soil forming materials. The quantitative content in the deposits determined the following order of the elements: Ce > La > Nd > Y > Pr > Sm > Gd > Dy, Er > Yb > Eu > Tb|Ho > Tm|Lu, according to the Oddo-Harkins rule (Laveuf and Cornu 2009). Compared with the data reported for sedimentary cover of Russian plate (Ronov and Migdisov 1995) concentrations of all REE were found to be somewhat lower with only exclusion for Ce concentration. We have calculated the mean and median concentrations of REE in the soil forming material of Northern Europe countries (exclude Norway and Sweden, since in it soil forming materials have magmatic origin). Generally, the median and mean concentrations of REE have similar values in the all countries (Table 2). Based on the REE content in the soil forming materials the considered countries are divided into two groups. The elevated quantities of REE are observed in Germany, Poland and Belarus. Lesser amounts of REE are found in the soil forming materials of Estonia, Finland, Germany, Latvia, Lithuania and Russia. T-statistic method that was applied to the samples from Russia permits to divide soil forming materials into four groups with significant differences in REE content.

**Table 2** Mean and median concentrations of REE and Y in the soil forming materials of series countries in northern Europe (mg/kg)

| Value | Y | La | Ce | Pr | Nd | Sm | Eu | Gd | Tb | Dy | Ho | Er | Tm | Yb | Lu |
|---|---|---|---|---|---|---|---|---|---|---|---|---|---|---|---|
| *Germany* ($n^a$ = 37) | | | | | | | | | | | | | | | |
| Mean | 10.6 | 14.7 | 30.5 | 3.5 | 13.6 | 2.6 | 0.5 | 2.3 | 0.3 | 1.8 | 0.3 | 1.0 | 0.1 | 1.0 | 0.1 |
| Median | 8.4 | 13.3 | 25.8 | 3.0 | 11.5 | 2.2 | 0.4 | 1.9 | 0.3 | 1.6 | 0.3 | 0.8 | 0.1 | 0.9 | 0.1 |
| *Poland* ($n$ = 136) | | | | | | | | | | | | | | | |
| Mean | 8.4 | 13.3 | 27.3 | 3.1 | 11.7 | 2.2 | 0.4 | 1.9 | 0.3 | 1.4 | 0.3 | 0.8 | 0.1 | 0.8 | 0.1 |
| Median | 7.1 | 12.2 | 24.6 | 2.7 | 10.1 | 1.9 | 0.3 | 1.6 | 0.2 | 1.2 | 0.2 | 0.7 | 0.1 | 0.7 | 0.1 |
| *Belarus* ($n$ = 31) | | | | | | | | | | | | | | | |
| Mean | 10.7 | 16.7 | 35.6 | 4.0 | 14.7 | 2.7 | 0.5 | 2.4 | 0.4 | 1.9 | 0.4 | 1.1 | 0.2 | 1.1 | 0.2 |
| Median | 9.4 | 14.0 | 34.3 | 3.7 | 12.2 | 2.3 | 0.4 | 2.1 | 0.3 | 1.7 | 0.3 | 0.9 | 0.1 | 0.9 | 0.1 |
| *Lithuania* ($n$ = 26) | | | | | | | | | | | | | | | |
| Mean | 15.3 | 21.0 | 44.7 | 5.2 | 19.5 | 3.7 | 0.7 | 3.5 | 0.5 | 2.7 | 0.5 | 1.5 | 0.2 | 1.4 | 0.2 |
| Median | 15.9 | 20.2 | 41.7 | 5.0 | 19.5 | 3.6 | 0.7 | 3.5 | 0.5 | 2.7 | 0.5 | 1.5 | 0.2 | 1.4 | 0.2 |
| *Latvia* ($n$ = 26) | | | | | | | | | | | | | | | |
| Mean | 17.0 | 23.1 | 47.5 | 5.8 | 21.2 | 4.1 | 0.8 | 3.9 | 0.6 | 3.0 | 0.6 | 1.6 | 0.2 | 1.5 | 0.2 |
| Median | 17.2 | 22.9 | 46.7 | 5.8 | 21.1 | 4.1 | 0.8 | 3.7 | 0.6 | 2.9 | 0.6 | 1.6 | 0.2 | 1.5 | 0.2 |
| *Estonia* ($n$ = 16) | | | | | | | | | | | | | | | |
| Mean | 16.2 | 21.9 | 46.2 | 5.3 | 19.7 | 3.8 | 0.7 | 3.5 | 0.5 | 2.8 | 0.5 | 1.5 | 0.2 | 1.5 | 0.2 |
| Median | 16.9 | 23.1 | 46.9 | 5.7 | 20.5 | 4.1 | 0.8 | 3.7 | 0.6 | 2.9 | 0.6 | 1.6 | 0.2 | 1.6 | 0.2 |
| *Russia* ($n$ = 81) | | | | | | | | | | | | | | | |
| Mean | 16.4 | 22.5 | 47.2 | 5.5 | 20.9 | 4.0 | 0.8 | 3.6 | 0.5 | 2.9 | 0.6 | 1.6 | 0.2 | 1.5 | 0.2 |
| Median | 15.3 | 21.7 | 45.0 | 5.5 | 20.9 | 3.9 | 0.8 | 3.5 | 0.5 | 2.7 | 0.5 | 1.5 | 0.2 | 1.5 | 0.2 |
| *Finland* ($n$ = 129) | | | | | | | | | | | | | | | |
| Mean | 15.8 | 23.9 | 48.6 | 5.7 | 21.4 | 3.9 | 0.8 | 3.7 | 0.5 | 2.8 | 0.5 | 1.5 | 0.2 | 1.4 | 0.2 |
| Median | 14.3 | 21.4 | 41.5 | 5.1 | 18.8 | 3.4 | 0.8 | 3.1 | 0.5 | 2.4 | 0.5 | 1.4 | 0.2 | 1.3 | 0.2 |

[a]Number of sample sites

The highest amount of REE is presented at the first group: in varve clays and carbonate moraines. Among the all soil forming materials varved clays were characterized by a maximum concentration of the all REE, which is due to formation of the clays in the deep glacial lakes, where fine-dispersed material accumulated. Increased amount of REE was observed in the second group: in non-carbonate moraine. Lowered amounts of REE are typical for the third group, which include lacustrine-glacial loams and sandy loams and fluvioglacial sands. The fourth group, consist of lacustrine-glacial sands and alluvial sands, characterized by small amount of REE.

Comparison of the REE content in soil forming materials of different soil geographical districts of the North-West Russia showed that limno-glacial clay sediments of Lovatsko-Volkhovsky district have the highest REE contents and the clays of local depressions (between stone ridges) of Baltic-Ladoga district have the lowest ones. Among non-carbonate moraine sediments the moraines of Chudskoy and Lugsky-Shelonsky districts are distinct in their elevated the REE content. The REE content in the carbonate moraines of different districts follows the descending order: moraines of Lugsky-Shelonsky district, Chudskoy district > Predvaldaysky district > Pskovsky district > Izhorsky district. In glaciofluvial sands the highest the REE contents were observed in Velikoluksky terminal moraine district and the lowest ones were observed in Lovat-Volkhov and Velikoretsky of lacustrine-glacial plains. Lacustrine-glacial sands were generally poor in REE.

The analytical data (Table 3) have proved that ploughing horizons of agricultural soils have less REE content than soil forming materials of all considered countries. Just as in the case of soil forming materials, ploughing horizons of Estonia, Finland, Germany, Latvia, Lithuania and Russia contain the increased amount of REE compared to the samples of the same horizons from Germany, Poland and Belarus.

T-statistic method was applied for the data of REE content in ploughing horizons from Russia. It was shown that higher REE concentration in varve clays and carbonate moraines is statistically significant in comparison with the low REE content in non-carbonate moraines, lacustrine-glacial loams, lacustrine-glacial sandy loams, lacustrine-glacial sands, fluvioglacial sands and alluvial sands.

Content of REE in ploughing soil horizons depends on their concentrations in soil forming materials. Comparison of REE concentrations in TOP and BOT horizons in samples from Russia shows the enrichment of soil forming materials in these elements. The factor is calculated as median TOP/median BOT: Ce, La, Tm—0.9; Dy, Er, Eu, Gd, Ho, Lu, Nd, Pr, Sm, Tb, Y, Yb—0.8.

Spatial distribution of REE in the investigated area (with Ce and La as an example) is shown in Figs. 3, 4, 5 and 6. The maps presented here demonstrate the geochemical regional specificity of the agricultural soils of the northwestern Russia. Overall the REE content depends on the amount of clay and fine-clay particles in deposits especially, when comparing deposits of the same origin. Deposits of the northeastern part of investigated area located within the moraine and fluvioglacial plains were found to be poorer in the REE than deposits of the southwestern hilly terrain—elevated part.

**Table 3** Mean and median concentrations of REE in the ploughing horizons of series countries in northern Europe (mg/kg)

| Value | Y | La | Ce | Pr | Nd | Sm | Eu | Gd | Tb | Dy | Ho | Er | Tm | Yb | Lu |
|---|---|---|---|---|---|---|---|---|---|---|---|---|---|---|---|
| *Germany* ($n^a = 37$) | | | | | | | | | | | | | | | |
| Mean | 7.3 | 11.8 | 23.2 | 2.6 | 9.7 | 1.8 | 0.4 | 1.6 | 0.2 | 1.2 | 0.2 | 0.7 | 0.1 | 0.7 | 0.1 |
| Median | 6.5 | 9.9 | 21.0 | 2.4 | 9.5 | 1.6 | 0.3 | 1.4 | 0.2 | 1.1 | 0.2 | 0.6 | 0.1 | 0.6 | 0.1 |
| *Poland* ($n = 136$) | | | | | | | | | | | | | | | |
| Mean | 7.9 | 12.2 | 25.7 | 2.9 | 10.8 | 2.0 | 0.4 | 1.8 | 0.3 | 1.4 | 0.3 | 0.8 | 0.1 | 0.8 | 0.1 |
| Median | 6.9 | 11.0 | 23.8 | 2.6 | 9.5 | 1.7 | 0.3 | 1.6 | 0.2 | 1.2 | 0.2 | 0.7 | 0.1 | 0.7 | 0.1 |
| *Belarus* ($n = 31$) | | | | | | | | | | | | | | | |
| Mean | 9.5 | 15.7 | 32.8 | 3.6 | 13.4 | 2.5 | 0.5 | 2.2 | 0.3 | 1.7 | 0.3 | 0.9 | 0.1 | 1.0 | 0.1 |
| Median | 8.2 | 14.8 | 30.2 | 3.3 | 12.0 | 2.2 | 0.4 | 2.0 | 0.3 | 1.5 | 0.3 | 0.8 | 0.1 | 0.8 | 0.1 |
| *Lithuania* ($n = 26$) | | | | | | | | | | | | | | | |
| Mean | 13.1 | 19.1 | 40.4 | 4.6 | 16.8 | 3.2 | 0.6 | 3.0 | 0.4 | 2.3 | 0.4 | 1.2 | 0.2 | 1.2 | 0.2 |
| Median | 13.0 | 18.9 | 40.7 | 4.5 | 16.8 | 3.2 | 0.6 | 2.9 | 0.4 | 2.2 | 0.4 | 1.2 | 0.2 | 1.2 | 0.2 |
| *Latvia* ($n = 26$) | | | | | | | | | | | | | | | |
| Mean | 11.5 | 18.6 | 38.1 | 4.3 | 15.8 | 2.9 | 0.5 | 2.6 | 0.4 | 2.0 | 0.4 | 1.1 | 0.2 | 1.1 | 0.2 |
| Median | 11.0 | 17.9 | 36.3 | 4.2 | 15.0 | 2.6 | 0.5 | 2.5 | 0.4 | 1.9 | 0.4 | 1.0 | 0.2 | 1.1 | 0.2 |
| *Estonia* ($n = 16$) | | | | | | | | | | | | | | | |
| Mean | 15.9 | 23.1 | 46.2 | 5.3 | 19.8 | 3.7 | 0.7 | 3.5 | 0.5 | 2.7 | 0.5 | 1.5 | 0.2 | 1.4 | 0.2 |
| Median | 15.7 | 22.6 | 45.4 | 5.2 | 19.4 | 3.6 | 0.7 | 3.2 | 0.5 | 2.6 | 0.5 | 1.5 | 0.2 | 1.4 | 0.2 |
| *Russia* ($n = 81$) | | | | | | | | | | | | | | | |
| Mean | 12.9 | 19.7 | 39.6 | 4.5 | 17.0 | 3.1 | 0.7 | 2.9 | 0.4 | 2.2 | 0.4 | 1.3 | 0.2 | 1.2 | 0.2 |
| Median | 12.4 | 19.4 | 38.9 | 4.3 | 16.7 | 3.1 | 0.6 | 2.8 | 0.4 | 2.1 | 0.4 | 1.2 | 0.2 | 1.2 | 0.2 |
| *Finland* ($n = 129$) | | | | | | | | | | | | | | | |
| Mean | 13.2 | 21.6 | 42.5 | 5.0 | 18.4 | 3.4 | 0.7 | 3.1 | 0.4 | 2.3 | 0.4 | 1.2 | 0.2 | 1.1 | 0.2 |
| Median | 11.1 | 18.2 | 35.3 | 4.2 | 15.1 | 2.8 | 0.7 | 2.5 | 0.4 | 2.0 | 0.4 | 1.0 | 0.2 | 1.0 | 0.2 |

[a] Number of sample sites

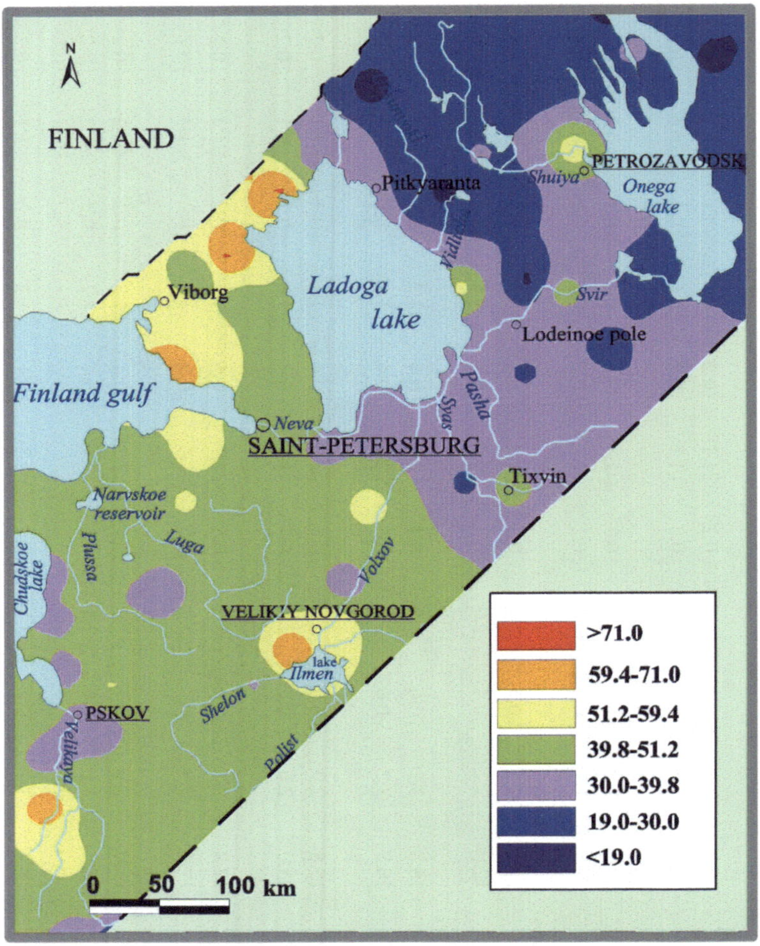


**Fig. 3** Ce content in ploughing horizons in North-West Russia (mg/kg)

The normal regional content of Ce in the ploughing horizons and soil forming materials in considered region is 39.8–512 mg/kg (Figs. 3 and 4). Near normal content intervals are: 30.0–39.8 and 51.2–59.4 mg/kg. The increased value of Ce content for this area is 59.4–71.0 mg/kg. The decreased values are: 19.0–30.0 mg/kg. Amounts more than 71.0 mg/kg are the higher values for Ce spatial distribution. Concentrations lower than 19.0 mg/kg are the low local regional values. It is natural phenomenon that Ce concentrations in upper soil horizons are lower than in forming materials. Ploughing soil horizons in the northeastern part of investigated area are characterized by decreased and low content of Ce. Central and

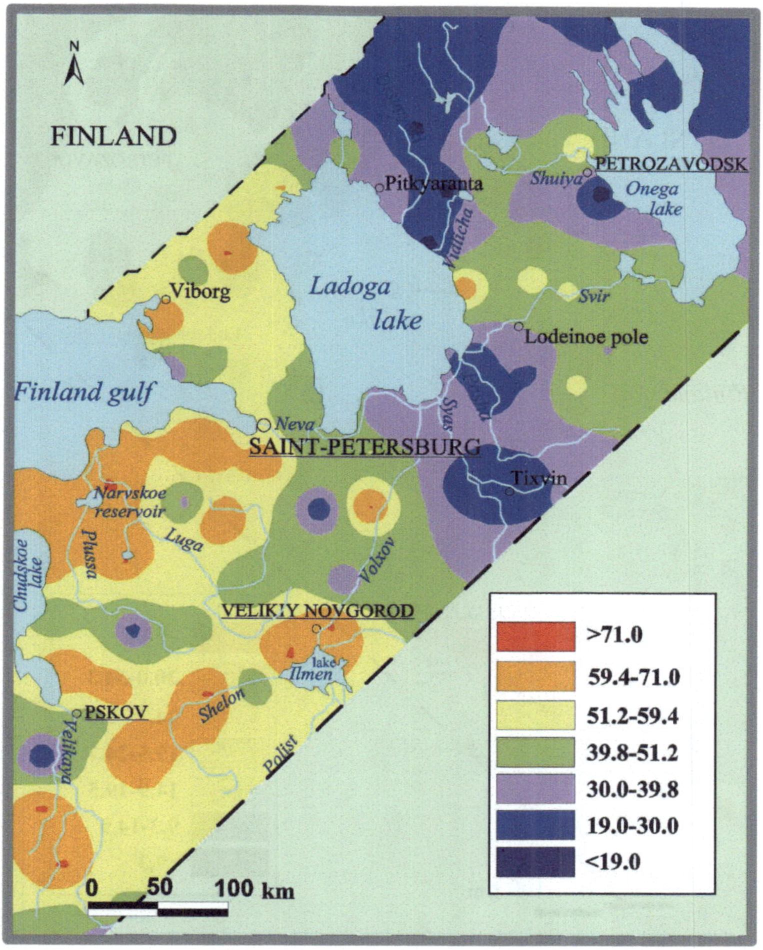


**Fig. 4** Ce content in soil forming materials in North-West Russia (mg/kg)

southwestern parts show normal and near normal Ce concentration, excluding five small areas near Lake Ladoga, The Gulf of Finland, Lake Ilmen and the Velikaya River. The samples of soil forming materials from northeastern part of investigated region are characterized by normal, near normal and decreased content of Ce. Opposite, the southwestern part of considered area contains normal, near normal and increased values of Ce. It seems to be determined by distribution of soil forming materials enriched in clay and fine-clay particles.

The normal content of La in the ploughing horizons and soil forming materials in investigated region is 19.5–24.7 mg/kg (Figs. 5 and 6). Near normal content

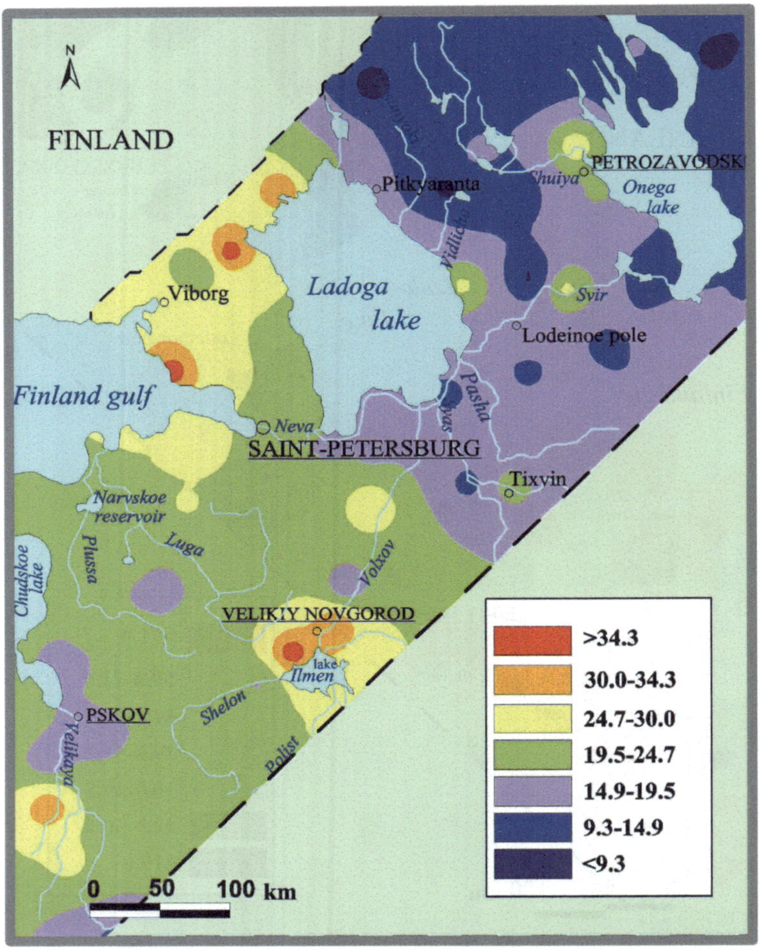


**Fig. 5** La content in ploughing horizons in North-West Russia (mg/kg)

intervals are: 14.9–19.5 and 24.7–30.0 mg/kg. The increased value of La concentration for this area is 30.0–34.3 mg/kg. The decreased values are: 9.3–14.9 mg/kg. Amounts above 34.3 mg/kg are the higher values for La spatial distribution. Amounts lower than 9.3 mg/kg are the low local regional values. Common way of La spatial distribution in soils and rocks of consideration area is similar to Ce distribution. However there are three points (near Lake Ladoga, Gulf of Finland and Lake Ilmen) with high content of La in ploughing soil horizons. The map of La distribution in soil forming materials shows eleven sites with a high concentration of this element that are most likely to depend on specific character of soil forming material.

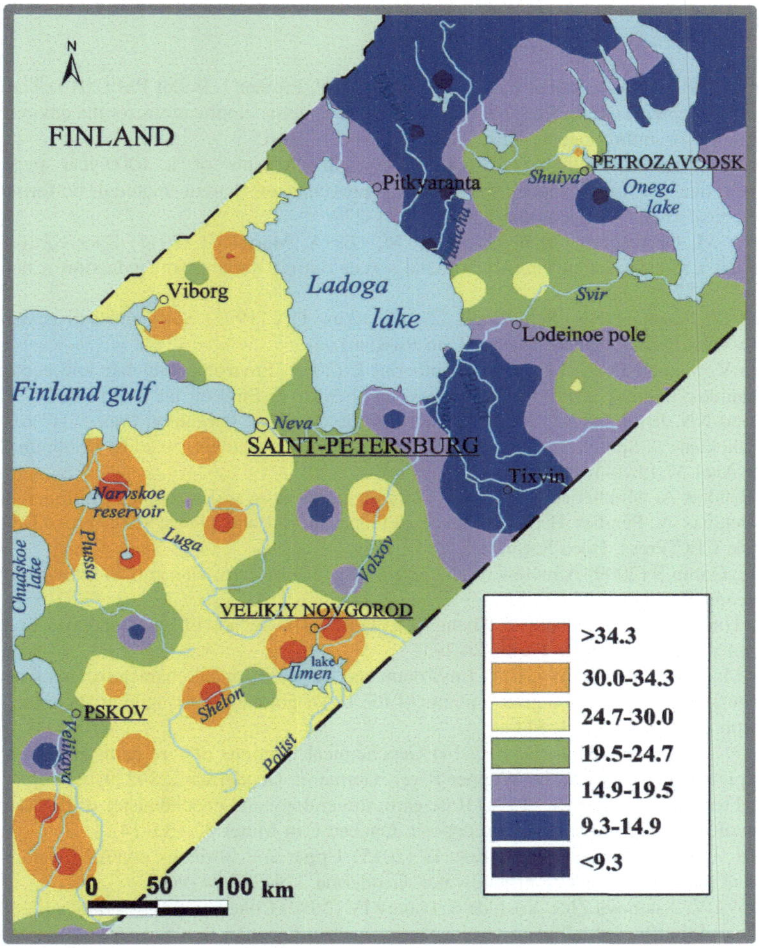


**Fig. 6** La content in soil forming materials in North-West Russia (mg/kg)

## 4 Conclusion

The content of rare earth elements in northwestern Russia soil forming materials is less than in Earth crust. Varve clays and carbonate moraines are enriched in REE content relatively another soil forming materials of northwestern Russia (non-carbonate moraines, lacustrine-glacial loams, sandy loams and sands, fluvioglacial sands and alluvial sands). Content of REE in soils depends on their concentrations in soil forming materials; therefore, spatial distribution of REE in ploughing horizons of agricultural soils reflects regional abundance of different soil forming materials.

## References

Brioschi L, Steinmann M, Lucot L, Pierret MC, Stille P, Prunier J, Badot PM (2013) Transfer of rate earth elements (REE) from natural soil to plant systems: implications for the environmental availability of anthropogenic REE. Plant Soil 366:143–163

Chen LM, Zhang GL, Jin JD (2014) Rare earth elements of a 1000-year paddy soil chronosequence: Implications for sediment provenances, parent material uniformity and pedological changes. Geoderma 230–231:274–279

Davranche M, Grybos M, Gruan G, Pedrot M, Dia A, Marsak R (2011) Rare earth element patterns: a tool for identifying trace metal sorces during wetland soil reduction. Chem Geol 284:127–137

Gagarina EI, Matinian NN, Schastnaya LS, Kasatkina GA (1995) Soils and soil cover of the North-West Russia, Saint Petersburg (in Russian)

Gonsalez V, Vignati DAL, Leyval C, Gamberini L (2014) Environmental fate and ecotoxity of lanthanides: are they a uniform group beyond chemistry? Environ Int 71:148–157

Greenwood NN, Ernshaw A (1984) Chemistry of the elements. Pergamon Press, Oxford

Hu Z, Haneklaus S, Sparovek G, Schnug E (2006) Rare earth elements in soils. Commun Soil Sci Plant Anal 37:1381–1420

Kabata-Pendias A, Mukherjee AB (2007) Trace elements from soil to human. Springer, Berlin

Kabata-Pendias A, Pendias H (2000) Trace elements in soil and plants. CRC Press/Taylor and Francis CEC Press, Boca Raton, USA

Laveuf C, Cornu S (2009) A review on the potentiality of rare earth elements to trace pedogenetic processes. Geoderma 154:1–12

Liang T, Ding S, Song W, Chang Z, Zhang C, Li H (2008) A review of fractionations of rare earth elements in plants. J Rare Earths 26:7–15

Miao L, Ma Y, Xu R, Yan W (2011) Environmental biogeochemical characteristics of rare earth elements in soil and soil = grown plants of the Hetai goldfield, Guangdong Province, China. Environ Earth Sci 63:501–511

Mihajlovic J, Stark HJ, Rinklebe J (2014) Geochemical fractions of rare earth elements in two floodplain soil profiles at the Wupper River, Germany. Geoderma 228–229:160–172

Pan JM, Fleet ME, Barnett RL (1994) Rare-earth mineralogy and geochemistry of the Mattagami lake volcanogenic massive sulfide deposit. Quebec Can Miner 32:133–147

Pedrot M, Dia A, Davranche M, Gruau G (2015) Upper soil horizons control the rare earths element patterns in shallow groundwater. Geoderma 239–240:84–96

Perelomov LV, Asainova ZhS, Yoshida S, Ivanov IV (2012) Concentrations of rare earth elements in soils of the Prioksko-Terrasnyi biospheric reserve. Eurasian Soil Sci 45(10):983–994

Reimann C, Siewers U, Tarvainen T, Bityukova L, Eriksson J, Gilucis J, Gregorauskiene V, Lukashev VK, Matinian NN, Pasieczna A (2003) Agricultural soils in Northern Europe: a geochemical atlas. Hannover

Ronov AB, Migdisov AA (1995) The abundance of rocks, minerals and chemical elements in the sedimentary cover of Russian plate and its change over time. In: Vinogradov AP (ed) The main directions of geochemistry: to 100-th jubilee of academician. Nauka, Moscow (in Russian)

Tyler G (2004) Rare earth elements in soil and plant systems—a review. Plant Soil 267:191–206

Tylor SR, McLennan SM (1985) The continental crust: its composition and evolution—an examination of the geochemical record preserved in sedimentary rocks. Blakwell Scientific Publication, Oxford

Tylor SR, McLennan SM (2003) Distribution of the lanthanides in the earth's crust. In: Sigel A, Sigel H (eds) The lanthanides and their interrelations with biosystems. Metal ions in biological systems. CRC, New York

Vodyanitskii YuN (2012) Geochemical fractionation of lanthanides in soils and rocks: a review of publications. Eurasian Soil Sci 45(1):56–67
Vodyanitskii YuN, Goryachkin SV, Savichev AT (2011) Distribution of rare-earth (Y, La, Ce) and other heavy metals in the profiles of the podzolic soil group. Eurasian Soil Sci 44(5):500–509
Zhu Z, Wang Z, Li J, Li Y, Zhang Z, Zhang P. (2012) Distribution of rare earth elements in sewage-irrigated soil profiles in Tianjin, China. J Rare Earths 30:609–613

# Phytotoxicity of Tailings Dam of the Dzhidinsky Tungsten–Molybdenum Combine (Western Transbaikalia)

**Svetlana G. Doroshkevich and Irina V. Bardamova**

**Abstract** The phytotoxicity of tailings dam of the Dzhidinsky tungsten–molybdenum combine has been studied. It is established that the content of toxic chemical elements (Pb, Zn, As, Cu, Co, Mo, Ni, W) in technogenic sands is extremely dangerous; the total contamination index (Zc) is 425–500. The phytotoxicity of the sands with respect to oats is different: the technogenic sands of the Barun-Naryn tailings refer to II (high) toxicity class, while the technogenic sands of the Modonkul deltoid deposit refer to V (normal) toxicity class. The index of total pollution by toxic elements (Zc) of plant oats grown on technogenic sands and their water extract is consistent to high level (aboveground part of plants—12–69, and underground parts—11–147).

**Keywords** Phytotoxicity · Sulfide-bearing tailings · Heavy metals · Oats

## 1 Introduction

Mining plays a significant role in the formation of technogenic landscapes and the extreme ecological situation in a surrounding area.

The Dzhidinsky molybdenum–tungsten combine that processed the Dzhidinsky molybdenum and sulfide-hubnerite ores of more than 60 years was one of the leading companies of the Russian mining industry. Production was conservated in 1997. The massifs of technogenic sands (waste of tailings) remained in the surrounding areas of Zakamensk town. The total volume of sands is more than 40 million tons. These sands are the main source of environmental pollution by heavy metals and other toxic elements (Yatsenko 1994; Smirnova et al. 2006). The bioavailability of toxic elements increases in tailings eventually (Smirnova and Dampilova 2010). The total pollution of soils within Zakamensk town is classified

S.G. Doroshkevich (✉) · I.V. Bardamova
Geological Institute of SB RAS, Ulan-Ude, Russia
e-mail: sv-dorosh@mail.ru

 
O.V. Frank-Kamenetskaya et al. (eds.), *Biogenic—Abiogenic Interactions in Natural and Anthropogenic Systems*, Lecture Notes in Earth System Sciences,
DOI 10.1007/978-3-319-24987-2_22

as: ecological disaster—25.5 %, crisis—26 %, tense—30 %, and satisfactory—18.5 % of the total area of the town (Smirnova and Plyusnin 2013).

Investigations at the present level assess the quality of the environment and biotic parameters using bioassay methods and bioindication (Salvatore et al. 2008; Neverova 2009; Branzini and Zubillaga 2010; Terekhova 2011; Cui and Du 2011; Dobrovolsky et al. 2013; Oros 2013; Shaikh et al. 2013) along with the determination of pollution by toxic elements. This approach allows to obtain complete information concerning the impacts of pollution on biological objects, as well as their response. The combination of the two approaches (abiotic and biotic) is the most informative in solving such problems currently (Shuysky et al. 2002).

This paper presents the results of phytotoxicity of tailings dam of the Dzhidinsky tungsten–molybdenum combine that included: (1) determination the degree of chemical contamination of tailings dam; (2) determination the level of toxicity of technogenic sands by bioassay methods; (3) assess the level of technogenic objects hazard.

## 2 Objects and Methods

The objects of the research are technogenic sands from waste of tailings dam of (1) the Barun-Naryn ores and (2) the Modonkul deltoid deposit formed by emergency breakthrough tailings dam (Fig. 1). The sand from Lake Baikal is taken as a control sample. The studied technogenic sands contain a significant amount of chemical elements of I–III hazard classes. The chemical composition of sands is presented in Table 1.

Determination of chemical elements bulk content in natural and technogenic sands was performed according to the methodology of measurement of mass concentrations of bulk forms of elements on x-ray energy dispersive polarization spectrometer EDPRS-1 (Zhalsaraev et al. 2010); in plants of oats (the sample of air-dry matter) by acid digestion method PND F 16.1:2.3:3.11-98 on the ICP-MS PerkinElmer Elan 9000 (USA) in Khabarovsk innovative-analytical center.

The level of chemical pollution of soil by elements of I–III hazard classes was estimated in accordance with the Sanitary rules and norms 2.1.7.1287-03. The phytotoxicity of tailings dam is determined by method of sprout on the state standards (GOST R ISO 22030-2009) in two directions: (1) the bioassay by a method of sprout on technogenic sands; (2) the bioassay by a method of seedlings on water extracts.

Conditions of the first experiment: sand samples of 100 g weight were placed in containers, then they were wetted by distilled water to 60 % of full capacity, 20 seeds of test-crops oats of Gerel sort were planted into containers with sands. Everyday, the control of water capacity of the samples carried out by the gravimetric method.

Conditions of the second experiment: 50 g of the sand was wetted by 100 ml of distilled water and kept for 7 days with periodic interfusion, T = 22 °C. The extract

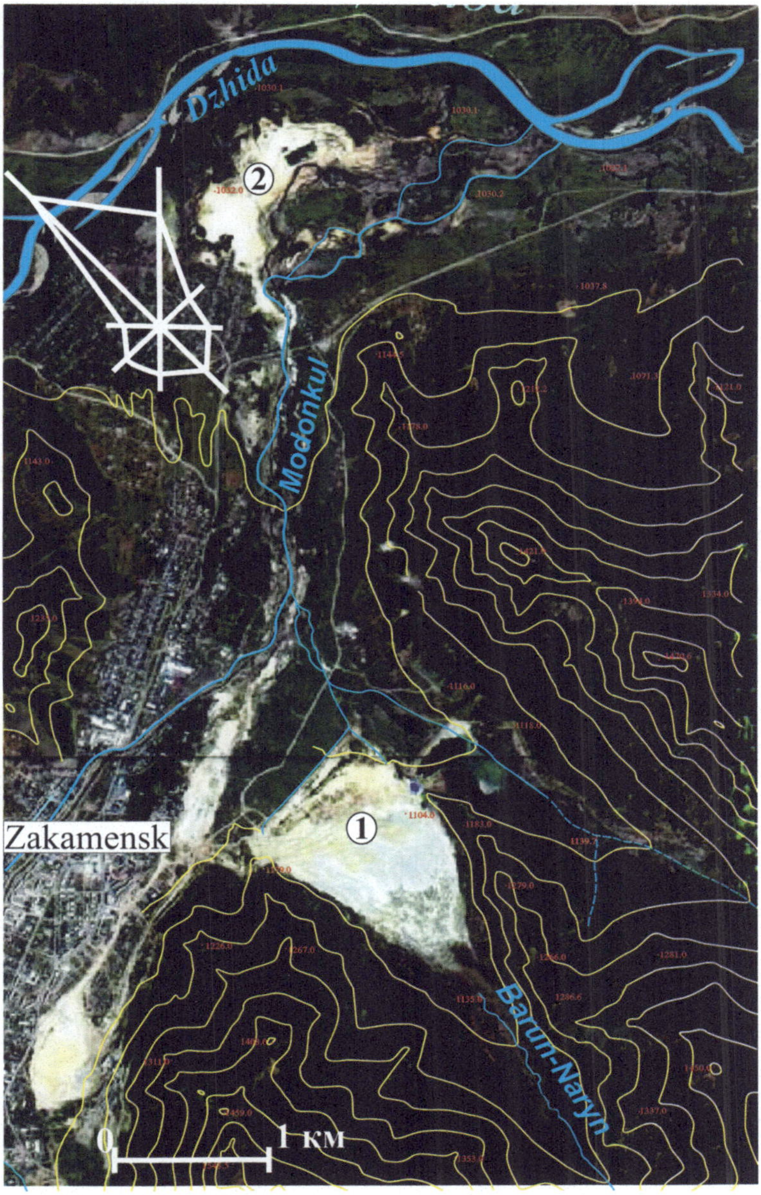


**Fig. 1** The allocation plan of the objects of study: *1*—the technogenic sands of waste sulfide-tungsten ores of the Barun-Naryn tailings; *2*—the technogenic sands of the Modonkul deltoid deposit

**Table 1** The amount of chemical elements of I–III hazard classes in the natural and technogenic sands, ppm

| Object | Zn | Pb | As | Cu | Co | Cr | Mo | Ni | W |
|---|---|---|---|---|---|---|---|---|---|
| Baikal sand (control) | 25 | 12 | 5.6 | 5 | 9 | 19 | 1.9 | 8.8 | 67 |
| Waste of tailings dam, the Modonkul deltoid deposit | 58 | 230 | 10 | 38 | 41 | 9.1 | 170 | 4.2 | 440 |
| The Barun-Naryn waste of tailings dam | 275 | 320 | 20 | 66 | 40 | 290 | 5.6 | 70 | 800 |
| APC (HN 2.1.7.2511-09) | 55 | 32 | 2 | 33 | – | – | – | 20 | – |
| The average amount in the world soils (Kabata-Pendias 2011) | 70 | 27 | 6.8 | 39 | 8 | 60 | 1.1 | 29 | 1.7 |

*Note* The dash—no data

through filter white ribbon was filtered. Germination was carried out in closed Petri dishes at room temperature. A filter paper and 20 seeds of plants were placed at the bottom of Petri dishes. The aqueous extract was added to fully wet the filter paper.

There were six replicates of experiments with the duration of germination 7 days. The amount of germinated plants was calculated and the length of seedlings was measured at end of the each experiment.

Germinating energy $B$ was determined by formula (1):

$$B = \frac{a}{b} \times 100\,\%, \tag{1}$$

where: $a$—amount of germinated seeds; $b$—total amount of seeds taken for experiment.

The length of sprouts was a parameter for determining the phytotoxic effect (PhE), calculated according to (2):

$$\mathrm{PhE} = \frac{\mathrm{Dc} - \mathrm{Dsp}}{\mathrm{Dc}} \times 100\,\%, \tag{2}$$

where: Dc—the average length of the sprouts in the control sample; Dsp—the average length of the sprouts in the test sample.

The toxicity of the studied technogenic sands was calculated using the toxicity index (ITF) by formula (3):

$$\mathrm{ITF} = \frac{\mathrm{TFe}}{\mathrm{TFc}}, \tag{3}$$

where: TFe—the value of the recorded test functions in the experiment; TFc—the value of the recorded test-function on the control.

The value of ITF varies from 0 to n, where n is any positive value (Volkova and Kondakova 2002). The scale of toxicity was estimated by ITF volume: VI class (stimulation) ITF > 1.10; V class (normal) ITF = 0.91–1.10; IV class (low toxicity) ITF = 0.71–0.90; III class (medium) ITF = 0.50–0.70; II class (high) ITF < 0.50; I class (high) total death of the test object.

Criteria of dangerous contamination of soils were decreasing of seed germination more than by 50 % and decreasing of length of seedlings by more than 30 % compared to control samples (Volkova and Kondakova 2002).

## 3 Results

The amount of toxic chemical elements in the technogenic sands of the Modonkul deltoid deposit and the technogenic sands of the Barun-Naryn tailings is extremely dangerous (Table 2).

As a result of the experiment on germination of oats on technogenic sands it is determined that the technogenic sands of the Modonkul deltoid deposit affected decrease in the length of oat seedling, but they had a more intense green color; the technogenic sands of the Barun-Naryn tailings strongly inhibited the seedling plants of oats (Fig. 2). The germination of oats in the experiments with technogenic sands of the Modonkul deltoid deposit remained at the level of the control; the length of the seedlings decreased to 1.37 times compared to the control variant (Fig. 3). The seed germination and the seedling plants of oats in the experiments with the technogenic sands of the Barun-Naryn tailings decreased by 2.18 and 7.09 times relative to the control variant, respectively (Fig. 3).

The water extracts from technogenic sands of the Modonkul deltoid deposit increased seed germination of oats by 1.04 times compared with the control variant; the length of the seedlings decreased from 17.1 to 16.4 cm. In case of water extracts from the Barun-Naryn technogenic sands the germination of seeds and the length of the sprout oats are decreased by 1.13 and 2.16 times relative to the control, respectively (Fig. 3).

Phytotoxic effect was 96.1 and 49.7 % for experiments with technogenic sands and water extracts from technogenic sands of the Barun-Naryn tailings, respectively (Table 3). Phytotoxic effect on the technogenic sands of the Modonkul deltoid

**Table 2** Category of pollution of technogenic sands on the amount of chemical elements of I–III hazard classes according to Sanitary rules and norms 2.1.7.1287-03

| Object | Chemical elements defining the classes of pollution (hazard classes) | Zc | Category of pollution |
|---|---|---|---|
| Waste of tailings dam, the Modonkul deltoid deposit | Pb and As (I), Co and Mo (II), W (III) | 425 | Extremely dangerous |
| The Barun-Naryn waste of tailings dam | Pb, Zn and As (I), Cu, Co, Mo and Ni (II), W (III) | 500 | Extremely dangerous |

**Fig. 2** The influence of technogenic sands (**a**) and water extracts from technogenic sands (**b**) on the appearance of the oats: *1*—control, the natural sand of lake Baikal, *2*—the Modonkul deltiod deposit, *3*—the Barun-Naryn tailings

deposit was 3.51 times lower than for technogenic sands of the Barun-Naryn tailings. Phytotoxic effect in cases with water extracts of technogenic sands of the Modonkul deltiod deposit was low (4.05 %).

In accordance with the scale of toxicity (Environmental 2002), the technogenic sands of the Modonkul deltoid deposit and their water extracts belong to V (normal) toxicity class. The technogenic sands of the Barun-Naryn tailings refer to II (high) toxicity class with ITF index of 0.35 (Table 3); water extract of technogenic sands of the Barun-Naryn tailings has IV (low) toxicity class.

The damaging effect of the technogenic sands and their water extracts is determined on the test-culture oats (Table 4) using guidelines (MI 1.2.2968-11): the technogenic sands of the Modonkul deltoid deposit are characterized by the absence of the damaging effect; while the technogenic sands of the Barun-Naryn tailings have average damaging effect on oats.

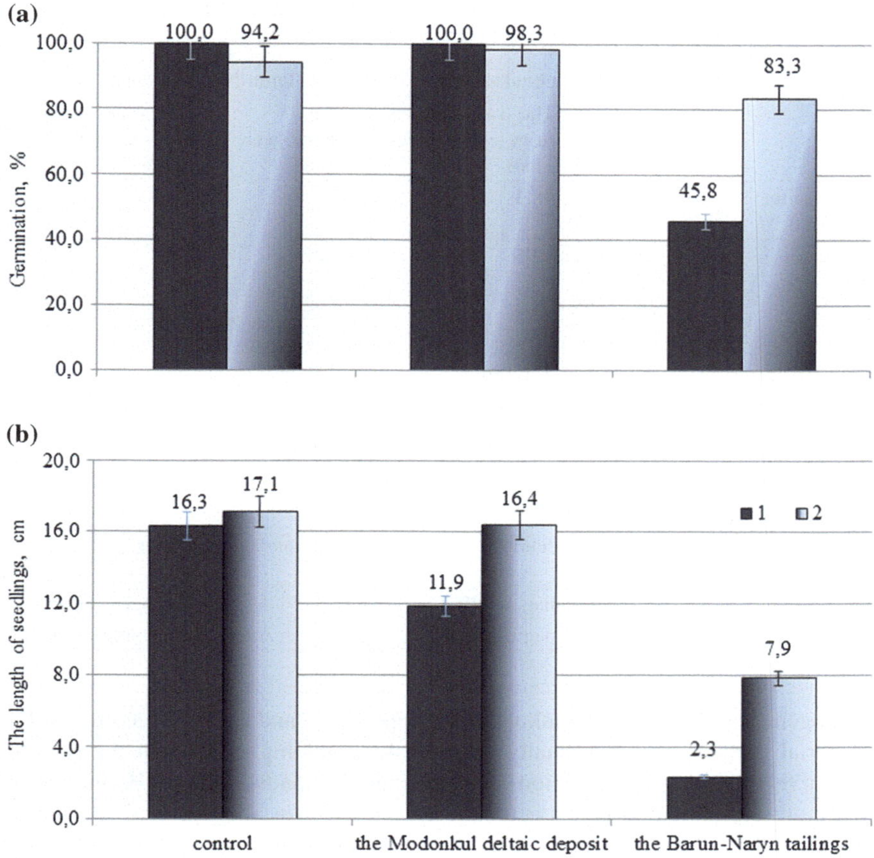


**Fig. 3** The influence of technogenic sands (*1*) and water extracts from technogenic sands (*2*) on the seed germination (**a**) and the length of sprouts (**b**) of oats

Due to the fact that the wastes of tailings dam with high degree of contamination by toxic elements were used in the experiment, plants of oats were analyzed in the presence of these toxic elements in the aboveground and underground parts (Table 5). Overall, during germination of oats on technogenic sands and their water extracts, the Zn, Cr, Ni, Co, and W concentrations in plants of oats exceed their worldwide average values. The contents of Zn, As, Cu, Cr, and Ni increase up to 5 times in aboveground and underground parts of oats compared to those in the control variant. Pb (in aboveground and underground parts 10 and 18 times higher than in the control, respectively), Mo (in the underground part 24 times higher than the control), and W (in aboveground and underground parts 57 and 95 times higher than in the control, respectively) were accumulated as much as possible during germination on technogenic sands of the Modonkul deltoid deposit. High accumulation of toxic elements was not observed during germination on water extract of

**Table 3** Phytotoxicity of tailings dam of Dzhidinsky Tungsten–Molybdenum combine

| Test function | Objects | | | |
|---|---|---|---|---|
| | The Modonkul deltoid deposit | | The Barun-Naryn tailings | |
| | The technogenic sands | The water extracts from technogenic sands | The technogenic sands | The water extracts from technogenic sands |
| Germinating energy, B, % | 100 | 98.3 | 35 | 83.3 |
| Phytotoxic effect, PhE, % | 27.4 | 4.05 | 96.1 | 49.7 |
| Toxicity index, ITF | 1.00 | 0.98 | 0.35 | 0.83 |
| The toxicity class on ITF | V (norm) | V (norm) | II (high) | IV (low) |

**Table 4** The damaging effect of technogenic sands (first experiment) and water extracts from technogenic sands (second experiment) on the oats test-culture

| Experiment | The Modonkul deltoid deposit | | The Barun-Naryn tailings | |
|---|---|---|---|---|
| 1 | – | No damaging effect | ++ | Average damaging effect |
| 2 | – | No damaging effect | ++ | Average damaging effect |

technogenic sands of the Modonkul deltoid deposit. Co and Pb were accumulated in the aerial parts of oats maximally (33 and 42 times higher than in the control, respectively) during germination on technogenic sands tailings Barun-Naryn. Maximum accumulation is characteristic only for Co (in aboveground and underground parts 29 and 53 times higher than in the control, respectively) during germination on water extraction of technogenic sands of the Barun-Naryn tailings. The index of total pollution of toxic elements is in the range from 11 (in the plant oats) to 85 (aboveground part) and 147 (underground part), which according to Kasatikov (1992) corresponds to the strong contamination in comparison with control.

Thus, the content of toxic chemical elements is extremely dangerous in the technogenic sands of waste sulfide-tungsten ores of the Barun-Naryn tailings and the Modonkul deltiod deposit. Although these technogenic sands belong to the extremely dangerous category of pollution with total pollution 425–500, the phytotoxicity with respect to oats are different: the technogenic sands of the Barun-Naryn tailings refer to II (high) toxicity class; the technogenic sands of the Modonkul deltaic deposit–V (normal) toxicity class. The water extract of technogenic sands of the Barun-Naryn tailings has lower phytotoxic effect than the technogenic sands and has IV (low) toxicity class. During germination on the technogenic sands of the Modonkul deltoid deposit of the maximum accumulated Pb, Mo, and W; on the technogenic sands tailings Barun-Naryn—Co and Pb. The plants oats are heavily

**Table 5** The amount of chemical elements of I–III hazard classes in the aboveground (numerator) and underground (denominator) parts of oats (air-dry weight), ppm

| Experiment | Variant | Zn | Pb | As | Cu | Co | Cr | Mo | Ni | W | Zc |
|---|---|---|---|---|---|---|---|---|---|---|---|
| 1 | Control | 35.16/38.20 | 0.12/1.78 | 0.89/1.06 | 6.61/13.13 | 0.05/0.46 | **1.17**/2.36 | 0.58/0.8 | 2.68/2.61 | 0.04/0.49 | |
| | The Modonkul deltoid deposit | **49.69**/200.8 | 1.19/31.13 | 0.98/0.69 | 8.32/51.72 | 0.08/0.76 | **1.46**/11.05 | 1.36/18.87 | **2.97**/5.36 | **2.29**/46.56 | 69.3/146.7 |
| | The Barun-Naryn tailings | **97.69**/– | 5.03/– | 1.33/– | 15.39/– | **1.66**/– | **1.93**/– | 0.77/– | **6.81**/– | **0.22**/– | 84.7/– |
| 2 | Control | 31.04/39.52 | 0.17/3.06 | 0.85/0.79 | 8.08/15.75 | 0.02/0.08 | **0.90**/1.18 | 1.20/0.97 | 2.73/2.35 | 0.03/0.31 | |
| | The Modonkul deltoid deposit | 45.0/86.7 | 0.43/4.11 | 0.88/1.29 | 8.38/15.75 | 0.09/1.18 | **1.02**/1.46 | 1.12/1.01 | **4.46**/5.73 | **0.16**/1.51 | 11.6/11.1 |
| | The Barun-Naryn tailings | **69.63**/153.35 | 0.65/4.53 | 0.93/0.48 | 10.0/31.71 | **0.58**/4.25 | **3.29**/2.5 | 0.63/0.50 | **4.76**/5.89 | 0.08/0.55 | 38.4/60.9 |
| The volatility of average amount in the worldwide plants (Kabata-Pendias 2011) | | 12–47 | 0.36–8.0 | 0.009–1.5 | 1.1–33.1 | 0.03–0.57 | 0.02–0.2 | 0.33–2.3 | 0.13–2.7 | 0.01–0.15 | |

*Note* The dash—no data, bold—the excess of average amount in the worldwide plants polluted with toxic elements in comparison with control. The obtained data indicate that along with the determination of chemical pollution at estimation of their impact on the ecosystems (in particular the waste ores tailings) it is necessary to study the phytotoxicity of techogenic objects.

**Acknowledgments** The studies were supported by Russian Found of Fundamental Research (RFFR), grant No. 13-05-01155 and grant No. 15-45-04123_r_siberia_a.

## References

Branzini A, Zubillaga MS (2010) Assessing phytotoxicity of heavy metals in remediated soil. Int J Phytorem 12(4):335–342

Cui Y, Du X (2011) Soil heavy-metal speciation and wheat phytotoxicity in the vicinity of an abandoned lead–zinc mine in Shangyu City, eastern China. Environ Earth Sci 62:257–264

Dobrovolsky GV, Terekhova VA, Dgebuadze YuYu (2013) Bioindication in the ecological assessment of soils and related habitats. Volga Ecol J 4:365–367 (in Russian)

Kabata-Pendias A (2011) Trace elements in soils and plants. CRC Press, Taylor and Francis Group, Boca Raton, London, New York

Kasatikov VA (1992) Criteria contamination of soil and plant trace elements, heavy metals when used as fertilizer precipitation of urban wastewater. Message 2. Criteria pollution plants. Agri Chem 5:110–115 (in Russian)

Neverova OA (2009) Application of phytoindication in the assessment of environmental pollution. Biosphere 1(1):82–92 (in Russian)

Oros V (2013) Aquatic phytotoxicity of heavy metals: Cu, Cd and Zn ecotoxicological tests with duckweed plants (Lemna minor). Environ Eng Manage J 12(2):343–350

Salvatore MDi, Carafa AM, Carratù G (2008) Assessment of heavy metals phytotoxicity using seed germination and root elongation tests: a comparison of two growth substrates. Chemosphere 73:1461–1464

Shaikh IR, Shaikh PR, Shaikh RA, Shaikh AA (2013) Phytotoxic effects of heavy metals (Cr, Cd, Mn and Zn) on wheat (*Triticum aestivum* L.) seed germination and seedlings growth in black cotton soil of Nanded, India. Res J Chem Sci 3(6):14–23

Shuysky VF, Maksimova TV, Petrov DS (2002) Bioindication of water environment quality, state of freshwater ecosystems and anthropogenic changes. Collection of scientific reports of the VII international conference "ecology and the development of North-West Russia" StP:441–451 (in Russian)

Smirnova OK, Dampilova BV (2010) Dynamic of lead, zincum, copper species and their biological accessibility in stale tailings after dressing of sulfide-tungsten ores. Labours of III all-russian symposium with international participation "mineralogy and geochemistry of landscape of mountain-ore territories" Chita:58–62 (in Russian)

Smirnova OK, Plyusnin AM (2013) Dzhidinsky ore region (problems of environmental condition). BSC Publishing House, Ulan-Ude (in Russian)

Smirnova OK, Khodanovich PYu, Yatsenko RI (2006) Heavy metals in technogenic landscapes of the Dzida mining-processing combine region. Labours of the fitst all-russian symposium with international participation "mineralogy and geochemistry of landscape of mountain-ore territories" Chita:82–87 (in Russian)

Terekhova VA (2011) Soil bioassay: problems and approaches. Eurasian Soil Sci 44(2):173–179

Volkova IN and Kondakova GV (2002) Environmental soil science. Laboratory classes for students-ecologists (bachelors). Methodical instructions. Publishing house of Yaroslavl University, Yaroslavl (in Russian)

Yatsenko RI (1994) The methods and intensity of the dispersion of pollutants from the concentrator galinsoga molybdenum-tungsten mining-processing combine. Ann Ed BSC SB RAS 1:71–75 (in Russian)
Zhalsaraev BZh, Kutovoy AN, Tsynguev VG (2010) X-ray spectrometer Patent 2397481. Inf Bull 23 (in Russian)

# Distribution of Organic Compounds in the System of Geochemically Linked Mires (the Spurs of Vasuygan Mire)

**Lidia I. Inisheva, Alla V. Golovchenko and Lech W. Szajdak**

**Abstract** Specific natural and geochemical conditions formed on Vasuygan Mire include a wide variety of vegetation, types of peat deposits, and peats composing them. The study of the Vasyugan Mire's biospheric functions and elaboration of the scientific bases of regional monitoring are important. Therefore, we investigated mire regimes under field conditions. The examined plot, which includes the biogeocenoses connected geochemically with the landscapes, is the model system for the Vasuygan Mire. According to our data, the age of this plot dates back to 2500–4800 years. Our work revealed peculiarities of biochemical processes that exert influence on the formation of hydrochemical runoff from the paludified territory in the peat deposits of the landscape profile. The chemical composition of mire water and the subsequently migrating stream are formed due to the mixture of atmospheric precipitation with swampy waters. The composition of swampy water is defined by the arrival of movable compounds from the peat deposit, which underwent a regular biochemical transformation. The total runoff of chemical elements during the course of the runoff were as follows: $Ca^{2+}$—1398, $Fe_{total}$—311, $SO_4^{2-}$—391, $NO_3^-$—236, $NO_2^-$—1, Pb—2.253 × $10^{-3}$, Mn—317.29 × $10^{-3}$, Zn—41.191 × $10^{-3}$, Ni—8.151 × $10^{-3}$, and Ti—29.651 × $10^{-3}$ kg/km$^2$. The annual runoff losses of dissolved organic matter were equal to 6945 kg/km$^2$. Our study of the concrete water objects and physical, chemical, and biological processes of the transformation of substance and energy flows on the catchment areas provide insight into the chemical constituents of georunoff.

**Keywords** Vasuygan mire · Landscape · Peat formation · Oligotrophic stage · Organic matter · Micromycete · Biomass · Migration · Mire water

L.I. Inisheva (✉)
Tomsk State Pedagogical University, Tomsk Oblast, Russia
e-mail: inisheva@mail.ru

A.V. Golovchenko
Faculty of Soil Science, Moscow State University, Moscow, Russia

L.W. Szajdak
Institute for Agricultural and Forest Environment, Polish Academy of Sciences, Poznan, Poland

 
O.V. Frank-Kamenetskaya et al. (eds.), *Biogenic—Abiogenic Interactions in Natural and Anthropogenic Systems*, Lecture Notes in Earth System Sciences,
DOI 10.1007/978-3-319-24987-2_23

# 1 Introduction

The Vasyugan mire is the largest mire in the world, with an area of about 5,269,437 ha. It captured the Ob—Irtysh watershed and stretches out between 55° 40′–58° 60′N and 75° 30′–83° 30′E (Fig. 1). Its natural and geochemical conditions formed a wide variety of vegetation, types of peat deposits, and peats composing them.

This mire is not only the biggest one, but it is also the most unique in the mire territory of Western Siberia; by concentration of mires, their arrangement, and the intensity of paludification, this area has no analogs on the globe. Vasyugan Mire occupies the highest part of Western-Siberian lowland (Vasyugan Plateau) and is located in two natural and geochemical subbelts: southern taiga and forest-steppe. More than 200 rivers flow down from Vasyugan Mire. The central part of the raised moor is above the mire's borders by 7.5–10 m.

Specific natural and geochemical conditions formed a wide variety of vegetation here, as well as types of peat deposits and the peats composing them. Vasyugan Mire is a unique place with a wide spread of transitional mires. The important factor here is the formation of the overmoistened Hypnum-sedge fen on the very top of the Vasyugan Ridge at the highest point for the West Siberian flat plain—146 m above sea level.

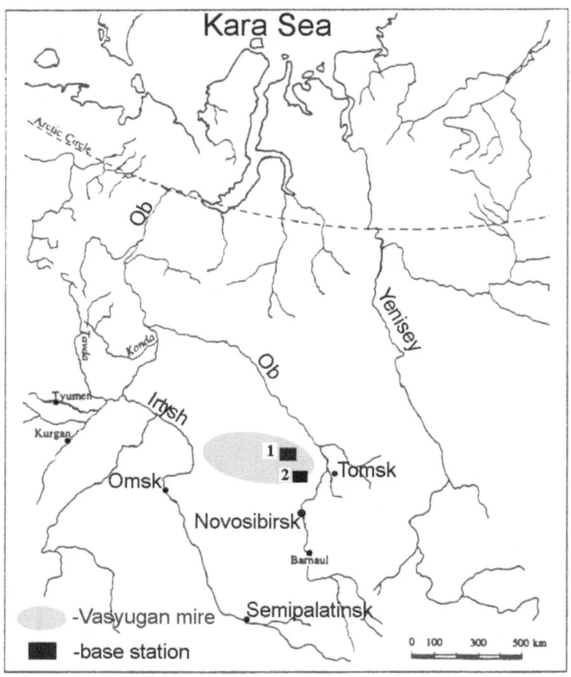


**Fig. 1** Schematic map of Vasyugan Mire: *1*—field research station of Polynyanka, *2*—field research station in sources of rivers Shegarka and Iksa

The study of its biospheric functions of the Vasyugan Mire and elaboration of the scientific bases of regional monitoring are important. Therefore, we investigated mire regimes under field conditions.

## 2 Materials and Methods

A special testing area was selected to carry out the regime observations, with several investigation objects on the plot of the northeastern part of Vasyugan Mire. Among these objects, the landscape profile (catena) was chosen as a model object in the basin of the River of Klyuch. The originality of the profile is in the accessibility to carry out the research on a small territory of the basin of a swampy river within the limits of 1 km, where there are practically all possible varieties of geomorphologic elements and biogeocenoses. This allowed us to carry out the studies of turnover by a balance method under Siberian conditions. The total area of swamp here reaches 50,000 ha.

The plot under study, which includes the biogeocenoses connected geochemically with the landscapes, is the model system for the Bakchar mire district (Liss et al. 2001). According to our data, the age of the mire dates back 2500–4800 years.

**Point 2** (pine-undershrub-sphagnum biocenoses, high ryam; Fig. 2) is a border of the peaty mire. The peat deposit of tall ryam is 90 cm thick; its structure is forest-marshy. It is formed by five kinds of peats. Only two of them—fen sedge and pine-cotton grassy—are 20 cm thick (each of them accounts for as much as 29 % of

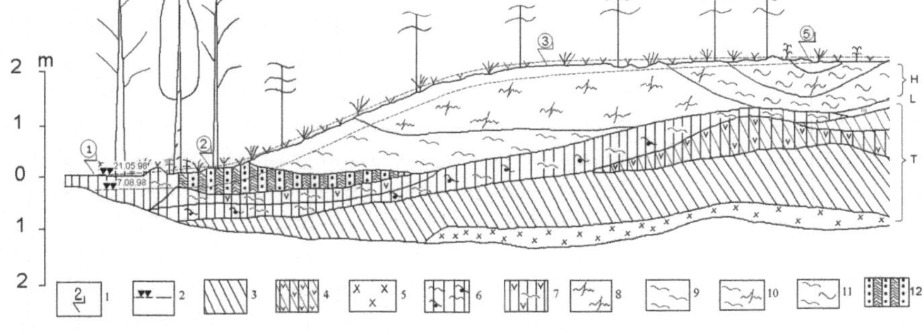


**Fig. 2** Points of description of phytocenoses on landscapes of the River Klyuch basin. Criteria for the placement of observation points within landscapes of the basin of the River of Klyuch: *1*—observation points; *2*—mire water table. Kind of peat: *3*—fen sedge, *4*—fen wood-sedge, *5*—fen horsetail, *6*—transitional wood-sphagnum, *7*—transitional wood-grass, *8*—fuscum peat, *9*—magellanicum peat, *10*—raised complex, *11*—peat of mire depressions, *12*—raised pine-cottongrass. Observation points on mire phytocenoses: *1*—birch-pine-green moss paludal forest; *2*—high ryam; *3*—low ryam; *5*—sedge-sphagnum swamp. Types of deposits: *H*—raised; *L*—transitional; *T*—fen

the structure of the peat deposit). The remaining peats reach up to 10 cm in thickness (each of them amounts to 14 % of the deposit).

At point 2, the layer of fen sedge peat with a high degree of decomposition (65 %) is located in its foundation. Later, the mesotrophic stage becomes apparent in the development of the mire. At this stage, the plat groupings formed the layer of well-decomposed transitional peat of three kinds: wood-sphagnum, cotton grassy, and wood-herbaceous. The mesotrophic stage of development of the mire vegetation is changed by the oligotrophic stage, during which raised peat of two kinds was formed: pine-cotton grassy and peat-magellanicum. The raised peat is notable for the medium and low degree of decomposition. The mean degree of decomposition of the peat deposits amounts to 39 %. Therefore, the stratigraphic features of the peat deposit define the development of the border of the peat tract where plant associations of forest-marshy type were predominant for a long period.

**Point 3** (spine-undershrub-sphagnum phytocenosis, low ryam) is spread along the profile (850 m) and is a typical ryam (Fig. 2). The structure of the peat deposit defines the most widely spread facies of the studied peat tract—i.e., ryam (see Fig. 2). The largest revealed thickness of the peat is equal to 3 m in this point; the deposit is of a mixed marshy type. The raised peat is represented by two kinds—fuscum of a weak degree of decomposition (40 % of participation) and magellanicum of a medium degree of decomposition (10 % of participation). These kinds of peat form a thicker layer of raised peat (1.5 m) as compared with the other points.

The layer of the fen peat is situated in the foundation of the peat deposit; its thickness is equal to 30 cm (10 % of the total) and a degree of decomposition accounts for 50 %. The thicker layer of the fen sedge peat (40–50 % of the total) is located above it. At the turn of two layers—raised and fen peats—the peat layer of the transitional type can be found, which was formed by mesotrophic plant communities that formerly existed here, such as wood-sedge and wood-sphagnum. The availability of most herbaceous remains among the fossil plants in the peat is the evidence of the decrease in moisture and climate warming at the mesotrophic stage of the mire development.

**Point 5** is a peripheral part of the open sedge-sphagnum marsh. The layer of the raised peat defines the oligotrophic stage of the mire development, which is revealed in the successive change of peat kinds: hollow, raised complex, and fuscum ones (Fig. 2). The replacement of plant groupings occurred very quickly for transfer of the mire from the stage of ground feeding to that of atmospheric feeding; furthermore, the transfer did not reflect on the peat deposit. Therefore, transitional peat is not found here. The hollow forms the underlying layers from a depth of 120 cm, mostly by sedge peats (34.6 %). The average degree of hollow peat decomposition accounts for 35 %, with fluctuations of 25–40 % in different layers.

Therefore, the history of evolution of the mire tract in the basin of the River of Klyuch is clearly pronounced by the stratigraphy of the peat deposit, beginning with the predominance of the eutrophic herbaceous phytocenosis, such as horsetail and then sedge. The essential predominance of the eutrophic and mesotrophic stages is worth noting. The transfer to the oligotrophic stage is accompanied by the formation of pine-undershrub-sphagnum communities. Nowadays, the biggest part of the peat deposit of oligotrophic landscapes of the basin of the River of Klyuch changed to the oligotrophic stage of evolution; the thickness of the raised peat reaches 120 cm.

The botanical composition and the degree of decay were determined in the samples (State Standard 28245-89). The group composition of peat was researched by Instrof's method. Humic acid (HA) separation was made on the following scheme: from air-dried samples, water-soluble components and lipids were removed one after another by extraction of chlorophorm (1:3). Humic matters were extracted by decimolar solution of natrium hydroxide. The total micromycete number and biomass were determined directly through luminescence microscopy. The cells were preliminarily desorbed by means of ultrasonic dispersant. White calcofuor was used to stain the fungal mycelium and spores (Zvyaginzev 1991). Eukaryote microbe biomass was counted, taking into account the measured diameter of fungal spores and mycelium by formula ($0.836r^3 \times 10^{-12}$ g for spores and $0.628r^2 \times 10^{-6}$ g for mycelium) (Polyanskaya 1996). The number and taxonomic composition of soil micromycete were studied by the inoculation of acidified agar-containing Czapek's medium (Dudka 1982). The identification of species was performed by generally accepted determinants (Domsch and Gams 1993; Ellis 1971; Pitt 1979). The number and composition of yeast fungi were studied by the inoculation of must-agar medium acidified by lactic acid. The isolates were obtained at room temperature and identified by standard methods (Babieva and Golubev 1979) with the help of a determinant (Kurtzman et al. 1998).

The mire water was sampled in every well of swampy phytocenosis (points 2, 3, 5), as well as in the River of Klyuch and River of Bakchar before the Kyuch falls into it. The macrocomponents were analyzed by generally accepted methods (Lurie 1973). HA and FA were examined by the method described in Bazin et al. (1992). The determination of heavy metal concentrations was performed by the certified method of quantitative atomic emission analysis with the preparation of ash residue, according to the State Standard 27784-88.

## 3 Results

The general technical characteristics of the peats are given in Table 1. The upper layer of the peat deposit (point 3) consists of mossy peats; below this, the peat deposit is represented by fen herbaceous species. In the peat deposit of point 2, the transitional wood-herbaceous peats are most common. The individual plant associations, such as sphagnum mosses (fuscum, magellanicum), cotton grass, and sedges, are mainly developed. These define the degree of decomposition, which increases with the depth of deposition in transition from mossy peats to arborescent peats from 5 to 65 %.

Ash content in raised slightly transformed peats is low and amounts to 2, 1–5, 2 %. Transitional and fen peats are notable for the normal ash content, which

**Table 1** Botanical composition of peats

| BGS, research points | Depth (cm) | Peat kind, deposit type | Botanical composition (%) | R/A (%) |
|---|---|---|---|---|
| High ryam, p. 2 | 0–25 | Pine-cotton grass-sphagnum, Rt | Pinus-35; Sph. Magellanicum-20; Sph. Angustifolium-10; Eriophorum-25; Undershrubs-10 | 35/5.8 |
| | 25–50 | Wood-cotton grass, T | Pinus, Betula-30; Undershrubs-10; Sph. Magellanicum-5; C. Lasiocarpa-rare; C. Rostrata-10; Eriophorum-45 | 55/4.7 |
| | 50–80 | Wood-cotton grass, T | Pinus, Betula-25; Undershrubs Sph. Magellanicum-5; Sph. Flexiosum-5; C. Rostrata-10; C. Lasiocarpa-5; C. Globularis-5; Eriophorum-40; Equisetum—rare. | 55/6.8 |
| Low ryam, p. 3 | 0–100 | Fuscum peat | Undershrubs-solitary; Sph. Fuscum-90; Sph. Magellanicum-10; Sph. Angustifolium-rare, | 5/1.5 |
| | 100–150 | Medium peat | Undershrubs-5; Sph. Fuscum-25; Sph. Magellanicum-40; Sph. Angustifolium-25 | 10/3.1 |
| | 150–200 | Pine-cotton grass-sphagnum, Rt | Pinus-25; Undershrubs-10; Sph. Magellanicum-15; Sph. Angustifolium-10; Sph. Fuscum-10; Eriophorum-30 | 50/4.6 |
| | 200–250 | Sedge, F | Wood-solitary; Eriophorum-10; C. Rostrata-10; C. Lasiocarpa-35; Menyanthes trifoliate-10; Fern-5; Equisetum-units, Sph. Centrale-10; C. Limosa-solitary; | 45/4.0 |
| | 250–300 | Herbaceous, F | Wood-5; C. Rostrata-5; C. Lasiocarpa-20; Fern-35; Menyanthes trifoliate-5; Drepanocladus sendtneri-10; Meesia triquetra-rare; Sph. Centrale-10; Sph. Riparium-5; Carex sp.-5 | 40/6.2 |

(continued)

**Table 1** (continued)

| BGS, research points | Depth (cm) | Peat kind, deposit type | Botanical composition (%) | R/A (%) |
|---|---|---|---|---|
| Sedge-sphagnum marsh, p. 5 | 0–50 | Fuscum peat | Sph. Balticum-10; Sph. Jensenii-rare; Sph. Flexuosum-60; Sph. Fallax-10; Sph. Magellanicum-10; Eriophorum-5 | 0/2.3 |
| | 50–100 | Complex, Rt | Sph. Majus-10; Sph. Jensenii-15; Sph. Obtusum-35; Sph. Papillosum-10; Sph. Flexuosum-10; C. Rostrata-5; C. limosa-solitary; Eriophorum-5; Equisetum-5 | 15/3.1 |
| | 100–150 | Sedge-sphagnum, T | C. Rostrata-20; C. Lasiocarpa-10; Scheuchzeria palustris-5; Menyanthes trifoliate-5; Equisetum-10; Sph. Majus-solitary; Sph. Jensenii-5; Sph. Obtusum-20; Sph. Magellanicum-5; Undershrubs, Pinus, Betula-15 | 35/5.5 |
| | 150–200 | Sedge, F | C. Rostrata-25; C. Lasiocarpa-15; Carex sp.-10; Wood-5; Menyanthes trifoliate-10; Equisetum, Fern-10; Sph. Fallax-5; Sph. Obtusum-10; Sph. Papillosum-5 | 50/5.8 |
| | 200–250 | Herbaceous, F | Equisetum-5; Fern-35; C. Rostrata-10; C. Lasiocarpa-15; Carex sp.-solitary; Sph. Fallax-5; Menyanthes trifoliate-20; Sph. Obtusum-5 | 50/9.2 |
| | 250–300 | Fern, F | Wood-10; Fern-60; Horsetail-5; Bogbean-10; Carex rostrata-5; C. asparatilis-5; Carex sp.-solitary; Sph. dusenii-solitary; Sph. apiculatum-5 | 40/6.8 |

Note: *Rt* raised type; *T* transitional type; *F* fen type; *R* degree of decomposition, %; *A* ash content, %; *BGS* biogeosystem reaches 10 and 9 %, respectively. Mineralization is greatly increased in the underlying grounds only.

The chemical composition of organic matter (OM) of the peats, which forms the peat deposit of the landscape profile, is also considered. The botanical characteristics of the peat exert an essential influence upon its composition. For example, the appearance of cotton grass defined the increase in bitumen content. If we comparatively analyze the composition of OM of the surface (oxidative conditions) and lower (reductive conditions) horizons, the essential differences can be noted in the content of HAs, lightly and sparingly hydrolysable substances (LH and SH). The process of humification is more clearly pronounced in the lower horizons. Therefore, the increase in LH and SH content is noted in the layer of 50–100 cm in point 3 (transit part of the profile) and point 2 (transaccumulative part) (Table 2).

At the same time, the high content of LH is observed in the water-unloading zone (point 2, geochemical barrier of the profile). The redistribution of water-soluble compounds can be seen in the horizon with the oxidative conditions; this fact is observed in most content of the transaccumulative part of the landscape throughout the whole landscape profile. This phenomenon also confirms the availability of the migration process of the substances.

Research on biogeochemistry of the peaty mires showed that the peat deposit contained considerable amounts of elements acting as a global sorbent (Table 3). If we compare the degree of element accumulation in the peat deposits of the landscape profile with the average background values that we obtained earlier for the oligotrophic peats (Inisheva and Tsybukova 1999), we can see the following: the contents of S, Br, and Sr are comparable to the content of the West-Siberian peats of the raised type. As for the content of Co, Zn, Sc, Cr, Sr, Hg, and La, a small excess is noted in comparison with the average background values. As a matter of fact, Ca and Fe are found considerably more in the studied peats. Their background values are 0.28 and 0.2 %, respectively, and the highest concentration is observed in the transaccumulative part of the landscape.

It is very interesting to examine the structure and stocks of microbe biomass in the peat deposits. This method, in contrast to that of inoculation, reveals the reserve of viable microbe biomass. The natural variability of the amount and concentration of microorganisms has been revealed throughout the entire depth of the peat deposit, but more clearly it is revealed in the active layer. The reserve of microbe biomass in the peat deposits of the landscape profile varies in the layer of 0–100 cm from 0.18 to 1.42 kg/m$^2$; it is also evidence of high microbiological activity in the peat deposits of the oligotrophic sequence. At the same time, the reserve of microbe biomass in the peat deposit of the sedge-sphagnum marsh is four times higher than in the deposit of the low ryam. The difference is noted in the main components of the microbe biomass in the surface and underlying horizons. The fungal mycelium is prevalent in the surface layer of the deposit (43–83 %), whereas fungal spores and yeast cell prevail down the profile (57–93 %). There are more spores and bacterial cells in the lower part of the peat deposit (9–42 %). Thus, the availability of active biochemical processes can be verified throughout the entire depth of the

**Table 2** Total technical and chemical characteristics of peat deposits of biogeocenoses of landscape profile

| Position in landscape profile | Layer (cm) | Botanical composition | R (%) | A (%) | Group composition of organic matter, % of mass | | | | | |
|---|---|---|---|---|---|---|---|---|---|---|
| | | | | | WS | LH | SH | HA | B | R |
| Autonomous part (sedge-sphagnum swamp) | 0–50 | Peat of Sphagnum hollows | 5 | 10.9 | 0.4 | 43.0 | 12.3 | 31.3 | 0.6 | 12.4 |
| | 50–100 | | 5 | 6.0 | 0.3 | 33.4 | 10.5 | 30.5 | 0.2 | 25.1 |
| Transit partryam) | 0–50 | Fuscum-peat | 5 | 2.7 | 0.9 | 30.6 | 15.2 | 25.1 | 2.6 | 25.6 |
| | 50–75 | | 5 | 2.0 | 1.2 | 16.6 | 16.4 | 25.2 | 1.7 | 38.9 |
| | 75–100 | | 5 | 2.1 | 0.4 | 32.6 | 14.3 | 19.8 | 0.6 | 32.3 |
| Transaccumulative part (raised ryam) | 0–25 | Pine-Eriophorum-sphagnum raised | 45 | 5.2 | 1.6 | 22.4 | 4.2 | 27.6 | 3.9 | 40.3 |
| | 25–50 | Wood-Eriophorum transitional | 55 | 6.5 | 1.6 | 28.3 | 8.1 | 22.7 | 3.6 | 35.7 |
| | 50–75 | | 60 | 8.0 | 1.2 | 36.6 | 9.2 | 23.3 | 2.6 | 27.4 |
| | 75–100 | | 60 | 9.8 | 0.4 | 32.7 | 11.0 | 39.8 | 0.9 | 15.2 |

Note: *R* degree of decomposition; *A* ash content; *WS* water-soluble; *LH* lightly hydrolysable; *SH* sparingly hydrolysable; *HA* humic acids; *B* bitumen; *R* residue

**Table 3** Element content in the peat of landscape profile, $10^{-4}$ %

| Biogeocenosis | Layer, cm | Ca[a] | Sc | Cr | Fe[a] | Co | Zn | Br | Sr | Hg | Ba | La | Ce | Sm |
|---|---|---|---|---|---|---|---|---|---|---|---|---|---|---|
| Sedge-sphagnum swamp | 0–50 | 0.8 | 4.7 | 21.5 | 1.0 | 4.3 | <6 | 27.9 | <50 | <20 | <20 | 6.8 | 17.8 | 1.3 |
| | 50–100 | 1.8 | 1.6 | 16.5 | 0.9 | 3.7 | 68.5 | 54.4 | 279 | 20.1 | 97.3 | 2.3 | 7.0 | 0.6 |
| Low ryam | 0–50 | 0.7 | 1.1 | 20.9 | 0.6 | 2.0 | 148.0 | 33.4 | 84 | 72.8 | 58 | 3.4 | 7.0 | 0.6 |
| | 50–75 | 0.9 | 0.8 | 15.7 | 0.5 | 1.3 | 124.1 | 27.6 | 72 | 54.9 | <30 | 1.4 | 3.8 | 0.3 |
| | 75–100 | 1.7 | 0.9 | 18.3 | 0.5 | 1.7 | 130.0 | 36.4 | 76 | 58.0 | 45 | 2.1 | 5.4 | 0.3 |
| High ryam | 0–25 | 1.4 | 2.4 | 18.1 | 1.1 | 2.8 | 41.4 | 25.2 | 303 | 92.8 | 129 | 3.4 | 9.9 | 0.3 |
| | 25–50 | 1.9 | 2.4 | 18.1 | 1.1 | 4.0 | 6.9 | 40.2 | 324 | 31.3 | 103 | 3.1 | 9.2 | 0.2 |
| | 50–75 | 2.6 | 3.0 | 18.3 | 1.5 | 6.2 | <6 | 56.5 | 320 | 284.7 | 103 | 5.5 | 15.3 | 1.3 |
| | 75–100 | 2.1 | 2.0 | 16.4 | 1.5 | 5.5 | 31.0 | 55.5 | 396 | <20 | 96 | 3.5 | 6.1 | 0.8 |
| Background content | | 0.28 | 0.56 | 7.6 | 0.2 | 1.1 | 6.6 | 39.4 | 79.4 | 0.4 | 46.9 | 1.5 | 4.8 | 0.5 |

[a]Less than 6 %

peat deposit, but the direction of the processes is different in the surface and lower horizons of the deposit.

The peat deposits of the landscape profile differ in their micromycete concentration and character of the depthwise distribution of the fungal biomass. The total concentration of eukaryote microorganisms is higher in a thin peat deposit of high ryam (point 2) than in the peat deposits of points 3 and 5 (shallow ryam and sedge-sphagnum marsh); the highest concentration is observed in the upper layer of 50 cm thick. The distribution of these microorganisms is even within the limits of the profile in the peat deposits of points 3 and 5. The stock of the eukaryote biomass ranges from 200 g to 1.3 kg/m$^2$ in the deposit of 1 m thickness and from 300 g to 1.2 kg/m$^2$ in the deposit of 3 m thickness (Table 4). The stock of microscopic fungi is rather high (2–13 tons per ha), but it is insignificant in comparison with their enormous reserve of vegetative mortmass. The portion of carbon of microbial origin (eukaryote microorganisms) is not more than 3 % in the layer of 0–50 cm, 2 % in the 1-m layer and 1 % in the 3-m layer.

The structure of eukaryote microbe biomass (ratio of active compounds of the complex of microscopic fungi [i.e. mycelium] and those of inactive compounds [i.e. spores]) has its specific features in each peat deposit. Thus, fungal spores prevail throughout the whole profile in the peat deposit of the low ryam (point 3). Furthermore, their proportion was higher by 10–30 % than that of mycelium in the upper layers; in the lower layers, the spore prevalence became absolute (100 %). The peat deposit of the sedge-sphagnum marsh (point 5) occupies an intermediate position between the peat deposits of points 2 and 3, according to the morphological structure of the micromycete complex. Fungal mycelium dominates here up to a depth of 1.5–2 m and fungal spores in the deeper layers. It should be noted that fungal spores could not be practically distinguished from yeast cells in the electron microscope. Therefore, it can be supposed that the high number of eukaryote cells is determined more likely by the yeast-like cell rather than the fungal spores. Oligotrophic peat deposits are a favorable ecotope for this group of micromycetes due to the high content of plant residues, low pH value, and capacity of the yeasts

**Table 4** Fluctuation in the concentration of microscopic fungi (A), fungal biomass reserve (B), and carbon of fungal biomass of total carbon pool (C) in peat deposits of landscape profile

| | Layer, cm | High ryam | Low ryam | Sedge-sphagnum swamp |
|---|---|---|---|---|
| A (mg per g) | 0–50 | 13–44 | 2–8 | 2–26 |
| | 50–100 | 16–56 | 10–21 | 5–30 |
| | 100–300 | –[a] | 13–24 | 11–36 |
| B (kg/m$^2$) | 0–50 | 0.2–0.9 | 0.05–0.1 | 0.05–0.7 |
| | 50–100 | 0.4–1.3 | 0.2–0.4 | 0.2–0.9 |
| | 100–300 | – | 0.3–0.5 | 0.5–1.2 |
| C (%) | 0–50 | 1–2 | 0.2–1 | 0.2–3 |
| | 50–100 | 0.5–1 | 0.4–0.8 | 0.3–2 |
| | 300 | – | 0.1–0.2 | 0.3–1 |

[a]Data are not given for the peatland thickness is equal to 1 m

for oxygenic metabolism. The yeast concentration in the peat deposit ranges from several hundreds to millions of colony-forming units per gram of the peat. The spreading of the yeasts is even in the deposit; their lowest number was noted only in the deepest layers of the peat deposit.

The yeasts are considered as a group, which is ecologically connected with the initial stages of the succession of plant residues. In fact, a negative dependence between the two parameters was discovered—that is, a yeast number and a degree of decomposition of the peat (correlation factor $r = -0.44$ at level of significance of $p < 0.05$). This dependence indicates the existence of the back connection (the lower the degree of decomposition, the higher the density of the yeast groupings).

However, 114 strains were singled out during the work. The collection was represented by 17 species (both banal cosmopolitan and those peculiar to the peat deposits (Golubev et al. 1981; Polyakova et al. 2001) from seven genera and a group of isolates of asporous yeasts of genera of *Candida, Cryptococcus,* and *Rhodotorula;* these are not identified in the composition of the known species (Table 5).

Three species, such as *Cryptococcus albidus* Skinner and "red" epiphyte *Rhodotorula glutinis* Harrison, *Sporobolomyces roseus* Kluyver et van Niel., prevailed in all communities in the study. It should be also noted that the isolation of the species of *Candida paludigena* Golubev et Blagodatskaya is considered to be ecologically connected with the raised mire ecosystems (Golubev et al. 1981).

The revealed peculiarities of the biochemical processes exert influence upon the formation of hydrochemical runoff from the paludified territory in the peat deposits of the landscape profile. It should be taken into account that chemical composition of mire water and subsequently migrating streams are formed due to a mixture of atmospheric precipitation with swampy waters. The composition of swampy water is defined by the arrival of movable compounds from the peat deposit, which underwent a regular biochemical transformation.

Geochemically, the influence of mires on the composition of the river and underground waters has not been practically studied. On the one hand, swampy ecosystems are a geochemical barrier (Glazovskaya 1983); they fix many

**Table 5** Occurrence of yeast genera in peat deposits in oligotrophically connected landscapes

| Genus | Number of species of the given genus | Occurrence |
|---|---|---|
| *Basidiomycetes affinity* | | |
| *Cryptococcus* | 5 | 67 |
| *Rhodotorula* | 6 | 71 |
| *Sporobolomyces* | 1 | 55 |
| *Trichosporon* | 1 | 19 |
| *Ascomycetes affinity* | | |
| *Candida* | 4 | 19 |
| *Debaryomyces* | 1 | 7 |
| *Pichia* | 1 | 7 |

contaminants from the atmosphere and remove them from the biological cycle due to their absorbing capacity. However, on the other hand, the complex chemical composition of the peats in the peat deposit of swampy ecosystems, their physical structures, and colloidal structures form the proper hydrochemical composition of swampy waters. Atmospheric precipitation passes through the stage of mire genesis before coming to underground water-bearing horizons. The groundwater feeding mire is also transformed in organogenic medium of the peat deposit. As a result, freshwaters rich in carbonic acid, methane, iron, manganese, and swampy components are formed. A special kind of mire water is formed in such a way that composition and the processes of interaction have been studied incompletely.

All the processes of the interaction of water with the products of vital activity of biogeocenoses can be considered, to some degree, as a special regional thermodynamic system (sun-basin unit by Kaznacheyev and Anshin 1986), where the living substance is predominant. Furthermore, the raised bogs are autonomous in eluvial and geochemical respect; this fact permits one to follow the migration flow of the substances in the balance variant.

The removal of the elements of the River of Klyuch runoff was estimated by daily intervals. The consumption of every element was estimated based on concentrations of the respective elements and the average daily water consumption as a product of concentration and water consumption. The removal for the longer time intervals was counted by the summation of daily removal values.

The runoff volume, its dynamics, and the conditions of formation define the migration flow of the substances from the oligotrophic landscapes into swampy rivers. The data of chemical element concentration in the River of Klyuch show considerable variability in different years and hydrological stages of the runoff (Table 6).

The swampy origin of the small waterway of the River of Klyuch determines the lowered content of average values of ions in water, such as $Ca^{2+}$, $HCO_3^-$, and $SO_4^{2-}$. However, to some extent, the increased content of total Fe and $NH_4^+$, as well as the appearance of intermediate reduction products, such as $NO_3^-$–$NO_2^-$, can appear in reductive conditions only.

The water of the River of Klyuch is rich in OM, as confirmed by the high values of chemical oxygen consumption (ChOC) and HA and FA (Table 7). The chemical composition of the River of Bakchar, which springs from the swamps and flows through them, is to a greater extent of the same chemical composition as that of the River of Klyuch.

The analysis of the obtained data indicates the influence of mire waters of the active layer of the peat deposit, which passed through the biochemical cycle of the exchange processes in the peat-water, on the hydrochemical composition of the river. The highest concentration of practically all compounds is noted on the mire border in the paludified forest (point 2), which is a transaccumulative part of the landscape profile of geochemically connected landscapes. It also serves as a geochemical barrier for swampy waters running off from the autonomous part of the profile.

**Table 6** Chemical composition of river and mire waters, mg per l

| Components | River of Bakchar | River of Klyuch | Mire waters | | | Solid precipitations |
|---|---|---|---|---|---|---|
| | | | High ryam, p. 2 | Low ryam, p. 3 | Open swamp, p. 5 | |
| PH | $\frac{6.5\text{–}7.5}{7.2}$ | $\frac{6.6\text{–}7.1}{6.7}$ | $\frac{4.6\text{–}5.6}{4.7}$ | $\frac{3.6\text{–}4.3}{4.0}$ | $\frac{3.9\text{–}4.5}{4.1}$ | 7.0 |
| $Ca^{2+}$ | $\frac{9.3\text{–}45.3}{32.5}$ | $\frac{8.0\text{–}28.0}{17.6}$ | $\frac{3.0\text{–}7.2}{5.2}$ | $\frac{0.6\text{–}6.2}{1.9}$ | $\frac{1.0\text{–}6.0}{2.6}$ | 0.7 |
| $Mg^{2+}$ | $\frac{7.3\text{–}18.8}{12.1}$ | $\frac{6.7\text{–}15.2}{9.9}$ | $\frac{0.5\text{–}6.3}{2.6}$ | $\frac{0\text{–}3.0}{1.6}$ | $\frac{0\text{–}4.3}{1.7}$ | 0 |
| $Na^{+}$ | $\frac{1.4\text{–}14.0}{5.3}$ | $\frac{1.7\text{–}2.6}{2.2}$ | $\frac{0.1\text{–}1.9}{1.2}$ | $\frac{0.4\text{–}0.7}{0.6}$ | $\frac{0.1\text{–}1.1}{0.7}$ | 1.5 |
| $NH_4^{+}$ | $\frac{0.2\text{–}3.8}{1.6}$ | $\frac{0.4\text{–}3.9}{1.8}$ | $\frac{0.4\text{–}4.1}{2.3}$ | $\frac{0.2\text{–}2.9}{1.4}$ | $\frac{0.2\text{–}2.3}{1.4}$ | 1.2 |
| $Fe_{total}$ | $\frac{1.5\text{–}4.9}{3.1}$ | $\frac{2.0\text{–}5.9}{3.6}$ | $\frac{1.6\text{–}7.0}{4.7}$ | $\frac{1.6\text{–}3.8}{2.8}$ | $\frac{1.0\text{–}3.8}{2.5}$ | 0.01 |
| $HCO_3^{-}$ | $\frac{54.3\text{–}160.0}{96.5}$ | $\frac{24.4\text{–}109.1}{56.6}$ | $\frac{4.8\text{–}24.5}{11.8}$ | $\frac{0.0\text{–}13.5}{3.6}$ | $\frac{0.0\text{–}12.8}{3.5}$ | 8.5 |
| $SO_4^{2-}$ | $\frac{0.0\text{–}0.8}{0.3}$ | $\frac{0.0\text{–}3.8}{1.8}$ | $\frac{0.0\text{–}1.0}{0.3}$ | $\frac{0.0\text{–}2.9}{0.5}$ | $\frac{0.0\text{–}3.3}{0.7}$ | 1.0 |
| $NO_3^{-}$ | $\frac{0.23\text{–}1.7}{0.63}$ | $\frac{0.3\text{–}2.4}{1.5}$ | $\frac{0.3\text{–}0.8}{0.5}$ | $\frac{0\text{–}1.9}{0.7}$ | $\frac{0.2\text{–}1.4}{0.7}$ | 0.3 |
| $NO_2^{-}$ | $\frac{0.002\text{–}0.07}{0.03}$ | $\frac{0.002\text{–}0.1}{0.03}$ | $\frac{0.0\text{–}0.01}{0.004}$ | нет | $\frac{0.0\text{–}0.009}{0.004}$ | 0 |

*Note* In numerator—extreme values, in denominator—average values

**Table 7** Content of organic substances, mg per l

| Components | River of Bakchar | River of Klyuch | Mire waters | | |
|---|---|---|---|---|---|
| | | | High ryam | Low ryam | Open swamp |
| Carbon (C) | 14–49 / 30 | 28–85 / 55 | 56–106 / 78 | 46–109 / 65 | 37–96 / 54 |
| ChCO | 94–200 / 146 | 81–293 / 181 | 108–269 / 175 | 60–222 / 142 | 103–216 / 166 |
| Humic acids | 3.5–16.7 / 9.2 | 5.8–25.1 / 17.8 | 10.6–27.0 / 16.8 | 4–26.1 / 12.5 | 6.1–20.0 / 11.0 |
| Fulvic acids | 28.6–32.7 / 30.6 | 41.5–61.6 / 51.3 | 81.0–92.4 / 87 | 51.4–60.2 / 56 | 42.6–53.3 / 49.6 |

*Note* In numerator—extreme values; in denominator—average values; ChCO—Chemical consumption of oxygen

The peculiarity of the organo-mineral type of swampy waters is also revealed in the elemental composition (Table 8). The peat deposit contains a considerable quantity of dispersed elements operating as a global sorbent, which regulates their content (Dobrovolsky 1983). It has been shown by different authors that finely dispersed particles, sesquioxides, and HAs in a large amount of the swampy water (Kalbitz and Geyer 2002; Frank and Heinrich 2008) participate in ion absorption.

As shown earlier (Inisheva and Tsybukova 1999), the content of trace elements in peat deposits is determined to a greater extent by their botanical composition; to a smaller extent, they are concentrated in the raised peats. Therefore, the element's content in the mire waters of the open marsh (point 5), deposited from the surface by the raised sphagnum peat of a weak degree of decomposition, is less than at the mire periphery (point 2, high ryam). The compounds of Ti and Sr—the contents of which are the same in the peat deposits of all geochemically connected landscapes of the river of Klyuch—are an exception. The high content of heavy metals is practically noted in the studied swampy waters of all phytocenoses compared to the data of other researchers (Kreshtapova 1993; Inisheva et al. 2005).

Water-soluble HAs, especially fulvic acids, play a decisive role in the formation of the hydrochemical composition of swampy waters. The waters of mires are characterized in this case by a yellowish color and the value of oxidability usually ranges on average within the limits of several tens of $O_2$/l; mean values amount to 200–300 mg $O_2$/l. For example, the content of carbon in snow is not more than 8 mg/l and oxidability is 12.3 $O_2$/l. The excess of FA content is 5–20 times (most often 10) with regard to HA amounts in the studied waters; this is in accordance with their content in peats. The high content of movable compounds in the peats and rather high microbiological activity throughout the whole profile of the peat deposit confirm that chemical composition of the runoff from the paludified catchment is determined not only by atmospheric precipitation, as some researchers assert, but also at the expense of transformation processes in the peat deposit.

Therefore, the general runoff of mineral and organic substances agrees with above-mentioned regularities of water removal and migration of the substances in the landscape profile. On the whole, the dynamics of the element runoff are mainly defined by the course of water runoff; this index is a synchronism of the distribution of water expenditure and chemical elements. The total runoff values for the

**Table 8** Content of elements in mire water and runoff of river of Klyuch, average values in the course of vegetation season, mcg per l

| Point of sampling | Pb | Cu | Mn | Zn | V | Cr | Ni | Yb | Ti | Ga | Hg | Sr |
|---|---|---|---|---|---|---|---|---|---|---|---|---|
| Open swamp, (p. 5) | 0.95 | 0.95 | 159.16 | 11.38 | 0.95 | 2.59 | 1.95 | 0.12 | 6.18 | 0.39 | 0.08 | 27.03 |
| Low ryam (p. 3) | 0.75 | 6.80 | 45.12 | 7.52 | 1.78 | 2.39 | 1.43 | 0.14 | 7.99 | 0.63 | 0.08 | 31.05 |
| High ryam (p. 2) | 1.06 | 4.62 | 52.87 | 5.95 | 2.44 | 2.52 | 1.42 | 0.20 | 8.35 | 0.63 | 0.13 | 49.51 |
| River of Klyuch (upper course) | 0.57 | 0.57 | 36.44 | 2.28 | 0.76 | 1.52 | 0.81 | 0.12 | 4.81 | 0 | 0.08 | 31.69 |
| River Klyuch (lower course) | 0.56 | 0.56 | 170.9 | 7.48 | 0.99 | 2.15 | 1.38 | 0.17 | 7.25 | 0.67 | 0.11 | 50.96 |
| Water from raised bog (Bakhnov 1986) | 0.60 | 0.50 | 41.90 | 5.70 | – | – | 0.70 | – | 2.60 | – | – | – |
| Water from raised bog (Nazarov et al. 1977) | – | 3.80 | 10.00 | 38.00 | – | – | – | – | – | – | – | – |
| Water from raised peats of Western Siberia (Bakhnov 1986) | 3.10 | 7.00 | 74.70 | 6.60 | 2.10 | 7.60 | – | 0.36 | – | – | 0.45 | 79.40 |

*Note* '–' Element is not found

chemical elements in the course of the runoff were as follows: $Ca^{2+}$—1398, $Fe_{total}$—311, $SO_4^{2-}$—391, $NO_3^-$—236, $NO_2^-$—1, Pb—2.253 × $10^{-3}$, Mn—317.29 × $10^{-3}$, Zn—41.191 × $10^{-3}$, Ni—8.151 × $10^{-3}$, and Ti—29.651 × $10^{-3}$ kg/km$^2$. The annual runoff losses of dissolved OM were 6945 kg/km$^2$ (Inisheva et al. 2011).

## 4 Conclusions

When the conditions of the formation of chemical composition of swampy water waters are considered, the hierarchy of the landscapes in the river basin and the participation of surface and inner-swampy flow during the vegetation season must first be taken into account. The botanical composition of the peat deposits of mire ecosystems are significant. Biochemical processes in the active and inactive horizons of the peat deposit exert influence upon the formation of the runoff composition.

The study of the concrete water objects, as well as the physical, chemical, and biological processes of transformation of substance and energy flows on the catchment areas typical for them, seems to influence the chemical constituent of the georunoff. When one considers the conditions of formation of the chemical composition of swampy, fluvial waters of the paludified catchments and the role of mires in this process, it is necessary to take into account the hierarchy of landscapes in the river basin and the organic constituent of swampy waters.

The role of OM as a final product of peat formation (peat genesis) on the territory of Vasyugan Mire and its part in the subsequent migration with lateral and inner swampy runoff into the rivers require further research in the future.

**Acknowledgements** This research was supported by grants from the Ministry of Education and Science RF (No. 174), RFFI (No. 15-05-00288), and Polish Ministry of Education (No. NN305 3204 36).

## References

Babieva IP, Golubev VI (1979) Methods of isolation and identification of yeasts. Food industry, Moscow (in Russian)
Bakhnov VK (1986) Biogeochemical aspects of paludification process. Nauka, Novosibirsk (in Russian)
Bazin ET, Kopenkin VD, Kosov VI (1992) Determination of humics content in water. Technical analysis of peat, Nedra, Moscow (in Russian)
Dobrovolsky VV (1983) Geography of trace elements. Global dispersion. Mysl, Moscow (in Russian)
Domsch KH, Gams W (1993) Compendium of soil fungi. Academic Press, London
Dudka IA (1982) Methods of experimental micology. Naykova dumka, Kiev (in Russian)
Ellis MB (1971) Dematiaceous Hyphomycetes. Inst., Kew, Commenw, Mycol

Frank S, Heinrich S (2008) Dissolved organic carbon concentration along carbon and water table gradients in raised bogs in Lower Saxony (Germany) under grassland. In: 14th International Peat Congress, Extended abstracts, No 232
Glazovskaya MA (1983) Foundations of classification of natural ecosystems by sustainability to man-caused impact and prediction of landscape and geochemical regioning. Sustainability of geoecosystems. Nauka, Moscow (in Russsian)
Golubev VI, Blagodatskayay VM, Manukyan AR, Liss OL (1981) Yeast flora of peats. In: Proceedings of academy of sciences of USSR, Biology—No 2 (in Russian)
Inisheva LI, Tsybukova TN (1999) Environmental and geochemical assessment of the peats of the south-western part of West-Siberian flat plain. Geography and natural resources, Soc Notes 1:105–111 (in Russian)
Inisheva LI, Yudina NV, Inishev NG, Golovchenko AV (2005) Distribution of organic compounds in the system of geochemically linked mire landscape. Geochem Int 43(2):168–176
Inisheva LI, Zemtsov AA, Novikov SM (2011) Vasyugan mire natural conditions, structure and functioning. Tomsk State Pedagogical University Press, Tomsk
Ilyin RS, Nature Territory (Rrlief, Geology, Landscapes, Soils) Materials on Reserch of Siberia. Tomsk, 1930, vol 2. ( in Russian), Saprykin F. Yu. Soil Geochemistry and Nature Protection. Leningrad: Nedra, 1984. (in Russian)
Kaznacheyev VP, Anshin FG (1986) Doctrine on transformation biosphere and human ecology. Znanie, Moscow (in Russian)
Kreshtapova VN (1993) Nrace elements in peat soils and peat landscapes of the European Russia. Rossel khozizdat, Moskow
Kurtzman CP, Fell JW et al (1998) The Yeasts, a taxonomic study. Fourth revised and enlarged edition, Elsevier Science B.V., Amsterdam
Liss OL, Abramova LI, Avetov NA, Berezina NA, Inisheva LI et al (2001) Mire systems of West Siberia and their nature-protective significance. Moscow (in Russian)
Lurie Y (1973) Unified methods of water analysis. Chemistry, Moscow (in Russian)
Nazarov AD, Rasskazov NM, Udodov PA, Shvartsev SL (1977) Hydrogeological conditions of mire formation. Scientific prerequisites of developing of West-Siberian mires. Nauka, Moscow (in Russian)
Pitt JI (1979) The Genus Penicillium and its teleomorphic states Eupenicillium and Talaromyces. Academic Press, Waltham
Polyakova AV, Chernov IY, Panikov NS (2001) Yeast biodiversity in hydromorphic soils by example of herb-sphagnum mire (Western Siberia) and hummocky tundra (Barrow, Alaska). Microbiology 70(5) (in Russian)
Polyanskaya LM (1996) Microbe successions in soil. Thesis of doctorate, science, boiling. Moscow (in Russian)
Zvyaginzev DG (1991) Methods of soil microbiology and biochemistry. Moscow state university publishing house, Moscow (in Russian)

# Spin Labeling EPR Analyses of Soil: A New Method to Investigate Biogenic and Abiogenic Interactions of Amines in the Environment of Natural Soils

Olga N. Alexanderova

**Abstract** A major focus of this paper is to investigate the environmental fate of chemical pollutants with the amine functionality in soils and its underlying molecular processes. Taking into account a key role of the amine in the structure of chemical pollutant and its reactivity, spin labels with different amine groups were applied to natural soils and investigated by means of electron paramagnetic resonance (EPR) analysis. As spin labels, stable nitroxide radicals modeled chemical pollutants with the amine and, meanwhile, afforded knowledge of the molecular environment of soil where they were located or bound via the amine. The first results of the Spin Labeling EPR investigation showed that chemical pollutants including the amine were differently partitioned among soil sites with the different water–octanol coefficients Kow in dependence on the amine base that allowed for accumulating the amines with a weak base by a non-hydrolysable part of soil organic matter, whereas the amines with a strong base were transformed or bound to small molecules (<5 kDa) of a hydrolysable part of soil organic matter. The uneven partitioning of the amines among soil sites of different hydrophobicity intensified a part of anoxic areas in soil that, in turn, resulted in a restriction of areas fit for soil aerobic microorganisms and a decrease in their magnitude.

**Keywords** Polluted soil · Amines · Electron paramagnetic resonance (EPR) analysis · Spin labeling · Humic substances; non-extractable residues (NER)

## 1 Introduction

Interaction of soil with organic chemical pollutants released into the soil environment poses a complex of some processes between biogenic and abiogenic components in soils that change the soil environment, influence on soil microorganisms,

O.N. Alexanderova (✉)
Ural Federal University, Chapaeva Street 20, Office 414 Ekaterinburg, Russia
e-mail: o.n.alexanderova@urfu.ru



O.V. Frank-Kamenetskaya et al. (eds.), *Biogenic—Abiogenic Interactions in Natural and Anthropogenic Systems*, Lecture Notes in Earth System Sciences,
DOI 10.1007/978-3-319-24987-2_24

and lead to contamination of plants and crop. To answer the questions whether chemical pollutants released into soil are dangerous for soil, soil microorganisms and human health, we need to understand the underlying processes.

A wide variety of emerging environmental pollutants, such as biocides, pesticides, pharmaceuticals, etc., have the amine as a key structural component. Many investigators emphasized a peculiar role of the amine functionality in the molecular structure of some xenobiotics at their interaction with biogenic and abiogenic components in soils (Bollag et al. 1992; Senesi 1992; Weber et al. 1996; Thron et al. 1996). Biogenic and abiogenic interaction of amines with soil brings to mineralization of organic xenobiotics as well as to formation of their bound residues or non-extractable residues (NER). Xenobiotics covalently linked to soil organic matter are often referred to as the 'the bound residues' of soil, and these complexes are very resistant to acid and base hydrolysis, thermal treatment and microbial degradation (Bollag et al. 1992; Senesi 1992). NER of biotic and abiotic origin were studied by a lot of investigators but their chemical structure is currently not understood. Weber et al. (1996) concluded the importance of covalent binding of aromatic amines to natural organic matter from numerous studies of the reaction of aniline and other aromatic amines with soils and sediments. They hypothesized that dissolved organic matter, and model compounds might be representative of monomeric units forming humic and fulvic acid polymers. Bollag et al. (1992) elucidated covalent binding of xenobiotics with aromatic compounds, such phenols and anilines, as oxidative coupling with the formation of C–C, C–O, C–N, and N–N linkages between the humic substances and pollutants. They put the accent on the mediation of this reaction by a number of biological and abiotic catalysts including microbial enzymes, inorganic chemicals, clays, etc. Senesi (1992) pointed at a charge-transfer complex between the electron-rich donating amine of xenobiotic molecule and humic quinone moiety.

Ample investigations used the laboratory extraction of pollutants from contaminated soil as a critical step in analyses of the structure of NER and their sorption properties but often it resulted in unsatisfactory yield (Nowak et al. 2011). A common finding of all studies was that recovery of pollutants from field soils and spiked soils amounted to about 90 % immediately after an application of pollutants to soil and decreased significantly with increasing contact time, regardless of which extraction method was applied (Schauss et al. 2009; Nowak et al. 2011). The NER formed immediately was assumed to originate from abiotic/non-biogenic processes. Therefore, abiotic/non-biogenic NER was estimated to about 10 %, whereas the NER formed during biodegradation of pollutants was estimated to 35–50 % of the added labeled carbon (Schauss et al. 2009; Nowak et al. 2011). However, these investigations did not provide any information about chemical structure of the NER from tested compounds and underlying processes (Nowak et al. 2011; Gulkowska et al. 2014).

Meanwhile, the NER was believed to effect soil biota. Mention has been made of a residual toxic effect of many chemical pollutants including pollutants with amine released into natural soils (Nowak et al. 2011). Tappe et al. (2008), Schauss et al. (2009), and Rosendahl et al. (2011) showed that an application of antibiotic sulfadiazine to soil via its fertilizing with manure from antibiotic-treated animals often results in clear reduction of microbial biomass in soil, change in the soil bacterial community structure, persistent inhibitory effects on the growth rate of soil microorganisms, etc. Also, it was speculated that NER pose a risk for human health because of possible remobilization and further distribution of active metabolites over the food web (Nowak et al. 2011).

This work focused on covalent binding of pollutants to soil constituents via the amine functionality and accompanying processes in the soil environment in dependence on the amine base. This consideration allowed for recognizing and systematizing the interaction of pollutants including the amine functionality as a structural unit with soil constituents, as well as for revealing the importance of nonuniformity of soil interacted with pollutants that can be described using the water–octanol coefficient Kow characterizing a different degree of hydrophobicity of soil compartments. The investigation was based on the effect of spectral anisotropy that is not revealed when label nitroxide group is tumbling rapidly and isotropically in the soil environment, whereas covalent binding of special nitroxide radicals with the sewed amine brings to the pronounced anisotropy of hyperfine splitting with respect to the magnetic field direction, and a complex broadened EPR spectrum is observed (Berliner 1998; Voinov and Smirnov 2011).

# 2 Materials and Methods

## *2.1 Soil and Humic Acids Samples*

Samples of the deep typical chernozems, rich ($C_{org}$ = 5.52 %) and depleted ($C_{org}$ = 2.79 %) in fresh organic matter, were obtained from the annually mown steppe of the Alekhin Central Chernozemic State Biosphere Nature Reserve and from the permanent fallow (since 1964) of the Dokuchaev Soil Science Institute field experiment, respectively (Kholodov et al. 2009; Kogut et al. 2012). Samples were taken from Horizon Ap. They are of black, well- and medium-humified (9.51 and 4.81 %, respectively), and have a highly dispersed structure of clay loam (33 % of sand, 30 % of silt, and 37 % of clay). Soil samples had neutral pH (pH = 6.3–6.8).

Samples of humic acids (HA) isolated by alkaline extraction from typical chernozems were obtained from Moscow State University (Kholodov et al. 2009). As shown by $^{13}C$ NMR spectroscopy, the most fraction of total carbon composed of carbon of O, N-substituted aromatic fragments and amounted to about 44 % (Kholodov et al. 2009).

Some samples of standard HA, such as Elliott Soil Humic Acids Standard (1S102H), Pahokee Peat Humic Acids Standard (1S103H), Leonardite Humic Acids Standard (1S104H), were purchased from International Humic Substances Society IHSS.

## 2.2 Chemicals

Spin labels 4-amino-2,2,6,6-Tetramethylpiperidin-1-oxyl (4-Amino-TEMPO) were purchased from Aldrich. 2,5,5-Trimethyl-2-(3-aminophenyl)pyrrolidin-1-yloxy (Anilino SL) was obtained from Kalman Hideg, University of Pécs, Hungary (Gadányi et al. 2000).

4-Amino-TEMPO and Anilino SL are nitroxide radicals whose EPR spectra are defined by unpaired electron density at the nitrogen atom: N–O˙ (Berliner 1998; Ferreira et al. 2001). The free electron relevant in EPR gives rise to the characteristic three-line hyperfine structure of the spin-label spectrum. The distance between neighboring spectral lines of freely tumbling in solution SL is defined as the hyperfine splitting constant $a_0$ (in magnetic field units).

Extracellular fungal laccase from Trametes versicolor (oxygen oxidoreductase E. C. 1.10.3.2.) with a reported activity of 22.4 U/mg was obtained from Sigma-Aldrich. A fresh enzyme stock solution of 1 mg/mL in 0.1 mM ammonium acetate buffer (pH = 5.6) was prepared before each set of experiments and diluted as required.

## 2.3 Soil–SL Samples

SL was dissolved in distill water and then shaken for at least one hour. To study spin-label behavior in soil, fresh 0.3 mM aqueous solution of SL was prepared, and next, added to soil to moisten and label it. Moistened samples were centrifuged for 3 min. For experiments, the moisture of soil sample was adjusted to 35 % by d.w. For the use in spectral measurements, 1-mm glass capillaries were filled with 20 mg of labeled soil. Prior to experiments, an effect of glass capillary on EPR measurements of soil was defined. To compare EPR spectra of labeled soil samples and samples of natural soil, natural soil samples were also moistened with only distilled water to 35 % by d.w. For experiments with laccase, a sample of soil/humic acids was moistened with a laccase solution and immediately stirred for 1–2 min. At the experiment of labeling of soil/humic acids influenced by laccase, the solution of SL was prepared using a laccase solution and then added to soil/humic acids and stirred.

## 2.4 Humic Acids-SL Samples

To study spin-label behavior in HA, fresh 3 mM aqueous solution of SL was prepared, and next, added to dry HA to moisten and label them. Moistened samples were centrifuged for 3 min. For experiments, the moisture of HA sample was adjusted to 35 % by d.w., as in experiments with soil.

## 2.5 EPR Spectroscopy at X-band

EPR spectra of soil samples were recorded at X-band frequency at room temperature using a spectrometer of the type Magnettech Mini Scope MS 200 (Magnettech GmbH, Analysenmesstechnik, Louis-Blériot-Str. 5, Berlin, Germany) at a microwave power of 10 mW, corresponding to a microwave attenuation of 10 db, and a modulation amplitude of 0.150 mT. All spectra were obtained at a receiver gain equal to 100. Sweep and sweep time were set to 6.775 mT and 25 s, respectively. For determination of the g-value and spin concentration, DPPH (2,2-diphenyl-1-picrylhydrazyl, $g = 2.0038$) and MTSL (S-(2,2,5,5-tetramethyl-2,5-dihydro-1H-pyrrol-3-yl)methyl methanesulfonothioate) were applied to soil (Aleksandrova et al. 2011; Aleksandrova 2013).

## 2.6 Spin Labeling EPR Analyses

The EPR spectrum of SL is very susceptible to the molecular environment of SL, as well as to their mobility. In this paper, close consideration has been given to a displacement of the side spectral lines away from or toward the central spectral line and to inhomogeneous broadening. These effects in the EPR spectra of SL deal with a transformation of SL molecules, and a change of their mobility, which can decrease from a slight degree to complete immobilization or covalent binding of SL to large molecules/particles of soil. Thus, $a_0$ is a polarity indicative of the soil environment in which the SL is located, as well as a line broadening in the spectrum evidences a slight slowing down of nitroxide radicals. Increasing spin-label immobilization or their binding to soil constituents lead to a differential line broadening in the spectrum. It is the anisotropy of hyperfine splitting with respect to the magnetic field direction that confers both structural and motional susceptibility on the spin-label method (Berliner 1998).

## 2.7 Washout of Samples of Labeled Soil

In the course of repeated washouts of labeled soil samples, use has been made of the ultrafiltration device Vivaspin 500 µl with a carrier of 2.2 ml and diameter of 11 mm. The polyethersulfone (PES) membrane of device was characterized with 5000 MWCO PES. The usage of the membrane of 5000 MWCO PES allowed for filtrating of all dissolved components that arise in soil mechanically mixed with some distilled water at room condition.

A soil sample was labeled, incubated for some hours, and then put into concentrator. Some water was added to sample to wash it out. Filled and assembled concentrator was fixed and centrifuged at the recommended spin speed of

15,000×*g*. After centrifuging, retentate and permeate were sampled and investigated using the EPR spectroscopy. Duration of the experiment did not exceed the lifetime of SL in soil (Aleksandrova 2013).

# 3 Results and Discussions

## *3.1 EPR Spectra of Soil and Soil–SL Samples*

The EPR spectrum, which spans a wide range of magnetic field (50–450 mT) at microwave power $P = 10$ mW and room temperature $t = 20$ °C, shows a wide variety of paramagnetic substances, paramagnetic defects, and impurities present in natural soil (Senesi and Loffredo 2005; Paul et al. 2006; Skrzypczak-Bonduelle et al. 2008; Aleksandrova 2013). Directly after application of aqueous solution of SL to soil, a total EPR signal poses a composition of three narrow lines of SL superimposed on the complex EPR spectrum of soil at spectral region around $g = 2$ (Fig. 1). The EPR examination of soil–SL samples showed a good reproducibility and relative stability of the soil signal at an expanded range of B-field (Fig. 1) except the spectral region around $g = 2$ (inset to Fig. 1). As shown by experiments, free OR of soil, originated from complex organic molecules with polyphenolic and quinoid units of humic substances (Paul et al. 2006), were substantially susceptible to SL 4-Amino-TEMPO applied to soil (Fig. 2), interacted with them in dependence on their concentration (Aleksandrova 2013), and provided a gradual reduction of nitroxide radicals in the soil environment donating H-atoms or electron–proton

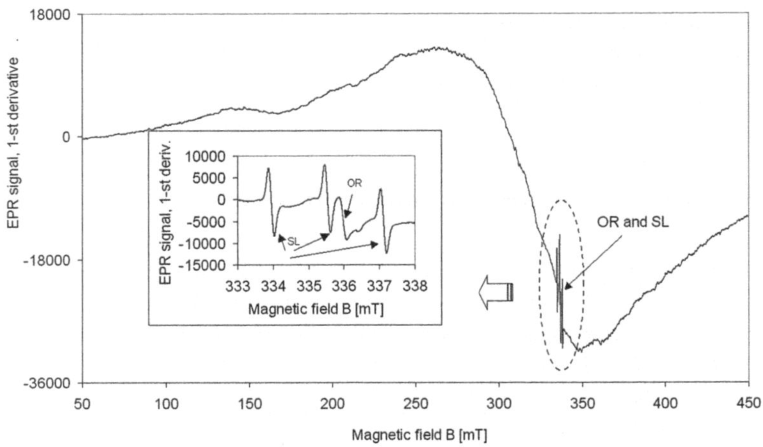


**Fig. 1** EPR spectra of soil–SL sample recorded at *X*-band in the *B*-field region between 50 and 450 mT and around $g = 2$ on an expanded *B*-field axis (*Inset* in Figure). *Symbols* are described in the text

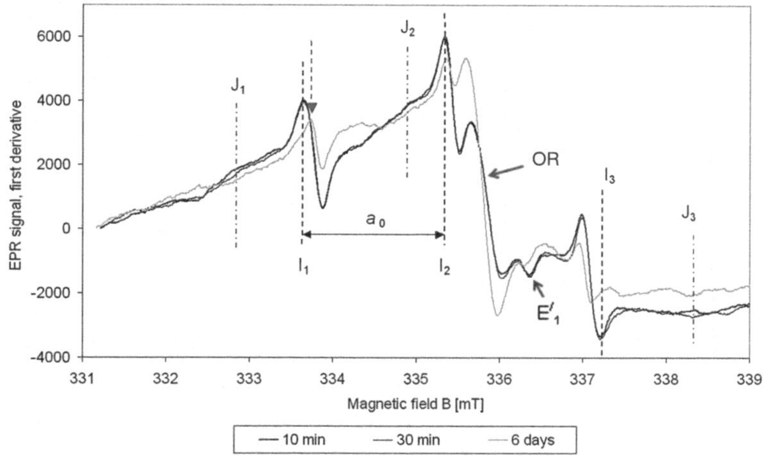


**Fig. 2** EPR spectra of 4-Amino-TEMPO incubated with depleted chernozem that were recorded at different time of incubation at *X*-band in the spectral region around $g = 2$ on an expanded *B*-field axis

pairs. At the same time, any change in OR of soil at their incubation with Anilino SL was not pronounced (Inset in Fig. 1). An EPR line of paramagnetic defects and impurities in silica lattice of soil $E_1^{/}$, observed in the high-field flank of the main signal of soil OR (inset in Figs. 1 and 2), was found mostly stable during all experiments with native soils.

## 3.2 *Abiogenic Interactions of Amines with Natural Soils in Dependence on Amine Base*

In SL EPR experiments, SL 4-Amino-TEMPO modeled pollutants with the aliphatic amine that allowed for considering the interaction and the environmental fate of the amine with a strong base in natural soils, whereas Anilino SL labeling soil modeled pollutants with the aromatic amine that highlighted their interactions and environmental fate of the amine with a weak base. Both SL were applied to two chernozems depleted and rich in fresh non-humified organic matter.

Figure 2 shows the complex EPR spectra of 4-Amino-TEMPO incubated with depleted chernozems, whereas Fig. 3 shows the EPR spectra of Anilino SL interacted with the same chernozem. While the EPR spectrum of 4-Amino-TEMPO posed a composition of three sub-spectra, namely a sub-spectrum of three narrow lines marked with dashed line and symbols $I_k$, $k = 1, 2, 3$, a sub-spectrum with shifted side spectral lines marked with dashed arrow, and a slightly broadened sub-spectrum marked with dash-dot lines and symbols $J_k$, $k = 1, 2, 3$, the EPR

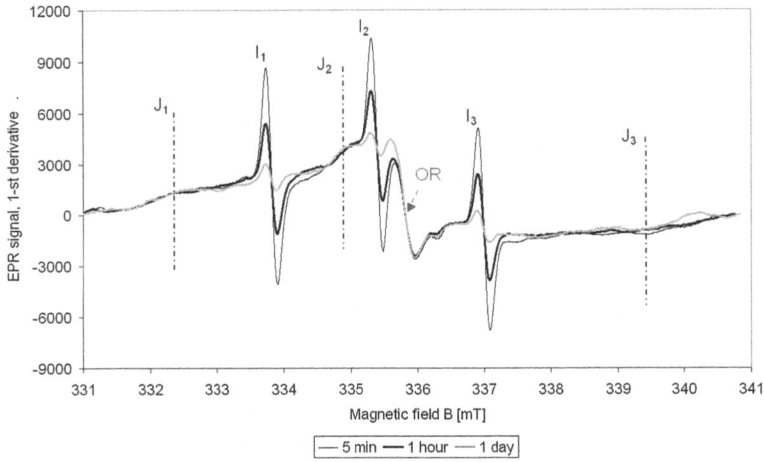


**Fig. 3** EPR spectra of Anilino SL incubated with depleted chernozem that were recorded at different time of incubation in the spectral region around $g = 2$

signals of Anilino SL interacted with both chernozems consisted only of two sub-spectra, such as a sub-spectrum of free SL (symbols $I_k$, $k = 1, 2, 3$) whose concentration decreased rapidly, and a sub-spectrum of bound SL (symbols $J_k$, $k = 1, 2, 3$). Three narrow lines characterized with $a_0$ indicated at a part of molecules of nitroxide radicals, which remained free in soil sites with the same polarity as in the aqueous solution. A shifted sub-spectrum pointed at a displacement of a part of spin-label molecules in the soil site with different polarity. A broadened sub-spectrum specified nitroxide radicals bound to small particles of soil. Figures 2 and 3 obviously showed different temporal behavior of considered spin labels that modeled molecules with the aromatic and aliphatic amines applied to natural soil. A most part of SL with the aliphatic amine was found in soil being not bound for more than 6 days. A second part of them were shifted into the soil environment with different polarity that might be provided by the transformation of spin-label molecules. At the same time, molecules with the aromatic amines were not found as shifted. They remained not bound and reduced by one–two days, or became bound immediately and observed longer than 6 days. These results elucidated the different partitioning of molecules with the aromatic and aliphatic amines among soil compartments. Overheating of both chernozems at 450 °C in oven for 1 h before their labeling provided burnout of hydrolysable part of soil organic matter, and caused the dying out of the bound sub-spectrum and a long-term observation of the shifted sub-spectrum of 4-Amino-TEMPO incubated with overheated chernozems. Incubation of 4-Amino-TEMPO with humic acids extracted from natural chernozems revealed the only sub-spectrum not shifted and not bound. At the same time, the spectral shape of Anilino SL incubated with overheated chernozems and humic acids extracted from natural chernozems did not change. These results highlighted that the most part of molecules with the aliphatic amines were found in the soil compartments of hydrolysable organic matter, whereas molecules with the aromatic amines were removed from a hydrolysable part of soil. SL EPR analyses also showed that in rich chernozem, more types of transformation of SL molecules were observed. Strong heating resulted in the dying out of the non-humified organic matter in a hydrolysable compartments of soil, and the EPR signals of overheated rich and depleted chernozems became the same.

As shown in Figs. 2 and 3, both spin labels became bound to soil constituent. Binding was observed immediately (in a minute) after the application of both SL to chernozems that put the accent on the abiotic nature of observed process. However, the application of both SL to humic acids extracted from natural chernozems revealed that binding of molecules to non-hydrolysable part of soil organic matter via the aromatic amine was well observed, whereas via the aliphatic amines, molecules were likely not to be bound to the same part of organic matter.

Examination of labeled soil samples using repeated washouts with distill water showed that the aliphatic amines were revealed mostly (>99 % at $n = 15$, $p < 0.05$) in the permeate, whereas the significant part of bound Anilino SL amounted to 25 % (at $n = 15$, $p < 0.05$) was found in the retentate that evidenced a reversible binding via the aliphatic amines and nonreversible binding of modeled pollutants via the aromatic amine to non-hydrolysable part of soil organic matter.

EPR investigation of Anilino SL incubated with standard HA, such as Elliott soil HA, Pahokee peat HA, leonardite HA, showed the same broadened spectrum in all cases. All measured ESR spectra changed the same way in shape (Fig. 4) but differently with time. The EPR spectrum of Anilino SL in leonardite that changed slower than other two spectra is shown in Fig. 4. All broadened EPR spectra of Anilino SL incubated with both chernozems, humic acids extracted from chernozems, standard humic acids have the same spectral parameters: $2 \cdot a_{//} \approx 7$ mT,

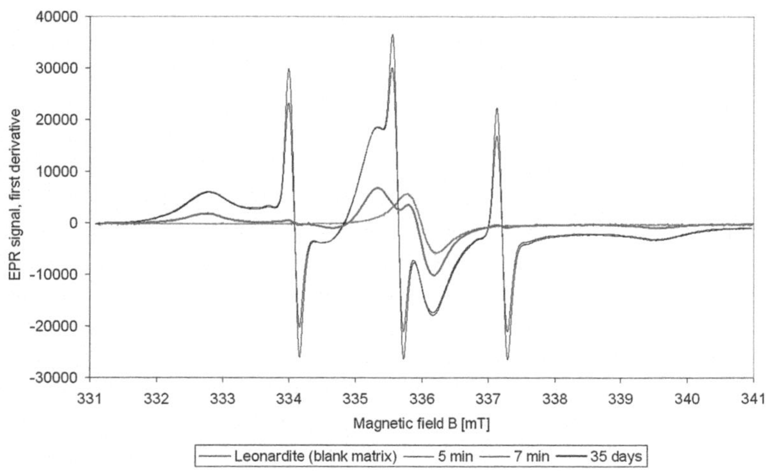


**Fig. 4** EPR spectra of Anilino SL incubated with leonardite HA that were recorded at *X*-band in the spectral region around $g = 2$

$2 \cdot a_{\perp} \approx 0.36$ mT, $g_{//} \approx 2.0035$, $g_{\perp} \approx 2.0078$, where $a_{//}, g_{//}$ and $a_{\perp}$, $g_{\perp}$ describe hyperfine splitting constants and g-factors with a parallel and perpendicular to an axis of symmetry, respectively. This finding hinted at the size of humic particles to which spin-label molecules were bound. Thus, particles of considered soils and HA, which bound SL, assumed larger proportions, which exceeded 50–100 kDa. At the same time, original particles substantially differ in the size, with them being less than 50 kDa. So, the averaged molecular size amounted to 2–5 kDa for leonardite, and to 15–30 kDa for soil HA (Perminova 2000). Therefore in the experiment, an enlargement of humic particles, which bound the aromatic amine, was observable, and hinted at the beginning of polymerization. It might well be that polymerized structures, which immediately followed binding of the aromatic amines, had been referred to as the NER of abiotic origin. Repeated washouts of leonardite labeled with Anilino SL and ultrafiltration using a membrane of 100 kDa showed a significant part of bound SL (>90 % at $n = 10$, $p < 0.05$) in retentate that confirmed the formation of polymerized units from humic particles binding the aromatic amines.

## 3.3 Biogenic Interactions of Amines with Natural Soils in Dependence on the Amine Base

An influence of biogenic factors on soil interacted with the amines were considered using laccase, i.e., copper-containing oxidases. Several types of redox enzymes were produced by fungi that degrade aromatic compounds in natural soils. Some investigations showed that laccase reduces dissolved oxygen to water and, at the same time, oxidize phenolic and non-phenolic substrates with the formation of quinones and cation radicals (Rabinovich et al. 2004). This model of an enzymatic effect on substrate allows for understanding the processes in soil influenced by laccase that was observed by EPR spectroscopy. An addition of laccase to chernozems led to dying out of many paramagnetic signals except the signal with g-factor around 2.0035–2.0038, whereas incubation of laccase with humic acids revealed any change in their paramagnetic signal. Moreover, the incubation of Anilino SL with rich chernozem influenced by laccase showed the EPR spectrum of the same spectral shape as in the case of the incubation of the same SL with depleted chernozem. The experimental results with a model of mechanism of laccase activity in the substrate hinted at the importance of the presence of oxygen in considered location. In the considered experiment, laccase oxidized only radicals, which were located in compartments sufficiently rich in oxygen, and could not oxidized radicals in the compartment depleted in oxygen as well as radicals of humic acids, i.e., radicals of insoluble (non-hydrolysable) part of soil organic matter. Therefore, hydrophobic compartments of soil depleted in oxygen seem to be characterized with a reduced activity of laccase. Thus, the uneven distribution of oxygen provided by the distribution of hydrophilic and hydrophobic compartments in natural soil well agreed with the activity of laccase.

However as shown above, an addition of molecules with the aromatic amines caused polymerization of humic particles, and in turn, the expanding of non-hydrolysable part of soil organic matter accompanied by restriction of its hydrolysable part. Thus, the crucial property of the aromatic amines to be incorporated into the hydrophobic soil sites, which were found as depleted in oxygen and inclined to expanding, might influence some beneficial soil-dwelling bacteria and the processes of nitrogen cycle in soil. The majority of the beneficial bacteria in soil need oxygen and/or moistened conditions (Zvyagintsev et al. 2005), which were lacking in the found specific hydrophobic compartments of soil, that point at a restriction of life space and a reduction of ample aerobic bacteria. Moreover by oxygen-free conditions, denitrifying bacteria convert nitrates and nitrites in soil into nitrogen gas or into gaseous compounds, such as nitrous oxide or nitric oxide, and therefore intensified nonuniformity of oxygen distribution in soil might bring to excess in denitrification, and in turn to overall losses of available soil nitrogen and subsequent loss of soil fertility.

## 4 Conclusions

Amines are attributed to typical functional groups of ample pollutants released into the soil environment. Applied method of spin labeling EPR analysis showed a close connection of abiogenic and biogenic interactions in natural soils contaminated with amines. A key role in these interactions plays the particular distribution of anoxic areas in soil microcompartments provided by hydrophobic sites of soil constituents that substantially depends on abiogenic and biogenic interaction in soil. In the soil environment, the aromatic amines are immediately attracted by hydrophobic sites of soil and bound to them, whereas the aliphatic amines remain in a hydrolysable part of soil and become partially transformed or reversibly bound. The formation of NER occurs via binding of tha aromatic amines to humic particles. The aromatic amine causes a transformation of biogenic processes in soil. They influences the distribution of anoxic areas in soil, intensifying their volume via polymerization of humic particles that leads probably to a restriction of areas fit for aerobic bacteria, and in turn a reduction of its magnitude.

**Acknowledgements** I gratefully acknowledge fruitful opportunity to carry out personally all experiments granted by Prof. Dr. Michael Matthies and Prof. Dr. Heinz-Juergen Steinhoff of University Osnabrueck, Germany. My sincere thanks are given to Kalman Hideg of University of Pe'cs (Hungary) for granted chemicals of Anilino SL.

I wish to thank Prof. Dr. Irina Perminova and Dr. V. Kholodov of Moscow State University for useful discussions.

# References

Aleksandrova ON (2013) Spin labeling ESR investigation of the molecular environment of soil interacting with chemical organic contaminants. J Geochem Explor 129:6–13

Aleksandrova ON, Steinhoff HJ, Klasmeier J, Matthies M (2011) ESR investigation of stable organic radicals in soils. Abstr Magn Reson Conf EUROMAR 2011:235

Berliner LJ (Ed) (1998) Spin labelling. The Next Millennium. Biological magnetic resonance V 14. Plenum Press, New York

Bollag JM, Myers CJ, Minard RD (1992) Biological and chemical interactions of pesticides with soil organic matter. Sci Total Environ 123(124):205–217

Ferreira JA, Nascimento OR, Martin-Neto L (2001) Hydrophobic interaction between spin-label 5-SASL and humic acid as revealed by ESR spectroscopy. Environ Sci Technol 35:761–765

Gadányi S, Kálai T, Jekı J, Berente Z, Hideg K (2000) Synthesis of 2-substituted pyrrolidine nitroxide radicals. Synthesis 14:2039–2046

Gulkowska A, Thalmann BD, Hollender J, Krauss M (2014) Nonextractable residue formation of sulfonamide antimicrobials: new insights from soil incubation experiments. Chemosphere 107:366–372

Kholodov VA, Konstantinov AI, Belyaeva EY, Kulikova NA, Kiryushin AV, Perminova IV (2009) Structure of humic acids isolated by sequential alkaline extraction from a typical chernozem. Eurasian Soil Sci 42(10):1095–1101

Kogut BM, Sysuev SA, Kholodov VA (2012) Water stability and labile humic substances of typical chernozems under different land uses. Eurasian Soil Sci 45(5):496–502

Nowak KM, Miltner A, Gehre M, Schaeffer A, Kaestner M (2011) Formation and fate of bound residues from microbial biomass during 2,4-D Degradation in soil. Environ Sci Technol 45:999–1006

Paul A, Stoesser R, Zehl A, Zwirmann E, Vogt RD, Steinberg CEW (2006) Nature and abundance of organic radicals in natural organic matter: effect of pH and irradiation. Environ Sci Technol 40:5897–5903

Perminova IV (2000) Analyses, classification and prediction of properties of humus acids. Habilitation dissertation, University of Moscow (in Russian)

Rabinovich ML, Bolobova AV, Vasil'chenko LG (2004) Fungal decomposition of natural aromatic structures and xenobiotics: a review. Appl Biochem Microbiol 40:1–17

Rosebdahl I, Siemens J, Groeneweg J, Linzbach E, Laabs V, Herrmann C et al (2011) Dissipation and sequestration of the veterinary antibiotic sulfadiazine and its metabolites under field conditions. Environ Sci Technol 45:5216–5222

Schauss K, Focks A, Heuer H, Kotzerke A, Schmitt H, Thiele-Bruhn S et al (2009) Analysis, fate and effects of the antibiotic sulfadiazine in soil ecosystems. Trend Anal Chem 28(5):612–618

Senesi N (1992) Binding mechanisms of pesticides to soil humic substances. Sci Total Environ 123(124):63–76

Senesi N, Loffredo E (2005) Metal ion complexation by soil humic substances. In: Tabatabai MA, Sparks DL (eds) Chemical processes in soils. Soil Science Society of America Inc., New York

Skrzypczak-Bonduelle A, Binet L, Delpoux O, Vezin H, Derenne S, Robert F et al (2008) EPR of radicals in primitive organic matter: a tool for the search of biosignatures of the most ancient traces of life. Appl Magn Reson 33:371–397

Tappe W, Zarfl C, Kummer S, Burauel P, Vereecken H, Groeneweg J (2008) Growth-inhibitory effects of sulfonamides at different pH: dissimilar susceptibility patterns of a soil bacterium and a test bacterium used for antibiotic assays. Chemosphere 72:836–843

Thron KA, Pettigrew PJ, Goldenberg WS (1996) Covalent binding of 2-Aniline to humic substances. 2. 15N NMR studies of nucleophilic addition reactions. Environ Sci Technol 30 (9):2764–2775

Voinov MA, Smirnov AI (2011) Spin Labels and spin probes for measurements of local pH and electrostatics by EPR. In: Gilbert BC et al (eds) Electron paramagnetic resonance. The Royal Society of Chemistry, Cambridge

Weber EJ, Spidle DL, Thron KA (1996) Covalent binding of aniline to humic substances. 1. Kinetic Studies. Environ Sci Technol 30(9):2755–2763

Zvyagintsev DG, Babieva IL, Zenova GM (2005) Soil biology. Moscow State University, Moscow (in Russian)

# Peculiarities of Migration of Some Biogenic Elements in System Soil-Infiltration Waters in the Aragats Mountain Massif (Armenia)

**Marieta H. Avetisyan, Levon A. Araratyan and Tatevik E. Poghosyan**

**Abstract** This article considers water-migration changes of some biogenic elements in system soil-infiltration waters and the role of organic matter during biogenic elements leaching into high mountain ecosystems. The researches covered alpine (3250 m a.s.l.) and meadow steppe (2085 m a.s.l.) belts of the Aragats mountain massif. To study a water migration flow of biogenic elements, a study was done of peculiarities of their migration in system atmospheric precipitation–soil-infiltration waters. The research was done employing a lysimetric method which provides information about soil infiltration rate, chemical composition, and migration of soil water elements. The research allowed to reveal high contents of biogenic elements in the studied compartments of meadow steppe belt. As indicated, migration intensity of the studied elements and rates of soil biomineralization in meadow steppe versus alpine belt are higher as in cold high mountainous conditions mineralization and humification rates of elements are comparatively lower. Consequently, this phenomenon brings about an increase in the contents of the study of biogenic elements and acceleration of their removal with infiltration runoff. Collation between main migration flows of biogenic elements beginning from their entering (atmospheric precipitation) to leaching into groundwater flow allows to conclude that the highest transformation of a migration flow of elements in the ecosystem of the Aragats mountain massif occurs in meadow steppe belt.

**Keywords** Biogenic elements · Soil water · Leaching · Lysimeter · Infiltration waters

M.H. Avetisyan (✉) · L.A. Araratyan · T.E. Poghosyan
The Center for Ecological-Noosphere Studies, NAS, Yerevan, Republic of Armenia
e-mail: marieta_0208@mail.ru



O.V. Frank-Kamenetskaya et al. (eds.), *Biogenic—Abiogenic Interactions in Natural and Anthropogenic Systems*, Lecture Notes in Earth System Sciences,
DOI 10.1007/978-3-319-24987-2_25

# 1 Introduction

Ecological tension in mountain meadow ecosystems of a number of Armenia's mountain regions has increased to such an extent that presently the latter are on the verge of destruction. In such conditions the rate of natural restoration of soil fertility is substantially lower than that of nutrient loss that finally brings disturbance of migration of biogenic elements in the ecosystem, a decrease of productivity of mountain meadow ecosystems, destruction of topsoil horizon, degradation of soil cover, and development of soil erosion. Degradation brings not only deterioration of some soil properties, but also to a decrease in the contents of organic matter, and disturbance of biogeochemical cycles of the most essential nutrients. The importance of soil organic matter to mountain meadows is that it improves soil structure as a result of organic matter transformation by microorganisms. Soil organic matter regulates soil temperature, influences moisture content, increases cation exchange capacity (CEC), and serves as nutrient reserves. A decrease of soil organic matter entails a number of ecological problems; first of all water and wind erosion, as water and wind carry away soil particles and nitrogen, phosphorus, and potassium contained in them or adsorbed on their surface (Evrendilek et al. 2004; Revazyan 2001; Tate 1987).

One of the factors determining leaching of biogenic elements is groundwater flow. Biogenic elements leaching data obtained through a lysimetric method are one of the most essential informative sources for complex landscape–biogeochemical investigations into the state and functioning of mountain ecosystems. Lysimetric solutions characterizing vertical migration of flow of elements not only partly reflect biogeochemical cyclicity, but also provide direct information about geochemical specificity and functioning of ecosystems (Revazyan 2001).

Migration of chemical substances combined with water flow leaches a considerable mass of substances. With respect to water flow, particularly essential is the ecosystem migration of substances, a considerable portion of which is leached and used by other ecosystems (Kovda 1985; Sakoyan et al. 2010). From such positions it is essential to study migration of biogenic elements in system atmospheric precipitation-soil-infiltration waters for mountain ecosystems of the Aragats massif by vertical zonality, which will allow to characterize migration processes. This system characterizes soil links between the ecosystem's compartments and leach of substances out of it. So, studying and assessing the current status and quality composition of a water migration flow is an essential issue.

The aim of this research was to study peculiarities of a water migration flow of biogenic elements in the ecosystem of the Aragats mountain massif.

To achieve the goal, studies were done of migration of biogenic elements in system atmospheric precipitation-soil-infiltration waters within meadow steppe and alpine belts of the Aragats massif.

This work deals with the mean values of research results for 2011–2013.

## 2 Research Materials and Methods

The studies were implemented in one of the most complex ecological physico-geographical subregions of the country: alpine (3250 m a.s.l.) and meadow steppe (2085 m a.s.l.) belts of the Aragats mountain massif in a summer–winter period.

Soil sampling and analyses were done by accepted methods of geochemical and agrochemical research (Arinushkina 1970; Yudin 1971).

Atmospheric precipitation and lysimetric waters were sampled from meteo stations of Aragats (3250 m a.s.l.) and Hamberd (2085 m a.s.l.) located in the mentioned belts.

Vertical soil water runoff was studied through the lysimetric method which helps assess nutrient losses caused by leaching in conditions close to natural (Shilova 1972). Flat fit-in lysimeters were installed 50 cm deep into the soil layer to minimally damage the soil structure.

The initial stage of analytical treatment of water samples was to separate solid and liquid constituents of solution phases. Biogenic elements in all the collected water samples were determined by standard hydrochemical methods (Alekin 1954; Reznikov et al. 1963).

## 3 Results and Discussion

Atmospheric precipitation being a component of the ecosystem and having biogeochemical activity can transport different substances into soil, plants, and infiltration waters.

The atmospheric precipitation research implemented in the Aragats mountain massif allowed to reveal large amounts of hydrocarbonates and sulfates (at pH-7.3–7.4). So, in alpine belt their contents were 146.2 and 45.1 and in meadow steppe belt 129.6 and 62.9 kg/ha, respectively (Table 1).

**Table 1** Mean values of biogenic elements in alpine and meadow steppe belts of the Aragats mountain massif, kg/ha[a]

| Sampling site | Major nutrients | $NH_4^+$ | $K^+$ | $Ca^{2+}$ | $Mg^{2+}$ | $Na^+$ | $HCO_3^-$ | $SO_4^{2-}$ | $NO_3^-$ | $HPO_4^{3-}$ | $Cl^-$ |
|---|---|---|---|---|---|---|---|---|---|---|---|
| Alpine belt (3250 m a.s.l.) | Entering (with atmospheric precipitation) | 4.20 | 7.50 | 51.85 | 8.70 | 6.10 | 146.2 | 45.1 | 5.20 | 2.20 | 18.35 |
| | Leaching (by lys. waters) | 0.75 | 2.50 | 38.50 | 5.50 | 0.57 | 144.5 | 16.5 | 5.60 | 0.10 | 12.1 |
| Meadow steppe belt (2085 m a.s.l.) | Entering (with atmospheric precipitation) | 6.80 | 18.70 | 24.0 | 22.8 | 6.50 | 129.6 | 62.9 | 10.2 | 1.85 | 22.80 |
| | Leaching (by lys. waters) | 3.20 | 13.80 | 42.8 | 18.5 | 3.20 | 92.5 | 33.4 | 13.1 | 0.49 | 14.6 |

[a]Calculations were done with regard to the annual amount of atmospheric precipitation

Basic nutrients entering with atmospheric precipitation were $NH_4^+$, $NO_3^-$, $HPO_4^{2-}$, and $K^+$.

Over the study period we established that both the contents of chemical substances in atmospheric precipitation and mineralization rate are increasing over the years. The mean of contents of cations and anions of biogenic elements contained in atmospheric precipitation were ranked in descending order (mg/L): *in meadow steppe* belt of the Aragats massif (at 2085 m a.s.l.)—$Ca^{2+}_{(25.7)} > K^+_{(12.8)} > Mg^{2+}_{(7.1)} > Na^+_{(6.4)}$ and $SO^{2-}_{4(119.7)} > HCO^-_{3(28.8)} > Cl^-_{(16.3)}$, *in alpine* (at 3250 m a.s.l.)—$Ca^{2+}_{(15.7)} > K^+_{(9.2)} > Mg^{2+}_{(5.9)} > Na^+_{(4.7)}$ and $SO^{2-}_{4(26.7)} > HCO^-_{3(19.3)} > Cl^-_{(8.8)}$.

Although cations and anions are ranked similarly in both the descending series, meadow steppe versus alpine belt exhibit high contents of chemical substances, and this may be connected with meadow steppe belt pollution.

Formation of chemical composition of atmospheric precipitation in the Aragats mountain massif is determined by the nature of regional processes of transportation of air masses owing to geographical position of the region; the latter is characterized by contrasting seasonal variations in the atmosphere with domination of air masses in a winter–spring period.

One should emphasize that in warm seasons formation of elemental composition of atmospheric precipitation is basically associated with activation of landscape–geochemical processes which induce a dramatic increase in the impact of biogenic factors. In winter, elemental composition of atmospheric precipitation is formed at low temperatures, under which many processes of functioning of the ecosystem run less intensively than in summer.

It is known that man-made factors produce a direct, active impact on soil cover. In soil, the flow of elements transported from the atmosphere is distributed between vegetation and ground waters. The intensity of chemical elements entering into migration flows in most cases depends on agrochemical indices of soils, forms of combination of chemical elements, the impact of a water factor on vertical migration of elements, the level of ion exchange, and so on (Kovda 1985).

According to the implemented studies, humus contents of meadow steppe soils vary between 5 and 10, and total nitrogen 0.2–0.5 %. The contents of total phosphorus are high, while those of potassium are low. These soils are not rich in available nitrogen and phosphorus and are moderately and well supplied with potassium. In mountain meadow, topsoils of alpine belt typical have high contents of humus (8–12 %), total phosphorus (0.2–0.4 %), and nitrogen (0.3–0.8 %), whereas the contents of total and available potassium in these soils are much lower compared with meadow steppe soils of meadow steppe belt. From meadow steppe to mountain meadow soils the contents of humus, total nitrogen, phosphorus, and soil acidity increase, whereas those of potassium decrease.

Mountain meadow soils as a basic feed resource enjoy a large share in Armenia's land reserves. It should be mentioned that heavy grazing results in disturbance of nutrient regime of soils and decrease of organic matter. The latter is determined first by disturbance of physical structure of topsoil and with regard to actively reacting excrements on the top layer of soil—this leads in turn to disturbance of stability of biochemical compounds, i.e., soil degradation followed by loss of nitric and carbon compounds in gaseous form and leaching of elements (Revazyan et al. 2011; Tate 1987).

Soil water migration and quantitative determination of leachable chemical elements are the most essential constituents of the capacity of a biogeochemical cycle.

Initially, atmospheric precipitation passes through a tree waste layer and litter toward soil cover in which release of mineral components and soluble organic matter and formation of humic and organic compounds result and biochemical processes of mineralization and leaching occur. In lysimetric litter solutions versus, total rainfall mineralization rate increases mainly at the expense of increasing concentrations of biogenic elements: potassium, calcium, magnesium, phosphorus, the entering of which into solutions is connected with transformation of tree waste and litter biomass (Revazyan et al. 2011).

It is known that in the case of atmospheric precipitation there is acid reaction, then infiltration waters are enriched with chemical substances at the expense of dissolution of soil salts and cation exchange. Under alkaline reaction an inverse process may occur—loss of chemical substances from soil solution and their accumulation in soil adsorption complex (Kovda 1985).

A study of water migration of chemical elements in alpine and meadow steppe belt soils have indicated that mean contents of chemical elements lysimetric waters (50 cm deep) vary within wide limits (Fig. 1).

Lysimetric waters of meadow steppe soils (at pH—7.40) are characterized by high concentration of chemical elements and high rate of mineralization (64.3 mg/L), which is explained by more active rates of tree waste decomposition and mineralization in the given belt (Fig. 1). Lysimetric waters of alpine soils have a low pH value (5.80) and comparatively low contents of chemical substances.

As seen from Fig. 1, lysimetric soil waters of both the belts are characterized by high contents of $HCO_3^-$, $Ca^{2+}$, $SO_4^{2-}$ and $Cl^-$. Low contents of nitrates in lysimetric

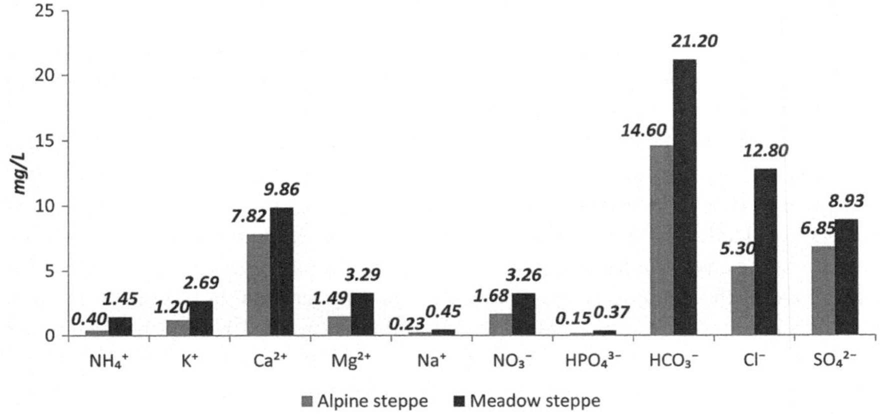


**Fig. 1** Chemical composition of lysimetric waters of Aragats mountain massif, mg/L

waters of alpine belt soils are explained by poor nitrification property of soils. Basically, this phenomenon is determined by acid reaction and low temperatures of the soils.

In spite of an altitude-dependent increase in the amount of atmospheric precipitation and the volume of lysimetric water runoff from different horizons, the volume of biogenic elements leached with vertical migration flows regularly decreases, which is explained by infiltration of solutions from upper organogenic to lower illuvial horizons and increasing absolute altitudes as meadow steppe belt changes into alpine (Sakoyan et al. 2010).

Under conditions of humid climate and mountain relief, one of the most important functions of mountain soils is their regulatory role in migration and limitation of organic matter and biogenic elements leach with a runoff (Kovda 1985). The alpine belt soil profile practically wholly retains the most essential biogens: nitrogen and phosphorus, whereas the contents of potassium and calcium are somewhat higher.

Basically, entering of sulfates and chlorides into lysimetric solutions is connected with atmospheric precipitation. Another source of sulfates in lysimetric waters is their entering during mineralization, leaching, and transformation of organic matter. It should be emphasized that high contents of meadow steppe belt sulfates are determined by aerosol emissions and atmospheric pollution with sulfur dioxide (Revazyan et al. 2011).

It is noteworthy that no absolute fixation of soluble organic matter and soluble forms of biogenic elements occur in the bounds of soil profile. Differences in the contents of soluble organic matter are associated with peculiarities of biogeochemical cyclicity of carbon, which is basically characterized by the volume and structure of phytomass, its productivity, composition of plants, diversity, and activity of microorganisms (Follett 2001). These factors contribute to occurrence of a weak alkaline reaction in meadow steppe belt in particular. Presumably, weak alkaline reaction and high contents of calcium in lysimetric waters create more favorable conditions for vital activity of microorganisms, this enhancing their ability to efficiently decompose organic residues.

Concluding, one may ascertain that in meadow steppe the soil cover of which is dominated by mountain black soils, the contents of chemical substances entering into soil with atmospheric precipitation are higher versus those leached out from soil. The contents of ammonia nitrogen ($NH_4^+$) transported with atmospheric precipitation into soils in alpine and meadow steppe belts make 4.20 and 6.80 kg/ha, respectively, those leached from soils by lysimetric waters—respectively 0.75 and 3.20 kg/ha (Table 1). Similar data were obtained for the remaining elements, too, whereas in the case of nitrates and meadow steppe calcium quite an opposite picture is observed as the contents of leached nitrates are higher than those transported into the soil with atmospheric precipitation. Migration activity of $HPO_4^{2-}$ in the studied soils is low unlike nitrates ($NO_3^-$) that are not absorbed by soil colloids, do not form hard soluble salts, and therefore—due to high mobility—are leached easily from a root-inhabited layer into ground waters, and finally become unavailable for plants and a possible cause for underground water pollution.

The researches indicate that among the studied elements, migration of $Ca^{2+}$, $Mg^{2+}$, $SO_4^{2-}$ and $Cl^-$ is comparatively more active.

## 4 Conclusion

The results obtained demonstrate that the higher the absolute altitudes (alpine belt), the higher, but not lower, the contents of organic compounds in lysimetric waters are, whereas lysimetric solutions of alpine versus meadow steppe belt soils are more acid and less mineralized.

Hence, a water migration flow of elements by vertical zonality of the Aragats massif obeys regularities connected with processes of chemical element migration in biogeochemical cycles. The nature of changes in biogeochemical cycles is determined by migration of elements and changes of their concentrations in soils, vegetation, and natural waters.

The results obtained explain a mechanism of a biogenic element leach from different types of soils and help get a better understanding of their migration ability, which is particularly essential for controlling environmental status and providing prognostic assessments and prevention of ground- and surface water pollution.

So, a complex research of water migration flow of biogenic elements with regard to their entering with atmospheric precipitation and leaching into infiltration (lysimetric) waters in mountain meadow soils may be used for balance calculations to define the dosage of fertilizers for improving fertility of mountain meadow soils and increase productivity of mountain meadows.

## References

Alekin OA (1954) Chemical analysis of continental waters. Gidromet Leningrad

Arinushkina EV (1970) Manual for chemical analyses of soils. Publ h MSU Moscow

Evrendilek F, Celik I, Kilic S (2004) Changes in soil organic carbon and other physical soil properties along adjacent Mediterranean forest, grassland, and cropland ecosystems in Turkey. J Arid Environ 59:743–752

Follett RF (2001) Soil management concepts and carbon sequestration in cropland soils. Soil Till Res 61:77–92

Kovda VA (1985) Biochemistry of soil cover. Nauka, Moscow

Revazyan RH (2001) Biospherology and environmental protection issues. Papers NAS RA 101 (2):158–163 (in Russian)

Revazyan RH, Sakoyan AG, Avetisyan MG (2011) Peculiarities of transformation of biogeochemical cycles in mountain meadows of alpine belt under a man-induced impact. Bioinert interaction in natural and man-made systems. Mat IV intern Symposium, Saint-Petersburg 500–504 (in Russian)

Reznikov AA, Mulikovskaya EP, Sokolov IYu (1963) Methods of analyses of natural waters. Gos geoltekhizdat Moscow

Sakoyan AG, Revazyan RH, Avetisyan MG, Araratyan LA (2010) Peculiarities of in-ecosystem links in the Aragats mountain massif. Biol J Armenia 62(4): 86–93 (in Russian)
Shilova EI (1972) Application of lysimetric methods in soil science, agrochemistry and landscape science. Nauka L 1–21
Tate RL (1987) Soil organic matter: biological and ecological effects. Wiley, Chichester
Yudin FA (1971) Methods of agrochemical research. Kolos Moscow

# Application of Raman Spectroscopy and High-Precision Geochemistry for Study of Stromatolites

**Pavel V. Medvedev, Svetlana Y. Chazhengina and Sergey A. Svetov**

**Abstract** Paleoproterozoic (2.0–2.2 Ga) stromatolites from drill cores of the ICDP Project FAR DEEP were studied by Raman spectroscopy and inductively coupled plasma mass spectrometry with laser ablation (LA-ICP MS) methods. Morphological features of the studied Paleoproterozoic stromatolites together with identification of synsedimentary carbonaceous matter by Raman spectroscopy are strong argument for biogenic origin of these carbonate buildups. Precision geochemical analysis of the stromatolite laminae composition links to sedimentary settings and environmental conditions of the microbialite formation.

**Keywords** Palaeoproterozoic · Stromatolites · Eastern fennoscandian shield · Raman spectroscopy · LA-ICP mass spectrometry

## 1 Introduction

Stromatolites are mineralized microbial buildups which vary in morphology, have carbonate composition and characteristic fine lamination. Stromatolites are fossilized laminated organosedimentary structures produced by carbonate sediment trapping, binding, and/or precipitation as a result of the growth and metabolic activity of microorganisms, principally Cyanophyceae (Awramik and Margulis 1974). They are the only macrofossils for Archean and Proterozoic and mark environmental conditions on the Early Earth. Morphology of the buildups (stratiform, bulbous, bimical, or columnar) is defined by microbial community as well as environmental conditions: water depth, currents, etc. Stromatolite formation is a result of complex biogenic and abiogenic processes (Filho and Fairchild 2011). Stromatolites, representing fossilized microbial mats, reflect microbial biomineralization processes, which have

P.V. Medvedev (✉) · S.Y. Chazhengina · S.A. Svetov
Institute of Geology, Karelian Research Centre, Russian Academy of Sciences, Petrozavodsk, Russia
e-mail: pmedved@krc.karelia.ru



O.V. Frank-Kamenetskaya et al. (eds.), *Biogenic—Abiogenic Interactions in Natural and Anthropogenic Systems*, Lecture Notes in Earth System Sciences,
DOI 10.1007/978-3-319-24987-2_26

been operating from the Archean to the Present. There are long-standing controversies about the participation of microorganisms in the formation of stromatolites (microbial buildups, microbialites), as well as about the definition of the term stromatolite itself (Riding 2008). Taking in account that stromatolites are microbial buildups, it is necessary to prove their biogenic origin. For this purpose, there are a number of criteria, one of them is presence of synsedimentary carbonaceous matter within individual laminae.

The purpose of the research work was to investigate Paleoproterozoic (2.2–2.1 Ga) dolomite stromatolites (fossilized microbial buildups) from Eastern Fennoscandian Shield, which had been formed under evaporate conditions (Melezhik et al. 1999, 2005b) by modern analytical techniques (Raman spectroscopy and LA-ICP mass spectrometry) to prove their biogenic origin and to reveal environmental conditions of formation.

## 2 Materials

The subject of the study was stromatolites in core samples from holes which were drilled within the frames of the project FAR-DEEP, International Continental Drilling Program (Melezhik et al. 2005a, 2010; McLoughlin et al. 2013) (Fig. 1). The samples were collected from a number of drill holes 5A, 8B (Pechenga Greenstone Belt) and 10A, B, and 11A (Onego Paleobasin), which intersected dolostones with stromatolites (Figs. 2, 3 and 4). Sampling was done by the author in

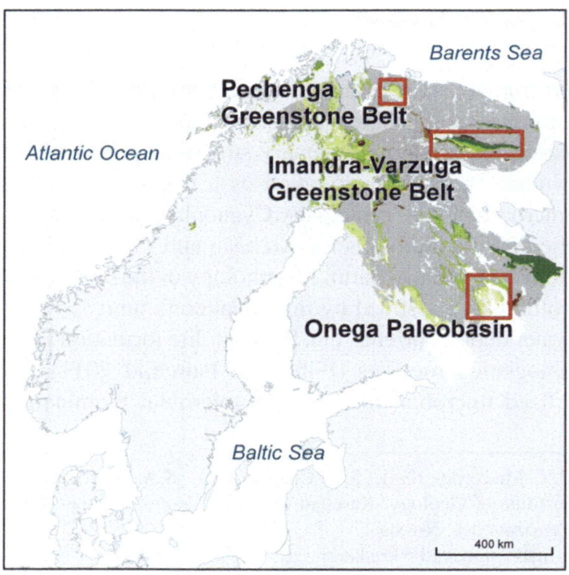


**Fig. 1** Geographic locations of the ICDP Fennoscandian Arctic Russia drilling early earth project (from Melezhik et al. 2010)

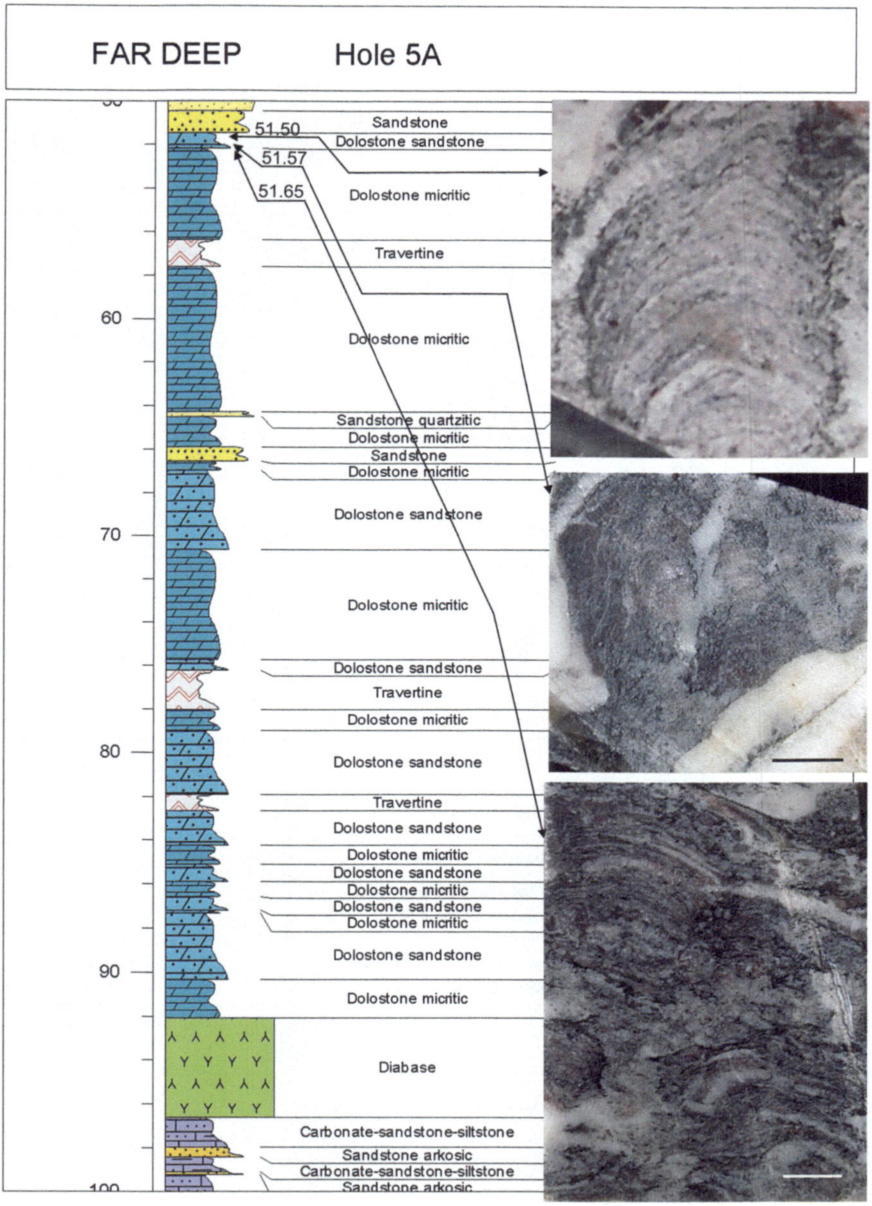


**Fig. 2** Stromatolites from the Kuetsjarvi formation in FAR-DEEP drill hole 5A (Pechenga Greenstone Belt). Scale bar is 1 cm. Lithological log adapted from FAR DEEP database (http://far-deep.icdp-online.org)

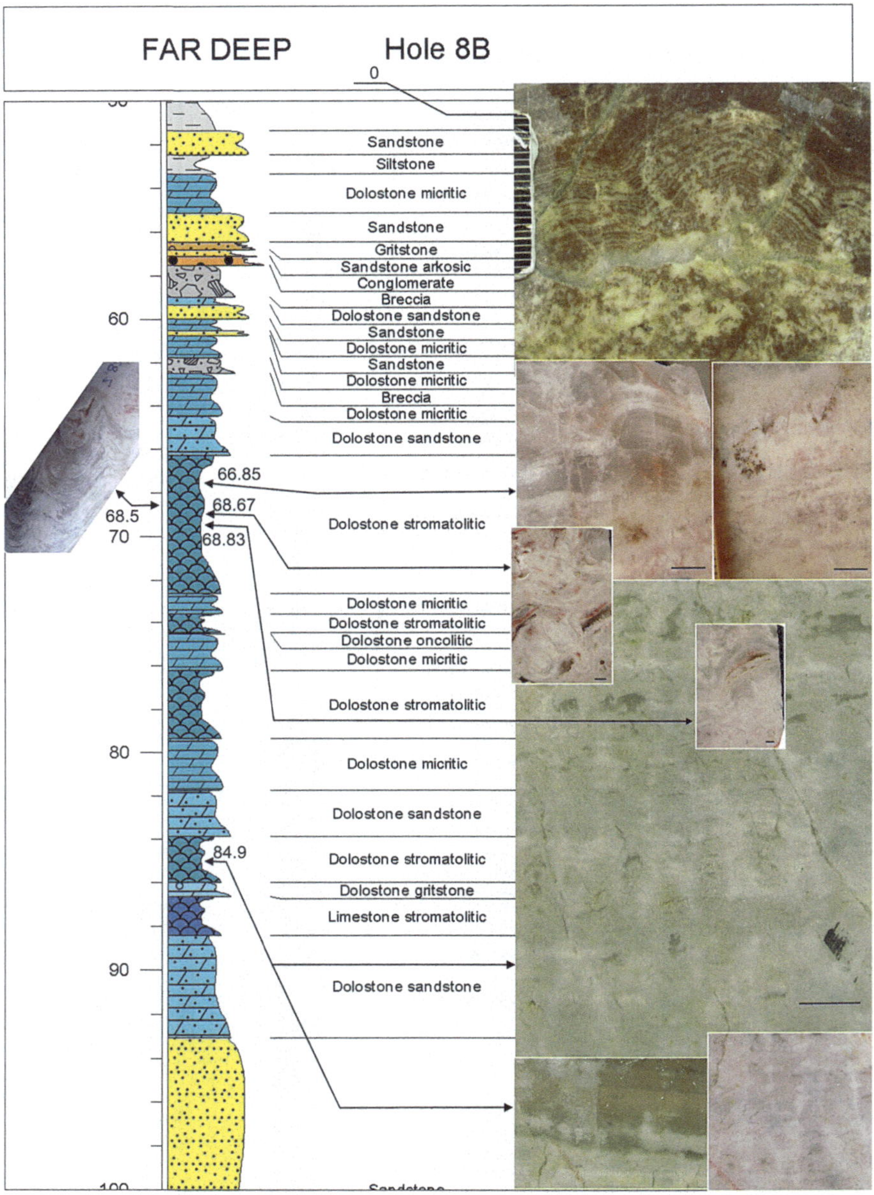


**Fig. 3** Stromatolites from the Kolasjoki formation in FAR-DEEP drill hole 8A (Pechenga Greenstone Belt). Scale bar is 1 cm. Lithological log adapted from FAR DEEP database (http://far-deep.icdp-online.org)

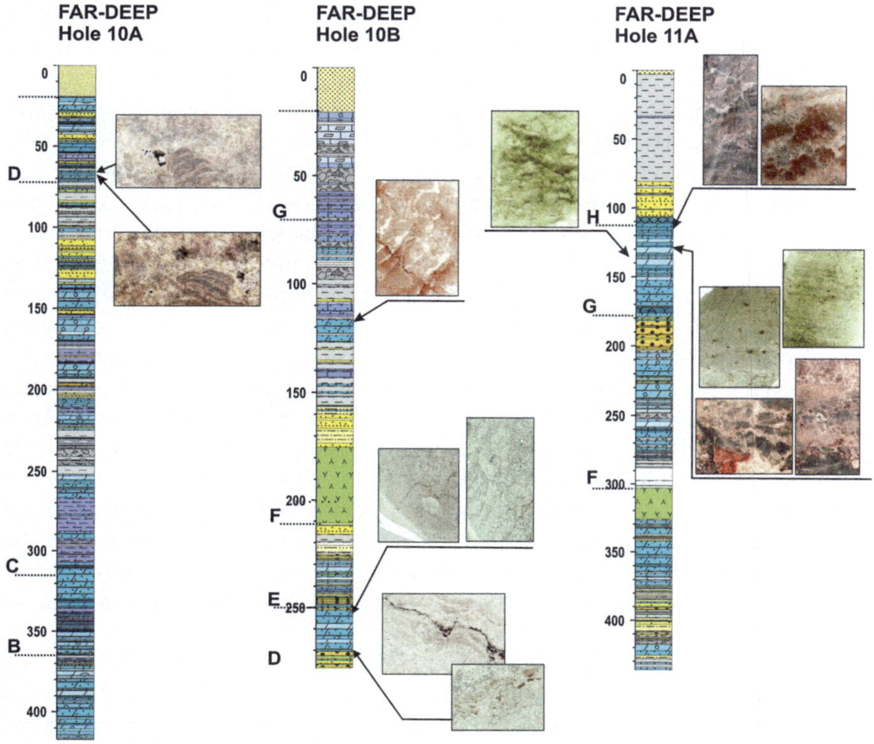


**Fig. 4** Stromatolites from the Tulomozero formation in FAR-DEEP drill holes 10A, 10B, 11A (Onega Paleobasin). Lithological log adapted from FAR DEEP database (http://far-deep.icdp-online.org)

2009 from ICDP FAR-DEEP drill cores stored at Norwegian Geological Survey in Trondheim. Although Paleoproterozoic stromatolites from Fennoscandian Shield vary in size and morphology, only cm-sized buildups (ministromatolites) can be distinguishable in the core of 5 cm diameter. Consequently, these buildups were the subject of our research. Few samples of Mesoproterozoic stromatolites from Sredny peninsula (Kochetkov et al. 2013) were used in Raman spectroscopy study as a comparative material.

## 3 Methods of Analysis

1. Light microscopy and photography of the thin sections from Paleoproterozoic stromatolitic dolostones have been done at the Institute of Geology in Petrozavodsk using VK-9710 Violet-Laser Confocal Scanning Microscope.

2. Scanning Electron Microscopy as well as determination of mineral composition of the dolostones has been done at the Institute of Geology in Petrozavodsk using SEM TESCAN VEGA II LSH equipment.
3. Raman spectroscopy (RS) analysis was carried out using a dispersive NicoletAlmega XR Raman spectrometer with a green laser (532 nm, Nd-YAG). The spectra were collected at 2 $cm^{-1}$ spectral resolution. The spectrometer was calibrated before each analytical session by 'zero-point' centering and by analyzing a Si-standard with a characteristic Si Raman band at 520.4 $cm^{-1}$. A confocal microscope with a 50× objective was used to focus an excitation laser beam on the sample and to collect a Raman signal from a 2 μm diameter area. The final laser power was about 5–10 mW at the sample surface. Raman spectra were collected on microprobe-quality petrographic thin section.
   Though the effect of polishing on the CM (carbonaceous matter) Raman spectra is still being debated (Beyssac et al. 2002, 2004; van Zuilen et al. 2007; Aoya et al. 2010), it has been reported that for poorly ordered CM it could be neglected (van Zuilen et al. 2007; Sforna et al. 2014).
   Raman spectral data, such as peak position, band area (i.e., integrated area) and band width (i.e., full width at half maximum, FWHM), were determined by using band decomposition based on a Gaussian–Lorentzian function in OMNIC software. Band-fitting protocols for five spectrums were described in detail in (Sforna et al. 2014).
4. Regions of stromatolitic laminae were identified on polished sections and marked for geochemical analysis using laser ablation (e.g., Fig. 4b). Elemental abundances of REY, and Th were determined by inductively coupled plasma mass spectrometry with laser ablation (LA-ICP MS) on a thermo elemental mass spectrometer at the Institute of Geology in Petrozavodsk following procedures outlined by Webb et al. (2009). In addition to pre-ablation of the analysed sites, data from each ablation site were time-resolved allowing removal of any erroneous values resulting from surface contamination. REY data were normalised to a shale average (Post-Archaean Australian shale—PAAS; McLennan 1989) for pattern analysis.

## 4 Results

Over the past decades Raman spectroscopy has been widely used to investigate the organic microfossils in Precambrian rocks (Schopf et al. 2005; Allwood et al. 2006; Schopf and Kudryavstev 2009; Marshall et al. 2010; Bower 2011). RS method gives the opportunity not only to determine the mineral composition of rocks in microscale, but also to identify the carbonaceous matter (CM) in ancient microfossils and derive Raman spectral properties of CM.

The Raman study of Paleo- and Mesoproterozoic stromatolites displayed that these rocks had similar mineral composition with dominant minerals of dolomite and quartz with the presence of rutile, anatase, muscovite, microcline, pyrite, hematite, and CM. The only exception is sample 2648 of Paleoproterozoic stromatolites, which is composed only of dolomite and CM with accessory mineral of pyrite.

In samples 6241, 6229 of Paleoproterozoic stromatolites CM was identified by Raman spectroscopy within individual dolomite laminae, and it consists of rare irregular grains with size ca. 30 μm, disseminated through a matrix of dolomite. For other studied samples of Paleo- and Mesoproterozoic stromatolites CM is concentrated in dark laminated rock fabrics of different thickness.

The Raman spectra for all samples of CM from studied Paleo- and Mesoproterozoic stromatolites are consistent with poorly ordered carbonaceous material (Wopenka and Pasteris 1993; Ferrari and Robertson 2000; Beyssac et al. 2002). The Raman spectra of stromatolite CM are composed of two intensive bands G ($\sim 1580$ $cm^{-1}$) and D1 ($\sim 1350$ $cm^{-1}$) in first-order region (Fig. 5). The graphite band G corresponds to in-plane vibration of aromatic carbons in the graphitic structure. The bands D1 and D2 ($\sim 1620$ $cm^{-1}$) appear in disordered carbons, and they are assigned to defects in the graphitic structure (Wopenka and Pasteris 1993; Ferrari and Robertson 2000; Beyssac et al. 2002).

For CM from Paleoproterozoic stromatolites (6241, 6229, 2648, and VI) the intensity of D1-band is higher than that of G-band, but FWHM of these bands are 50–55 $cm^{-1}$. In contrast, CM spectra of Mesoproterozoic stromatolites (905 A, B, C, and D) differ from Paleoproterozoic stromatolites, because the intensity of G-band is higher than that of D1-band (Fig. 5). Besides, the FWHM of D1-band is significantly

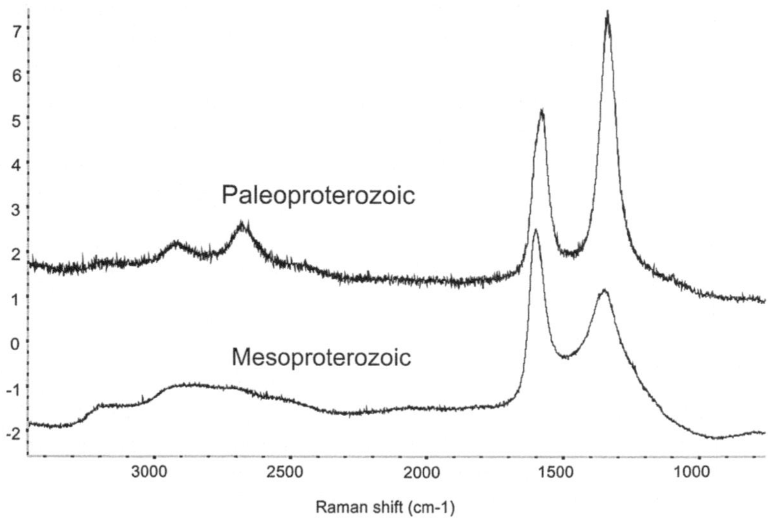


**Fig. 5** Representative spectra for CM of Paleo- (**a**) and Mesoproterozoic (**b**) stromatolites showing first and second-order bands

higher than FWHM of G-band, and exceeds the FWHM of D1-band of CM from Paleoproterozoic stromatolites.

In poorly ordered CM at least two additional peaks D3 ($\sim 1530\ cm^{-1}$) and D4 ($\sim 1260\ cm^{-1}$) can be decomposed from the Raman spectrum. Broad D3-band ($\sim 1500\ cm^{-1}$) results from out-of-plane defects, related to tetrahedrally coordinated carbons, dangling bonds, and heteroatoms, occur in natural CM (Wopenka and Pasteris 1993; Beyssac et al. 2002). When the crystallinity of CM is very low the D4-band appears like a shoulder of D1-band which has been tentatively attributed to $sp^2$–$sp^3$ bonds or C–C and C=C stretching vibration of polyene-like structure (Dippel and Heintzenberg 1999). CM of Paleoproterozoic stromatolites displays either no (sample 2648) or slight (samples 6241, 6229 и VI) intensity of the D3- and D4-bands comparing with the Mesoproterozoic stromatolites that indicates the higher degree of structural order of Paleoproterozoic CM.

The degree of CM structural order is also reflected by the bands in the second-order region. Spectra of CM of Paleoproterozoic stromatolites display well-developed second-order bands at $\sim 2700\ cm^{-1}$ and $\sim 2900\ cm^{-1}$ (Fig. 5). The former one is attributed to overtone scattering of the D1-band and is absent in poorly ordered CM, while in pure crystalline graphite it occurs as a doublet of two partially overlapping peaks (Wopenka and Pasteris 1993). The later one is attributed to the combination modes (G+D) and appears only in poorly ordered CM (Wopenka and Pasteris 1993; Beyssac et al. 2002). CM from Mesoproterosoic stromatolite differs from Paleoproterozoic since it does not contain bands in the second-order spectrum (Fig. 5). This is the additional argument to the higher structural order of CM from Paleoproterozoic stromatolites comparing with the Mesoproterosoic stromatolites.

Profiling geochemical microsampling of Paleoproterozoic stromatolites has revealed considerable variations in the concentrations of trace elements induced by the internal heterogeneity of the structures. Many elements display multi-order cyclicity indicated by variation in the step length between the peaks on the variation curves (Fig. 6). Strong correlations in the group HFSE—Zr, Ti, Y, Nb, La, Th and U and those of zirconium with Ge, Cd, Sb and Bi suggest contamination of chemically precipitated carbonate matter by acid crustal and exhalative material.

## 5 Discussion

The Raman spectroscopy of CM in ancient microfossils cannot provide definitive evidence of biological or abiological origin of CM. However, it not only confirms the presence of CM that is of great importance for ancient stromatolites, but also is used to determine structural characteristics that may provide information on the origin and history of the carbon.

CM is characterized by the degree of structural ordering, which can be quantified by different Raman parameters. Traditionally, the Raman parameters R1 (defect peak height/ordered peak height ratio) and R2 (defect band area/ordered graphite

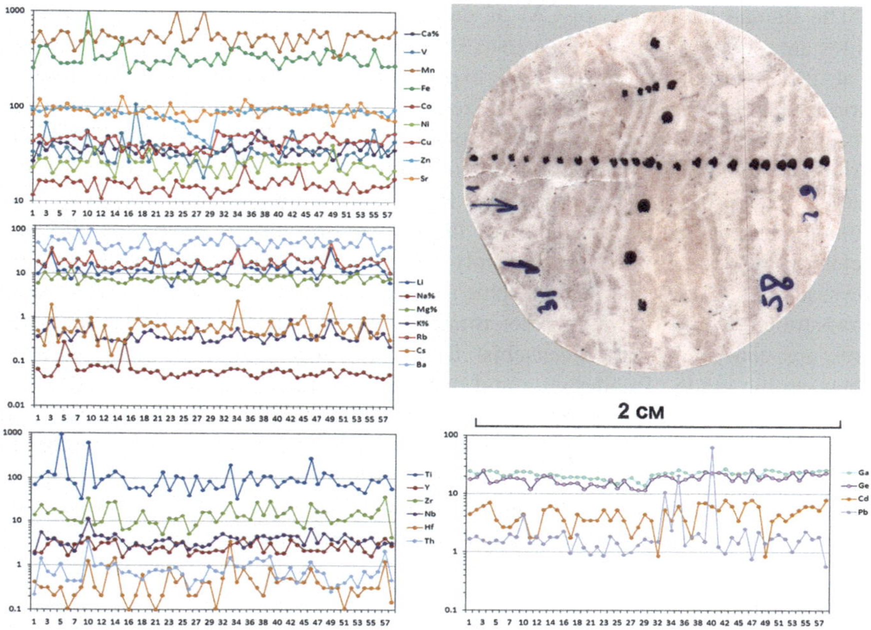


**Fig. 6** Position of laser microsampling points in a Paleoproterozoic stromatolite (sample 6229 from the Kolasjoki Formation in FAR-DEEP drill hole 8A (Pechenga Greenstone Belt) and variation diagrams of petrogenic and trace elements. All concentrations, except Ca, Mg, Na and K, are in ppm

band area ratio) are used to evaluate the degree of structural order (Wopenka and Pasteris 1993; Beyssac et al. 2002, 2003, 2004). The increasings RI and R2 values correspond to the decreasing of structural order of CM (Wopenka and Pasteris 1993; Beyssac et al. 2002; Jehlicka et al. 2003). However, these parameters are conclusive for metamorphosed CM, but is not sufficient for the CM experienced either no or slight metamorphic transformations (~250 °C) (Beyssac et al. 2002; Rahl et al. 2005). For the latter one it is necessary to consider not only R1 and R2 parameters, but also FWHM of D1- and G-bands (Bower 2011; Bower et al. 2013; Sforna et al. 2014).

As it has been shown above, the CM of Paleoproterozoic stromatolites has the higher degree of structural order comparing with the Mesoproterozoic stromatolites. Though the R1 parameter is higher than that of Mesoproterozoic stromatolites, the lower FWHM of D1-band and either no or slight intensity of D3- and D4-bands corresponds to the higher degree of structural order of Paleoproterozoic stromatolites. The additional signature of its higher structural ordering is the well-developed second-order bands which are absent in CM spectra of Mesoproterozoic stromatolites.

The Raman parameters of CM of studied Paleo- and Mesoproterozoic stromatolites are plotted on the the R1 versus FWHM-D1 diagram constructed by (Bower et al. 2013) for CM of microfossils experienced metamorphic transformations of various degree (Fig. 7). Paleoproterozoic stromatolites fall into cluster with the Precambrain stromatolites, including Apex chert, Strelley Pool chert and Tumbiana carbonate, which are hydrothermally overprinted (T > 300 °C). Thus, for the studied Paleoproterozoic stromatolites it is possible to apply an empirical thermometer based on the correlation of the Raman spectra parameters of CM and metamorphic temperature. If we apply the RS thermometer, e.g., proposed by Beyssac et al. (2002), the metamorphic alteration temperatures of Paleoproterozoic stromatolites would be roughly determined to be around 400 °C. These temperatures are consistent with greenschist facies, and this is in agreement with the metamorphic grade of underlying and overlying basalts (Melezhik et al. 2013).

Unlike Paleoproterozoic stromatolites, Mesoproterozoic stromatolites fall into cluster (Fig. 7) with the samples with the low degree of metamorphic transformations (T < ~150 °C) and where precursor carbonaceous material was compositionally complex. This cluster includes both Phanerozoic (Rhynie chert) and Archean (Duoshanto phosphorite and Gunflint chert) stromatolites.

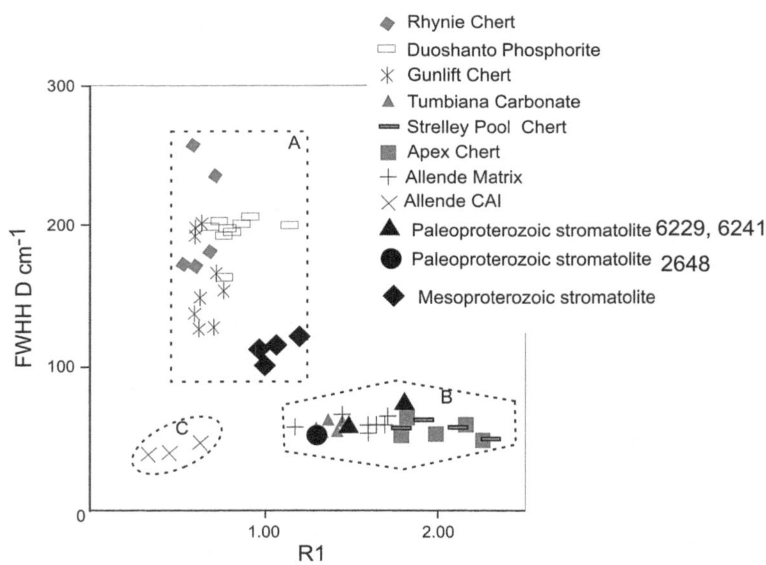


**Fig. 7** CM of Paleo- and Mesoproterozoic stromatolites, showed in FWHH-D versus R1 diagram (after Bower et al. 2013). **a** Cluster of samples where precursor carbonaceous material was compositionally complex: Rhynie chert, Duoshanto phosphorite, Gunflint chert, Murchison chondrite, ~27 °C > T < 150 °C. **b** Cluster of samples where precursor materials may have been less compositionally complex and hydrothermally overprinted: Apex chert, Strelley Pool chert, Tumbiana carbonate, Allende matrix, ~300 °C > T < ~1000 °C. **c** Purely crystalline graphitic carbon in Allende CAI, T > 1000 °C

The most of RS studies have been carried out for the CM of microfossils, including that of the Precambrian age, but only few RS studies have been done for CM of stromatolites with carbonate matrix (e.g. Allwood et al. 2006). Schopf et al. (2005) have introduced the Raman Index of Preservation (RIP) to determine the organic geochemical maturity of the preserved organic matter. The studied Paleo- and Mesoproterozoic stromatolites are characterized by RIP values, determined according to Schopf et al. (2005), of 1 and 0.5 correspondently that evidence of their high degree of degradation. These values are equal or lower the RIP minimum values determined by Schopf et al. (2005) for the stromatolites of various ages. But according to the Schopf et al. (2005) data some of the Proterozoic stromatolites, for example, from Gunflint Formation, have relatively high RIP (8.8) constituent to relatively high degree of preservation. But if we estimate the grade of metamorphic transformations using the R1 versus FWHM-D1 diagram (Fig. 7), it would be similar for CM of studied Paleoproterozoic stromatolites and of Gunflint Formation. The low RIP values determined for CM of studied stromatalites may be caused by the fact that CM of studied stromatalites are preserved directly in carbonate matrix rather than in the chert as for the CM of stromatolites, studied by Schopf et al. (2005), where the resistance to degradation is higher. Thus the RIP parameter seems to be not applicable to characterize the CM, preserved in carbonate matrix.

## 6 Conclusion

Morphological features of the studied Paleoproterozoic stromatolites together with identification of synsedimentary carbonaceous matter by Raman spectroscopy are strong argument for biogenic origin of these carbonate buildups. Precision geochemical analysis of the stromatolite laminae composition links to sedimentary settings and environmental conditions of the microbialite formation.

**Acknowledgments** Dr. Igor Chikirev from Geological Institute in Apatity for providing samples of Mesoproterosoic stromatolites is grateful acknowledged.

This research forms part of the Russian Academy of Science Programm "Origin and evolution of the Biosphere" and was partly supported by Russian Fund of Basic Research (grant 15-05-08705 A).

## References

Allwood AC, Walter MR, Marshall CP (2006) Raman spectroscopy reveals thermal palaeoenvironments of c. 3.5 billion-year-old organic matter. Vib Spectrosc 41:190–197

Aoya M, Kouketsu Y, Endo S, Shimizu H, Mizukami T, Nakamura D, Wallis SR (2010) Extending the applicability of the Raman carbonaceous-material geothermometer using data from contact metamorphic rocks. J Metamorphic Geol 28:895–914

Awramik SM, Margulis L (1974) Definition of stromatolite. Stromatolite News 2:1–5

Beyssac O, Goffé B, Chopin C, Rouzaud J-N (2002) Raman spectra of carbonaceous material in metasediments: a new geothermometer. J Metamorphic Geol 20:859–871

Beyssac O, Bollinger L, Avouac JP, Goffe B (2004) Thermal metamorphism in the lesser Himalaya of Nepal determined from Raman spectroscopy of carbonaceous material. Earth Planet Sci Lett 225:233–241

Beyssac O, Goffé B, Petitet J-P, Froigneux E, Moreau M, Rouzaud J-N (2003) On the characterization of disordered and heterogeneous carbonaceous materials by Raman spectroscopy. Spectrochim Acta A Mol Biomol Spectrosc 59:2267–2276

Bower DM (2011) Micro Raman spectroscopic investigations of mineral assemblages in parallel to bedding laminae in 2.9 Ga sandstones of the Pongola Supergroup, South Africa. J Raman Spectrosc 42:1626–1633

Bower DM, Steele A, Fries MD, Kater L (2013) Micro Raman spectroscopy of carbonaceous material in microfossils and meteorites: improving a vethod for life detection. Astrobiology 13:103–113

Dippel B, Heintzenberg J (1999) Soot characterization inatmospheric particles from different sources by NIR FT Raman spectroscopy. J Aerosol Sci 30 (Suppl. 1):S907–S908

Ferrari AC, Robertson J (2000) Interpretation of Raman spectra of disordered and amorphous carbon. Phys Rev B 61:14095–14107

Filho WS, Fairchild TR (2011) Proterozoic stromatolites of the Itaiacoca Group, Southeast Brazil. In: Tewari VC and Seckbach J (eds) Stromatolites: interaction of microbes with sediments. Springer Science+Business Media B, Berlin

Jehlicka J, Urban O, Pokorny J (2003) Raman spectroscopy of carbon and solid bitumens in sedimentary and metamorphic rocks. Spectrochim Acta A Mol Biomol Spectrosc 59:2341–2352

Kochetkov OS, Mihailenko YV, Ivanov NF (2013) Stromatolites Stratifera flexurata Kom of Sredny peninsula, north part of Kola peninsula. In: Materials of X Fersman conference «Geology and mineral resources of Kola peninsula», Apatity, pp 85–88

Marshall CP, Edwards HGM, Jehlicka J (2010) Understanding the application of Raman spectroscopy to the detection of traces of life. Astrobiology 10:229–243

McLoughlin N, Melezhik VA, Brasier AT, Medvedev PV (2013) Palaeoproterozoic Stromatolites from the Lomagundi-Jatuli interval of the Fennoscandian Shield. In: Melezhik VA (ed) Reading the archive of earth's oxygenation. Springer, Berlin

McLennan SM (1989) Rare earth elements in sedimentary rocks: influence of provenance and sedimentary processes. In: Lipin BR, McKay GA (eds) Geochemistry and mineralogy of rare earth elements. Reviews in mineralogy. Mineralogical Society of America

Melezhik VA, Fallick AE, Medvedev PV, Makarikhin VV (1999) Extreme 13Ccarb enrichment in ca. 2.0 Ga magnesite-stromatolite-dolomite-'red beds' association in a global context: a case for the world-wide signal enhanced by a local environment. Earth Sci Rev 48:71–120

Melezhik VA, Fallick AE, Hanski E, Kump L, Lepland A, Prave A, Strauss H (2005a) Emergence of the Modern Earth System during the Archean-Proterozoic. Trans GSA Today 15:4–11

Melezhik VA, Fallick AE, Rychanchik DV, Kuznetsov AB (2005b) Palaeoproterozoic evaporites in Fennoscandia: implications for seawater sulphate, δ13C excursions and the rise of atmospheric oxygen. Terra Nova 17:141–148

Melezhik VA, Lepland A, Romashkin AE, Rychanchik DV, Mesli M, Finne TE, Conze R, The FAR-DEEP Scientists (2010) The great oxidation event recorded in Paleoproterozoic Rocks from Fennoscandia. Sci Drilling 9:23–29

Melezhik VA, Medvedev PV, The Svetov SA (2013) Onega Basin. In: Melezhik VA (ed) Reading the archive of earth's oxygenation. Springer, Berlin

Rahl JM, Anderson KM, Brandon MT, Fassoulas C (2005) Raman spectroscopic carbonaceous material thermometry of low-grade metamorphic rocks: calibration and application to tectonic exhumation in Crete, Greece. Earth Planet Sci Lett 240:339–354

Riding R (2008) Marrying stromatolite perspectives: 3500 million year of history and a century of research. In: Reitner J, Quéric NV and Reich M (eds) Geobiology of stromatolites. In: International Kalkowsky-symposium. Abstract volume and field guide to excursions, Universitätsverlag, Göttingen, pp 29–30

Schopf JW, Kudryavstev AB (2009) Confocal laser scanning microscopy and Raman imagery of ancient microscopic fossils. Precambrian Res 173:39–49

Schopf JW, Kudryavstev AB, Agresti DG, Czaja AD, Wdowiak TJ (2005) Raman imagery: a new approach to assess the geochemical maturity and biogenicity of permineralized Precambrian fossils. Astrobiology 5:333–371

Sforna MC, van Zuilen MA, Philippot P (2014) Structural characterization by Raman hyperspectral mapping of organic carbon in the 3.46 billion-year-old Apex chert, Western Australia. Geochim Cosmochim Acta 124:18–33

van Zuilen MA, Chaussidon M, Rollion-Bard C, Marty B (2007) Carbonaceous cherts of the Barberton greenstone belt, South Africa: isotopic, chemical and structural characteristics of individual microstructures. Geochim Cosmochim Acta 71:655–669

Webb GE, Nothdurft LD, Kamber BS, Kloprogge JT, Zhao J-X (2009) Rare earth element geochemistry of scleractinian coral skeleton during meteoric diagenesis: a sequence through neomorphism of aragonite to calcite. Sedimentology 56:1433–1463

Wopenka B, Pasteris JD (1993) Structural characterization of kerogens to granulite-facies graphite: applicability of Raman microprobe spectroscopy. Am Mineral 78:533–557

# Part IV
# Bioweathering and Destruction of Cultural Heritage Monuments

# Granite Weathering in Urban Environments

**Elena G. Panova, Dmitry Yu. Vlasov, Hannu Luodes, Alexey D. Vlasov, Tatijana A. Popova and Marina S. Zelenskaya**

**Abstract** Various aspects of the granite destruction in urban environments were studied in cities of Northern Europe (Russia and Finland). Granite is commonly used to create monuments, buildings, and embankments. Rock degradation is associated with physical, chemical, and biological factors. In this chapter, the main forms of granite destruction are described and a classification of biofouling granite is developed. Features of granite biological colonization were established. Rapakivi granite is most strongly subjected to damage by abiotic weathering processes as well as biological colonization, which can be explained by the peculiarities of the mineral composition and texture of this type of rock.

**Keywords** Granite destruction · Weathering processes · Biogenic–abiogenic interactions · Granite crust · Anthropogenic influence · Biological colonization

## 1 Introduction

Cultural heritage monuments have become victims of the "ecological aggression" of modern industrial production, urbanization, and other anthropogenic and natural factors. Granite is traditionally used as a stone decoration in cities, especially near granite mining areas. It is one of the most widespread stones used in the architecture of St. Petersburg, Vyborg, Priozersk, and Primorsk in Russia, as well as Helsinki, Lappeenranta, Kotka, Hamina, and Kuopio in Finlan (Bulakh et al. 2010). Granites from different quarrying locations have been used in historical buildings and

E.G. Panova (✉) · D. Yu. Vlasov · M.S. Zelenskaya
Saint-Petersburg State University, Saint-Petersburg, Russia
e-mail: e.panova@spbu.ru

H. Luodes
Geological Survey of Finland, Espoo, Finland

A.D. Vlasov · T.A. Popova
Herzen State Pedagogical University, Saint-Petersburg, Russia



O.V. Frank-Kamenetskaya et al. (eds.), *Biogenic—Abiogenic Interactions in Natural and Anthropogenic Systems*, Lecture Notes in Earth System Sciences,
DOI 10.1007/978-3-319-24987-2_27

monuments of the central parts of these cities. It is necessary to know how to cure granite from destruction in order to save our cities and our history.

Rapakivi granite is the most famous of these types of rocks. Typical rapakivi granite has a unique appearance: large ovoid clusters of K-feldspar with a diameter of 3–8 cm surrounded by a rim of greenish-grey plagioclase are set into a fine-grained matrix of feldspars, quartz, and biotite. However, compared with other types of granite, the weathering processes in this rock are more intensive. Hence, more attention has been paid to rapakivi granite in studies of the weathering processes. The most essential problems include evaluation of long-term changes in the stone, the influence of air quality, and biodeterioration processes on the stone.

In recent decades, the problem of rock monument preservation has drawn the attention of geologists, biologists, architects, and restoration specialists. The greatest attention has been given in the literature to the processes of destruction of limestone and sandstone (Fitzner et al. 1995; Dakal and Cameotra 2012); granite destruction has not been given sufficient attention. Only a few publications exist on the biological colonization of granite (Silva et al. 1999; Favero-Longo et al. 2011).

## *1.1 Weathering Factors and Process*

Weathering is a process of mineral or rock destruction under the influence of physical, chemical, biotic, and anthropogenic factors. *Physical weathering* is the destruction of rocks without considerable changes in the composition of the fragments. Physical weathering mainly takes place under the influence of temperature fluctuations, freezing and thawing of water, crystallization of salts contained in the capillary water, wind, the impact of biotic communities, and the root system of plants.

*Chemical weathering* is a process of chemical alteration of minerals and rocks under the influence of water, oxygen, carbon dioxide, organic acids, and biochemical processes. The main factors affecting the intensity of the chemical weathering are the surface area and particle size, rock type, location, climate, and time. The main agents of chemical weathering are water, carbonic acid, sulphuric acid, nitric acid, organic acid, oxygen, hydrogen sulphide, methane, and ammonia, among others. There is also a considerable amount of data on heavy metals accumulated in dust (Rampazzi et al. 2010; Sverlova 2009). The main chemical processes of weathering include dissolution, leaching, oxidation, hydration, carbonization, and hydrolysis. In the course of weathering, the removal of chemical elements, oxides, and hydroxides occurs in the form of true and colloidal solutions and of suspensions of clay particles.

*Biogenic weathering* includes the impact of different types of biological objects on minerals and rocks. Biofouling is usually understood as the development (accumulation) of living organisms (microorganisms, plants, fungi, animals) on a solid substrate. Often, this term stands for "biological colonization." Biofouling takes place in both air and water. This process may have a varying duration and be accompanied by the gradual destruction of the substrate.

*Biodestruction* is a special type of destruction of rocks and materials by living organisms or their metabolic products. It usually caused by a group of organisms. The development of destructive processes may lead to the loss of the essential properties of the material, resulting in its consistent and complete destruction. Destructors of granite are bacteria, microalgae and fungi, mosses, lichen, mosses, seed plants, invertebrates, and vertebrates. However, most researchers consider that the most dangerous destructor for granite buildings are microorganisms (Berthelin 1983; Warscheid and Braams 2000). The mechanisms of their influence on granite are very diverse.

Biogenic weathering is connected with the physical and chemical influence of flora and fauna on the rock. The biochemical impact on the rock starts with the colonization of the rock surface by microorganisms, lichen, and mosses. The biochemical components have a strong influence on mineral substances. Microorganisms produce chemically active metabolites, which influence the extraction of a mineral substance from the minerals and contribute to the destruction of the rock. Microorganisms and lichens are capable of leaching elements of stone (Štyriaková et al. 2012; Adamo and Violante 2000). Their action leads to conditions for further colonization of the stone surface by higher plants.

The dynamics of the development of organisms on granite were determined by two groups of factors. The first group includes the ecological and physiological features of biological objects and their ability to grow on the rock. There is also a complex of abiotic factors that play a crucial role at every stage of colonization of the substrate.

This chapter examines the characteristics of the physical, chemical, and biological destruction of granite under urban conditions in Fennoscandia, which can be employed by different specialists and will help in preserving architectural monuments for further generations.

# 2 Materials and Methods

## *2.1 Sampling*

For our study, we examined granites with different structure–texture characteristics and investigated the weathering of the rocks in buildings of different ages, which allowed us to trace the processes of stone destruction during 300 years. In the course of our observations in nature, samples were taken of rapakivi granite in Vyborg, St. Petersburg, Helsinki, and Kuopio, among others. Objects for study included granite embankments, roads, buildings, bridges, and monuments.

In some cases, the stone samples were taken from the damaged parts of the buildings. More than 1000 samples were taken. The sites of sampling were photographed and the types of destruction were determined. In total, more than 2000 photos were taken. In some samples, the weathered crust was sawed away from the unchanged granite. Due to weathering, granite decomposes into separate grains, ultimately transforming into a dust, which was also taken as an object of our investigation. The main attention in detecting the damage to granite was focused on changes in the colour and texture of the surface layer of the stone, different forms of fouling, and new formations. In describing the nature of the damage, we noted films and stains of various colours, cracks, chips, and pits, then evaluated the degree of destruction of the surface layer of the stone.

In characterizing the biofouling of the granite, we noted the color of biofilms, their number, thickness, density, connection with certain minerals, cracks, or weathered fragments of the stone (selective biofouling). We evaluated biofouling in relation to the surface relief. At first visually, then in the laboratory, we defined the types of biofilm according the dominant species: biofilms dominated by algae, cyanobacteria, microscopic fungi, microlichens, and other organisms. In characterizing the macrofouling of granite, we recorded the presence of macrolichens, mosses, and seed plants, paying attention to their association with specific components of rocks and structural spaces on the granite. We estimated the total spatial distribution of biodamage to the object. A compulsory element of the study was a highly detailed description of the visible signs of damage, as well as photography (documentation).

The samples were divided into 2 groups: samples of collapsing materials and samples taken by nondestructive methods from the surface of the object under study. In cases where there were significant violations of the integrity of the damaged surface with fragmentation, flaking, and shedding of the stone material, the samples were taken into special sterile containers. In addition, scrapings were taken from damaged (colonized) parts of the stone surface in a sterile container or directly onto a nutrient medium in Petri dishes. Sampling of the macrofouling species (lichens, mosses, and vascular plants) was carried out according the rules of herbarium collection. Soil samples on the border of the studied objects were taken in order to determine the possible pathways of distribution and accumulation of destructive microorganisms. Samples of primary soil were collected from under the turf moss that had developed on sites of granite destruction. They were used for geochemical studies.

## 2.2 Analytical Methods

In the course of laboratory analysis of the samples, various modern analytical methods can be applied to studying the mineral and chemical composition of unchanged and weathered rocks. The following research methods were applied in this study: petrographic study of thin sections, scanning electron microscopy and microanalysis, confocal microscopy, granulometric analysis of the dust, X-ray phase analysis, X-ray spectral silicate analysis, X-ray spectral analysis, Mass spectrometry with inductively coupled plasma (ISP MS), calculation of weathering indices, and water extraction experiments.

### *2.3 Biological Methods*

Microbiological research was mainly focused on micromycetes and heterotrophic bacteria. Traditional mycological and microbiological methods have been applied for the detection and identification of microorganisms in biofilms on the surface of the granite. In the course of bacteriological studies, the microorganisms were isolated on solid nutrient medium—enzymatic hydrolyzate meat—to detect a wide range of heterotrophic bacteria and determine the total microbial numbers (Tepper et al. 2005). The following nutrient media were used for primary isolation, sustenance in culture, and the identification of micromycetes: the classic Chapek-Doks medium and modifications of it (by adjusting the content of glucose and some salts), agarosed oat broth with the addition of glucose, Saburo agar, potato-glucose agar, water agar, and 2 % Malz-Agar.

Identification of the microorganisms took place after their extraction into pure culture. The duration of cultivation varied depending on microorganisms. For example, in the course of identification of micromycetes, the obtained cultures were incubated in a thermostat for 2–3 weeks at 25 °C until sporulation, followed by the study of colonies using light microscopy. Bacteria were cultivated for 7 days at 28 °C. Typical forms of bacteria were analyzed by a molecular method in the resource center of development of cell and molecular technology of St. Petersburg State University (SPbSU). Lichens were identified according to standard techniques at the Department of Botany SPbSU under the leadership of D.E. Gimelbrant, and mosses were identified at the Botanical Institute of the Russian Academy of Science under the leadership of L.E. Kurbatova.

## 3 Results and Discussion

The processes of the destruction of granite in an urban environment are accelerated by the complex impact of closely interrelated physical, chemical, biological, and anthropogenic factors. The main characteristics of granite weathering forms in the urban environment were determined based on our observations. They correlate with known classifications (Fitzner and Heinrichs 2004) but have many peculiarities connected with the granite's properties.

*Surface roughening* (Fig. 1) is the most widely represented type of mechanical destruction, which is typical for open surfaces of granite. It develops with different intensity depending on the position of the surface relative to the direction of the wind, as well as the vertical or horizontal position. Roughening of the surface is the process of grain loss. According to their resistance to weathering, the main minerals

**Fig. 1** Surface roughening of the granite rapakivi (St. Petersburg, Facade of Russian museum building)

of granite are divided as very stable (quartz), stable (K-feldspar plagioclase [albite], muscovite), instable (plagioclase [andesine], pyroxene, amphibole), and very instable (plagioclase [anorthite], biotite, chlorite).

*Hollows and deepenings* appear on the surface of rock due to deep weathering and the loss of numerous mineral grains. Typical rounded deepenings in the rock are explained by the loss of large ovoids. *Exfoliation* leads to the appearance of thin plates on the rock surface, which consist of granite and are the products of its destruction. Exfoliation occurs as a result of the heating of the stone surface to a certain depth, differences in the coefficients of expansion of the minerals, and the formation of weathered crust (plate) that is in parallel to the surface of the bedrock (Fig. 2).

*Fissures* are caused by the textural and mineral heterogeneity of granite. Fracturing first affects feldspars and amphiboles due to their cleavage. The quantity of quartz crystals, which are less susceptible to mechanical destruction, thus increases in the crust. *Loss of fragments* can be observed at the site of fissures or chips. Loss of granite fragments causes aesthetic damage to architectural buildings, which is amplified by careless restoration.

**Fig. 2** Exfoliation of granite (St. Petersburg, Facade of Russian museum building)

*Chemical weathering* can be observed as a change of colour caused, first of all, by the decomposition of sulphides with the formation of iron hydroxides (Fig. 3). The presence of sulphides in granite leads to their oxidation and the appearance of brown spots on the surface of the stone. On vertical walls, the brown colour flows downwards from sulphide accumulations, forming vertical stripes. This effect can be avoided by the appropriate choice of stone for buildings (i.e. without sulphides).

Mechanical links between the parts of stone are the first to be destroyed under the influence of water, wind, and temperature fluctuations. Water penetrates the cracks and micro-cracks in the rock, creating a favorable environment for chemical reactions. Gases and substances from the air and water have harmful chemical impacts. Micas, pyroxenes, amphiboles, and feldspars are transformed into clay minerals and leached from the rock. The processes of oxidation and substitution of minerals lead to micro-cavities and fissures and, as a result, to an increase in mechanical destruction.

The dust of aerosols may take part in the chemical destruction of granite by accumulating in the irregularities of the surface and reacting with moisture. Dust particles settling on the weathered surface of stone can change its chemical composition. In the dust of Petersburg, the following chemical elements have been found (average, ppm): Zr (515), Zn (458), Sr (388), V (172), Cr (151), Rb (135), La (82), Pb (65), Ni (45), Cu (42), and As (13). Concentrations of some elements in the dust are 2–13 times higher than in the lithosphere.

The data indicate that the crust of altered granite is enriched with silica, aluminum oxide, and sodium and is poor in potassium and calcium compared to unaltered granite. This is explained by the higher proportion of clay minerals in altered granite, its disintegration into minerals, and the removal of clay particles by the wind. Based on the micro- and macro-element composition of samples, we can conclude that the crust accumulates 1.5–3 times more chemical elements in comparison with fresh granite. The accumulation of Sb, As, Pb, Cu, Mo, Zn, and S is typical for the crust (Table 1), which can be explained by the influence of the urban environment.

**Fig. 3** Brown spots (iron hydroxides) on the surface of the stone (Facade of Saint Petersburg Exchange Building)

**Table 1** Average concentrations (ppm) of chemical elements in granites and crust

| Elements | Granite ($n$ = 13) | Crust ($n$ = 13) | Elements | Granite ($n$ = 13) | Crust ($n$ = 13) |
|---|---|---|---|---|---|
| Zn | 95.35 | 144.6 | Ba | 119.5 | 149.0 |
| As | 11.72 | 14.83 | Sr | 12.82 | 19.16 |
| Cu | 6.28 | 7.31 | La | 100.3 | 130.6 |
| Pb | 17.66 | 30.65 | Ce | 181.3 | 230.1 |
| Bi | 0.29 | 0.34 | Y | 54.52 | 57.90 |
| Cd | 0.24 | 0.32 | U | 7.37 | 6.49 |
| Co | 3.30 | 2.99 | Th | 47.09 | 42.77 |
| Mo | 1.66 | 2.30 | W | 1.79 | 2.06 |
| Sb | 0.09 | 0.14 | Cr | 2.92 | 2.94 |
| Ag | 0.13 | 0.15 | Be | 0.99 | 1.25 |
| S | 102.0 | 126.6 | Sc | 4.88 | 4.96 |
| Se | 3.19 | 3.88 | Li | 38.28 | 38.64 |

*Biogenic weathering* is one of the permanent factors in granite destruction. The classification of biofouling is presented in Fig. 4. The surfaces of the granite monuments, buildings, and embankments at St. Petersburg are usually covered by green biofilms with a predominance of algae belonging to the phylum Chlorophyta. High humidity creates optimal conditions for the colonization of the granite surface by microscopic aerophilic algae. Algae can show selectivity with respect to the mineral composition of the granite. Hence, there is a tendency for algal distribution over the inclusions of biotite and around feldspar ovoids. Graffiti contributes to the development of green biofilms. Algae hold moisture, which accelerates the destruction of granite. In addition, their development contributes to the accumulation of organic matter on the surface of granite, which is used by more aggressive destructors, such as by micromycetes. Scanning electron microscopy studies

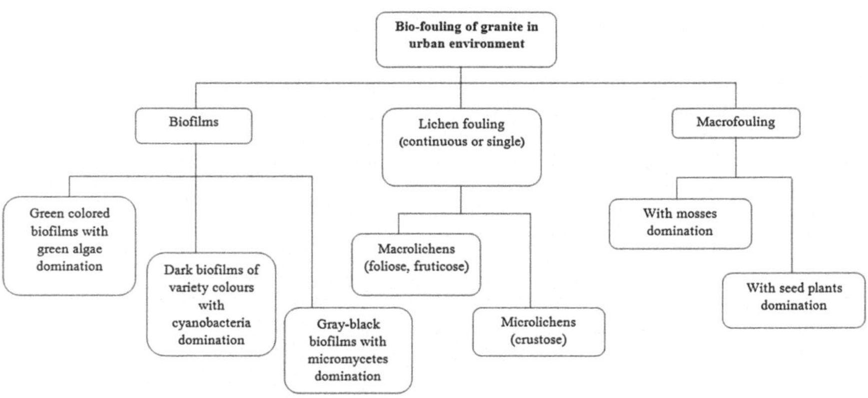


**Fig. 4** Classification of biofouling of granite in urban conditions

indicate that unicellular and multicellular (filamentous) forms of algae develop on the granite, and the cells of diatoms can be often found.

Microscopic fungi and bacteria, as a rule, are found in biofilms that often have a grey-black color. Such biofilms are usually formed in the direction of moisturization. In many cases, dark-colored filamentous fungi, as well as black microcolonial yeast-like fungi, are the dominating group in grey-black biofilms. They cover some parts of embankments, the basements of historical buildings, and the surface of monuments in St. Petersburg. Atmospheric pollution contributes to the development of dark-colored fungi. The number of micromycetes in such biofilms can reach 10,000 colony-forming units per gram of sample.

Often, dark films are associated with the development of cyanobacteria. Such films can be observed in places with a constant high humidity. They were found on many buildings in Vyborg and the Peter and Paul fortress in St. Petersburg. Their color varies from dark green to almost black. In areas with an intensive development of cyanobacteria, they create the conditions for the development and accumulation of saprophytic bacteria, especially from the genus *Bacillus*.

An SEM study of granite covered with dark biofilms demonstrated that black microcolonial yeast-like fungi occupy "microzones" in the cracks and caverns of the rock surface (Fig. 5). They are able to penetrate under the loose flakes of stone. Dark-coloured fungi settle in the space around the crystals of the rock, as well as in the cracks and caverns.

On contaminated areas of granite, the growth of fungal mycelia and formation of biofilms with micromycetes domination was observed most actively. It is noteworthy that the colonization of granite by fungi and crustose lichens is connected with peculiarities of the texture and mineral composition of the rock. Hence, the colonization of quartz by fungi is mainly observed along microcracks. In the process of mica colonization, the hyphae of micromycetes develop between the plates of this mineral. The accumulation of microcolonies on feldspar is observed in zones of crystal cleavage. Colonies are often formed in the contact zones between

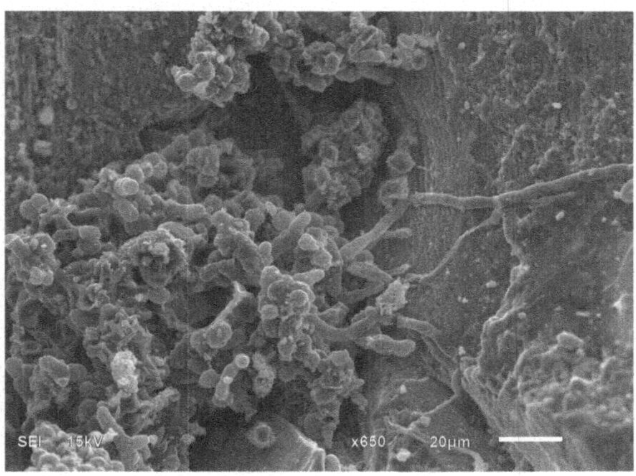


**Fig. 5** Microcolonies of fungi with branching and penetrating hyphae in a zone of feldspar destruction (Lieutenant Schmidt embankment)

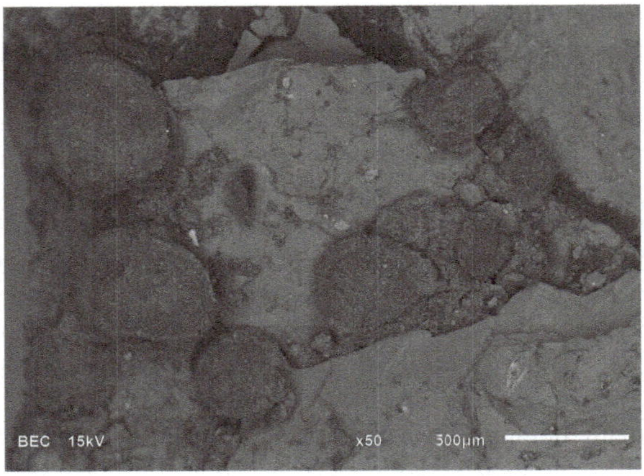


**Fig. 6** Apothecia of crustose lichens in microcracks of granite (Robespierre embankment)

different minerals with significant heterogeneity of the surface layer and fracturing. Crustose lichens can also develop here, forming apothecia in the micro-cracks of the rocks (Fig. 6).

Spore-bearing (mosses, horsetails, lycopods, ferns) and seed-bearing plants (herbaceous, shrub and woody) take part in the biological colonization of granite embankments in St. Petersburg. As a result, 110 species of plants have been detected in these habitats. The plants mainly occur in the gaps between the granite blocks. Moreover, they are ubiquitous on the protruding parts of the embankments: borders, pedestals, and their joints in cast-iron railings, carved ornaments, and figure images. The greatest number of species occurs near parks and gardens, which are potential sources of seeds and spores, as well as around bridges. Mosses of 3 species are ubiquitous: *Ceratodon purpureus* (Hedw.) Brid., *Pohlia nutans* (Hedw.) Lindb., and *Physcomitrium* sp. (definition by L.E. Kurbatova). They trap moisture and create conditions for the gradual destruction of granite (Fig. 7). Under the

**Fig. 7** The moss *Pohlia nutans* (Hedw.) Lindb. on granite embankment

mosses, it is possible to observe the formation of a thin primary layer of soil, which is composed of dead fragments of mosses, particles of damaged granite, and sand and dust particles from the external environment. The large number of mica and feldspar particles (granite components) in the samples, which are accumulated in rhizoids, points to the destructive role of mosses.

Among the identified species of plants, 72 % are indigenous and 28 % are alien. Analysis of the ratio of life forms revealed that herbaceous perennials predominate (50 %). On the embankments, plants can grow singly as well as in communities. The investigations carried out have demonstrated that granite embankments in the central part of Saint Petersburg are subjected to considerable biological colonization (macro- and microfouling), which induces their gradual destruction.

## 4 Conclusions

The processes of the destruction of granite in urban environments are accelerated by the complex impact of closely interrelated physical, chemical, biological, and anthropogenic factors. The development of biofouling accelerates the processes of the physical-chemical destruction of granite caused by fluctuations in temperature and moisture, as well as by the influence of salt removal, which is notable on the vertical walls of the embankments. Rapakivi granite is most strongly subjected to damage, and its colonization is explained by the peculiarities of the mineral composition and texture. Products of stone destruction often remain in biofilms. Crust on the granite surface consists of products of granite destruction, air pollutants, cells, and metabolites of living organisms. The diversity of rock fouling in the urban environment is poorer than in natural ecosystems (outcrops, abandoned quarries, and others). In the city, the most adapted species dominate in the lithobiontic communities. The composition and structure of these communities can be used as a bioindicator of the state of urban ecosystems.

**Acknowledgments** We thank L. Kurbatova and D. Himelbrandt for their help identifying mosses and lichens. Our investigations were supported in part by Saint-Petersburg State University (research grant 1.37.151.2014) and the project of South-East Finland—Russia ENPI CBC 2007–2013.

## References

Adamo P, Violante P (2000) Weathering of rocks and neogenesis of minerals associated with lichen activity. Appl Clay Sci 16:229–256

Berthelin J (1983) Microbial weathering processes. In: Microbial geochemistry. Blackwell, London, pp 223–262

Bulakh AG, Abakumova NB, Romanovsky JV (2010) St Petersburg: a history in stone. St Petersburg (In Russian)

Dakal TS, Cameotra CC (2012) Microbially induced deterioration of architectural heritages: routes and mechanisms involved. Environ Sci Eur 24–36

Favero-Longo SE, Gazzano C, Girlanda M, Castelli D, Tretiach M, Baiocchi C, Piervittori R (2011) Physical and chemical deterioration of silicate and carbonate rocks by meristematic microcolonial fungi and endolithic lichens (Chaetothyriomycetidae). Geomicrobiol J 28:732–744

Fitzner B, Heinrichs K (2004) Photo atlas of the weathering forms on stone monuments. www. Stone.rwth-aachen.de

Fitzner B, Heinrichs K, Kownatzki R (1995) Weathering forms—classification and mapping. Denkmalpflege und Naturwissenschaft, Natursteinkonservierung I:41–88

Rampazzi L, Giusanni B, Rizzo B, Corti C, Pozzi A, Dossi C (2010) Monuments as sampling surfaces of recent traffic pollution. Environ Sci Pollut Res 18:184–191

Silva B, Rivas T, Prieto B (1999) Effects of lichens on the geochemical weathering of granitic rocks. Chemosphere 39(2):379–388

Štyriaková I, Štyriak I, Oberhänsli H (2012) Rock weathering by indigenous heterotrophic bacteria of Bacillus spp. at different temperature: a laboratory experiment. Miner Petrol 105:135–144

Sverlova LI (2009) Scientific bases of modern approach to an assessment of level of pollution of atmospheric air of the cities. Achievements Mod Nat Sci 7:20–22 (in Russian)

Tepper EZ, Shilnikova VK, Pereversev GI (2005) Workshop on microbiology. Moscow (In Russian)

Warscheid Th, Braams J (2000) Biodeterioration of stone: a review. Int Biodeterior Biodegradation 46:343–368

# The Crystallization of Calcium Oxalate Hydrates Formed by Interaction Between Microorganisms and Minerals

**Aleksei V. Rusakov, Aleksei D. Vlasov, Marina S. Zelenskaya, Olga V. Frank-Kamenetskaya and Dmitry Yu. Vlasov**

**Abstract** The work is devoted to the research of the interactions between microorganisms, rocks and minerals in simulation experiment conditions. Bacteria and microscopic fungi isolated from different types of rocks were studied as agents of biomineralization processes on the mineral surface in moist chamber and in liquid medium. The formation of calcium oxalates under the influence of fungi (*Aspergillus* and *Penicillium*) and bacteria (*Bacillus*) was observed on the surface of different calcium-bearing minerals: carbonates, phosphates, silicates. The obtained results allow to compare the role of microorganism metabolism, peculiarities of the stone substratum and the experimental conditions (the role of the environment) in calcium oxalate crystallization. It was shown that it is a complicated multifactor process. The metabolic activity of microorganisms affects the morphogenesis of the forming oxalates which can be explained by the difference in solubility of stone substrates in various metabolites. The results of simulation experiments showed that the metabolism of the microbial community is an important factor of secondary mineralization on the surface of rocks and minerals. The present results give a scientific basis for creating new methods for cultural heritage stone monument preservation with the use of modern biotechnologies.

**Keywords** Oxalate crystallization · Whewellite · Weddellite · Microbial community · Microscopic fungi · Bacteria · Bioweathering · Biotransformation

A.V. Rusakov (✉) · M.S. Zelenskaya · O.V. Frank-Kamenetskaya · D.Yu.Vlasov
Saint Petersburg State University, Saint Petersburg, Russia
e-mail: alex.v.rusakov@gmail.com

A.D. Vlasov
Herzen State Pedagogical University of Russia, Saint Petersburg, Russia

 
O.V. Frank-Kamenetskaya et al. (eds.), *Biogenic—Abiogenic Interactions in Natural and Anthropogenic Systems*, Lecture Notes in Earth System Sciences,
DOI 10.1007/978-3-319-24987-2_28

# 1 Introduction

Microbes play a significant role in bioweathering and geomicrobiological processes. Microorganisms (bacteria, fungi and etc.,) deteriorate rocks and minerals and cause biotransformation of natural rock and building materials (Ehrlich 1999; Sterflinger 2000; Burford et al. 2003; Adeyemi and Gadd 2005; Vlasov and Frank-Kametskaya 2006; Gadd 2007; Frank-Kamenetskaya et al. 2009). Microorganisms also play a significant role in the global carbon cycle as well as in biotechnological applications in the bioremediation processes of xenobiotic-, metal- and radionuclide-contaminated soils and wastes (Gadd 1999; Burford et al. 2006). Last but not least, the study of the biochemical activity of microorganisms is important for development of effective procedures for the protection of stone monuments influenced by urban environments (Frank-Kamenetskaya et al. 2012b; Bonaventura et al. 1999).

Calcium oxalate hydrates, particularly whewellite $CaC_2O_4{\cdot}H_2O$ and weddellite $CaC_2O_4{\cdot}2H_2O$, are commonly present in biofilms on carbonate rocks (marble, limestone, dolomite) surfaces in association with lichen thalli, fungal hyphae and bacteria (Pinna 1993; Bonaventura et al. 1999; Burford et al. 2003; Magnuson and Lasure 2004; Gadd 2007; Rousakov et al. 2010). Calcium oxalates were also found in association with lichens on the surface of granite (Prieto and Silva 2003; de Oliveira et al. 2001). Whewellite and weddellite occurs also in biofilms on other rocks and minerals (serpentinite, cupriferous rocks, andesite and volcaniclastite, basalt, gabbro, dolerite) (Burford et al. 2003; Gadd 2007). Signs of dissolution found on the surface of fossil ivory (apatite–organic composite) (Pinzari et al. 2013) allowed to propose that oxalates can form on the surface of apatite as well.

The results of laboratory experiments (Monte 2003a, b; Kolo et al. 2007; Frank-Kamenetskaya et al. 2012a, b; de la Torre et al. 1993a, b; Pinzari et al. 2013) showed that the formation of calcium oxalates on the surface of calcium-bearing rocks and minerals can take place as the result of the action of oxalic acid produced by microscopic fungi (Barinova et al. 2010) on a mineral surface.

Bacteria producing organic acids and in particular oxalic acid (Kuis and Markevich 2008; Binbin and Bin 2011; Girgis et al. 2008) can also take part in oxalate formation on the surface of rocks and minerals. It has been investigated that *Bacillus subtilis* contributes to the formation of quasi-hexagonal calcium oxalate monohydrate (Chen et al. 2007). However, experimental research on the formation of calcium oxalate under the influence of bacteria is scarce.

In general, the crystallization of calcium oxalate hydrates crystals under the influence of microorganisms and factors regulating this process have not yet been well investigated.

The present work continues our researches of this problem (Frank-Kamenetskaya et al. 2012a, b; Rosseeva et al. 2015) and is devoted to the crystallization laws of calcium oxalate hydrates formed by interaction between calcium-bearing minerals (calcite, apatite, feldspar) and microorganisms (fungi and bacteria). In particular, the sequence and the morphogenesis of oxalate crystals, the comparative role of microorganisms, peculiarities of the stone substratum and the environment in crystallization processes will be studied.

# 2 Materials and Methods

## 2.1 Experimental Conditions

To study the effect of microorganisms on the stone substratum in experimental conditions several species of microscopic fungi and bacteria were chosen (Table 1), these species excrete oxalic acid and are active stone destructors (Kuis and Markevich 2008; Barinova et al. 2010). The strains of fungus *Aspergillus niger* (Ch4/07) were previously isolated from the surface of a damaged proconesos marble of a monument column of the «Basilica in Basilica» (Tauric Chersonesos, Crimea); strain of fungus *Penicillium* sp.—from the carrara marble monument in Lasarevskaya chamber of Eighteenth century Necropol of Urban Sculpture museum (Saint Petersburg), strains of bacteria *Bacillus subtilis*—from the eroded rapakivi granite from the quarry (Finland, Virolahti area) which was used for mining the block for the Alexander column on Dvortsovaya square (Saint Petersburg) and from granite near Russian «Progress» polar station (Antarctica).

For the experiments aimed to study the effect of microorganisms on calcite homogeneous calcite marble from Carrara quarry (Italy) was used, on hydroxyapatite $Ca_5(PO_4)_3(OH)$—a fossil mammoth tooth fragment (hydroxyapatite–protein composite), as well as the hydroxyapatite powder obtained by the precipitation

**Table 1** The cultivation times of microorganisms on different mineral substratums (days)

| Underlying substratum | *Aspergillus niger* | *Bacillus subtilis* | *Bacillus subtilis + Aspergillus niger* |
|---|---|---|---|
| *Experiment in moist chamber* | | | |
| Marble | 30, 120 | 90, 150 | 90[a] |
| Mammoth tooth | 14, 60 | Not conducted | Not conducted |
| Granite | 90 | 150 | 90 |
| *Experiment in liquid nutrient medium* | | | |
| Marble | 13, 16, 21, 45 (Rosseeva et al. 2015) | 60 | 13, 30, 60 |
| Mammoth tooth | 7, 16, 20, 23, 60, 180 | Not conducted | Not conducted |
| Granite | 60, 90 | 60, 90 | 90 |
| Synthetic hydroxyapatite powder | 13, 18, 25, 60, 180 | Not conducted | Not conducted |

[a]The cultivation time of *Bacillus subtilis* + *Penicillium* sp. in experiment on marble was 90 days

method. In addition rapakivi granite from the quarry (Finland, Virolahti area) was used as a stone substratum, which contains acid plagioclase—calcium-bearing feldspar (Na, Ca)$AlSi_3O_8$. Marble and granite were cut into blocks (1 × 1 × 0.5 cm), tooth fragments were used as shards from 0.5 cm to 1 cm in size. To estimate and compare the effect of microorganisms on the stone substratum the inoculation of the substratum was carried out in three different ways: with micromycetes, bacteria, and both micromycetes and bacteria (Table 1).

The experiments were carried out at room temperature in moist chambers in the lack of nutrients (the conditions close to natural) and in liquid Czapek–Dox medium of the following composition (g/l): $NaNO_3$—3.0; $KH_2PO_4$—1.0; $MgSO_4 \cdot 7H_2O$—0.5; KCl—0.5; $FeSO_4 \cdot 7H_2O$—0.015; glucose—30.0 (additionally—$CaCO_3$—2.0 g/l).

Several drops (3–4 drops) of microorganism spore suspension of a fixed concentration (mixed with liquid Czapek–Dox medium) were applied to the surface of stone samples held in moist chambers. No nutrient solution was applied later. The incubation period for the stone samples was between 16 and 150 days.

In the experiments with liquid medium stone samples were put on the bottom of glass weighing bowls in the Czapek–Dox medium (15 ml) and inoculated by microorganisms. The inoculation was carried out by fungal conidia and bacteria cells, using fungus culture taken from Czapek–Dox agar after 10 days of cultivation and bacteria culture taken from the agar media HMF (hydrolyzate meat fermentative) after 2-day cultivation. The cultivation period for microorganisms for different samples varied from 7 to 180 days (Table 1). The number of spores and bacteria cells in 1 ml of the initial suspension of Czapek–Dox medium was measured as $10^4$–$10^5$, $10^8$–$10^9$, respectively.

## 2.2 *Laboratory Methods*

The resulting products of the crystallization experiments were visualized by optical and scanning electron microscopy. In addition, elemental analysis and X-ray powder diffraction were used. X-ray diffraction was carried out only when the amount of crystalline material was high enough. In general, the identification of calcium oxalate hydrate crystals was based on their morphological features as reported in previous investigations (Zuzuk 2003; Thomas 2009; Thomas et al. 2012). To estimate the relative development of dipyramidal and prismatic faces of tetragonal weddellite crystals the [100] to [001] edges ratio was estimated.

### 2.2.1 Scanning Electron Microscopy (SEM) and Energy-Dispersive X-ray Spectroscopy (EDXS)

Morphological investigation and elemental analysis (using EDXS) of stone samples (calcite marble and granite blocks, pieces of mammoth tooth and apatite powder) inoculated by microorganisms were performed. The research was made by means of the Tescan MIRA3 LMU system operated in a low vacuum (60 Pa) mode and at an acceleration voltage of 15 kV and Jeol JCM-5000 Neoscope system operated in a low vacuum (60 Pa) mode and at an acceleration voltage of 15 kV. Samples were mounted on carbon tapes adhered to aluminum or brass holders. To increase the quality of the SEM images (to avoid charging effect) the samples were coated by a thin gold layer of 20 nm for 1 min (Gressington 108 Auto Sputter Coater). The coating was performed on a high vacuum coating machine «Leica EM SCD500». For EDXS measurements the microscope was additionally equipped with the system Oxford Aztec X-Max 80. The EDX spectra were analyzed by means the EDAX Genesis software package (semiquantitative analysis was performed by standardless method which is in general reliable for elements with $Z > 10$).

#### 2.2.2 X-Ray Powder Diffraction (XRD)

The determination of the phase composition of the crystallization products was carried using a Bruker «D2 Phaser» and Rigaku «MiniFlex II» powder diffractometers ($CuK_{\alpha}$ radiation, $\lambda = 1.54178$ Å). X-ray diffraction patterns were collected at room temperature in the range of $2\theta = 5–50°$ with a step of 0.02° $2\theta$ and a counting time of half second per data point. The JCPDS PDF-2 database was used for phase identifications.

## 3 Results

### *3.1 Calcite Transformation*

#### 3.1.1 The Experiments in Moist Chamber

Small tetragonal crystals of dipyramidal–prismatic habit were found on the surface of calcite marble in moist chamber in the presence of *Aspergillus niger* fungus after one month of experiment (Fig. 1a; Tables 2 and 3). Such crystals are typical for calcium oxalate dihydrate (weddellite $CaC_2O_4 \cdot 2H_2O$), and it was approved by EDXS analysis (Fig. 1b). Prismatic faces were developed much less than dipyramidal faces (Table 3). Bigger crystals with more developed prismatic faces with the dissolution traces were found on the surface of marble after 4 months of experiment (Fig. 1c).

Tetragonal weddellite crystals of similar size and morphology were found on the surface of marble in moist chamber after 3 months of cultivation with the presence of bacterium *Bacillus subtilis* (Tables 2 and 3). Crystals without prismatic faces were found as well (Fig. 2a). Only dipyramidal weddellite crystals were found after 5 months of cultivation, their size and quantity increased.

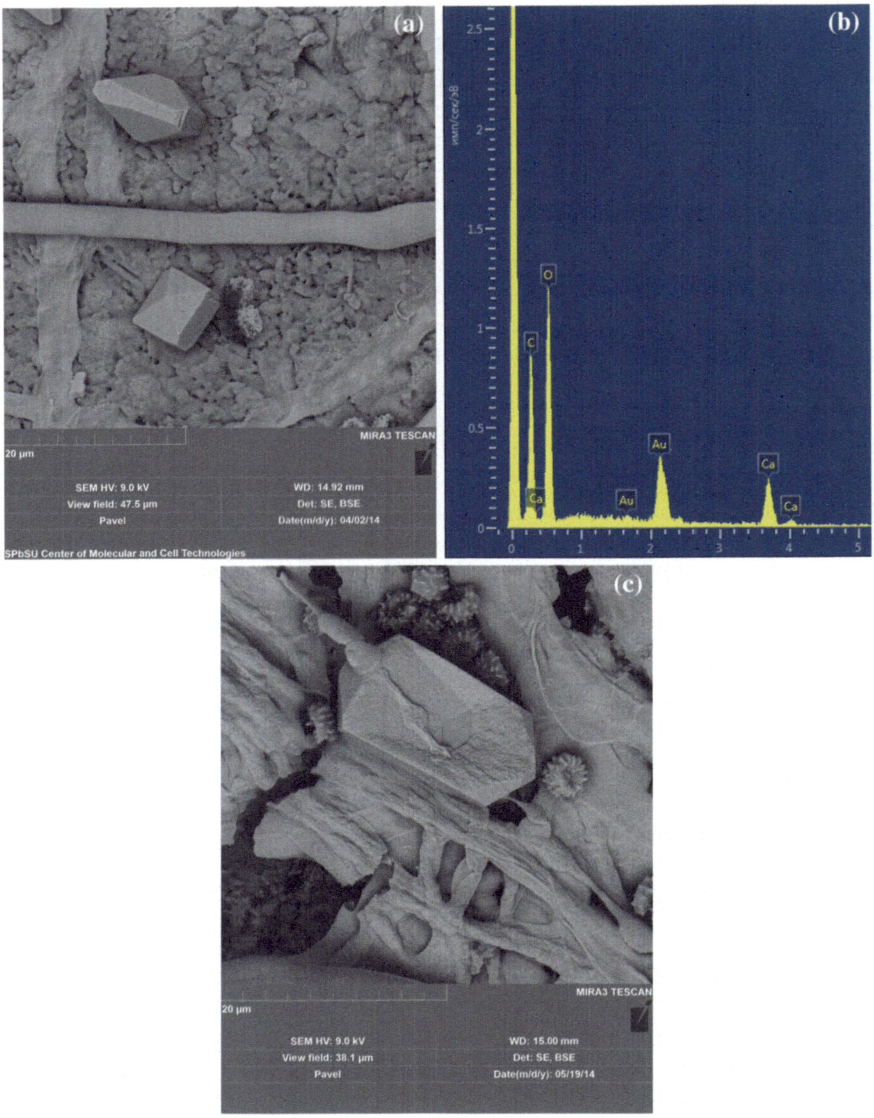


**Fig. 1** Dipyramidal-prismatic tetragonal weddelite crystals formed on marble surface by fungal *Aspergillus niger* treatment (moist chamber experiment): **a** SEM image after 1 month; **b** EDX spectrum after 1 month; **c** SEM image after 4 months

When both bacterium *Bacillus subtilis* and microscopic fungi (*Aspergillus niger, Penicillium* sp.) were inhabiting marble surface the size of the crystals was the same but their phase contents and morphology was different to the previous case. Weddellite crystals with prismatic faces considerably bigger than dipyramidal faces

**Table 2** Fixation time (days) of calcium oxalate crystals (weddellite—Wd, whewellite—Wh) under different experiment conditions (mc—in moist chamber, lm—in liquid nutrient medium)

| Substratum | Calcium oxalates | *Aspergillus niger* | | *Bacillus subtilis* | | *Bacillus subtilis* + *A. niger* | |
|---|---|---|---|---|---|---|---|
| | | mc | lm | mc | lm | mc | lm |
| Marble | Wd | 30 | 13[b] | 90 | No crystals (till 60 days) | 90 | 13 |
| | Wh | No crystals (till 120 days) | $13 < x \le 16$[b] | No crystals (till 150 days) | | No crystals (till 90 days)[a] | No crystals (till 60 days) |
| Mammoth tooth fragment | Wd | 14 | $7 < x \le 16$ | No data | | | |
| | Wh | $14 < x \le 60$ | $16 < x \le 20$ | | | | |
| Synthetic hydroxyapatite powder | Wd | No data | $13 < x \le 18$ | | | | |
| | Wh | | | | | | |
| Granite | Wd | No crystals (till 90 days) | No crystals (till 90 days) | 150 | No crystals (till 90 days) | 90 | No crystals (till 90 days) |
| | Wh | | | No crystals (till 150 days) | | No crystals (till 90 days) | |

[a]A polycrystalline pseudomorphosis of secondary calcite over spherulitic whewellite intergrowths in experiments with *Bacillus subtilis* + *Penicillium* sp. after 90 days was found

[b]According to Rosseeva et al. (2015)

**Table 3** The characteristics of dipyramidal-prismatic weddellite crystals formed in experiments with different conditions (mc—in moist chamber, lm—in liquid nutrient medium)

| Substratum | *Aspergillus niger* | | | | *Bacillus subtilis* | | | | *Bacillus subtilis* + *A. niger* | | | |
|---|---|---|---|---|---|---|---|---|---|---|---|---|
| | mc | | lm | | mc | | lm | | mc | | lm | |
| | Median size, μm (exposure, days) | [100]:[001] | Median size, μm (exposure, days) | [100]:[001] | Median size, μm (exposure, days) | [100]:[001] | Median size, μm (exposure, days) | [100]:[001] | Median size, μm (exposure, days) | [100]/[001] | Median size, μm (exposure, days) | [100]:[001] |
| Marble | 14 (30)<br>20 (120) | 1:0.20 (30)<br>1:1.5 (120) | 10–20[b] (13–16) | 1:0.14 (13)<br>1:0.14–0.50 (16)[b] | 5–15 (90)<br>10–40 (150) | 1:0.2 (90)<br>No prismatic faces (90–150) | No crystals | | 5–20 (90)[a] | 1:1.5 (90) | 30–50 (13) | 1:1.5 (13) |
| Granite | No crystals | | No crystals | | 25–40 (150) | No prismatic faces | No crystals | | 20–25 (90) | No prismatic faces | No crystals | |
| Mammoth tooth fragment | 10–15 (14–60) | No prismatic faces | 30–120 (16–60)<br>100–130 (180) | 1:0.20 (16)<br>1:0.25 (20)<br>No prismatic faces (23–180) | No data | | | | | | | |
| Synthetic hydroxyapatite powder | No data | | 90–120 (18)<br>100–150 (60) | 1:0.20 (18)<br>1:0.40 (25)<br>No prismatic faces—60 | | | | | | | | |

[a]In experiments with *Bacillus subtilis* + *Penicillium* sp. the median crystal sizes were 10 μm (90 days); [100]:[001]—1:0.4
[b]According to Rosseeva et al. (2015)

**Fig. 2** SEM images of weddellite crystals with varying prismatic face size on the surface of marble, formed during first three months of experiment (moist chamber experiment): **a** prism faces almost absent (under the influences of bacterium *Bacillus subtilis*); **b** prism faces are more prominent than dipyramidal faces (under the influence of bacterium *Bacillus subtilis* and fungus *Aspergillus niger*); **c** prism faces are less prominent than dipyramidal faces (under the influences of bacterium *Bacillus subtilis* and fungus *Penicillium sp.*)

(Fig. 2b; Table 3) as well as crystal intergrowths were found when bacterium *Bacillus subtilis* and fungus *Aspergillus niger* grew together. Single small dipyramidal-prismatic weddellite crystals with prismatic faces smaller than dipyramidal faces (Fig. 2c; Table 3) were found on marble under the influence of bacterium and fungus *Penicillium* sp. after the same period of incubation. The prismatic faces were still bigger than the ones of the crystals formed under the influence of bacterium *Bacillus subtilis* only. Besides that multiple crystal intergrowths (under 60 μm) similar to spherulitic whewellite intergrowths (Fig. 3) were found under the influence of bacterium and fungus *Penicillium* sp. after 3 months. Only calcite peaks can be observed on the X-ray diffraction pattern which leads to a conclusion that a polycrystal pseudomorphosis of secondary calcite over whewellite formed in this experiment as a result of weddellite dehydration. A hole can be observed in the center of the aggregate (Fig. 3b). It could form as a result of the dissolution of weddellite crystal during the experiment. It is known that bacteria of different trophic groups, in particular of *Bacillus* specie, can use oxalates as the source of carbon, electrons and energy by oxidizing them to calcium carbonate. Oxalate–carbonate pathway under the influence of oxalotrophic bacteria is described by different authors (Verrecchia et al. 2006; Martin et al. 2012). This result

**Fig. 3** SEM images of calcite polycrystalline pseudomorphose grown over the spherulitic whewellite intergrowth, formed under the influence of bacterium *Bacillus subtilis* and fungus *Penicillium* sp. (moist chamber experiment): **a** general view; **b** spherulitic crystal intergrowth

approves the data (Sahin 2003; Martin et al. 2012) that bacteria act as oxalotrophic organisms in the presence of fungi. Thus it can be stated that the process of the transformation of calcium oxalates into secondary calcium carbonate is catalyzed by microorganisms.

### 3.1.2 The Experiments in Liquid Medium

*In Czapek–Dox liquid medium* the process of calcite–oxalate pathway under the influence of organic acids, produced by microscopic fungi, is faster (Table 2), while under the influence of bacterium *Bacillus subtilis* no oxalate crystallization was observed after 5 months.

As was reported earlier (Rosseeva et al. 2015), in experiment with fungus *Aspergillus niger* there were small dipyramidal–prismatic weddellite crystals (with very small prismatic faces) on the surface of marble on the 13th day of experiment (Tables 2 and 3) and multiple spherulite-like intergrowths of platelike whewellite crystals already on the 16th day. Some of the individuals reveal more pronounced prismatic faces. According to the results of XRD studies the amounts of weddellite and whewellite crystals at this stage are nearly equal (Fig. 4a).

In experiment with both bacterium *Bacillus subtilis* and micromycete *Aspergillus niger* in liquid nutrient medium there were dipyramidal–prismatic weddellite crystals (weddellite presence approved by XRD, Fig. 4b) on the surface of marble on the 13th day of experiment (same as in the experiment with fungus

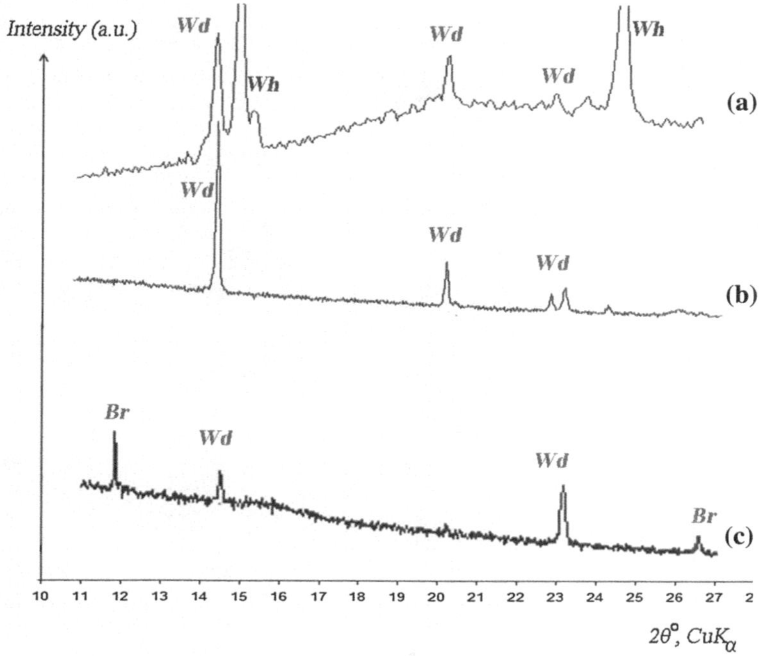


**Fig. 4** XRD pattern of calcium oxalates in experiment on marble surface in liqiud Czapek medium under the influence of: **a** fungus *Aspergillus niger* after 16 days of reaction (Rosseeva et al. 2015); **b** fungus *Aspergillus niger* and bacterium *Bacillus subtilis* after 13 days of reaction; **c** fungus *Aspergillus niger* and bacterium *Bacillus subtilis* after 60 days of reaction. Key: *Wh* whewellite, *Wed* weddelite, *Br* brushite

*Aspergillus niger* only). They are bigger in this experiment and the prismatic faces are more pronounced than dipyramidal faces (Tables 2 and 3; Fig. 5a). The crystals are surrounded by biofilm which consists of fungal hyphae and spores, bacterium cells, and metabolites. With the increase of cultivation time (up to 2 months) the size and the morphology of the crystals remained the same (Fig. 5b). There were no whewellite peaks on the XRD patterns, only weddellite and calcium phosphate (brushite) after 2 months (Fig. 4c).

## 3.2 Hydroxyapatite Transformation

### 3.2.1 The Experiments in Moist Chamber

There were small dipyramidal weddellite crystals on the surface of fossil mammoth tooth (hydroxyapatite–protein composite) formed under the influence of fungus *Aspergillus niger* after 14 days of experiment (weddellite presence approved by

**Fig. 5** SEM images of dipyramidal-prismatic tetragonal weddellite crystals on marble surface under the influence of bacterium *Bacillus subtilis* and fungus *Aspergillus niger* (liquid Czapek medium) after: **a** 13 days; **b** 2 months

XRD, Fig. 6a). Single plate-like whewellite crystals (13–15 µm) appear after 60 days, the size and shape of weddellite crystals didn't change (Tables 2 and 3).

### 3.2.2 The Experiments in Liquid Medium

In liquid Czapek–Dox medium with fungus *Aspergillus niger* there were calcium oxalate crystals after 16 days on the surface of mammoth tooth and after 18 days on the surface of hydroxyapatite powder (Figs. 6b and 7a, Table 2). According to XRD patterns there were more weddellite than whewellite crystals on the surface of mammoth tooth, and there were same quantities of both oxalates on the surface of hydroxyapatite (Fig. 7a).

The size of weddellite crystals formed on the surface of mammoth tooth in liquid medium was bigger than in moist chamber (Table 3). And even bigger crystals were

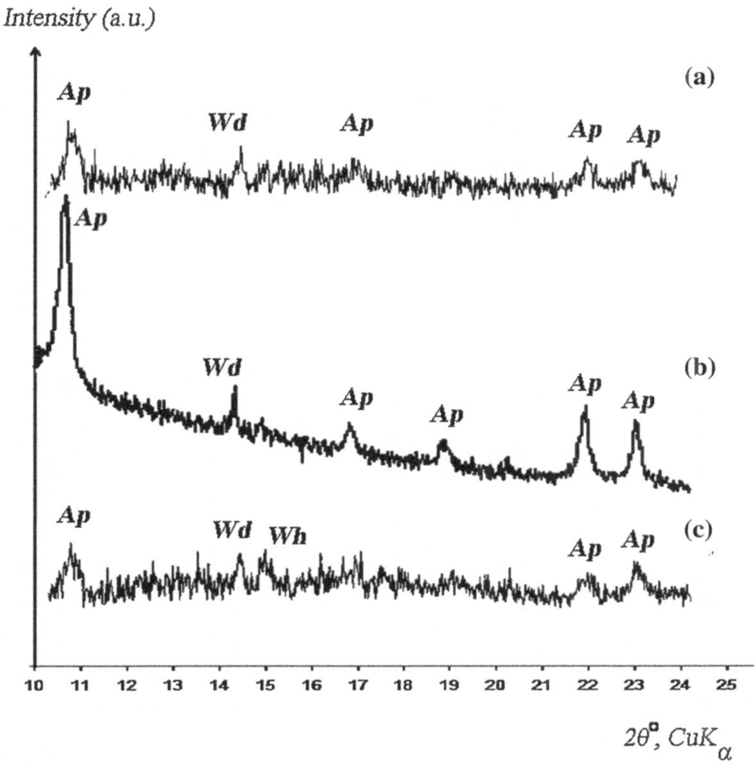


**Fig. 6** XRD pattern of crystallization products on a surface of a mammoth tooth under the influence of fungus *Aspergillus niger*: **a** in moist chamber after 60 days of reaction; **b** in liqiud Czapek medium after 16 days of reaction; **c** in liqiud Czapek medium after 60 days of reaction. Key: *Wh* whewellite, *Wed* weddelite, *Ap* apatite

found under the influence of the same fungus on the surface of synthetic hydroxyapatite powder. In both cases there were dipyramidal-prismatic crystals with small prismatic faces (Figs. 8a and 9a).

Whewellite was found as single plate-like crystals as well as their rose shaped intergrowths (15–30 μm) (Fig. 8), which tended to grow on fungal hyphae or on the dipyramidal faces of weddellite crystals.

The ratio between monohydral and dihydral oxalate quantities on mammoth tooth flattened out with time (Fig. 6c), and in case of the synthetic hydroxyapatite powder the amount of whewellite was larger than weddellite after 25 days (Fig. 7b). The size of weddellite crystals was gradually increasing with time. Their morphology changed the same way as was described in the experiment on marble. Their prismatic faces grew with time and then disappeared (Table 3; Figs. 8b and 9b). Starting from 25 days there are multiheaded crystals and crystals with marks of dissolution which gives evidence for the oversaturation of the solution. The

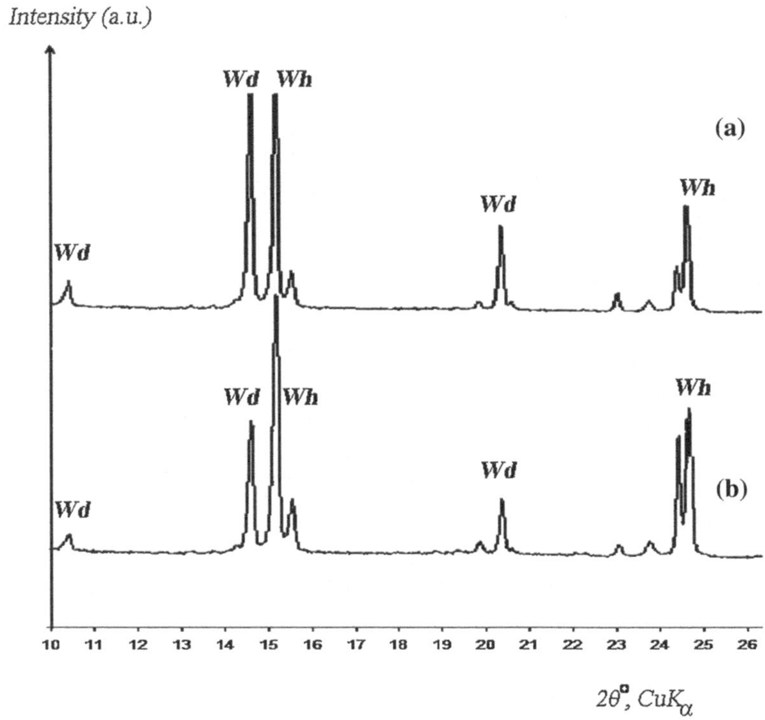


**Fig. 7** XRD pattern of crystallization products on hydroxiapatite powder (liqiud Czapek medium experiment) under the influence of fungus *Aspergillus niger*: **a** 18 days of incubation; **b** 60 days of incubation. Key: *Wh* whewellite, *Wed* weddelite

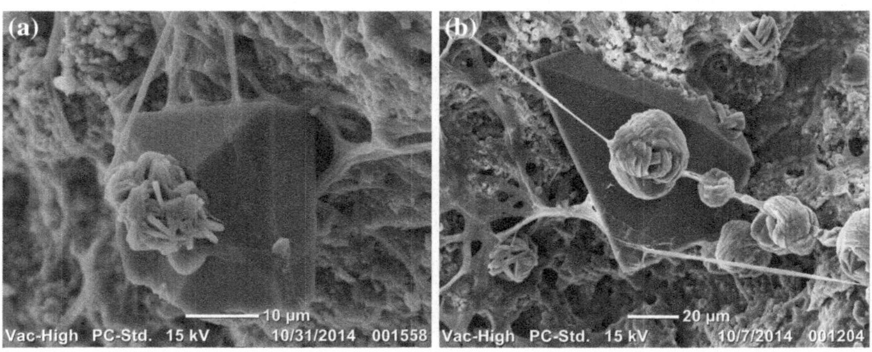


**Fig. 8** SEM images of oxalate crystals formed on the mammoth ivory surface by fungal *Aspergillus niger* treatment (liquid Czapek medium experiment): **a** whewellite rose on top of the weddellite crystal after 20 days; **b** multiple whewellite roses on fungal hyphae and on dipyramidal crystal of weddellite after 23 days

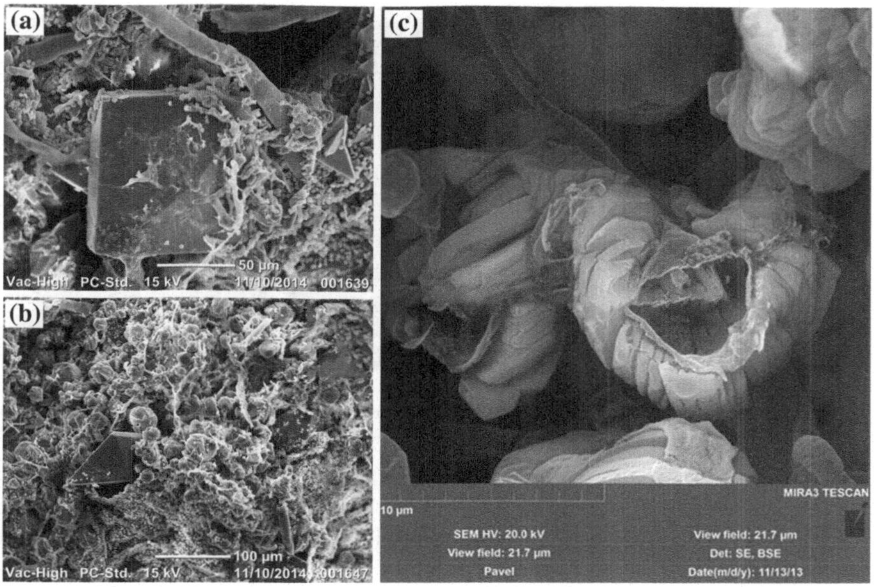


**Fig. 9** SEM images of oxalate crystals formed on synthetic hydroxyapatite by fungal *Aspergillus niger* treatment (liquid Czapek medium experiment): **a** dipyramidal-prismatic weddelite crystals after 18 days; **b** dipyramidal crystals of weddellite after 60 days; **c** crystals surrounding a fungal hyphae after six months

intergrowths of plate like crystals (30–40 µm) formed a thick coat on the surface of fungal hyphae after 6 months (Fig. 9c).

## 3.3 Feldspar Transformation

The oxalate crystallization took place only in moist chamber under the influence of bacterium *Bacillus subtilis* (Tables 2 and 3).

It's presence lead to the formation of dipyramidal oxalate dihydrate crystals after 5 months of cultivation (Tables 2 and 3), which is approved by their typical crystal morphology and EDXS patterns (Fig. 10). The weddellite appearance is far from ideal, there are multiple signs of dissolution as well as cracks (Fig. 10a).

The mix of bacterium *Bacillus subtilis* and fungus *Aspergillus niger* on the surface of granite lead to the formation of one dimensional wisp-like structures formed on fungal hyphae after 3 months of the experiment (Fig. 11a). Oxalate crystals are confined to those structures which is approved by their morphology and EDXS patterns (Fig. 11b). Multiple signs of dissolution and cracks can be seen on

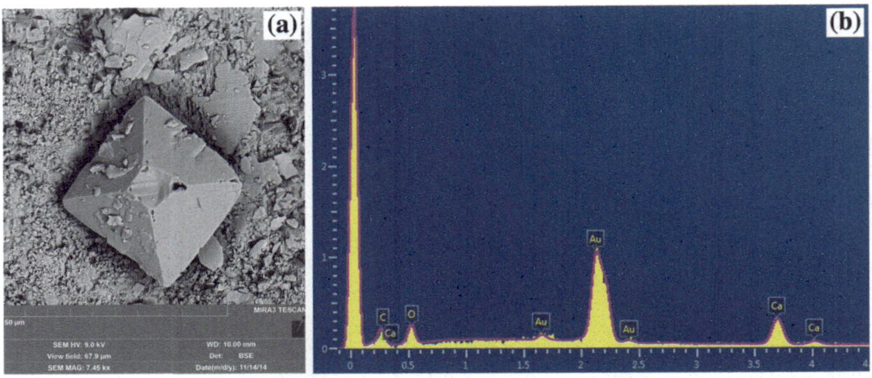


**Fig. 10** Dipyramidal tetragonal weddelite crystals formed on granite surface by bacterial *Bacillus subtilis* treatment (moist chamber experiment) after 5 months: **a** SEM image; **b** EDX spectrum of weddellite crystal

weddellite crystals. They are pierced by multiple needle-like and plate-like calcium oxalate monohydrate crystals which formed due to the dehydration.

The elemental contents of stone substratum near oxalate intergrowths is close to feldspar—acid plagioclase (Fig. 11c), which gives evidence that the oxalate formation took place due to the bacterium effect on the feldspar.

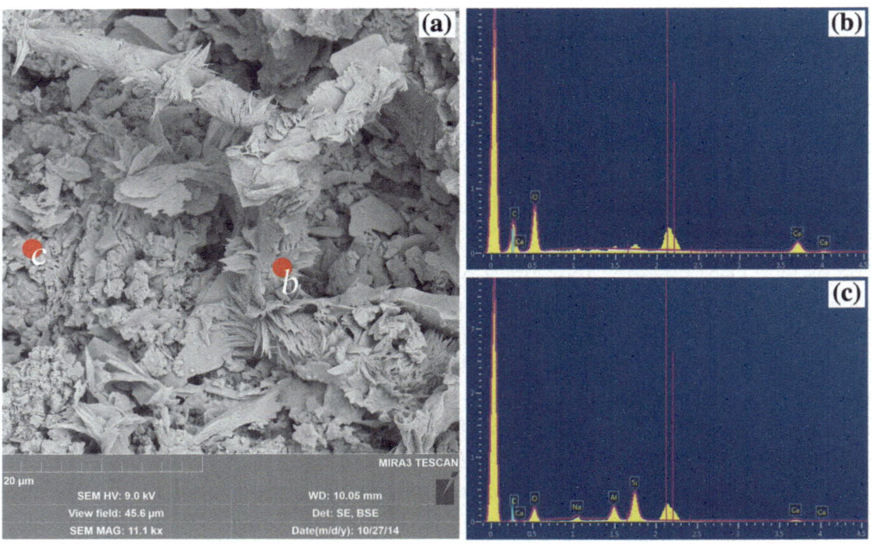


**Fig. 11** Dipyramidal tetragonal weddelite and needle-shaped whewellite crystals formed on granite surface by bacterial Bacillus subtilis and fungal *Aspergillus niger* treatment (moist chamber experiment) after 3 months: **a** SEM image (*red dots* show points where EDX spectra were measured); **b** EDX spectrum of oxalate crystals (*red dot b*); **c** EDX spectrum of granite surface near crystals (*red dot c*)

# 4 Discussion

The formation of calcium oxalates under the influence of fungi and bacterium was observed on all studied stone substrates (Table 2). Regardless of the stone substrates, metabolic activity of microorganisms and experiment conditions oxalate crystallization starts from the formation of almost ideal dipyramidal and dipyramidal–prismatic (with dominant {101} pyramidal faces) tetragonal weddellite crystals (observed on all stone substrates, Figs. 1a, 2a, 8a, 9a, and 10a). Then with the increase of $Ca^{2+}$ cation content in the crystallization medium the prismatic faces {001} of weddellite crystals became more pronounced (Figs. 1c, 2b and 5). The stabilization of metastable weddellite can be explained by the presence organic molecules in the crystallization medium which are likely to be adsorbed by calcium-bearing biominerals (Thomas 2009; Thomas et al. 2012; Izatulina et al. 2014; Rosseeva et al. 2015). Thomas et al. (2012) assumed that the elongation of the weddellite crystals along [001] is due to the interaction of organic molecules with the prismatic faces of weddellite which leads to an inhibition of the growth rate of these faces along [100]. When the saturation of the solution becomes high enough, weddellite becomes unstable, starts to dissolve and calcium oxalate monohydrate, whewellite, appears in the medium as splitted (branched) spherulite-like aggregates (Rosseeva et al. 2015, Fig. 8). Then if the crystallization takes place under the influence of oxalotrophic bacteria, including *Bacillus subtilis* according to our research, the whewellite to secondary calcite transformation starts which can lead to a formation of polycrystal pseudomorphosis of calcite over whewellite (Fig. 3). In our experiment this transformation was reproduced under the synergizing joint action of bacterium *Bacillus subtilis* and fungus *Penicillium* sp. on the calcite marble which resulted in the increase of their biochemical action on the stone substratum. With the aging of cultures the microorganism metabolite effect on the stone substratum as well as the contents of $[C_2O_4]^{2-}$ and $Ca^{2+}$ ions in the crystallization medium significantly decreased. At this stage there were almost ideal dipyramidal weddellite crystals of the new generation (Figs. 8b and 9b).

The obtained results allow to compare the role of microorganism metabolism, peculiarities of the stone substratum to the experiment conditions (the role of the environment) in calcium oxalate crystallization.

Under the influence of bacterium *Bacillus subtilis* the formation of weddellite took place only in the moist chamber. The comparison between the experiments of the fungus *Aspergillus niger* effect on marble in moist chamber and in liquid medium (Tables 2 and 3) gives evidence that the crystallization in liquid medium is faster than in the moist chamber. These examples give evidence for a significant role of the environment on the biogenic crystallization.

The role of the microorganism species and of the stone substratum peculiarities is well described by the following examples. On the surface of granite oxalates formed only in the experiments with bacterium *Bacillus subtilis*, or with bacterium *Bacillus subtilis* and fungus *Aspergillus niger*. We can propose that this effect is caused by the different dissolution rates (rates of $Ca^{2+}$ leaching) of silicate minerals, in particular calcium-bearing acid plagioclase, under the influence of the metabolites from fungus *Aspergillus niger* and bacterium *Bacillus subtilis*. The $Ca^{2+}$ and $[C_2O_4]^{2-}$ ions need to be on the surface of the stone substratum for the crystallization process to start. The fungus *Aspergillus niger* excretes oxalic acid but the $Ca^{2+}$ ions leaching does not occur so the crystallization does not start. With the appearance of bacterium *Bacillus subtilis* on the surface of granite the dissolution of granite mineral components starts, and while there is oxalic acid among their metabolites the $Ca^{2+}$ and $[C_2O_4]^{2-}$ ions appear in the surface solution and the oxalate crystallization starts (beginning with the weddellite formation). If the surface of granite is inhabited by both fungi and bacteria the concentration of oxalate ions in the surface solution increases even more which leads to weddellite dehydration and whewellite formation (Table 2; Fig. 11).

On the surface of marble calcium oxalates formed under the influence of microscopic fungi, bacteria as well as under the joint action of both. The metabolic activity of microorganisms affects the morphogenesis of the forming oxalates which can be explained by the difference in solubility of marble in different metabolites. In moist chamber the crystallization of weddellite under the influence of bacterium *Bacillus subtilis* starts later than under the influence of microscopic fungi (Table 2). As the result at the same stages of cultivation the dehydrate oxalate crystal morphology (the relative sizes of dipyramidal and prismatic faces) differs depending on the microorganism (Table 3; Fig. 2). The cooperative action of bacterium *Bacillus subtilis* and fungus *Penicillium* sp. leads to weddellite instability. Spherulite-like whewellite crystal intergrowths form as a result of dehydration and after that they transform into calcite (Fig. 3).

## 5 Conclusion

The results of simulation experiments showed that crystallization of calcium oxalate hydrates by participation of microscopic fungi and bacteria can take place on the surface of different calcium-bearing minerals: carbonates, phosphates, silicates. It was shown that it is a complicated multifactor process which is connected not only with the metabolic activity of microorganisms, but with the environment and stone substratum peculiarities. The necessary condition for the beginning of the oxalate crystallization is not only the oxalate ion presence on the surface of mineral substratum but the presence of $Ca^{2+}$ cations leached from the stone substratum. The detected general morphogenetic laws of oxalate crystallization under the influence of microorganisms are true for all stone substrates and experimental conditions.

The experiments showed that bacterium *Bacillus subtilis* act as an oxalotrophic agent. Their combined action with the microscopic fungus *Penicillium* sp. on the calcite marble leads to a substantial change in biochemical activity of the microbial community and initiates the transformation of calcite monohydrate (whewellite) into calcite.

Evidently, the metabolism of the microbial community is an important factor of secondary mineralization on the surface of rocks and minerals. The transformation of oxalates into carbonates is potentially one of the ways of the formation of massive carbonate deposits and probably regulates the atmospheric $CO_2$ as it is followed by the excretion of carbon dioxide into the atmosphere (Martin et al. 2012). Also the present results give a scientific basis for creating new methods for cultural heritage stone monument preservation with the use of modern biotechnologies.

The laboratory researches were carried out in the Research Resource Centers of Saint Petersburg State University: SEM investigations—in the center "Development of molecular and cell technologies" and XRD measurements—in the center for X-ray Diffraction Studies.

This research was partially supported by the Russian Foundation for Basic Research (Projects 13-05-00815-a) and Saint Petersburg State University (project 1.37.151.2014).

## References

Adeyemi AO, Gadd GM (2005) Fungal degradation of calcium-, lead- and silicon-bearing minerals. Biometals 18:269–281

Barinova KV, Vlasov DYu, Schiparev SM, Zelenskaya MS, Rusakov AV, Frank-Kamenetskaya OV (2010) Production of organic acids by micromycetes from the rock substrates. Mycol Phythopathology 44(2):137–142 (in Russian)

Binbin MJ, Bin L (2011) Interactions between *Bacillus mucilaginosus* and silicate minerals (weathered adamellite and feldspar): weathering rate, products, and reaction mechanisms. Chin J Geochem 30:187–192

Bonaventura MP, Gallo MD, Cacchio P, Ercole C, Lepidi A (1999) Microbial formation of oxalate films on monument surfaces: bioprotection or biodeterioration. Geomicrobiol J 16:55–64

Burford EP, Kierans M, Gadd GM (2003) Geomycology: fungi in mineral substrata. Mycologist 17:98–107

Burford EP, Hillier S, Gadd GM (2006) Biomineralization of fungal hyphae with calcite ($CaCO_3$) and calcium oxalate mono- and dihydrate in carboniferous limestone microcosms. Geomicrobiol J 23:599–611

Chen L, Xie A, Jia R, Shen Y, Tang W, Li C (2007) Influence of *Bacillus subtilis* on the growth of calcium oxalate. Cryst Res Technol 42(9):881–885

de la Torre MA, Gomez-Alarcon G, Palacios JM (1993a) "In vitro" biofilm formation by Penicillium frequentans strains on sandstone, granite, and limestone. Appl Microbiol Biotechnol 40:408–415

de la Torre MA, Gomez-Alarcon G, Vizcaino C, Garcia MT (1993b) Biochemical mechanisms of stone alteration carried out by filamentous fungi living in monuments. Biogeochemistry 19:129–147

de Oliveira PB, de la Rosa JM, Miller AZ, Saiz-Jimenez C, Gomez-Bolea A, Sequeira Braga M, A Dionısio A (2001) An integrated approach to assess the origins of black films on a granite monument. Environ Earth Sci 63: 1677–1690

Ehrlich HL (1999) Microbes as geologic agents: their role in mineral formation. Geomicrobiol J 16:135–153

Frank-Kamenetskaya OV, Vlasov DYu, Zelenskaya MS, Knauf IV, Timasheva MA (2009) Decaying of the marble and limestone monuments in the urban environment. Case studies from Saint Petersburg, Russia. Studia Universitatis Babes-Bolyai Geologia 54(2):17–22

Frank-Kamenetskaya O, Rusakov A, Barinova K, Zelenskaya M, Vlasov D (2012a) The formation of oxalate patina on the surface of carbonate rocks under influence of microorganisms. In: Broekmans MATM (ed) Proceedings of the 10th international congress of applied mineralogy (ICAM-2011). Springer, Berlin

Frank-Kamenetskaya OV, Vlasov DY, Shilova OA (2012b) Biogenic crystals genesis on a carbonate rock monument surface: the main factors and mechanisms, the development of nanotechnological ways of inhibition. In: Krivovichev S (ed) Minerals as advanced materials II. Springer, Berlin

Gadd GM (1999) Fungal production of citric and oxalic acid: importance in metal speciation, physiology and biogeochemical processes. Adv Microb Physiol 41:47–91

Gadd GM (2007) Geomycology: biogeochemical transformations of rocks, minerals, metals and radionuclides by fungi, bioweathering and bioremediation. Mycol Res 111:3–49

Girgis MGZ, Khalil HMA, Sharaf MS (2008) In vitro evaluation of rock phosphate and potassium solubilizing potential of some bacillus strains. Aust J Basic Appl Sci 2(1):68–81

Izatulina A, Gurzhiy V, Frank-Kamenetskaya OV (2014) Weddellite from renal stones: structure refinement and dependence of crystal chemical features on $H_2O$ content. Am Mineral 99:2–7

Kolo K, Keppens E, Preat A, Claeys P (2007) Experimental observations on fungal diagenesis of carbonate substrates. J Geophys Res 112:G01007

Kuis LV, Markevich RM (2008) The organic acid accumulation in the cultural liquid of bacterium of *Bacillus* genera. Trudi BGTU. Seria 4: Himia I tehnologia organucheskih veshestv 4 (1):195–198

Magnuson JK, Lasure LL (2004) Organic acid production by filamentous fungi. In: Tkacz JS, Lange L (Eds) Advances in fungal biotechnology for industry, agriculture, and medicine. Springer Science+Business Media, New York

Martin G, Guggiari M, Bravo D, Zopfi J, Cailleau G, Aragno M, Job D, Verrecchia E, Junier P (2012) Fungi, bacteria and soil pH: the oxalate-carbonate pathway as a model for metabolic interaction. Environ Microbiol 44:2960–2970

Monte M (2003a) Oxalate films formation on marble specimens caused by fungus. J Cult Heritage 4:255–258

Monte M (2003b) Biogenesis of oxalate patinas on marble specimens in fungal culture. Aerobiologia 19:271–275

Pinna D (1993) Fungal physiology and the formation of calcium oxalate films on stone monuments. Aerobiologia 9:157–167

Pinzari F, Tate J, Bicchieri M, Rhee YJ, Gadd GM (2013) Biodegradation of ivory (natural apatite): possible involvement of fungal activity in biodeterioration of the Lewis Chessmen. Environ Microbiol 15(4):1050–1062

Prieto B, Silva B (2003) Neoformed calcium minerals in granite colonized by lichens. Nova Acta Cientifica Compostelana (Bioloxia) 13:35–45

Rosseeva E, Frank-Kamenetskaya O, Vlasov D, Zelenskaya M, Sazanova K, Rusakov A, Kniep R (2015) Crystallization of calcium oxalate hydrates by interaction of calcite marble with fungus *Aspergillus Niger*. Am Mineral, 100(8)

Rousakov AV, Frank-Kamenetskaya OV, Zelenskaya MS, Vlasov DYu, Gimelbrand DE, Knauf IV, Plotkina JV (2010) The first find of calcium oxalates in biofilms on the surface of archeological monuments of Tauric Chersonesos limestone (Crimea, Ukraine). Russian Mineral Soc Notes 5:100–108 (in Russian)

Sahin N (2003) Oxalotrophic bacteria. Res Microbiol 154:399–407

Sterflinger K (2000) Fungi as geological agents. Geomicrobiol J 17:97–124

Thomas A (2009) Biomimetic growth and morphology control of calcium oxalates. Dissertation (PhD thesis), TU Dresden, Germany

Thomas A, Rosseeva E, Hochrein O, Carrillo-Cabrera W, Simon P, Duchstein P, Zahn D, Kniep R (2012) Mimicking the growth of a pathologic biomineral: shape development and structures of calcium oxalate dihydrate in the presence of polyacrylic acid. Chem Eur J 18:4000–4009

Verrecchia EP, Braissant O, Cailleae G (2006) The oxalate-carbonate pathway in soil storage: the role of fungi and oxalotrophic bacteria. In: Fungi Biogeochem Cycles, Cambridge University Press, Cambridge

Vlasov DY, Frank-Kametskaya OV (2006) Natural rock decaying in the urban environment. In: Gavrilenko VG, Panova EN (eds) Transactions of Saint Petersburg naturalist society, vol 96. St. Petersburg State University publisher, Saint-Petersburg

Zuzuk FV (2003) Urinary calculus mineralogy. Volynsk State Universit Luzk 2:507 (in Ukranian)

# Significance and Regulation of Acids Production by Rock-Inhabited Fungi

**Katerina V. Sazanova, Dmitry Yu. Vlasov, Natalia G. Osmolovskay, Sergei M. Schiparev and Alexey V. Rusakov**

**Abstract** According to the results of screening acidification activity of 24 species of micromycetes in vitro, 18 of them produced organic acids in artificial medium. The fungi from genera *Penicillium* and strains of *Aspergillus niger* have the most active production of organic acids. The ratio of acids formed by fungi on a liquid glucose-containing medium changes during ontogenesis: gluconic acid and acids of the Krebs cycle dominated in a log phase of growth; oxalic acid accumulated in a stationary phase while concentrations of other organic acids reduced. Oxalic acid is the main acid produced by fungi in a culture and apparently in a natural habitat. Formation of gluconate improves the efficiency of other carbon substrates consumption by fungi. Citric, succinic, malic, and fumaric acids are produced only at higher concentrations of carbohydrate and high ratio of C/N. Production of oxalic acid by micromycetes increases on the media containing only nitrate nitrogen. Zn in concentration up to 2 mM induces secretion of oxalic acid on nitrate media. Copper has stimulation effect on oxalic acid production only in very low concentration (25 μM). Presence of Zn in culture leads to formation of oxalate Zn crystals on the surface of mycelium. On ammonium-containing culture Zn and Cu does not stimulate oxalic acid production, but induce significant morphological changes.

**Keywords** Organic acids · Fungi · Rock substrate · Oxalates · Adaptation · Trophic conditions · Heavy metals

K.V. Sazanova (✉) · D.Yu.Vlasov · N.G. Osmolovskay · S.M. Schiparev · A.V. Rusakov
Saint Petersburg State University, St. Petersburg, Russia
e-mail: KSazanova@binran.ru

D.Yu.Vlasov
e-mail: dmitry.vlasov@mail.ru

A.V. Rusakov
e-mail: alex.v.rusakov@gmail.com

K.V. Sazanova · D.Yu.Vlasov
Komarov Botanical Research Institute, St. Petersburg, Russia

 
O.V. Frank-Kamenetskaya et al. (eds.), *Biogenic—Abiogenic Interactions in Natural and Anthropogenic Systems*, Lecture Notes in Earth System Sciences,
DOI 10.1007/978-3-319-24987-2_29

# 1 Introduction

Organic acid production by fungi, to a great extent, determines the geochemical role of micromycetes and their importance in rock weathering and primary soil formation. The growth of rock-inhabited fungi usually occurs in biofilms, including autotrophic and heterotrophic organisms (Dutton and Evans 1996; Gadd 1999; Grbic et al. 2010). Changing the availability of individual elements or being directly involved in the trophic chain, organic acids effect on the biotic interactions in microbial communities. Despite the attention to the relationship of rock-inhabited fungi with minerals, many ecological and physiological aspects of acid-producing activities of micromycetes on rocks substrate, until recently remained little studied.

A lot of species of fungi from different taxonomic and ecological groups can produce organic acids. Acidification has been described for filamentous microscopic fungi of the genera *Acremonium, Acrodontium, Aphanocladium, Aspergillus, Aureobasidium, Botrytis, Cladosporium, Fusarium, Hormonema, Oidiodendron, Paecilomyces, Penicillium, Trichoderma, Verticillium* (Braams 1992; Carlile et al. 2001; Magnuson and Lasure 2004; Papagianni 2007), and for the yeast *Candida, Hansenula, Pichia, Debariomyces, Torula, Torulopsis, Kloekera, Saccharomyces, Zygosaccharomyces* и *Yarrowia* (Papagianni 2007). Some species of *Zygomycota* produce mainly lactic acid (Braams 1992; Carlile et al. 2001; Magnuson and Lasure 2004; Papagianni 2007). Traditionally *Aspergillus niger* is the main object for study of organic acids production by fungi. But the data about organic acids production by other fungi are few and often contradictory.

The ***aim*** of this study is to establish the main regularities of organic acids formation by microscopic fungi and identify the functional role of acidification in different growth conditions of fungi.

# 2 Methods

In experimental study, 24 fungal species were used: *A. niger* Tiegh., *A. ustus* (Bain.) Thom., *A. versicolor* (Vuill.) Tirab., *Aureobasidium pullulans* (de Bary) G. Arnaud., *Botrytis cinerea* (De Bary) Whetzel; *Cladosporium sphaerospermum* Penz., *Fusarium oxysporum* Schlecht., *Geomyces pannorum* (Link) Sigler and J.W. Carmich., *Mucor heterosporum* Wehmer., *Penicillium brevicompactum* Dierckx., *P. chrysogenum* Thom., *P. citrinum* Thom., *P. funiculosum* Thom., *P. griseo-purpureum* Dierckx., *P. olivaceum* Biourg, *P. olivino-viride* Biourg, *P. oxalicum* Currie et Thom., *P. raperi* G Sm., *P. spinulosum* Thom., *P. vitale* Pidopl. et Bilai, *Rhodotorula colostri* (T. Castelli) Lodder, *Scopulariopsis brevicaulis* (Sacc.), *Trichoderma viride* Pers., and *Ulocladium chartarum* (Preuss.). *A. niger* was represented in the twelve strains isolated in different regions (Antarctica, Egypt, Sevastopol, and St. Petersburg) and from various substrates.

Cultivation of fungi was carried out on agar and liquid nutrient media. Czapec medium ($NaNO_3$—3.0; glucose—30.0; $KH_2PO_4$—1.0; $MgSO_4 \times 7\ H_2O$—0.5; KCl —0.5; $FeSO_4 \times 7\ H_2O$—0.015 g/l) and Rollen medium ($NH_4NO_3$—3.0; glucose—50.0; $KH_2PO_4$—1.0; $MgSO_4 \times 7\ H_2O$—0.5; KCl—0.5; $FeSO_4 \times 7\ H_2O$—0.015 g/l) were used as basic nutrient media. In the experiments of mineral nitrogen sources influence on acid formation by fungi ammonium nitrogen source—$NH_4Cl$ (30 g/l) also was used. To study the organic acid production by fungi under influence of zinc and copper, $ZnSO_4 \times 7H_2O$ and $CuSO_4 \times 5H_2O$ were added in concentration from 25 μM to 2 mM/l.

Cultures micromorphology was studied using Axio Scope A1 light microscope, as well as Tescan MIRA3 LMU scanning electronic microscope with a R-BSE detector and Advanced Aztec Energy (IE350)/X-max80 energy-dispersive microanalysis system. Zn and Cu concentrations in fungi mycelium were measured for 10-day fungi cultures, grown in liquid media with pH 5.6, by MGA-915MD Zeeman atomic absorption spectrometer with electro thermal atomization. The determination of phase composition of crystallization products was carried out using Rigaku powder diffract meter (CuKα radiation). The measurement of glucose concentration in cultural liquid was performed using a standard set «GlucoStar» on a spectrophotometer SF-46.

To determine the total content of organic acids in the cultural fluid, test samples were treated with 0.1 M HCl up to pH 1.0 in order to dissolve oxalic acid salts. In case of cultivation on solid media, agar was preliminary and dissolved in hot water in the ratio of 1 ml of agar: 10 ml of water then treated with 0.1 M HCl and filtered. To determine soluble oxalate forms exclusively, no treating of media with HCl was done. Further, an aliquot of a filtrate was passed through cationic exchanger (KU—2–8) to get free acids from their salts.

In order to eliminate excessive sugars, the samples after cationic exchanger were passed through anionic exchanger (AN-2FN), the bound acid anions were replaced with 0.2 M NaOH and these samples were passed through the cationic exchanger again. The obtained water solution of organic acids were evaporated; dry residue was dissolved in pyridine. TMS-derivatives of organic acids were obtained using N, O-bis-(trymethylsilyl) trifluoroacetamide (BSTFA).

Carbon acids extracted from the cultural fluid were analyzed using chromatography and mass spectrometry (GC-MS) on Agilent device with MSD5975 mass selective detector, column HP-5MS, 30 m × 0.25 mm. Chromatography was carried out with linear temperature programming from 70 to 320° at a speed of 4 °C/min. Data were collected using Agilent ChemStation software. Mass spectrometric information was processed and interpreted using AMDIS program (http://www.amdis.net/index.html) and standard NIST2005 library. Quantitative interpretation of chromatograms was carried out with hydrocarbon using UniChrom program http://www.unichrom.com/unichrome.shtml.

Statistical analysis was performed using programs Origin Pro and Microsoft Excel.

## 3 Results

According to the results of acidification activity, screening of 24 species of fungi, the most active acid production was found by *A. niger* and all fungi from genera *Penicillium*. A weak production of organic acids was typical for *Aspergillus terreus, A. versicolor, Aureobasidium pullulans, Geomyces pannorum, Rhodotorula colostri, Scopulariopsis brevicaulis,* and *Trichoderma viride*. The investigated strains of *Aspergillus ustus, Cladosporium sphaerospermum, Fusarium oxysporum, Mucor heterosporum* и *Ulocladium chartarum* did not produce organic acids. Oxalic acid was dominated among the acids produced by fungi (Fig. 1). All 18 species of fungi producing organic acids excreted oxalates. Several species produced are also gluconic, citric, malic, succinic, and fumaric acids. In the cultural fluid of some micromycetes, glyceric and threonic acids were found.

Organic acid production activity often varies among different strains of fungi belonging to one species (Ali et al. 2002). We analyzed an acid production activity of twelve strains of *A. niger*, isolated from different habitats. The results of this study indicate that the acid formation intensity by fungi is independent on the geographical location where strain was isolated and reduced during storage in a subculture in vitro (Table 1). Among the strains isolated at approximately the same time, the greatest ability to secrete oxalic acid had the strains isolated from plaster, marble, and gneiss.

Composition of organic acids in the cultural fluid of fungi was changed during the growth (Table 2). Gluconic acid was formed predominantly on initial stages of ontogenesis. On the third day of growth, *A. niger* and *P. citrinum* 10–12 % of glucose was converted to gluconic acid. On the seventh day in cultural fluid of *P. citrinum,* gluconate amount was 24 % of the initial amount of glucose in the medium, while in the culture of *A. niger* it reduced to 2 %. At the tenth day of cultivation, amount of gluconic acid in the cultural fluid decreased in both species of fungi.

The greatest amount of citric, malic, fumaric, and succinic acids were detected at the end of the log-phase of mycelium growth. Oxalic acid accumulated in the

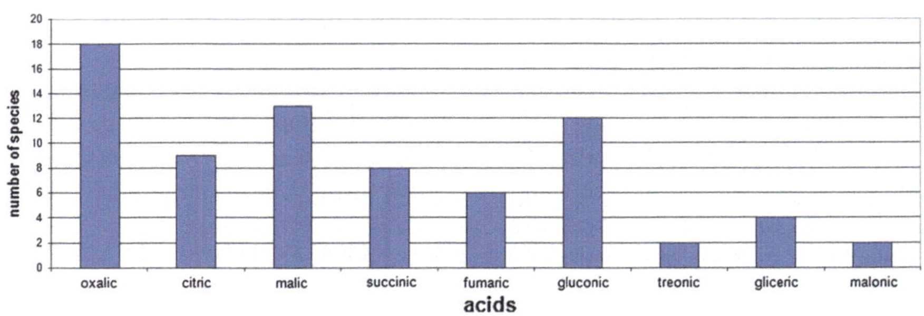


**Fig. 1** Organic acids produced by micromycetas

**Table 1** Oxalic acid production by different strains of *Aspergillus niger* (10-th days of cultivation)

| Strain number | Region | Substrate | Year of isolation | Oxalic acid (mg/g of mycelium) |
|---|---|---|---|---|
| Ch 4/07 | Sevastopol | Marble | 2007 | 246.3 ± 28.1 |
| S 1/11 | Saint Petersburg | Plaster | 2011 | 512.5 ± 78.6 |
| P 2/08 | Saint Petersburg | Cardboard | 2008 | 122.6 ± 125.0 |
| A 1/10 | Antarctica | Rock substrate | 2010 | 201.5 ± 38.0 |
| A 2/12 | Antarctica | Gneiss | 2012 | 255.2 ± 34.1 |
| E 1/04 | Egypt | Marble | 2004 | 140.2 ± 12.4 |
| S 2/12 | Saint Petersburg | Plaster | 2012 | 516.0 ± 39.1 |
| P 1/12 | Saint Petersburg | Paper | 2012 | 140.4 ± 83.1 |
| B 1/10 | Saint Petersburg | Wood | 2010 | 143.8 ± 23.1 |
| Bm-53 | Saint Petersburg | Paper | 2010 | 133.3 ± 32.8 |

**Table 2** Organic acids produced by fungi *A. niger* and *P. citrinum* during ontogenesis

| Day of growth | Organic acids, mg/g of mycelium | | | | | |
|---|---|---|---|---|---|---|
| | Gluconic | Citric | Malic | Fumaric | Succinic | Oxalic |
| *A. niger* Ch 4/07 | | | | | | |
| 3 | 167.5 ± 18.2 | 2.1 ± 0.5 | – | – | – | 40.7 ± 6.6 |
| 7 | 37.2 ± 8.2 | 15.4 ± 6.5 | 0.1 ± 0.2 | 0.1 ± 0.3 | 0.1 ± 0.3 | 344.4 ± 40.4 |
| 10 | 9.8 ± 2.5 | 5.0 ± 0.7 | – | – | – | 380.6 ± 35.2 |
| *P. citrinum* L 4/09 | | | | | | |
| 3 | 123.1 ± 12.9 | 0.7 ± 0.2 | – | – | – | 8.4 ± 0.5 |
| 7 | 498.0 ± 106.1 | 12.7 ± 0.4 | 5.8 ± 1.1 | 8.6 ± 2.0 | 8.4 ± 1.3 | 56.0 ± 8.7 |
| 10 | 125.4 ± 24.1 | 2.8 ± 0.4 | 0.7 ± 2 | 1.1 ± 0.3 | 4.0 ± 1.2 | 86.4 ± 18.1 |

medium in the stationary phase of growth, while the amount of other organic acids was decreased.

The intensity of glucose consumption and growth rates varied in different species of fungi. The fungi *A. niger* and *P. citrinum* grew more active and more quickly consumed the glucose than the nonacid-producing *U. chartarum* (Fig. 2). The coefficient of efficiency use of carbon substrate ($Y_{Glc}$ = biomass, g/substrate consumed, g) was the highest for the most active gluconate produced strain *P. citrinum*, slightly lower for *A. niger*, and the lowest for *U. chartarum*, which did not produce the acids (sixth day of growth: $Y_{Glc} = 5$ for *P. citrinum*, $Y_{Glc} = 1.2$ for *A. niger*, and $Y_{Glc} = 0.3$ for *U. chartarum*).

Gluconic acid can easily be consumed by fungi as single carbon sources (Elshafei 1989; Sazanova et al. 2014). Based on these data, we suggested that intensive oxidation of sugars into acids, especially gluconic acid, promotes their assimilation and therefore the rapid growth of mycelium.

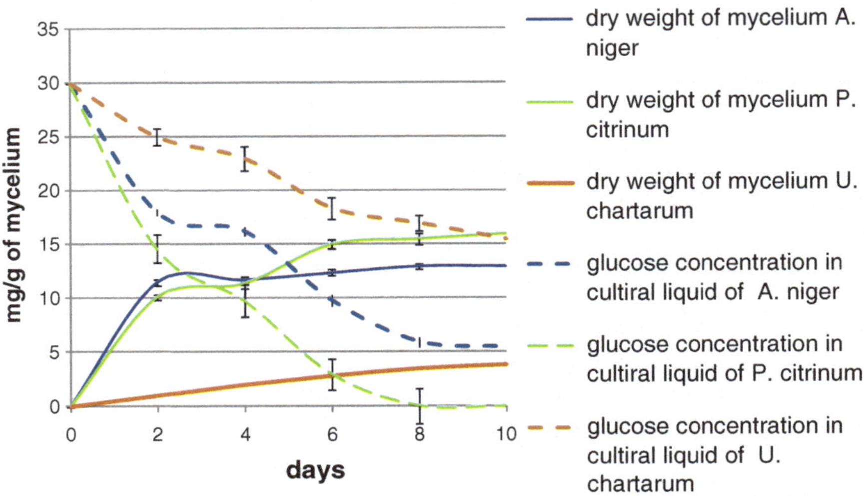


**Fig. 2** Dynamic of growth of mycelium and glucose consumption during ontogenesis of *A. niger*, *P citrinum* and *U. chartarum*

Carbon and nitrogen are the main trophic elements for fungi. Fungi can produce organic acids during the growth on monosaccharides and disaccharides, sugar alcohols and salts of other acids (Sazanova et al. 2014). In addition to the carbon source, nitrogen is very important for the development and metabolism of fungi. To investigate the effect of nitrogen source on growth and acid production by micromycetes nutrient medium containing $NaNO_3$, $NH_4NO_3$, or $NH_4Cl$ were used. Medium differed also by the amount of sucrose (30 and 50 g/l). Thus, total five variants of the media differing by nitrogen sources, and the ratio C/N have been used: $NaNO_3$ (sucrose 30 and 50 g/l), $NH_4NO_3$ (sucrose 30 and 50 g/l), or $NH_4Cl$ (sucrose 30 g/l).

The growth rate of *P. citrinum* on media containing sucrose in concentration 50 g/l has been the fastest on $NH_4NO_3$ in comparison with $NaNO_3$ or $NH_4Cl$ (Fig. 3). In the case of limited supply (30 g/l) of carbohydrates, the most suitable conditions for growth have been on the medium containing nitrate as a nitrogen source as compared with the $NH_4Cl$ form and $NH_4NO_3$.

Citric, succinic, fumaric, and malic acids have been produced only at high sugar concentration (50 g/l). Nitrate nitrogen source has been favorable to their production (Fig. 4). Amount of oxalic acid also was significantly higher on nitrate medium but less depended on sugars concentration in comparison the production of the Krebs cycle acids (Fig. 5).

Among the environmental factors that interact with fungi, as in their natural habitat, as well as in urban areas heavy metals are one of the most significant. In low concentrations, many metals stimulate growth of fungi as essential microelements being involved in various metabolic processes. However, high concentrations of metals may cause cell membrane damage, induce lipid peroxidation, and formation of reactive oxygen forms, suppress respiration, modify activity of certain enzyme, and damage DNA and proteins (Munir et al. 2005).

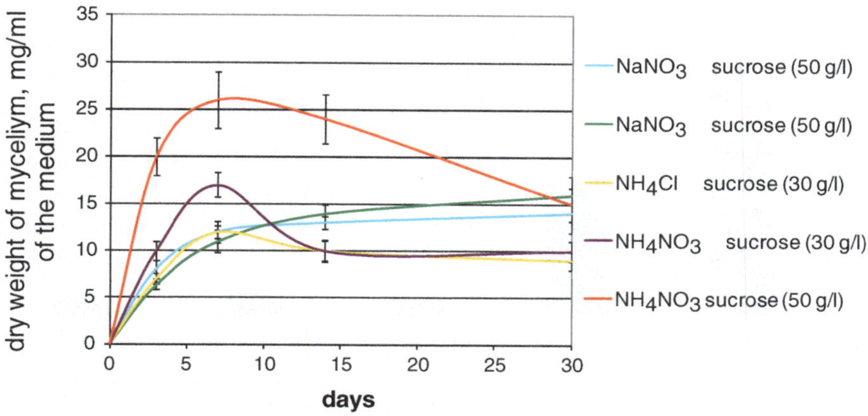


**Fig. 3** Growth of *Penicillium citrinum* on different nutrient media

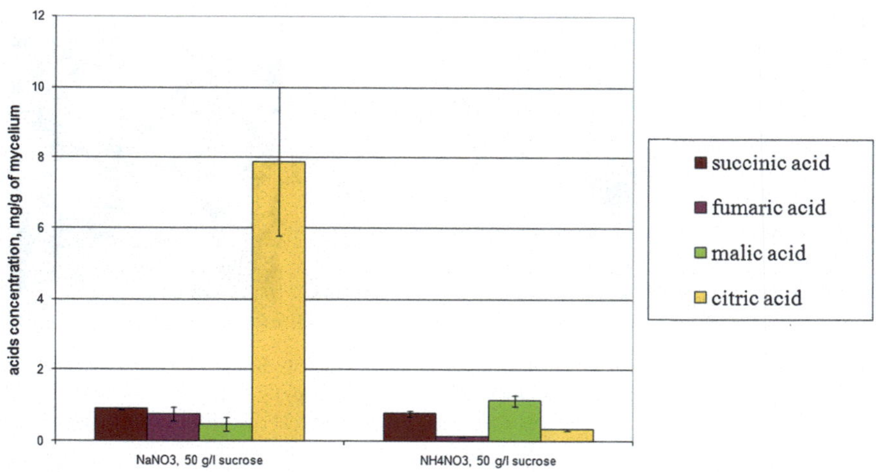


**Fig. 4** Citric, succinic, fumaric and malic acids production by *Penicillium citrinum* on agar-medium with different nitrogen sources

In our experiments, Zn and Cu have had different influence depending on nutrient medium and concentration. On Czapek medium, these metals in concentrations from 25 μM to 2 mM had no significant effect on micromorphology of mycelium. But on Rollen medium, addition of Zn and Cu resulted to formation of «swelled» cells, often constituting a main part of mycellium (Fig. 6).

Under influence of Zn in all concentration (25 μm–2 mM) on nitrate medium, the fungi *A. niger* and *P. citrinum* produced more oxalic acid than in control medium. Maximum of oxalic acid (five times higher than in control) was detected under 0.5 mM of Zn. At the same time, the formation of succinic acid and citric

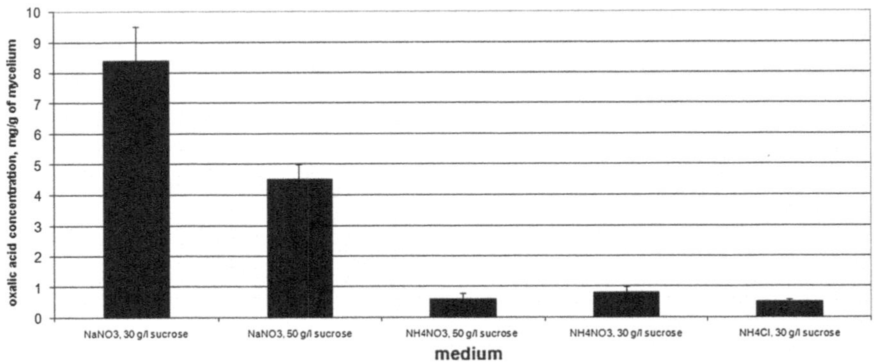


**Fig. 5** Oxalic acid production by *Penicillium citrinum* on agar-medium with different nitrogen and sugar content

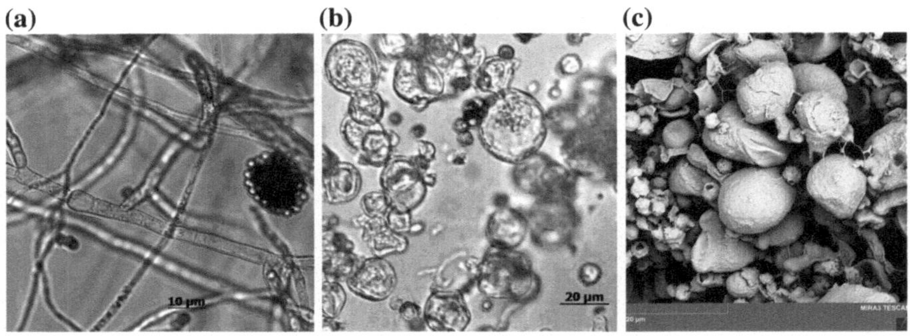


**Fig. 6** Mycelium of *A. niger* on Rollen medium: **a** control; **b** under Zn-influence (light microscopy); **b** under Zn-influence (scanning electron microscopy)

acid, which took place in the control culture of the *A. niger* and *P. citrinum*, in the presence of Zn is completely suppressed. On ammonium nitrate medium (Rollen) zinc did not have significant effect of production of oxalic acid and other organic acids.

Copper have had a little stimulation effect on oxalic acid production only in low concentration (25 µM) and only in Czapek medium. At higher concentrations (up to 0.5 mM) on both media, oxalic acid content in the culture fluid of fungi significantly reduced relative to control cultures (Table 3). But during one year of subculturing on Cu-containing medium (from 25 to 500 µM concentrations) *P. citrinum*, was obtained a Cu-resistant culture, the growth of which was not inhibited by copper in concentration of 500 µM. An oxalic acid produced by this strain was higher by more than two times than the original cultures and was 55.8 ± 6.30 mg/g of mycelium in control and 62.8 ± 5.7 mg/g of mycelium in the presence of 0.5 mM Cu.

By X-ray phase analysis, it was established that during cultivation *A. niger* and *P. citrinum* in the presence of Zn, zinc-containing crystals were formed on fungal hyphae and in a medium (Fig. 7). In *A. niger* these crystals were identified as dehydrate zinc oxalates ($ZnC_2O_4 \times 2H_2O$). Crystals formed on *P. citrinum* mycelium also had a plate crystal habit, but formed other splices. On media with copper, no Cu-containing crystals were identified for *A. niger* and *P. citrinum*.

Analysis of Zn and Cu content in mycelium of fungi showed that both metals were accumulated in the mycelium. The concentration of Zn was much higher in mycelium grown on a nitrate medium as compared to nitrate and ammonium (for *A. niger*-3 times and for *P. citrinum*-2.4 times), while for Cu an opposite dependence was observed.

## 4 Discussion

The data demonstrate that the main acids produced by fungi are gluconic, malic, succinic, citric, fumaric, and oxalic acids. In the cultural fluid of some micromycetes glyceric and threonic acids were found. The appearance of these acids may result from peroxidation of erythro-ascorbic acids to oxalic and glyceric acids, or by oxidation of ascorbic acid to form oxalic acid and threonic acids (Chupakhina 1997; Khan 1995). However, in our experiments, the presence of glyceric and threonic acids in the cultural fluid did not change constantly and the relationship between the intensity of their formation and oxalic acid formation did not observe. The greatest amount of glyceric acid was present in the cultures of psychrophilic fungi *Geomyces pannorum* and *Rhodotorula colostri*. But the amount of oxalic acid produced by these fungi has been low.

Unlike other acids, gluconic acid is synthesized mainly extracellular (Soldatenkov 1971; Magnuson and Lasure 2004; Ramachandran et al. 2006). According to our data, *A. niger* and *P. citrinum* are converted to gluconic acid up to 25 % from initial glucose content. The activity of glucose oxidation to gluconate, apparently promotes to more efficient use of carbon substrate and more active growth of fungi. It can be also assumed that gluconic acid formed by one species of fungi may be consumed as carbon source by other species of fungi. It is also likely that the gluconic acid forms complexes with sugars, which are preferably used by growing mycelium as a carbon substrate. Complexes, acids with sugars, are synthesized in vitro in the twentieth century and later they were isolated from germinating seeds of wheat, maize, and legumes (Soldatenkov 1971). Carbohydrate component in these complexes were glucose and acidic–gluconic, citric or malic acid. The physiological significance of these complexes in plants and fungi are still unknown. It is possible that they are more effectively used by these organisms due to lower energy costs for its absorption and transport.

The cause of excretion by fungi and the acid intermediates of Krebs cycle in vitro is probably an excess concentration of sugars in the nutrient medium. Apparently, for the production of citric, malic, fumaric, and succinic acids were not so important

**Table 3** Influence of Zn and Cu on organic acids production by *P. citrinum* on liquid nutrient media with different nitrogen sources (10-th days of growth)

| Nitrogen sources | Added metal | Accumulation of metals in mycelium mg/g of mycelium | Final pH | Organic acids, mg/g mycelium | | | | |
|---|---|---|---|---|---|---|---|---|
| | | | | Oxalic | Citric | Succinic | Malic | Gluconic |
| $NaNO_3$ | No (control) | – | 6.95 ± 0.11 | 24.95 ± 5.74 | 5.4 ± 1.15 | 0.69 ± 0.10 | – | 5.14 ± 0.95 |
| | $ZnSO_4$ 2 mM | 4265 ± 51 | 7.09 ± 0.09 | 36.90 ± 3.72 | – | – | – | – |
| | $CuSO_4$ 0.5 mM | 24 ± 4 | 7.28 ± 0.08 | 10.56 ± 2.30 | 1.43 ± 0.45 | – | 0.21 ± 0.02 | 21.95 ± 4.31 |
| $NH_4NO_3$ | No (control) | – | 2.13 ± 0.26 | 0.16 ± 0.03 | 0.86 ± 0.09 | 0.84 ± 0.08 | 0.18 ± 0.05 | – |
| | $ZnSO_4$ 2 mM | 1740 ± 390 | 2.10 ± 0.22 | 0.12 ± 0.02 | 0.42 ± 0.12 | 0.04 ± 0.01 | 0.14 ± 0.01 | – |
| | $CuSO_4$ 0.5 mM | 884 ± 334 | 2.29 ± 0.13 | – | 0.26 ± 0.04 | – | 0.31 ± 0.11 | – |

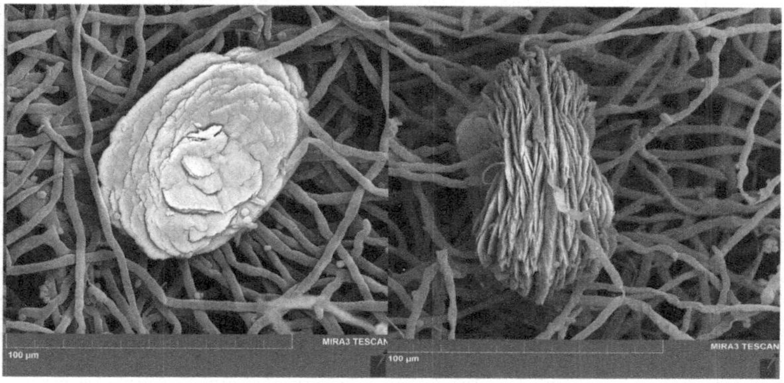


**Fig. 7** The crystals of $ZnC_2O_4 \times 2H_2O$ in cultures of *A. niger*

form of nitrogen as the ratio of carbon to nitrogen. Their production by stone-inhibited fungi in nature is unlikely. High C/N ratio favors their production. It is possible that in condition of an excess of sugar in the medium, these organic acids cannot be fully utilized for the synthesis of other compounds, e.g., amino acids are secreted into the medium. One of the probable causes for the low production of these acids on ammonium in compared to nitrate nitrogen source, most likely due to the great need of carbon for disposal of toxic ammonia. It should however be kept in mind that the total concentration of nitrogen in $NH_4NO_3$ in the standard Rollen medium should be twice higher than that in the standard Czapek (nitrogen) medium.

Production of oxalic acid is less dependent on the concentration of sugars than the formation of the Krebs cycle acids, but significantly reduced in condition of growth on ammonium and ammonium nitrate nitrogen source when compared to nitrate. Stimulation of oxalic acid production by nitrate is explained with increase of intracellular pH during $NO_3^-$ assimilation that activates carboxylating enzymes and oxaloacetate formation as an oxalate precursor in fungi and plants (Felle 2005).

Oxalic acid production is stimulated by $Ca^{2-}$ and $CO_3^{2-}$ (Braams 1992; Barinova et al. 2010). According to our data about strain differences of *A. niger*, isolated from varies substrates, it may be believed that oxalic acid production by fungi is a physiologically determined adaptation to inhabiting on substrates with a high concentration of calcium.

Oxalic acid production also is stimulated by stress factors such as higher content of zinc and copper in the medium. Zinc on a nitrate medium has in general less inhibiting effect than copper. Formation of swelled cells on a nitrate-ammonium medium is probably induced by metal ions accumulation inside the mycelium. This assumption is confirmed by the high concentration of copper in the mycelium of *P. citrinum* on the nitrate-ammonium medium. Formation of oxalic acid on the nitrate media is induced by zinc and leads to formation of insoluble zinc oxalates on the surface of mycelium, which decreases Zn toxicity. Oxalic acid production by Cu-adapted culture of *P. citrinum* was more intensive in comparison with the initial strain, but it was not stimulated with presence of Cu. Obviously, in this case, oxalate is produced to bind excessive Cu amounts, but is still not a primary factor, which determines culture resistance to copper. Thus, under influence of Zn and production of oxalic acid, Cu increase only under conditions enabling its formation, i.e., nitrate nitrogen source.

In addition to cultivation conditions and chemical stressors, physical influence also can change the products of organic acids by fungi. Ultraviolet radiation (UV) in sublethal doses leads to increase of oxalate production by fungi (Sazanova and Kirtcideli 2014). There is evidence that the treatment of plant tissues by oxalic acid maintains the stability of membranes, increases the activity of antioxidant enzymes (superoxide dismutase, peroxidase, and ascorbate oxidase), reduces the activity of lipoxygenase, inhibits the formation of reactive oxygen species and reduces lipid peroxidation (Zheng et al. 2007). In this case, increase of oxalate production under the UV-radiation can be regarded as an element of protection of the membrane against damaging influence.

Our experimental data are summarized in a scheme of organic acid's metabolism regulation in fungi (Fig. 8). Reactions of organic acid's biosynthesis are presented according to the literature (Ma et al. 1981; Kubicek et al. 1988; Munir et al. 2001; Magnuson and Lasure 2004; Papagianni 2007; Pel et al. 2007).

In practical terms, production of organic acids by fungi is regarded as one of the most important factors in the destruction of monuments and building. At the same time, in response to number of ecological factors, biodeterioration acidification activity of fungi changes and is often worse. Ability to overproduction of organic acids stimulated metals are often present in rock substrates and maintained with strain.

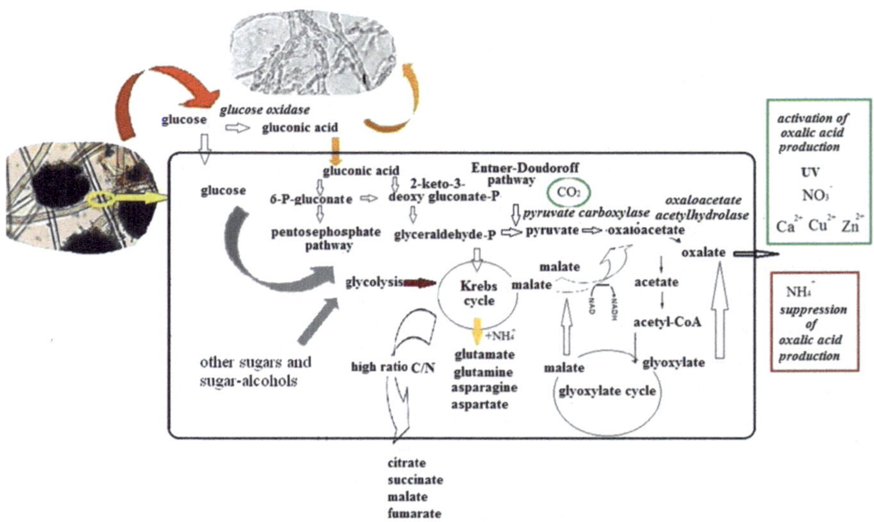


**Fig. 8** The scheme of regulation of metabolism of organic acids of fungi (detailed explanations see text)

**Acknowledgments** Biochemical analysis was carried out within the framework of the institution research projects (01201255617 and 01201255607) of Komarov Botanical Research Institute RAS.

Financial support was provided in part by Saint Petersburg State University (research grant 1.37.151.2014) and Russian Foundation for Basic Research (research grants 13.04.00843; 14-04-01795; 13-05-00815). The research was supported by research resource center «Molecular and cell technologies» of St. Petersburg State University.

## References

Ali S, Haq I, Iqbal J (2002) The role of $Mn^{++}$ ions for high and consistent yield of citric acid in recycling fed-batch bioreactor system and its novelty on kinetic basis. Electronic J Biotechnol 5 (2): http://www.ejbiotechnology.info/index.php/ejbiotechnology/article/view/v5n2-3/958

Barinova KV, Schiparev SM, Shavarda AL, Vlasov DYu (2010) Effect of calcium carbonate on acidification activity of micromycetes. Bull St. Petersburg State Univ 3(3):93–98 (in Russian)

Braams J (1992) Ecological studies on the fungal microflora inhabiting historical sandstone monuments. Dissertation, University of Oldenburg

Carlile MJ, Watkinson SC, Gooday GW (2001) The Fungi. Academic Press, San Diego

Chupakhina GN (1997) System ascorbic acid plants. Publishing House of the Kaliningrad State University, Kaliningrad (In Russian)

Dutton MV, Evans CS (1996) Oxalate production by fungi: its role in pathogenicity and ecology in the soil environment. Can J Microbiol 42:881–895

Elshafei AM (1989) Degradation of some sugars and sugar acids by the nonphosphorylated D-gluconate pathway in *Aspergillus ustus*. Acta Biotechnol 9(5):485–489

Felle HH (2005) pH regulation in anoxic plants. Ann Bot 96:519–532

Gadd GM (1999) Fungal production of citric and oxalic acid: importance in metal speciation, physiology and biogeochemical processes. Adv Microb Physiol 41:47–91

Grbic MVL, Vukojevic J, Gordana C, Simić S, Krizmanić J, Stupar M (2010) Biofilm forming cyanobacteria, algae and fungi on two historic monuments in Belgrade, Serbia. Arch Biol Sci Belgrade 62(3):625–631

Khan SR (1995) Calcium oxalate in biological systems. CRC Press, Chicago

Kubicek CP, Schreferl-Kunar G, Wöhrer W, Röhr M (1988) Evidence for a cytoplasmic pathway of oxalate biosynthesis in *Aspergillus niger*. Appl Environ Microbiol 54:633–637

Ma H, Kubicek CP, Rohr M (1981) Malate dehydrogenase isoenzymes from *Aspergillus niger*. FEMS Microbiol Lett 12:147–151

Magnuson JK, Lasure LL (2004) Organic acid production by filamentous fungi. Advances in fungal biotechnology for industry, agriculture, and medicine, pp 307–340

Munir E, Yoon JJ, Tokimatsu T, Hattori T, Shimada M (2001) New role for glyoxylate cycle enzymes in wood-rotting basidiomycetes in relation to biosynthesis of oxalic acid. J Wood Sci 47:368–373

Munir E, Hattori T, Shimada M (2005) Role for oxalate acid biosynthesis in growth of copper tolerant wood-rotting and pathogenic fungi under environmental stress. The 55th meeting of the Japan wood research society, pp 1–7

Papagianni M (2007) Advances in citric acid fermentation by *Aspergillus niger*: Biochemical aspects, membrane transport and modeling. Biotechnol Adv 25:244–263

Pel JH, de Winde JH, Archer DB et al (2007) Genome sequencing and analysis of the versatile cell factory *Aspergillus niger* CBS 513.88. Nat Biotechnol 25(2):221–231

Ramachandran S, Fontanille P, Pandey A, Larroche C (2006) Gluconic acid: a review. Food Technol Biotechnol 44(2):185–195

Sazanova KV, Kirtsideli IYu (2014) Effect of ultraviolet radiation on microscopic fungi isolated from Antarctic habitats. Mycol Phytopathol 48(5):315–321 (in Russian)

Sazanova KV, Schiparev SM, Vlasov DYu (2014) Formation of organic acids by fungi isolated from the surface of stone monuments. Microbiology 83(5):1–9
Soldatenkov SV (1971) Metabolism of organic acids in plants. Nauka, Leningrad (in Russian)
Zheng X, Tian S, Meng X, Li B (2007) Physiological and biochemical responses in peach fruit to oxalic acid treatment during storage at room temperature. Food Chem 104:156–162

# Development of Rock-Inhabiting Microfungi on Artificial (Synthetic) Marble Sculptures in the Summer Garden (St. Petersburg)

**Irina Yu. Kirtsideli, Anna V. Kazanova, Pavel A. Lazarev and Tatijana V. Pashkovskaya**

**Abstract** During the reconstruction of the Summer Garden from 2009 to 2011, all marble statues were moved to the exhibition halls of a Russian museum. Replicas of monuments from artificial marble were installed in the Summer Garden. Artificial marble consists of marble chips and polyester resin. It is known that microorganisms can degrade natural marble, but the influence of microfungi on artificial marble has been less studied. This chapter examines the dynamics of microfungi on different stone materials (granite, artificial marble and natural marble) in local ecosystems of the Summer Garden. We found that the artificial marble was covered by microfungi despite the presence of toxic resins in its composition. During the three time periods (spring–summer–autumn) in the study, 45 microfungi species from 21 genera and 3 subdivisions were identified. Of these, 16 typical rock-inhabiting species were observed on the surface of the statues and pedestals. More than 50 % of the species have dark pigmented mycelium or spores. Microfungi species composition varied at all rocky substrates. The greatest similarity was observed between the artificial and natural marble, and the smallest similarity was found between the artificial marble and granite. Mycobiota had a lower species diversity, whereas natural marble had the highest species diversity. The dynamics of microfungi on artificial marble surfaces were studied, and increases of species diversity were noted during the spring–summer season. There were no changes in the dominant species.

**Keywords** Microfungi · Artificial marble · Stone monuments · Microbial damage

The damage and demolition of marble sculptures is a real problem for many museums of the world. The Summer Gardens was laid out in the city of Saint Petersburg by the personal order of Peter I. According to his decree, a collection of

I.Yu.Kirtsideli (✉)
Komarov Botanical Institute, St. Petersburg, Russia
e-mail: microfungi@mail.ru

A.V. Kazanova · P.A. Lazarev · T.V. Pashkovskaya
The Russian Museum, St. Petersburg, Russia

 
O.V. Frank-Kamenetskaya et al. (eds.), *Biogenic—Abiogenic Interactions in Natural and Anthropogenic Systems*, Lecture Notes in Earth System Sciences,
DOI 10.1007/978-3-319-24987-2_30

marble sculptures from Italy was delivered for the first Russian urban park. A number of these works were purchased from the Venetian masters of late seventeenth and early eighteenth centuries. During the long period of time since the reign of Peter I up to today, many things have changed in the Summer Gardens. Nevertheless, not all the relics have been saved. By the beginning of the twenty first century, there were 92 exhibits of the sculpture and natural marble in the Summer Gardens in Saint Petersburg.

Extreme climatic conditions and the influence of aggressive biogenic and technology-related factors on the sculptures of the Summer Gardens have resulted in the destruction of the surface of exhibits, such as the destruction of polishing. Splits and destruction of the external layers of the marble have appeared. In addition to the physical and chemical erosion, numerous restoration interventions, acts of vandalism, and active development of microorganisms have negatively affected the preservation of the relics. The destruction of the material is due to both mechanical influence and exoenzymes and organic acids as a result of the vital activity of the microscopic fungi. Fairly recently, uncoordinated efforts to research the fungi populating the stone have received a common name of *geomichology* (Vlasov 2008).

One of the solutions to this problem is the substitution of the authentic marble sculptures with replicas. With an aim of preserving the authentic works during the reconstruction of the Summer Gardens in 2009–2011, all of the marble sculptures, marble busts, some of the marble pedestals, and four sculptural groups were substituted with replicas of artificial marble. The replicas were made by filling the silicon form with a mass of polyester resin and 80 % of the marble aggregate. This material imitates natural marble and possesses good physical properties for exhibits in the open air. In comparison with natural marble, the artificial material is more resistant to the influence of aggressive factors and possesses the best corrosion-protection characteristics. Dust is easy to remove from the surface of exhibits made from artificial marble with the help of polyester resin.

For the removal of such fouling, the Russian museum service conducted the following acts: weekly visual inspection of the sculpture in the Summer Gardens, spot removal, and, if required, washing the surfaces of exhibits in early May and late October. Washing a sculpture was done with a ready-made aqueous solution of Russian production. The aqueous solution of a biosurfactant substance has a weakly acidic reaction (pH 6), hydrogen peroxide, quaternary ammonium combinations, and is active in terms of bacteria, viruses, and fungi. The composition is applied to a somewhat damp surface with brushes and sits for 15 min. After that, it is washed with running water with the help of brushes.

Currently, 115 artificially marbled sculptures, the real marble sculptural group "World and Abundance," and pedestals made of granite and marble are on display in the Summer Gardens. Microorganisms have been long known to be able to destruct the real marble. At the same time, research on fungi destructors of the artificial marble demonstrates much less activity.

This chapter examines the dynamics of the development of microscopic fungi on artificial marble in different ecological conditions for exhibiting statues in the Summer Gardens.

# 1 Materials and Methods

The age of the artificial marble monuments at the beginning of the experiment was less than 15 days. Visually, the microbe colonization was not marked. A total of 20 probes of artificial marble, 4 probes of real marble, and 6 probes of granite were taken from the fixed parts of monuments during the open display of the statues in May, July, and September 2012–2014 on days with no major rainfall.

The probes from the surfaces were taken using a scrape and washout by a wad of cotton wool on a Petri dish with Czapek agar and malt agar (with the addition of antibiotics for suppression of bacteria growth), as well as by the reprint method. Next, there was a specific identification and calculation of colony-making units of fungi (CUF). Standard static programs and coefficients were used for the analysis of the data (Megarran 1992).

# 2 Results and Discussion

While researching artificial marble on open display, it was noted that the development of microfungi on the surface of the material often starts with the formation of microcolonies in the hollows, resulting from beads of air during the process of moulding statues from artificial marble (Fig. 1). In these areas, a single CFU of microfungi and microcolonies is able to remain for a long time. Apparently, a slight whitewash of the surface of sculptures made from artificial marble does not result the annihilation of propagula in lacunas; therefore, a source of infection takes place and results in the quicker reduction of mycelium on the surface of artificial marble. In addition, a microclimate could form in these cavities, which, in turn, promotes the faster growth and conservation of microscopic fungi.

Overall, on the surfaces of statues and pedestals, 45 species from 21 genera were found. The distribution of species according to substance are shown in Table 1. The specific content differs among all the stony substances. The majority of species were marked for genus *Penicillium*. This genus is traditionally represented by a large number of species when analyzing the fouling of various substances in the moderate zone. Nevertheless, it played a rather small part.

The results in Table 1 agree with the reported data about the species of stony substances (Sterflinger and Prillinger 2001; Ruibal et al. 2005; Zelenskaya and Vlasov 2006; Gorbushina et al. 2002; Gorbushina 2007; Vlasov et al. 2004, 2006; Gueidan et al. 2008; Vlasov 2008, 2014). We found 27 species of microfungi on the artificial marble, 29 species on the natural marble, and 13 species on the granite. Thus, the largest species variety was confirmed on the natural marble and the least on the granite, with artificial marble falling in the middle. Interestingly, more than 50 % of the species singled out have dark-pigmented mycelium and spores (shown in bold in Table 1).

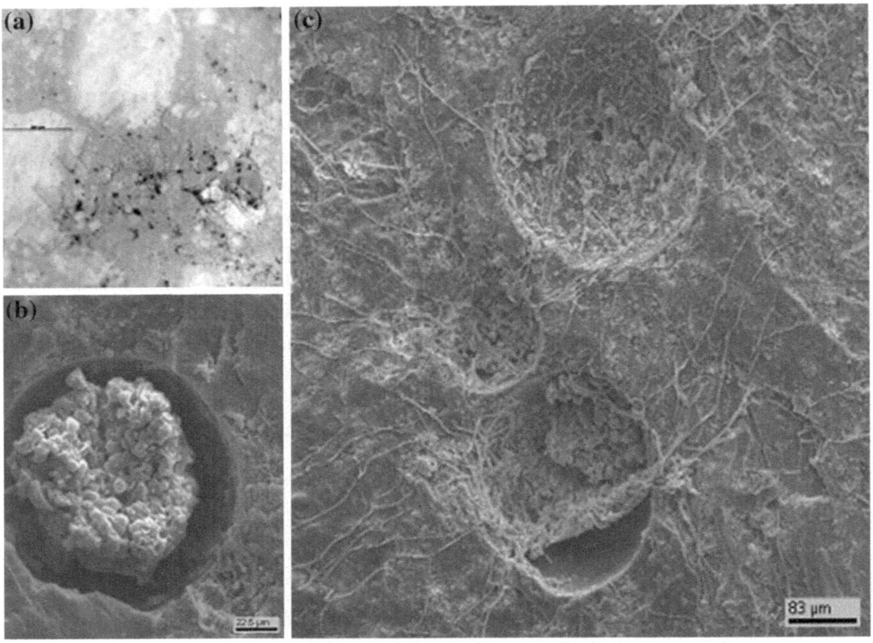


**Fig. 1** Development of microfungi on the surface of artificial marble. **a** Magnification of 500×, where the spots of the marble aggregate and the development of microfungi both on the surface as well as on the polyester resin are visible. **b–c** Development of colonies of microfungi in the hollows and their expansion on the surface of artificial marble

As mentioned before, the dynamics of the development of microfungi in the process of settling the surfaces of artificial marble have been investigated. Figure 2 shows the dynamics of the species diversity. In the first year of exhibition at the end of May, 10 species of microfungi were noted; in July, 13 species were found; and at the end of October, there were 17 species. Hence, during the summer period, the number of species in the complexes of micromycetes growing on the surface of the artificial marble increased by more than 1.5 times. At the same time, there was no change of the dominant species complex. The increase of the common specific diversity took place on the statues made from artificial marble during the three years. This can be explained by the gradual settlement by the new substance of microfungi from the outer environment (airborne fungi, microfungi settling on the surface of sculptures from soils and plant). The reason for the decrease of specific diversity after winter is connected with the statue processing (including biocide) before their conservation for the winter period.

Most similarity of the specific content is found between the natural and artificial marble, whereas the least similarity is found between the artificial marble and granite. The similarity of Sorensen/Jaccard's base on the analysis of the species list is represented in Table 2. However, using the similarity coefficients that include not

**Table 1** List of fungal species isolated from different stones

| | Artificial marble | Natural marble | Granite |
|---|---|---|---|
| 1. *Acremonium charticola* (Lindau) W. Gams | + | + | |
| 2. *Acremonium* sp. | + | | |
| 3. *Alternaria alternata* (Fr.) Keissl. | | + | |
| 4. *Alternaria tenuissima* (Kunze) Wiltshire | + | | |
| 5. *Alternaria* sp. | + | + | |
| 6. *Aspergillus niger* Tiegh. | + | | |
| 7. *Aspergillus terreus* Thom | + | | |
| 8. *Aspergillus waksmanii* Hubka, S.W. Peterson, Frisvad and M. Kolařík | | + | |
| 9. *Aspergillus* sp. | | | + |
| 10. *Aureobasidium pullulans* (de Bary and Löwenthal) G. Arnaud | + | + | + |
| 11. *Aureobasidium* sp. | + | + | + |
| 12. *Boeremia exigua* (Desm.) Aveskamp, 1. Gruyter and Verkley | | + | + |
| 13. *Chaetomium globosum* Kunze | + | | + |
| 14. *Cladosporium cladosporioides* (Fresen.) G.A. de Vries | + | + | + |
| 15. *Cladosporium herbarum* (Pers.) Link | + | + | + |
| 16. *Coniothyrium* sp. | | + | |
| 17. *Cryptococcus* sp. | | + | |
| 18. *Exophiala jeanselmei* (Langeron) McGinnis and A.A. Padhye | + | + | + |
| 19. *Exophialia* sp. | + | + | |
| 20. *Fusarium anthophilum* (A. Braun) Wollenw | | + | |
| 21. *Fusarium* sp. | | + | |
| 22. *Humicola grisea* Traaen | + | + | |
| 23. *Mucor hiemalis* Wehmer | + | | |
| 24. *Mucor* sp. | + | | |
| 25. *Penicillium aurantiogriseum* Dierckx | + | + | |
| 26. *Penicillium brevicompactum* Dierckx | | + | |
| 27. *Penicillium canescens* Sopp | + | | |
| 28. *Penicillium citrinum* Thom | | + | |
| 29. *Penicillium chrysogenum* Thom | | | + |
| 30. *Penicillium diversum* Raper and Fennell | + | | |
| 31. *Penicillium hirsutum* Dierckx | | | + |
| 32. *Penicillium thomii* Maire | + | | |
| 33. *Penicillium* sp. | + | + | + |
| 34. *Phialophora* sp. | | + | |

(continued)

**Table 1** (continued)

| | Artificial marble | Natural marble | Granite |
|---|---|---|---|
| 35. *Phialemonium atrogriseum* (Panas.) Dania García, Perdomo, Gené, Cano and Guarro | | + | |
| 36. *Phoma glomerata* (Corda) Wollenw. and Hochapfel | | + | |
| 37. *Phoma* sp. | + | | |
| 38. *Rhodotorula* sp. | + | | |
| 39. *Stachybotrys chartarum* (Ehrenb.) S. Hughes | + | + | |
| 40. *Torula herbarum.* (Pers.) Link | | + | + |
| 41. *Torula sp.* | + | + | |
| 42. *Trichoderma viride* Pers | | + | |
| 43. *Trichoderma* sp. | + | | |
| 44. *Ulocladium chartarum* E.G. Simmons | + | + | + |
| 45. *Ulocladium consortiale* (Thüm.) E.G. Simmons | | + | |

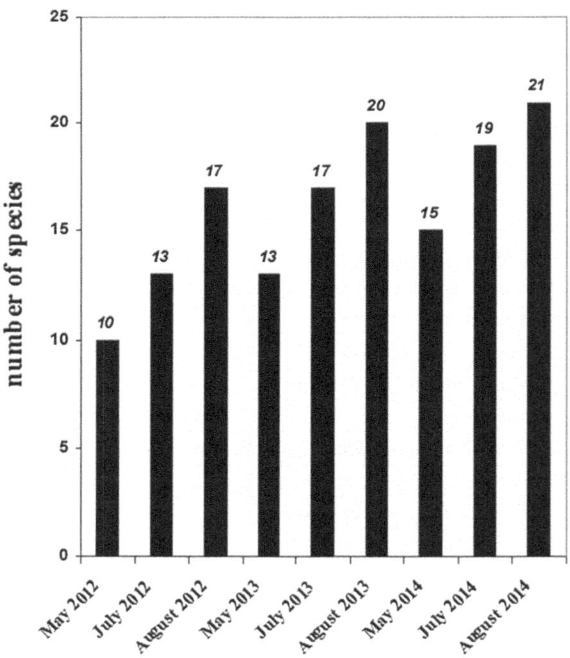


**Fig. 2** The dynamics of species diversity

only the list of species but also the CFU of species, these coefficients sharply increase. Thus, it is shown that the complexes of microscopic fungi growing on the surface of statues from artificial and natural marble are similar and different in single (random) species.

**Table 2** Coefficients of similarity to microfungi species composition on the different substrates

| | Artificial marble | Natural marble | Granite |
|---|---|---|---|
| Artificial marble | + | 0.33[b] | 0.25[b] |
| Natural marble | 0.50[a] | + | 0.20[b] |
| Granite | 0.40[a] | 0.33[a] | + |

[a]The coefficient of Sorensen
[b]The coefficient of Jaccard

At the same time, domineering and superdomineering species were noted on the surfaces of artificial marble statues at the moment of formation of stable complexes of microscopic fungi. The domineering species was *Aureobasidium pullulans* according to the CFU number, followed by *Cladosporium herbarum.* The number of all other species was minimal, although there was a variety of special diversity as indicated in Fig. 3.

Figure 4 shows the change of occurrence of the domineering species in the process of artificial marble exposure. The occurrence of microfungi *Aureobasidium pullulans* was quite high initially at 80 %; in the summer and autumn period, it

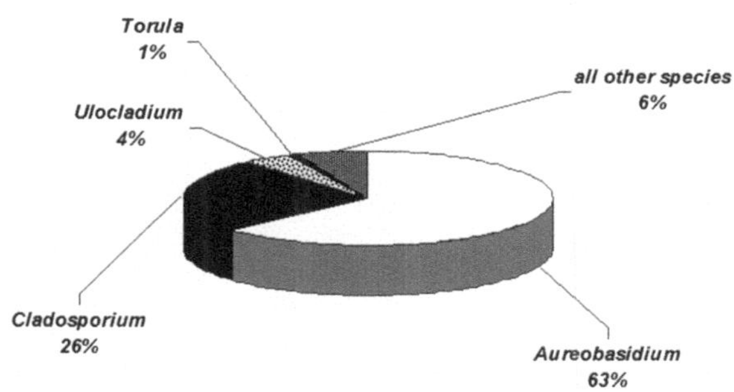


**Fig. 3** Proportion of different genera isolates

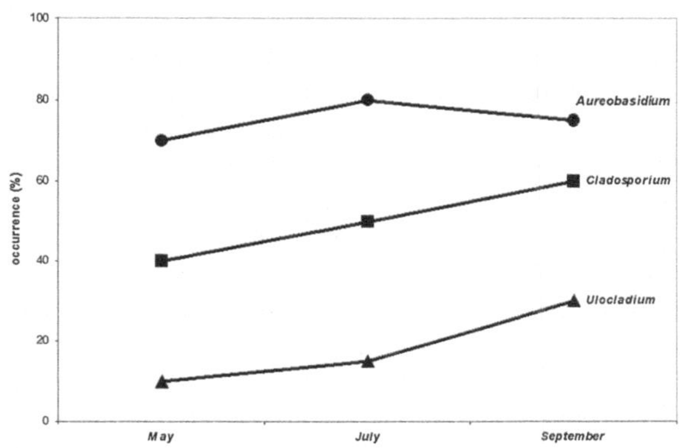


**Fig. 4** Occurrence of microfungi on the marble statues

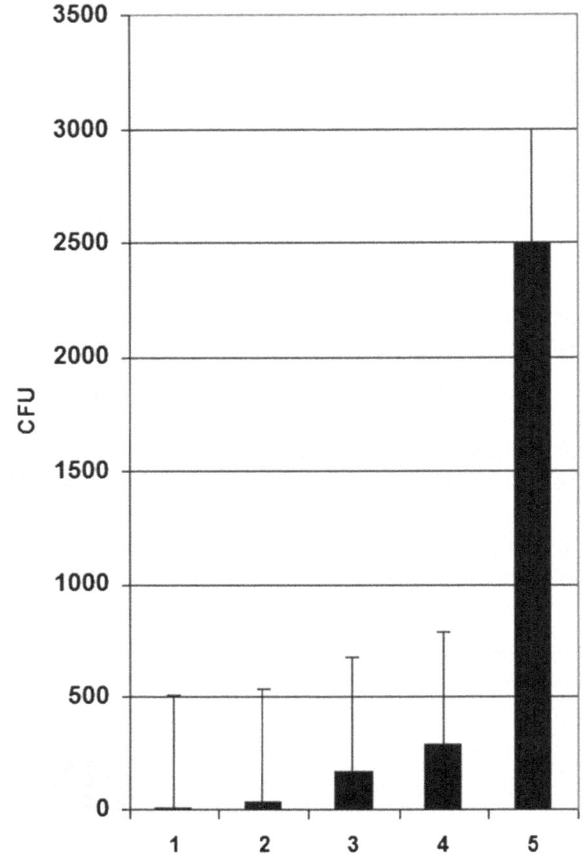


**Fig. 5** CFU number on different elements (parts) of one statue from artificial marble (*Ceres—copy*)

increased to 92 %. The occurrence of microfungi *Cladosporium herbarum* gradually increased from 55 % in spring to 73 % in early autumn. Microfungi *Ulocladium chartarum* were less in numbers, yet in autumn their occurrence grew up to 30 %.

The number of CFU on the surface of one sculpture could vary from 410 to $3.5 \times 10^4$ CFU/square sm (Fig. 5). Apparently, the probability of fixing the propagula of microfungi is less on the smooth polished surfaces, but dust bedaubing and the probability of propagula growth accumulate in the hollows of statues and fix on the surfaces. The further destruction of statues is significantly higher. In addition, the difference in the number of microscopic fungi between similar parts of statues located in different places of the garden seemed to be affected by how far the water and vegetative cover (trees) are located.

Thus, assemblages of microfungi on artificial and natural marble have a very high similarity among themselves; a narrower list of species might be determined by the initial stages of formation of the assemblages of litobiontic fungi. The monuments from artificial marble in the Summer Gardens were predominated by

*Aureobasidium*. The degree of monument colonization depends on the terms of exposition, conditions of exposition, the monument relief, and the protective service. A total of 45 species of fungi were found on the monuments from artificial marble. The majority of them are anamorphous forms of microfungi (*Ascomycota*). A group of domineering species is represented by dark-tinted anamorphous forms of fungi, which are more adapted to existing on a stony substance. Initially, on the surface of artificial marble, micromycetes are mainly fixed in fissures and fossas that are characteristic of artificial marble, which seems to depend on the quality of casting.

## References

Gorbushina AA (2007) Life in the rock. Environ Microbiol 9(7):1613–1631

Gorbushina AA, Lyalikova NN, Vlasov DY, Khizhnyk TV (2002) Microbial communities on the monuments of Moscow and St. Petersburg: biodiversity and trophic relations. Microbiology 71 (3):350–356 (in Russian)

Gueidan C, Ruibal Villaseñor C, de Hoog GS, Gorbushina AA, Untereiner WA, Lutzoni F (2008) A rock-inhabiting ancestor for mutualistic and pathogen-rich fungal lineages. Stud Mycol 61:111–119

Megarran E (1992) Ecological diversity and its measurement. Moscow: Mir (in Russian)

Ruibal C, Platas G, Bills G (2005) Isolation and characterization of melanized fungi from limestone formations in Mallorca. Mycological Progress 4(1):23–38

Sterflinger K, Prillinger H (2001) Molecular taxonomy and biodiversity of rock fungal communities in an urban environment (Vienna, Austria). Antonie Van Leeuwenhoek 80:275–286

Vlasov DY (2008) Rock-inhabiting micromycetes in communities: diversity, ecology, evolution and significance. Dissertation, University of St. Petersburg (in Russian)

Vlasov DY (2014) Modern problems of geobiology. Biogenic—abiogenic interactions in natural and anthropogenic systems, pp 32–34

Vlasov DY, Zelenskaya MS, Safronova EV (2004) Mycobiota rocky substrate in an urban environment. Mycol Phytopathol 38(4):13–22 (in Russian)

Vlasov DYu, Gorbunov GA, Krilenkov VA, Lukin VV, Safronov EV, Sienkiewicz YI (2006) Micromycetes location of the areas of the Antarctic polar stations (West Antarctica). Mycol Phytopathol 40(3):202–211 (in Russian)

Zelenskaya MS, Vlasov DYu (2006) Micromycetes on the monuments of the National Reserve “Chersonese” (Sevastopol, Crimea). Mycol Phytopathol 40(5):370–376 (in Russian)

# Lichen Diversity on Carbonate Stone Substrates in St. Petersburg, Russia: A Review

**Irina S. Stepanchikova, Oksana A. Kuznetsova, Dmitry E. Himelbrant and Ekaterina S. Kuznetsova**

**Abstract** The review of published records of lichens on carbonate stone substrates (marbles, limestones, limestone tufa, and concrete) in Saint Petersburg has been presented. Altogether, 66 species of lichens from 35 genera and 16 families have been found in more than 40 papers. The most diverse genera are *Verrucaria*, *Physcia*, and *Lecanora*. Most of the species inhabiting carbonate stone materials in Saint Petersburg form episubstratic thalli, only the species of genus *Verrucaria*, often have endolithic or mostly endolithic thalli and could cause pitting of carbonate materials. Other lichen inhabitants of these substrates could be agents of substrate piling. The majority of taxa were not reported in Saint Petersburg from natural carbonate materials (marbles, limestones, or limestone tufa), but only from concrete; species inhabiting exclusively carbonate stones in Saint Petersburg are very few; lichens strictly confined to natural carbonate substrates are almost absent.

**Keywords** Carbonate stones · Lichens · Saint Petersburg

Lichens are one of the well known and active agents in biodeterioration of stone substrates, especially carbonates (Wilson 1995; Seaward 1997; Chen et al. 2000; Nimis 2001; Clair and Seaward 2004). Their destructive influence can cause serious damage of historical and cultural heritage monuments.

Saint Petersburg is situated on the contact zone of the crystalline Baltic Shield of 1.6–1.8 billion years old and sedimentary Russian Plate. On most part of the city territory, crystalline basement (gneiss and granites) is covered by quaternary sediments 20–30 m thick in average and 100–130 m maximally. Pre-quaternary carbonate formations are extremely rarely exposed on the territory of Saint Petersburg

I.S. Stepanchikova (✉) · O.A. Kuznetsova · D.E. Himelbrant · E.S. Kuznetsova
St. Petersburg State University, Universitetskaya Emb., 7–9, 199034 St. Petersburg, Russia
e-mail: stepa_ir@mail.ru

I.S. Stepanchikova · D.E. Himelbrant · E.S. Kuznetsova
Komarov Botanical Institute, Prof. Popov Street, 2, 197376 St. Petersburg, Russia



O.V. Frank-Kamenetskaya et al. (eds.), *Biogenic—Abiogenic Interactions in Natural and Anthropogenic Systems*, Lecture Notes in Earth System Sciences,
DOI 10.1007/978-3-319-24987-2_31

for exception of small areas in river valleys in the southern part of the city (Khramtsov et al. 2013). Thus, all natural carbonate materials—marbles, limestones, and carbonate (limestone) tufa—in the city limits are incoming and anthropogenic. Moreover, all concrete constructions (alumosilicates) also could be classified as carbonate materials on the base of their surface chemical characteristics.

Saint Petersburg is one of most famous historical and cultural centers in Russia with more than 310 years of history. It is really strange but lichens on carbonate stone substrates in this megalopolis were not specially investigated. The aim of this review is to summarize all data on the lichen diversity on this type of substrates in Saint Petersburg.

The lichen investigations on the territory of Saint Petersburg and suburbs were started about 300 years ago. First, data was obtained by efforts of Johann Christian Buxbaum (1728, 1732), David de Gorter (1761), Johann Gottlieb Georgi (1790), Sobolewskj (1799, 1802), and other botanists. However, in the manuscripts of this period any information about lichens on carbonate substrates was absent.

Probably, the first records of lichens on carbonate stones in our region were published by Johann Anton Weinmann (1837). Unfortunately, there is no certain information about localities in this manuscript, so it is impossible to understand if the lichens were collected within the city limits or not. The first specific record of lichen from carbonate stones in Saint Petersburg was made by Elenkin (1922), who collected and described *Leptogium Issatschenkoi* Elenk (now, *Enchylium limosum*), from carbonate tufa in Imperial Botanical Garden (nowadays, Botanical Garden of the Komarov Botanical Institute RAS).

Many papers on Saint Petersburg lichens were published by Malysheva (1993, 1994, 1995a, b, 1997a, b, c, 1998a, b, 1999a, b, 2001a, b, c, 2002, 2003, 2004, 2005a, b, 2006a, b), Malysheva and Simachev (1994), Malysheva et al. (1995), Malysheva and Sviazeva (1995, 2009). Altogether 25 species of lichens from carbonate stones were reported by this author.

The latest lichen investigations in the limits of Saint Petersburg have been conducted by the lichenological group of Saint Petersburg State University and Komarov Botanical Institute RAS. During our inventory, 54 lichen taxa from carbonate stones were published (Himelbrant 2005; Himelbrant et al. 2006, 2007, 2013; Stepanchikova et al. 2008, 2009, 2010a, b, 2011, 2014, 2015; Stepanchikova and Himelbrant 2009; Stepanchikova and Kataeva 2010; Vondrák et al. 2010; Kuznetsova et al. 2012; Pykälä et al. 2012).

Finally, almost all data on lichen diversity from carbonate stone substrates in Saint Petersburg were obtained for the past 20 years only (since 1993). Altogether, 66 species of lichens from 35 genera and 16 families are known from carbonate stone substrates in Saint Petersburg; the records are present in more than 40 papers. The most diverse genera are *Verrucaria* (13 species), *Physcia* (5), and *Lecanora* (4). Alphabetic list of species is presented in Table 1, where nomenclature is given according to actual Scandinavian lichen checklist (Nordin et al. 2011). To reveal the most common lichen taxa in the city area, the occurrence was estimated roughly on the base of literature records. For one finding, we accepted one record of the species

**Table 1** The list of lichen species collected from carbonate stone substrates in the limits of Saint Petersburg (based on published data only)

| Species | Type of substrate | | | | |
|---|---|---|---|---|---|
| | Marbles | Limestones | Limestone tufa | Concrete | Total amount of records |
| *Acarospora glaucocarpa* (Ach.) Körb. | | | | 6 | 6 |
| *Acarospora moenium* (Vain.) Räsänen | | | | 10 | 10 |
| *Arthonia fusca* (A. Massal.) Hepp | | | | 5 | 5 |
| *Athallia holocarpa* (Hoffm.) Arup, Frödén et Søchting | | 5 | 2 | 29 | 36 |
| *Athallia pyracea* (Ach.) Arup, Frödén et Søchting | | | | 1 | 1 |
| *Bacidina arnoldiana* (Körb.) V. Wirth et Vězda | | | | 3 | 3 |
| *Bacidina chloroticula* (Nyl.) Vězda et Poelt | | | | 2 | 2 |
| *Caloplaca albolutescens* (Nyl.) H. Olivier | | | | 1 | 1 |
| *Caloplaca chlorina* (Flot.) H. Olivier | | | | 1 | 1 |
| *Caloplaca saxicola* (Hoffm.) Nordin | | | | 1 | 1 |
| *Candelariella aurella* (Hoffm.) Zahlbr. | 2 | 19 | 1 | 48 | 70 |
| *Candelariella vitellina* (Hoffm.) Müll. Arg. | | | | 2 | 2 |
| *Circinaria contorta* (Hoffm.) A. Nordin, S. Savić et Tibell | | | | 3 | 3 |
| *Clauzadea monticola* (Schaer.) Hafellner et Bellem. | | 1 | | | 1 |
| *Enchylium limosum* (Ach.) Otálora, P.M. Jørg. et Wedin | | | 2 | | 2 |
| *Flavoplaca citrina* (Hoffm.) Arup, Frödén et Søchting | | | | 7 | 7 |

(continued)

**Table 1** (continued)

| Species | Type of substrate | | | | |
|---|---|---|---|---|---|
| | Marbles | Limestones | Limestone tufa | Concrete | Total amount of records |
| *Flavoplaca dichroa* (Arup) Arup, Frödén et Søchting | | | | 1 | 1 |
| *Hypogymnia physodes* (L.) Nyl. | | | | 2 | 2 |
| *Lecania sylvestris* (Arnold) Arnold | | | | 3 | 3 |
| *Lecanora albescens* (Hoffm.) Branth et Rostr. | | 2 | 1 | 4 | 7 |
| *Lecanora crenulata* Hook. | 1 | 14 | | 20 | 35 |
| *Lecanora dispersa* (Pers.) Sommerf. | 1 | 13 | | 46 | 60 |
| *Lecanora hagenii* (Ach.) Ach. | | 3 | | 8 | 11 |
| *Lecidella stigmatea* (Ach.) Hertel et Leuckert | | | | 7 | 7 |
| *Lemmopsis arnoldiana* (Hepp) Zahlbr. | | | | 4 | 4 |
| *Lepraria incana* (L.) Ach. | | | 1 | | 1 |
| *Leproplaca chrysodeta* (Räsänen) J.R. Laundon | | | | 1 | 1 |
| *Parmelia sulcata* Taylor | | | | 1 | 1 |
| *Peltigera didactyla* (With.) J.R. Laundon | | 1 | 2 | 1 | 4 |
| *Peltigera praetextata* (Flörke ex Sommerf.) Zopf | | | | 1 | 1 |
| *Phaeophyscia nigricans* (Flörke) Moberg | | | | 9 | 9 |
| *Phaeophyscia orbicularis* (Neck.) Moberg | 1 | 6 | | 24 | 31 |
| *Phaeophyscia sciastra* (Ach.) Moberg | | | | 2 | 2 |
| *Physcia adscendens* H. Olivier | | 1 | | 10 | 11 |

(continued)

**Table 1** (continued)

| Species | Type of substrate | | | | |
|---|---|---|---|---|---|
| | Marbles | Limestones | Limestone tufa | Concrete | Total amount of records |
| *Physcia caesia* (Hoffm.) Fürnr. | | 3 | 2 | 8 | 13 |
| *Physcia dubia* (Hoffm.) Lettau | | | | 7 | 7 |
| *Physcia stellaris* (L.) Nyl. | | | | 2 | 2 |
| *Physcia tenella* (Scop.) DC. | | | | 15 | 15 |
| *Physconia detersa* (Nyl.) Poelt | | | | 1 | 1 |
| *Polycauliona phlogina* (Ach.) Arup, Frödén et Søchting | | | | 1 | 1 |
| *Polycauliona polycarpa* (Hoffm.) Frödén, Arup et Søchting | | | | 1 | 1 |
| *Porpidia crustulata* (Ach.) Hertel et Knoph | | | | 1 | 1 |
| *Protoparmeliopsis muralis* (Schreb.) M. Choisy | | 3 | 2 | 4 | 9 |
| *Psorotichia schaereri* (A. Massal.) Arnold | | | | 1 | 1 |
| *Rhizocarpon lavatum* (Fr.) Hazsl. | | | | 1 | 1 |
| *Rusavskia elegans* (Link) S.Y. Kondr. et Kärnefelt | | | | 3 | 3 |
| *Sarcogyne regularis* Körb. | | 1 | | 4 | 5 |
| *Scoliciosporum chlorococcum* (Graewe ex Stenh.) Vězda | | | | 2 | 2 |
| *Scytinium lichenoides* (L.) Otálora, P.M. Jørg. et Wedin | | | 1 | | 1 |
| *Scytinium subtile* (Schrad.) Otálora, P.M. Jørg. et Wedin | | | 1 | | 1 |
| *Verrucaria boblensis* Servít | | 3 | | 3 | 6 |

(continued)

**Table 1** (continued)

| Species | Type of substrate | | | | |
|---|---|---|---|---|---|
| | Marbles | Limestones | Limestone tufa | Concrete | Total amount of records |
| *Verrucaria cambrini* Servít | | | | 1 | 1 |
| *Verrucaria christiansenii* Servít | | 1 | | | 1 |
| *Verrucaria dolosa* Hepp | | 1 | | 4 | 5 |
| *Verrucaria elevata* (Nyl.) Zschacke | | | | 1 | 1 |
| *Verrucaria epilithea* Vain. | | | | 1 | 1 |
| *Verrucaria helsingiensis* Vain. | | | | 1 | 1 |
| *Verrucaria muralis* Ach. | 1 | 12 | 3 | 22 | 38 |
| *Verrucaria nigroumbrina* Servít | | | | 1 | 1 |
| *Verrucaria pilosoides* Servít | | 2 | | | 2 |
| *Verrucaria polystictoides* Vain. | | 1 | | | 1 |
| *Verrucaria tectorum* (A. Massal.) Körb. | | | | 1 | 1 |
| *Verrucaria tornensis* H. Magn. | | | | 1 | 1 |
| *Xanthocarpia crenulatella* (Nyl.) Frödén, Arup et Søchting | | | | 1 | 1 |
| *Xanthoparmelia conspersa* (Ach.) Hale | | | | 1 | 1 |
| *Xanthoria parietina* (L.) Th. Fr. | 1 | 4 | | 24 | 29 |

per certain type of carbonate stone substrates per paper (see Table 1). Moreover, the revision of genus *Verrucaria* in Saint Petersburg and Leningrad Region (Pykälä et al. 2012) showed that the majority of earlier records of *Verrucaria* spp. are unreliable because of the difficulties in identification of species; therefore, the *Verrucaria* records before 2012 were excluded from the analysis. Altogether, 88 records of this genus are present in literature; of them 60 are taken into account in the present paper.

Totally, 497 records of taxa were accepted from literature. The most common taxa are *Candelariella aurella* (70 records), *Lecanora dispersa* (60), *Athallia holocarpa* (36), *Lecanora crenulata* (35), *Phaeophyscia orbicularis* (31), and *Xanthoria parietina* (29). *Verrucaria* spp. are the most common in general (88 records), but it is impossible to estimate frequency of occurrence of separate species of the genus due to identification problems (see above).

All the most common taxa listed above inhabit more than one type of carbonate stone substrates (see Table and Fig. 1). All of them also show more than 50 % of occurrence on anthropogenic type of substrate (concrete). Concrete is especially important as a substrate for *Athallia holocarpa* and *Xanthoria parietina*—both species were recorded on concrete in more than 80 % of cases. It is very characteristic that among the most common taxa there are no lichens strictly or predominantly confined to natural carbonate materials.

The majority of taxa (42 species or 64 %) were not reported from natural carbonate materials (marbles, limestones, or limestone tufa), but only from concrete. However, these taxa in our opinion have ecological potential to inhabit natural substrates in Saint Petersburg. A few taxa were found only on natural materials (five species or 9.3 %) with one or two records—*Clauzadea monticola* (one record from limestone), *Enchylium limosum* (two records, limestone tufa), *Lepraria incana*, *Scytinium lichenoides*, and *S. subtile* (one record per species, all from limestone tufa). There are no species confined to marbles alone.

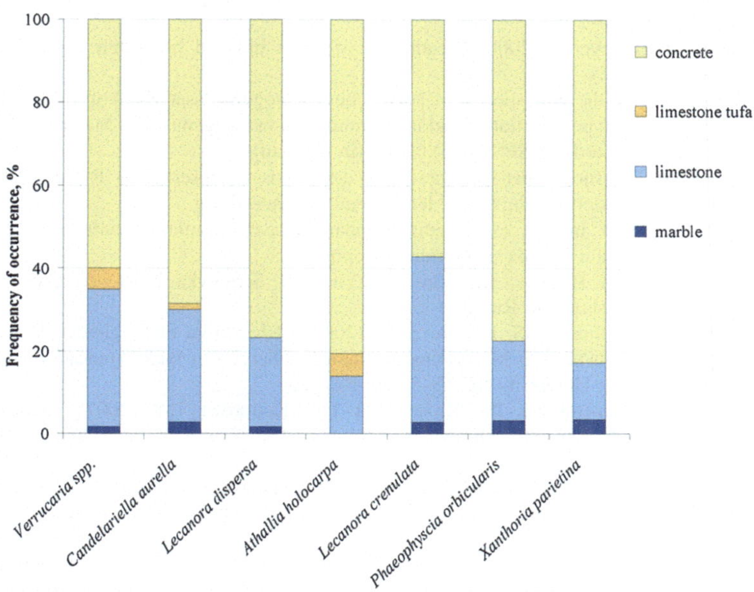


**Fig. 1** Frequency of occurrence (%) of the most common taxa on different types of carbonate stone substrates (based on published data only)

Carbonate stone substrates in the limits of Saint Petersburg make a little enrichment for city lichen diversity—13 species (2 % only from ca. 570 species known in the city) inhabit these types of substrates exclusively. Most of them have single occurrence and were reported from concrete. Other 53 species occurred on carbonate materials could inhabit at least one more group of substrates—bark of trees and shrubs, lignum, soil, siliceous stones, mica, bones, mosses, or anthropogenic materials (slate, bricks, iron, etc.).

Most of the species inhabit carbonate stone materials in Saint Petersburg form episubstratic thalli (epilithic lichens). Only the species of genus *Verrucaria* (e.g., *Verrucaria pilosoides*) often have endolithic or mostly endolithic thalli and could cause pitting of carbonate materials. Other lichen inhabitants of these substrates could be agents of substrate piling.

**Acknowledgments** The authors acknowledge Saint Petersburg State University for a research grant 1.37.151.2014, Russian Foundation for Basic Research (grant 14-04-01411) and Program of Basic Research of the Presidium of RAS «Wildlife: Current Status and Problems of Development».

## References

Buxbaum JC (1728) Plantarum minus cognitarum. Centuria II. Complectens plantas circa Byzantium et in oriente observatas. Petropoli

Buxbaum JC (1732) Observationes circa quasdam plantas Ingricas. Comment Acad Sc Petropolit III, pp 270–273

Chen J, Blume HP, Beyer L (2000) Weathering of rocks induced by lichen colonization—a review. Catena 39:121–146

Elenkin AA (1922) On the new species of jelly lichen Leptogium Issatschenkoi mihi in Central Botanical Garden and new section Pseudomallotium mihi in this genus. Bot Mater Inst Cryptog Plants Central Bot Garden RSFSR 1(5):65–69 (in Russian)

Georgi JG (1790) Versuch einer Beschreibung der russisch kaiserlichen Residenzstadt St. Petersburg und der Merkwurdigkeiten der Gegend. St. Petersburg

Gorter D (1761) Flora Ingrica ex schedis Stephani Krascheninnikow confecta et propriis observationibus aucta a Davide de Gorter. Petropoli

Himelbrant DE (2005) Lichens. In: Vlasov DYu (ed) Sergievka Park—complex natural monument. St. Petersburg (in Russian)

Himelbrant DE, Stepanchikova IS, Konoreva LA (2006) Lichen flora of Sergievka Park and its features. In: Vlasov DYu (ed) Monitoring of Nature of Sergievka Park. Transactions of Biol Inst of St. Petersburg University 52: 58–89 (in Russian)

Himelbrant DE, Stepanchikova IS, Kuznetsova ES, Konoreva LA (2007) Lichens. In: Volkova EA, Isatchenko GA, Khramtsov VN (eds) The Nature of the Elagin Island. St. Petersburg (in Russian)

Himelbrant DE, Motiejūnaitė J, Pykälä J, Schiefelbein U, Stepanchikova IS (2013) New records of lichens and allied fungi from the Leningrad Region, Russia. IV. Folia Cryptogamica Estonica 50:23–31

Khramtsov VN, Kovaleva TV, Natsvaladze NYu (eds) (2013) Atlas osobo okhranyaemykh prirodnykh territoriy Sankt-Peterburga [Atlas of the protected natural areas of Saint-Petersburg]. St. Petersburg (in Russian)

Kuznetsova ES, Motiejūnaitė J, Stepanchikova IS, Himelbrant DE, Czarnota P (2012) New records of lichens and allied fungi from the Leningrad Region, Russia. III. Folia Cryptogamica Estonica 49:31–37
Malysheva NV (1993) Lichens in vicinities of Leningrad. 1. Changes in species diversity of lichens in vicinity of Olgino station (Leningrad Region) for 72 years. Novosti Sist Nizsh Rast 29:119–124 (in Russian)
Malysheva NV (1994) The lichens of historical parks in the environs of St. Petersburg. Bot Zhurn 79(11):29–35
Malysheva NV (1995a) Lichens in arboretums of Saint Petersburg and surroundings. Bot Zhurn 80 (8):54–64 (in Russian)
Malysheva NV (1995b) Lichens of historical necropolis of Saint Petersburg. Bot Zhurn 80(10):74–78 (in Russian)
Malysheva NV (1997a) Lichens of historical gardens and parks of Saint Petersburg (founded in XVIII—beginning of XX century). Bot Zhurn 82(7):56–67 (in Russian)
Malysheva NV (1997b) Lichens of modern gardens and parks of Saint Petersburg (founded in 1920–1980). Bot Zhurn 82(8):77–84 (in Russian)
Malysheva NV (1997c) Lichens in vicinities of Saint Petersburg. 3. Specific features of lichens distribution in the Ekaterininsky Park in Tsarskoe Selo. Novosti Sist Nizsh Rast 32:58–64 (in Russian)
Malysheva NV (1998a) Lichens of the Saint Petersburg embankments. Bot Zhurn 83(2):40–47 (in Russian)
Malysheva NV (1998b) Lichens of the Petropavlovskaya Fortress Museum-Reserve territory (Saint Petersburg). Bot Zhurn 83(11):43–48 (in Russian)
Malysheva NV (1999a) Lichens of Menshikov's Homestead in Saint Petersburg and their comparison with lichens of historical territories beginning of XVIII century. Novosti Sist Nizsh Rast 33:138–142 (in Russian)
Malysheva NV (1999b) Lichens in vicinities of Saint Petersburg. 4. Present-day state and changes in species composition of lichens for 90 years in Repino settlement (former Kuokkala) and surroundings. Novosti Sist Nizsh Rast 33:142–153 (in Russian)
Malysheva NV (2001a) Lichens of the Letnij garden in Saint Petersburg and specific features of its distribution on the territory of garden. Novosti Sist Nizsh Rast 34:154–162 (in Russian)
Malysheva NV (2001b) Lichens in vicinities of Saint Petersburg. 5. Changes in species composition of lichens in the Osinovaya Grove Park for 200 years. Novosti Sist Nizsh Rast 34:162–166 (in Russian)
Malysheva NV (2001c) Lichens of gardens and cottage territories of Saint Petersburg and surroundings. Bot Zhurn 86(2):41–50 (in Russian)
Malysheva NV (2002) Lichens of the scientific town of Pulkovskaya observatory (Saint Petersburg). Novosti Sist Nizsh Rast 36:169–181 (in Russian)
Malysheva NV (2003) Lichens of Saint Petersburg. Transactions of Saint Petersburg Society of Naturalists. Series 3. 79:1–100 (in Russian)
Malysheva NV (2004) Lichens of the traffic arteries of Saint Petersburg and surroundings. Bot Zhurn 89(9):1453–1455 (in Russian)
Malysheva NV (2005a) Lichens in vicinities of Saint Petersburg: 6. Present-day state and changes in lichen flora of Dudergofskie Heights for the period of 1799–2003 years. Novosti Sist Nizsh Rast 38:226–237 (in Russian)
Malysheva NV (2005b) Lichens in vicinities of Saint Petersburg: 7. Lichens of the Zelenogorsk. Novosti Sist Nizsh Rast 39:213–219 (in Russian)
Malysheva NV (2006a) Lichens. In: Volkova EA, Isatchenko GA, Khramtsov VN (eds) Dudergofskie Heights—complex natural monument. St. Petersburg (in Russian)
Malysheva NV (2006b) Lichens on the man-made substrates in the Russian towns. Bot Zhurn 91 (11):1659–1672 (in Russian)

Malysheva NV, Simachev VI (1994) Lichens of the Botanical Garden of Saint Petersburg State University. Vestn Sankt-Peterb Univ. Series 3 (Biology) 3(17):48–52 (in Russian)

Malysheva NV, Sviazeva OA (1995) Lichens of the Park of Komarov Botanical Institute RAS (Saint Petersburg). Bot Zhurn (80)1:108–118 (in Russian)

Malysheva NV, Sviazeva OA (2009) Short-term biomonitoring of the lichen flora in the Park of Komarov Botanical Institute RAS (Saint Petersburg). Novosti Sist Nizsh Rast 43:190–198 (in Russian)

Malysheva NV, Nikolaev PM, Neshataev VYu, Reiman AL, Vlasov DYu (1995) History of creation and present-day state of Dubki Park in Sestroretsk (on the base of lichen indication, phytopathological and phytosociological analyses). Vestn Sankt-Peterb Univ. Series 3 (Biology) 2(10):46–51 (in Russian)

Nimis PL (2001) Artistic and historical monuments: threatened ecosystems. Living World 4:557–569

Nordin A, Moberg R, Tønsberg T, Vitikainen O, Dalsätt Å, Myrdal M, Snitting D, Ekman S (2011) Santesson's Checklist of Fennoscandian Lichen-forming and Lichenicolous Fungi. Ver. April 29, 2011. http://130.238.83.220/santesson/home.php. Accessed on 22 Dec 2014

Pykälä J, Stepanchikova IS, Himelbrant DE, Kuznetsova ES, Alexeeva NM (2012) The lichen genera *Thelidium* and *Verrucaria* in the Leningrad Region (Russia). Folia Cryptogamica Estonica 49:45–57

Seaward MRD (1997) Major impacts made by lichens in biodeterioration processes. Int Biodeterior Biodegradation 40(24):269–273

Sobolewskj G (1799) Flora Petropolitana. Petropoli

Sobolewskj G (1802) Sankt-Peterburgskaya flora. II [Flora of Saint Petersburg. Part II]. Saint Petersburg (in Russian)

St Clair L, Seaward M (eds) (2004) Biodeterioration of stone surfaces. Lichens and biofilms as weathering agents of rocks and cultural heritage. Springer, Netherlands

Stepanchikova IS, Himelbrant DE (2009) Lichens in vicinity of Lindenovsky Pond (Saint Petersburg). Vestn Tver Gosud Univ. Series Biologiya i Ekologiya 15(34):155–160 (in Russian)

Stepanchikova IS, Kataeva OA (2010) Lichens of Novoorlovsky Forest Park (Saint Petersburg). Vestn Tver Gosud Univ. Series Biologiya i Ekologiya 19(27):69–82 (in Russian)

Stepanchikova IS, Himelbrant DE, Konoreva LA (2008) Lichens of Severo-Primorsky Park in Saint Petersburg. Vestn Sankt-Peterb Univ. Series 3 (Biology) 3:55–66 (in Russian)

Stepanchikova IS, Himelbrant DE, Kuznetsova ES (2009) Lichens of protected area «Vicinities of the Schuche Lake» (Saint Petersburg). Vestn Tver Gosud Univ. Series Biologiya i Ekologiya 12(6): 123–139 (in Russian)

Stepanchikova IS, Himelbrant DE, Kukwa M, Kuznetsova ES (2010a) Additions to the lichen flora of the protected areas on the Gulf of Finland shores (in the limits of Saint Petersburg). Novosti Sist Nizsh Rast 44:237–244 (in Russian)

Stepanchikova IS, Himelbrant DE, Kuznetsova ES (2010b) Lichens of geological outcrops on the Popovka River (Saint Petersburg). Vestn Tver Gosud Univ. Series Biologiya i Ekologiya 18 (18):118–128 (in Russian)

Stepanchikova IS, Himelbrant DE, Kuznetsova ES (2011) Lichens. In: Volkova EA, Isatchenko GA, Khramtsov VN (eds) Nature ot Sestroretskaya Lowland. St. Petersburg (in Russian, English Summary)

Stepanchikova IS, Himelbrant DE, Konoreva LA (2014) The lichens and allied fungi of the Gladyshevsky protected area (Saint Petersburg). Novosti Sist Nizsh Rast 48:291–314

Stepanchikova IS, Himelbrant DE, Kuznetsova ES (2015) Lichens. In: Isatchenko GA, Khramtsov VN, Volkova EA (eds) Schuche Lake—Nature State Protected Area. St. Petersburg (in Russian, in print)

Vondrák J, Redchenko O, Himelbrant D, Stepanchikova I, Kuznetsova E (2010) Some sterile Caloplaca crusts identified by molecular data from the Leningrad Region (Russia). Folia Cryptogamica Estonica 47:97–99

Weinmann JA (1837) Enumeratio stirpium in agro Petropolitano sponte crescentium secundum systema sexuale linneanum. Petropoli

Wilson MJ (1995) Interactions between lichens and rocks: a review. Cryptogamic Botany 5:299–305

# Development of Stone Monuments Monitoring System Using Computer Technology

**Valeriy M. Grishkin, Stanislav B. Shigorets, Dmitry Yu. Vlasov, Elena A. Miklashevich, Alexey P. Zhabko, Alexander M. Kovshov and Alexey D. Vlasov**

**Abstract** Monitoring of the cultural heritage state is significant factor for their preservation, the timely restoration, and conservation. The aim of this work is to develop a system for monitoring cultural heritage using new computer technology. The proposed method of image comparison is based on the spectral characteristics of the studied monument. Photographing of the objects is carried out in visible and invisible spectra. Most information on the biological objects can be obtained, for example, when shooting is in the infrared range. Primary image processing, recognition of weathering features, and assessment of their potential hazards as well as evaluating of the overall state of the monument is made using complex computer programs, and the results are accumulated in a special database. This system works as the expert one. It has been shown that the possibility of classification of the main groups of organisms on monuments and assessment of the biofouling area in time is possible by using this method. This method was tested during the survey of ancient petroglyphs in Republic of Khakassia, Tomsk petroglyphs, as well as on a number of historical monuments in St. Petersburg.

**Keywords** Monitoring · Cultural heritage · Biofouling · Computer technology · Mapping

---

V.M. Grishkin (✉) · S.B. Shigorets · D. Yu. Vlasov · A.P. Zhabko · A.M. Kovshov
Saint-Petersburg State University, Saint-Petersburg, Russia
e-mail: valery-grishkin@yandex.ru

E.A. Miklashevich
Kemerovo State University, Kemerovo, Russia

A.D. Vlasov
Herzen State Pedagogical University, Saint-Petersburg, Russia



O.V. Frank-Kamenetskaya et al. (eds.), *Biogenic—Abiogenic Interactions in Natural and Anthropogenic Systems*, Lecture Notes in Earth System Sciences,
DOI 10.1007/978-3-319-24987-2_32

# 1 Introduction

Monitoring of the cultural heritage state is significant factor for their preservation, the timely restoration, and conservation. Known approaches to monitor monuments in the open air are based on the combination of visual assessments, including the method of qualimetric examination and analysis of damaged materials using complex instrumental methods (Frank-Kamenetskaya and Vlasov 2014). This approach gives a fairly complete characterization of the object, but it requires a lot of time. However, it often requires a rapid assessment of the changes that occur on the surface of the monument. Especially it concerns biofouling, which develop very quickly. It is well known that the main factors of destruction of stone monuments include external factors as well as the properties of the rock (Frank-Kamenetskaya and Vlasov 2014). Biological impact is also often seen as an important factor influencing the stone. Type of biofilm and species composition can determine the degree of impact on the monument. Biofouling can affect the durability of the stone physically and chemically. As result of the impact of living organisms can be bioweathering (biodeterioration) of rock monuments, changing their aesthetic value (Cutler and Viles 2010; Dakal and Cameotra 2012). It is important to note that all the above-mentioned factors influencing the stone are closely interrelated and are often have a synergistic nature.

Recently, a new methodological approach with the use of computer technology to investigating biofouling on stone monuments has been released (Shchigorets et al. 2009). **The aim** of this work is to develop a system for monitoring of cultural heritage using new computer technology that allows to assess the current state of the monument and the impact of environmental factors (to evaluate changes in the state of the monument in time).

**The objectives** of the research include:

- Development of a method for monitoring of stone monuments state with the use of computer technology (with using of group of indexes NDVI, ENDVI);
- Common determination (characteristic) of biofouling on the basis of their spectral characteristics;
- Mapping of the monument biofouling on the basis of typical biofilms characteristics.

# 2 Materials and Methods

The main objects of the research include:

- Monuments of Cultural Heritage (sculpture, architecture, petroglyphs, and others);
- The base material—stone (granite, limestone, sandstone, marble, etc.);

- Biofouling objects—bacteria, algae, fungi, lichens, mosses, plants, biofilm of complex composition;
- Contamination of different nature: organic, inorganic, anthropogenic;
- Mechanical damage—chipping, cracks, etc.

The proposed system is based on the idea identifying changes in the state of the monument by processing series of images obtained during the periodic observations of the monument. This approach can significantly reduce the effort required to monitoring and get information about the current state of a large number of monuments, concentrated in one place.

The first step in creating this system is the development of the program CONSOLIDATOR-BIO, which allows to assess the biological fouling of surfaces of stone monuments on the basis of the color characteristics of objects (Shchigorets et al. 2009). Later, it was shown that spectral and spatial characteristics of objects are their significant informative features (Grishkin et al. 2014). For development of stone monuments monitoring we use a group of indexes NDVI, ENDVI. These indexes have been generated as the basis for the monitoring system of cultural heritage. Testing of the system was carried out by the example of the historical monuments of St. Petersburg and Khakassia.

## 3 Results and Discussion

***The system of receiving and storing information about the monument.*** Initial information about the monument consists of the important data (location, material, historical information, etc.) and its images. These images are obtained with a digital camera. The input of the camera is all the reflected light spectrum from the monument. Outputs are digital images obtained in the visible spectrum, as well as a monument images obtained in the near infrared (NIR) and ultraviolet (UV) spectral range. For the selection of the spectral range, we use a system of filters in front of the camera, which allows to define the necessary range (RGB, NIR, or UV). Each object has its own set of characteristics. Property to absorb and reflect light rays depends on the material surface and its condition, presence of biofouling, or other contaminants.

Thus, at each photographing of the monument, there are, at least, two original images obtained: the first—normal RGB picture in the visible range, and the second—a picture in NIR or UV range. As a result of a periodic photographing of monument, series of its images accumulates. All information is recorded and stored in the database of monitoring system. The database also stores the information about the spectral characteristics of the monument materials, biological objects, and other contaminants. These characteristics are based on the well-known groups of formal indexes, such as NDVI, ENDVI, and similar, indirectly characterizing the spectral properties of biological and nonbiological materials. It should be noted that the results of image processing are also accumulate in the database.

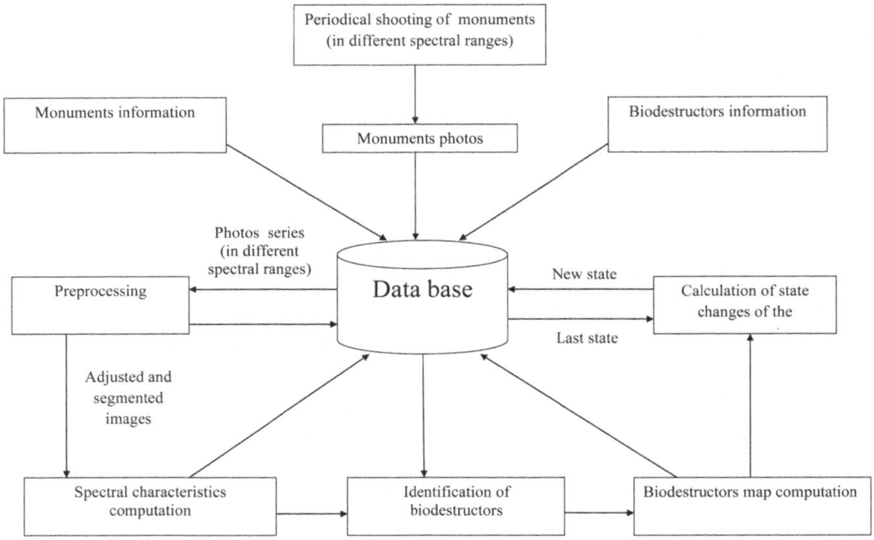


**Fig. 1** Structure of the computer monitoring system

Database and information processing system constitute a computer system monitoring. The structure of the monitoring system is shown in Fig. 1.

***Information processing system***. Based on the monument images series the computer system calculates the actual distribution (at the moment of shooting) biological and other contaminants on the surface of monument. Using previously known spectral characteristics of objects it carries out the identification of types of contamination or biofouling. As result of identification biofouling and surface contaminants the cartogram of monument could be constructed. This cartogram reflects the percentage ratio of main types of surface changes. The last stage of information processing is the identifying changes between the current and previous, already received, cartogram. Parameters of current cartograms and its changes compared to the previous cartogram characterize the state of the monument.

***Preprocessing***. Necessity of images preprocessing of monuments is related to the fact that under the proposed monitoring technology it is impossible to provide identical shooting conditions, even for the same monument. The images can be different due to shift of the shooting point and foreshortening. Therefore, one of the preprocessing functions—bringing images to the same shooting point and to the same foreshortening. This operation possible to carry out if all of the images lead to the same base image in the series. The base image in the series is selected by an expert. For each image in the series on the basis of control points it could be calculated the transformation matrix, by which the current picture leads to the shooting point and foreshortening of the base image. Another function of the preprocessing is separation of monument image from the background. This procedure is performed using interactive segmentation algorithm for the base image.

Then, the result is distributed to other images in the series. As a result of preprocessing, the new series of images containing only the image of the monument without a background are obtained.

***Determination of the spectral characteristics***. At this stage presented and segmented normal and infrared (NIR) current images of the monument are used. On the basis of these images indirect spectral characteristics of each pixel are calculated. These spectral characteristics are constructed on the basis of a group of indexes, such as NDVI and ENDVI, and their calculation is based on determining the difference in the appropriate channels in normal and infrared images. Thus, each pixel is associated with vector of characteristics—indexes. The result is a two-dimensional distribution of indirect spectral characteristics of the image on the surface of the monument.

***Identification of biodestructors and other contaminants***. The identification of the types of contamination or type of biofouling carried out for each pixels of image by comparing its vector of indexes, obtained by defining the indirect spectral characteristics, with the characteristics of a pollutant (biofouling) that are stored in the database. The comparison algorithms are based on the discriminant functions obtained previously with expert assistance.

***The construction of cartograms***. Cartogram images formed from two-dimensional vector distribution index by replacing each pixel of the image is the vector of spectral indexes corresponding index result of the identification of the vector. Each index is the result of identification posed and applied to the original image is a certain color expert. The resulting image undergoes a smoothing filter to eliminate interference. The result is a cartogram of the monument with a particular reference to the names of the colors of objects from the classifier: lichen biofilm, biofilm with algae domination, biofilm with micromycetes domination, area pollution with oil paint, salt precipitation zone, etc. Each object in this cartogram correlated with certain standard set of characteristics and properties inherent only to this object. This process involves expert experience. Process can go in automatic mode but if the system could not be automatically identify the object the result could be corrected by the expert hand.

***Assessment of the monument state***. Using cartogram of monument the percentages of areas occupied by various pollutants or organisms, could be calculated. These parameters define the current state of the monument. Determination of changes in the state of the monument obtained by comparing the current cartograms and its parameters with the data obtained with the cartograms reflecting the previous observation. Then expert can initiate specific actions about the necessity of any work to preserve the monument. System can be used for assessment of some cleaning actions (Fig. 2).

***Maintaining a database***. Large number of diverse data, which should be saved for comparing and processing, is received during the monitoring. For this purpose, a database is developed. In addition to the results of data in the database, there should be stored information about the objects as well as data relating to different scientific aspects, such as history, physics, chemistry, biology, mineralogy, meteorology. The data contains a standard set of numeric and string values, as well as original

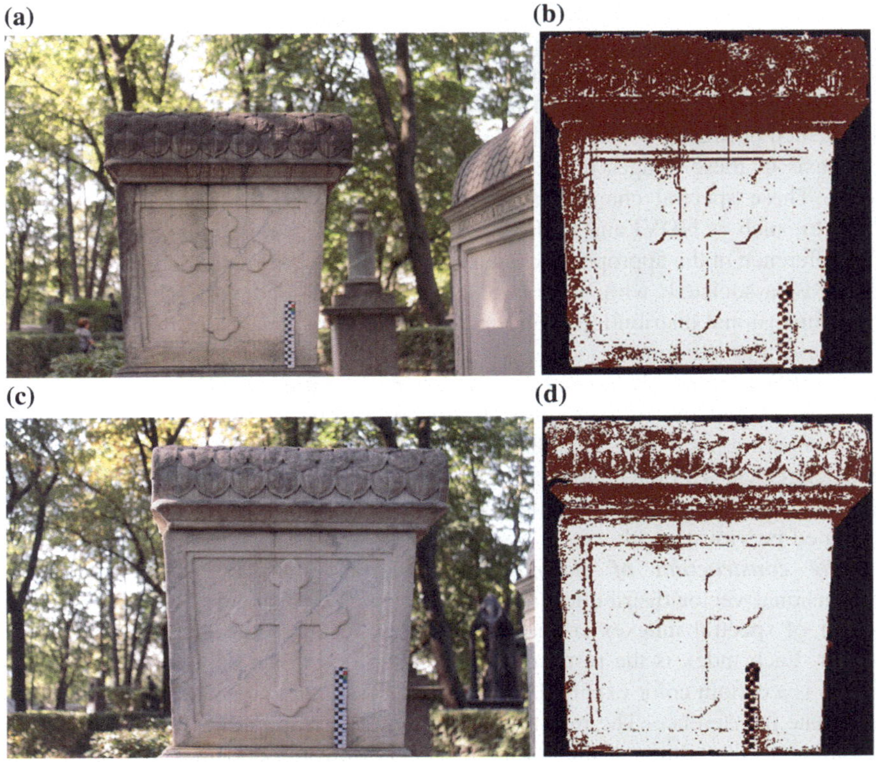

**Fig. 2** Marble monument state before and after preliminary cleaning action in the Museum Necropolis of Alexander Nevsky Lavra (St. Petersburg): **a** photo before cleaning, **b** map before cleaning: black biofilm (with micromycetes domination)—39.30 %, clean stone—60.70 %, **c** photo after preliminary cleaning, **d** map after cleaning: black biofilm (with micromycetes domination)—22.06 %; clean stone—77.94 %

documents containing text and graphics. The database allows users to structure the data so that in the future user can easily find the results of long surveys, compare them with the current state of the monument, follow the dynamics of changes of some parameters of the object state. Database is accessible through the Internet, which enables several users enter data. The organization of all of the data is such a thing as an event. For example, event may be a planned examination of an object, restoration or rehabilitation work, a scientific expedition, or something else. This event can be part of another event such as photographing a particular object may be part of a planned survey of the cultural reserve. Other data attached to their events that you can quickly find all the necessary information related to a particular set of events such as input shootings, measurements and samples, and the results of their processing, conclusions, recommendations for further action.

# 4 Conclusion

The whole monitoring system works as an expert one. This approach allows us to distinguish the main types of biofouling on the basis of color and spectral characteristics of biological objects. This method could be used for biofouling classification and quantitative assessment of the monument biodeterioration as well as for getting cartograms of biological colonization of the rock surface. The developed technique can be useful in the study of biofouling in time (for monitoring).

Currently, it has been shown the possibility of identification the main biodestructors groups and also the possibility of determining changes of biofouling over time. The system was tested during the survey of ancient petroglyphs in Khakassia, Tomsk, as well as a number of historical monuments of St. Petersburg.

**Acknowledgments** The authors acknowledge Saint-Petersburg State University for a research grants 1.37.151.2014, 9.37.157.2014.

## References

Cutler N, Viles H (2010) Eukaryotic microorganisms and stone biodeterioration. Geomicrobiol J 27(6–7):630–646

Dakal TS, Cameotra CC (2012) Microbially induced deterioration of architectural heritages: routes and mechanisms involved. Environ Sci Eur 24–36.doi: 10.1186/2190-4715-24-36

Frank-Kamenetskaya OV, Vlasov DY (2014) Monitoring of stone monuments state. St Perepsburg State University (in Russian)

Grishkin V, Zhabko A, Vlasov D, Schigorec S (2014) Multiple segmentation of the image series. In: International conference on computer technologies in physical and engineering applications (ICCTPEA), pp 59–60

Shchigorets SB, Doos AA, Grishkin VM, Zhabko AP, Vlasov DY (2009) Computer-aided tools for creating electronic atlases of the condition of historical monuments. In: Materials of the III international conference on remote sensing in archaeology. Tiruchirappalli, India, pp 70–72

# Part V
# Mineral Formation in Living Organisms and Biomimetic Materials

# Ion Substitutions, Non-stoichiometry, and Formation Conditions of Oxalate and Phosphate Minerals of the Human Body

**Olga V. Frank-Kamenetskaya, Alina R. Izatulina and Mariya A. Kuz'mina**

**Abstract** The pattern of ion substitutions and non-stoichiometry of oxalate and phosphate human body minerals (weddellite, whewellite, hydroxyapatite, struvite, and brushite) and their synthetic analogs was investigated by complex X-ray diffraction, spectroscopic, and chemical methods. All the studied biological minerals are characterized by variable non-stoichiometry composition that reflects the non- stationarity of formation conditions. Non-stoichiometry of calcium oxalates results from variations in quantities of disorderly distributed water molecules. The water amount in weddellite significantly influences the crystal structure parameters. In particular, there is positive correlation between the value of $a$ parameter and the occupancy of "zeolitic" water sites. Variations in composition of calcium and magnesium phosphates are caused by substitutions at all the crystallographic sites. Ion replacements in apatite and struvite are more prominent in comparison with those in brushite. Non-stoichiometry of apatite and brushite is due to the presence of vacancies at the Ca-sites, and that of struvite is due to the vacancies at the Mg-site. Two types of non-stoichiometry in carbonated apatites of B-type realize due to replacements $PO_4^{3-} \leftarrow CO_3^{2-}$ or $OH^- \leftarrow H_2O$. Maximum concentrations of impurity ions in magnesium and calcium phosphates are mainly limited by the content of these ions in the physiological solution. In case of Mg replacement with Ca in struvite and Ca with K in brushite and apatite, it is also essentially limited by the size differences of the host and admixture ions.

**Keywords** Biological minerals · Ion substitutions · Non-stoichiometry · Weddellite · Whewellite · Hydroxyapatite · Struvite · Brushite

O.V. Frank-Kamenetskaya (✉) · A.R. Izatulina · M.A. Kuz'mina
Saint Petersburg State University, Saint Petersburg, Russia
e-mail: ofrank-kam@mail.ru



O.V. Frank-Kamenetskaya et al. (eds.), *Biogenic—Abiogenic Interactions in Natural and Anthropogenic Systems*, Lecture Notes in Earth System Sciences,
DOI 10.1007/978-3-319-24987-2_33

# 1 Introduction

There are many uncertainties about the ionic substitutions and non-stoichiometry of minerals formed in living organisms. The presence of impurities in formula of biological minerals is usually ignored. At the same time the complex multicomponent composition of physiological solutions in the human body (Table 1) determines the probability of solid state formation. These work summarizes the regularities of complicate substitutions in phosphate and oxalate minerals—the main crystalline phases of a human body (Fig. 1). Such researches have not only the fundamental importance. They produce the scientific base for the creation of new biocompatible materials and technologies for medical purposes.

# 2 Experimental Section

## 2.1 Objects

Objects of the research were minerals of physiogenic and pathogenic aggregates (dental hard tissues, renal and salivary stones, dental and mitral valve calculi) formed in humans, and also synthetic analogs of biominerals obtained under experimental conditions in the presence of impurities were added to physiological solutions. Table 2 represents the ideal formulae of the studied minerals.

**Table 1** Inorganic components, ionic force, and pH of the human's physiological liquids

| Ion | Content (mmol/l) | | |
|---|---|---|---|
| | Urea (Severin 2003) | Mixed saliva (Rosseeva et al. 2009) | Blood (Kalyuzhniy 2003) |
| $Na^+$ | 67–133 | 4.80–30.40 | 135–150 |
| $K^+$ | 33–47 | 12.80–35.60 | 3.5–5.0 |
| $Ca^{2+}$ | 1.7–5.0 | 0.75–3.00 | 2.0–2.5 |
| $Mg^{2+}$ | 5.3–11 | 0.38–0.85 | 0.45–0.75 |
| $NH_4^+$ | 20–50 | 2.70–6.70 | |
| $Cl^-$ | 67–167 | 11.30–28.30 | 97–108 |
| $CO_3^{2-}$ | Oxalates 0.2–0.42 | 19.37–24.81 | |
| $PO_4^{3-}$ | 13–33 | 4.81–6.47 | 1–2 (inorganic phosphorus) |
| $SO_4^{2-}$ | 27–80 | | |
| $F^-$ | | 0005–0100 | |
| Ionic force | 0.3 | 20.92–76.81 | 0.15 |
| pH | 4.5–8.0 | 5.00–7.95 | 135–150 |

**Fig. 1** SEM images of Human body minerals: **a** bipyramidal weddellite crystals on the whewellite surface of renal stones; **b** sheaf aggregates of tabular whewellite crystals of renal stones; **c** epitaxial overgrowth of brushite fine crystals on the weddellite faces of salivary stones; **d** globular apatite aggregates of salivary stones; **e** rosette-like aggregates of brushite crystals of salivary stones

**Table 2** Objects consideration

| Biominerals | Ideal formula |
|---|---|
| *Oxalates* | |
| Whewellite | $CaC_2O_4{\cdot}H_2O$ |
| Weddellite | $CaC_2O_4{\cdot}2H_2O$ |
| *Phosphates* | |
| Hydroxyapatite | $Ca_5(PO_4)_3OH$ |
| Struvite | $NH_4MgPO_4{\cdot}6H_2O$ |
| Brushite | $CaHPO_4{\cdot}2H_2O$ |

## 2.2 Synthesis

To investigate the regularity of ion substitutions in calcium and magnesium phosphates of human body, synthesis of their analogs was realized by precipitation method (Frank-Kamenetskaya et al. 2011; Kus'mina et al. 2009, 2013). Concentrations of the main components included the values typical of physiological solutions. The additional ions of various impurities characteristic of urine and saliva were introduced into the initial solution in the form of Mg, Na, K, and $NH_4$ chlorides, Na and K sulfates, and Na hydrogen carbonate. Synthesis of *apatite* have been performed by both direct and reverse methods from aqueous solution of calcium acetate (0.02 M/L) or nitrate (0.02 M/L) (A) and hydrogen phosphate (0.01 M/L) with hydrogen carbonate of Na, K or $NH_4$ (B), at the pH ranges from 8 to 10. *Brushite* was synthesized from aqueous solution by mixing warm dilute solutions of calcium chloride (30–220 mmol/L) with Na, K, or $NH_4$ hydrogen phosphate solutions (30–220 mmol/L of phosphate ions) at the pH range 3–6. *Struvite* was synthesized from aqueous solution by mixing warm solution of magnesium chloride (10 mmol/L) and phosphoric acid (60 mmol/L) at the pH ranges from 6.3 to 9.0.

After completion of precipitation, solution was kept at different temperatures ($T = 90$–$100$ °C for apatite, $T = 110$ °C for brushite and 37 °C for struvite), than the precipitate was filtered, washed with distil water and dried at room temperature.

## 2.3 Analytical Procedures

Biological minerals and their synthetic analogs were studied by various methods including X-ray single crystal and powder diffraction, infrared spectroscopy and different chemical analyzes.

*Single crystal X-ray analysis* was used for detailed crystal structure investigations. The measurements were carried out using Bruker Smart APEX II CCD, STOE IPDS II, and Nicolet R3 diffractometers. Crystals (20 calcium oxalates from renal stones and synthetic Ca-struvite) were studied at a room temperature using monochromated Mo$K\alpha$ radiation. The unit cell parameters were refined by least square techniques using reflections in the 2θ range of 4.00–60.00°. The structures have been refined in the anisotropic approximation of thermal parameters for non-hydrogen atoms using SHELX-97 and CSD programs ($R$ = 0.024–0.057) (Sheldrick 2004, 2008; Akselrud et al. 1993). Positions of H atoms were localized objectively and refined without restrictions with individual isotropic temperature parameters.

*X-ray powder diffraction* measurements (phase identification, unit cell parameters determination) were carried out using the DRON and Rigaku MiniFlexII diffractometers (Cu$K\alpha$ or Co$K\alpha$ radiation, graphite monochromator; 2θ ranges 10–100° with a 0.05° steps and speed equal to 1°/min). Unit cell parameters were calculated by the least squares technique using 10–14 reflections and germanium powder as an internal standard.

*Infrared (IR) spectroscopy* was used during apatite investigations for localization of $CO_3^{2-}$ ions, detection of $OH^-$ and $HPO_4^{2-}$ ions, and detection of molecular water. The spectra were collected using a Bruker Vertex 70 and with modified two-wave (UR-20) spectrometers in the range of 400–4000 $cm^{-1}$ with a frequency resolution of 1 $cm^{-1}$. The samples were prepared as standard pellets in KBr. The interpretation of absorption bands in Frank-Kamenetskaya et al. (2011) is given.

*Chemical* analyzes. Concentrations of Ca, P, Mg, Na and K in phosphate crystals were determined by electron probe microanalysis using the scanning electron microscopes Camscan-4 and ATB-55 equipped with X-ray energy microanalyzers operating at 25 kV acceleration voltage and 2 lm electron beam diameter. SEM Calibration Specimens (registered standard number 1413) from MICROANALYSIS CONSULTANTS Ltd. were used as standards. The concentrations of nitrogen in monophase synthetic samples were measured using the gas-volumetric method. Samples were heated at 1100–1300 °C in an oxygen flow.

# 3 Results and Discussion

## 3.1 Variations of Crystal Lattice Parameters

X-ray Powder diffraction data demonstrates that the variations of crystal lattice parameters of mineral's majority (except brushite) are significantly greater than four standard deviations (Table 3; Fig. 2). Hence, the oxalates and phosphates of human body are characterized by variable composition.

## 3.2 Non-stoichiometry of Calcium Oxalate Hydrates

Calcium oxalates (weddellite and whewellite) are very common minerals of human body (Fig. 1a–c). The major part of the human urinary system stones consists of calcium oxalates both mono- and dihydrates (the amount of calcium oxalate stones is up to 75 % depending on the region) (Korago 1992; Izatulina and Yelnikov 2008; Benramdane et al. 2008; Izatulina et al. 2014). These minerals are found also in the salivary stones and other human pathogenic formations.

*Weddellite* $Ca(C_2O_4)\ (2 + x)H_2O$

The results of the crystal structure refinements of 17 weddellite crystals ($I4/m$, $a$ = 12.329–12.378 Å, $c$ = 7.345–7.366 Å, $V$ = 1117.8–1128.6 Å$^3$) which were taken from the oxalic renal stones of the St. Petersburg (Russian Federation) citizens of both sexes aged from 24 to 65 (Izatulina et al. 2014), as well as data of

**Table 3** Range of unit cell parameters of the urinary stone minerals (on powder data)

| Mineral | Sp.gr. | a | Δ/σ | b | Δ/σ | c | Δ/σ |
|---|---|---|---|---|---|---|---|
| Whewellite | $P2_1/c$ | 6.286–6.292(1) | 12 | 14.560–14.588(4) | 10 | 10.094–10.132(4) | 9.5 |
| Weddellite | $I4/m$ | 12.329–12.378 | | | 21 | 7.345–7.366 | 14 |
| Apatite | $P6_3/m$ | 9.395–9.457(4) | | | 15 | 6.849–6.885(4) | 9 |
| Struvite | $Pmn2_1$ | 6.933–6.955(2) | 11 | 11.193–11.224(4) | 8 | 6.127–6.139(2) | 6 |
| Brushite | $Cc$ | 6.364–6.366(1) | 2 | 15.188–15.193(2) | 2 | 5.813–5.814(1) | 1 |

*Designations* Δ—interval of parameter variation, σ—standard error

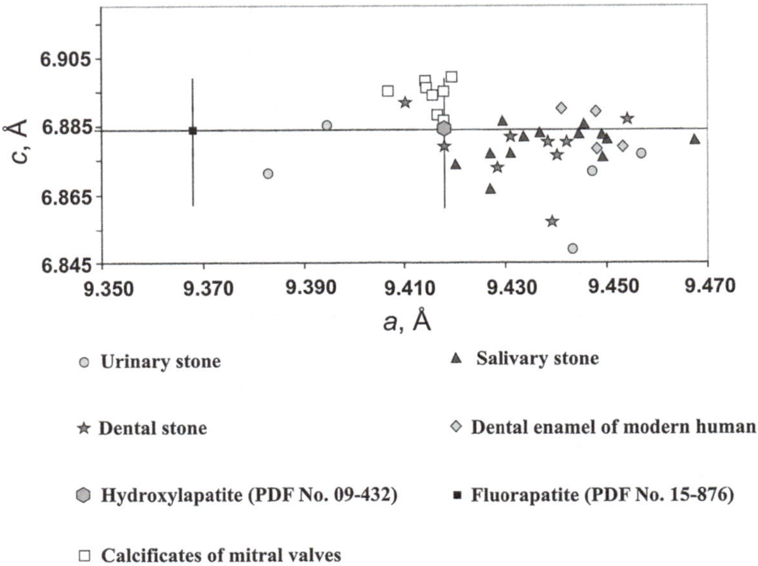


**Fig. 2** Unit cell parameters of apatites formed in humans

previous structural studies (Izatulina and Yelnikov 2008; Tazzoli and Domeneghetti 1980) indicate that no significant substitutions occur at calcium and other crystallographic sites.

The main "building" block of the weddellite structure is the Ca-polyhedron—slightly distorted square antiprism (Fig. 3a). Each Ca-polyhedron share common O1–O1 edges with two neighbor polyhedra to form infinite chains of square antiprisms arranged along the [001] direction. Rotation of Ca polyhedra chains and

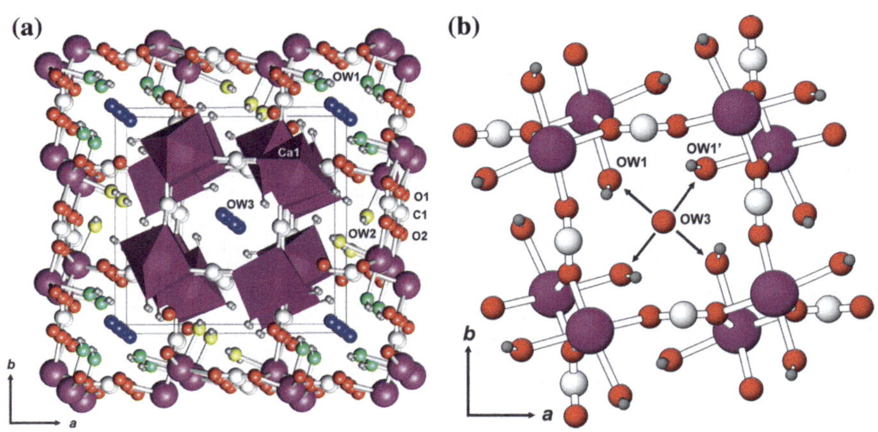


**Fig. 3** Crystal structure of weddellite (sp.gr. *P*4/*m*): **a** general view. Section almost parallel to (001); **b** displacement of the OW1 atoms with the increase of *x* value

water—oxalate complexes around the fourfold axis results in formation of large channels in the structure of weddellite arranged along the *c* axis. The oxygen atoms of water molecules occupy three independent crystallographic sites OW1, OW2, and OW3. The walls of the channels passing through the origin and center of the unit cell, formed by water molecule OW1. Four OW1 atoms arranged in the plane parallel to (001) to form a regular square with an edge OW1–OW1 ~ 3.2 Å (Fig. 3b). At the center of the large channels there are "zeolitic" water molecules OW3 arranged on the fourfold axis. The position of the oxygen atom of which OW3 is split into two independent similar and substantially vacant positions (OW3 and OW31), separated by a distance ~0.6 Å. The hydrogen atoms of the "zeolitic" water molecules were not localized at the difference Fourier synthesis which allows us to suggest that the protons are distributed statistically.

The distance between the oxygen atoms of the "zeolitic" water (OW3 and OW31) and oxygen atoms OW1 located in the same channel (~3.5–3.1 Å) are close to the distance OW1–OW1 (~3.2–3.3 Å), thus it could be described as the presence of slightly distorted octahedral groups (with the bases formed by OW1 atoms, and apical vertices by OW3/OW31) as well as the square complexes of oxygen OW1 atoms in the large channels. The OW3–OW1 distance (~3.5 Å) is significantly longer than the distance OW31–OW1 (3.1 Å) presumably because the structure tend to implement sustainable clusters of water molecules due to the "zeolitic" water position splitting (the appearance of OW31 position). The presence of a prohibited distance between oxygen atoms OW3, linked through the inversion center (1.9–2.2 Å), indicate that the centrosymmetricity of the weddellite structure realizes only statistically.

The "zeolitic" water amount (*x*) significantly influences the weddellite structure (in particular on the unit cell parameters). As the *x* increases (from 0.13 to 0.37 p.f.u) the value of $(OW1)_4$:$(OW1)_4(OW3/OW31)_2$ ratio decreases from ~7 to ~2, according to the ratio of the respective position's occupation. There is a significant positive correlation between amount of "zeolitic" water molecules in the structure of weddellite and the interatomic distance OW1–OW1 (3.211–3.279 Å)—edge of the OW1-based square complex (Fig. 4a). Changes in the interatomic distances and, especially, the increase in the bond lengths in the (001) plane, leads to significant variations in the values of the unit cell parameters. The *a* parameter increases significantly than others (from 12.329 to 12.378 Å) and considerable positive correlation observed between the value of the *a* parameter and the occupancy of "zeolitic" water positions (Fig. 4b). Obtained linear regression equation: $x = 5.43\,a - 66.80$, can be used for determination of the "zeolitic" water amount (*x*) using the known unit cell *a* parameter with mean-root-square error ±0.03 p.f.u (if standard error of *a* parameter determination ≤0.001 Å).

The *x* value for the crystals from mono-weddellite stones fall in the range of 0.13–0.24 p.f.u. In the recent paper, Conti et al. (2010) suggests that there is a range of water whereby weddellite crystals are stable, while the increase and decrease of the water amount leads to the destruction of crystals. It is also known (Rusakov et al. 2014) that for the weddellite crystals formed by the influence of microscopic

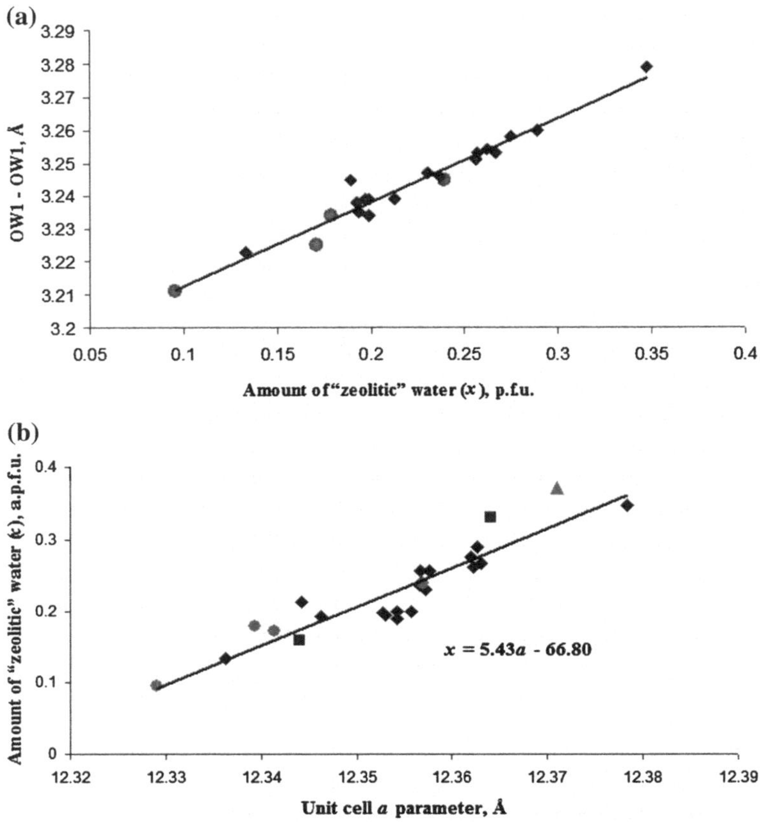


**Fig. 4** Correlation of "zeolitic" water molecules amount ($x$) with OW1–OW1 distance (*rhombuses*—Izatulina et al. 2014, *circles*—Rusakov et al. 2014) (**a**) and with unit cell a parameter (*rhombuses*—Izatulina et al. 2014, *circles*—Rusakov et al. 2014, *squares*—Izatulina and Yelnikov 2008, *triangular*—Tazzoli and Domeneghetti 1980) (**b**)

fungi (*Aspergillus niger* strain, active producer of organic acids) $x = 0.10$–$0.24$ p.f. u. Such weddellite could be considered relatively stable, since it is stored in the dried sludge. Whereas crystals obtained in our experiments, transforms to whewellite after drying the precipitate. The $x$ value determined for the weddellite crystals obtained in the presence of bacteria and viruses from the equation suggested above varies from 0.21 to 0.28 p.f.u. Thus, perhaps the weddellite in oxalate renal stones, substituted for calcium oxalate monohydrate (Fig. 5), was formed during inflammatory processes.

*Whewellite* $Ca(C_2O_4)\ (1 + x)H_2O$

The results of the crystal structure refinements of 3 whewellite crystals ($P2_1/c$, $a = 6.296$–$6.293$, $b = 14.598$–$14.590$, $c = 10.115$–$10.125$ Å, $V = 1117.8$–$1128.6$ Å$^3$)

**Fig. 5** Oxalate renal stone: *central part*—whewellite, *outer part*—weddelite. Picture taken in polarized light

which were taken also from the renal stones of the St. Petersburg citizens (Izatulina and Yelnikov 2008), as well as data of previous structural studies (Tazzoli and Domeneghetti 1980) indicate that as well as for weddellite, no significant substitutions occur at calcium crystallographic sites.

The crystal structure of whewellite consists from two types of alternating layers, parallel to (100) (Fig. 6a). The calcium cations are also at the center of a distorted square antiprism. Each antiprism paired by edges to three next, as result the hexagon motive of the first type of layers is formed. The second type of layers consists from the ribbons which formed by alternating oxalic ions and molecules of water bonding by hydrogen bonds laying in a plane (010) and directing along the

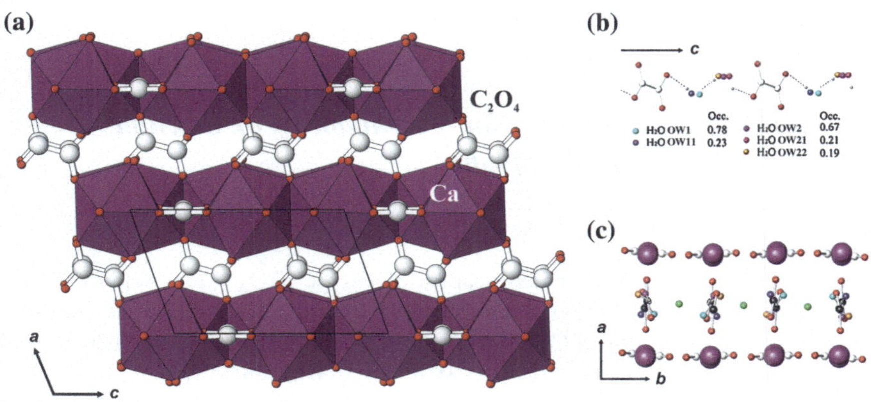


**Fig. 6** Crystal structure of whewellite (sp.gr. $P2_1/c$): **a** general view. Section parallel to (010); **b** disordered distribution of water molecules; **c** possible sites of additional water molecules

**Table 4** Occupancies of water–oxygen sites in whewellite structure

| Water oxygen sites | | Reference |
|---|---|---|
| OW1<br>OW11 | OW2<br>OW21<br>OW22 | |
| 0.78<br>0.23 | 0.67<br>0.21<br>0.19 | Izatulina and Yelnikov (2008) |
| 0.85<br>0.17 | 0.80<br>0.14<br>0.08 | |
| 0.85<br>0.15 | 0.86<br>0.14 | Tazzoli and Domeneghetti (1980) |

*c* axis. The oxygen atoms of water molecules occupy two independent crystallographic sites OW1 and OW2. On differential Fourier synthesis maps the positions of water molecules oxygens are disordered on OW1, OW11 and OW2, OW21, OW22 (Fig. 6b; Table 4) The sums of occupancies of both whewellite splitting sites are close to 1.0.

At the level of hypothesis, variations of crystal lattice parameters of whewellite can be explained by the presence of additional water molecules which are disorderly distributed and therefore cannot be found in the different Fourier synthesis maps. Disordered distribution of water molecules in structure of whewellite allows to assume the entering of additional water molecules OW3 in sites ½00, ½0½, and ½¼0—in the cavities between chains «$C_2O_4$–OW1–OW2–$C_2O_4$» (Fig. 6c). Distances between additional and other water molecules: OW3–OW1/OW11 = 2.60–2.65, OW3–OW21/OW22 = 2.86–2.69 Å. The maximum number of these molecules (0.07 p.f.u.) was estimated on the basis of residual electron density in the final difference Fourier synthesis maps.

## 3.3 *Ion Substitutions and Non-stoichiometry of Calcium and Magnesium Phosphates*

*Hydroxyapatite*

The study of synthetic carbonated apatites of B-type ($CO_3^{2-}$ ions replace $PO_4^{3-}$ ions) obtained by precipitation method revealed that its lattice constants are influenced by the concentration of $CO_3^{2-}$ ions, water molecules and also to vacancies and monovalent alkaline cations at Ca-sites (R+ = $K^+$, $NH_4^+$, $Na^+$) (Fig.7). In order to explain the compositional variations in carbonated apatites of B-type the following substitution equations should be considered (Elliott 1994, 2002; Frank-Kamenetskaya 2008a, b; Frank-Kamenetskaya et al. 2011):

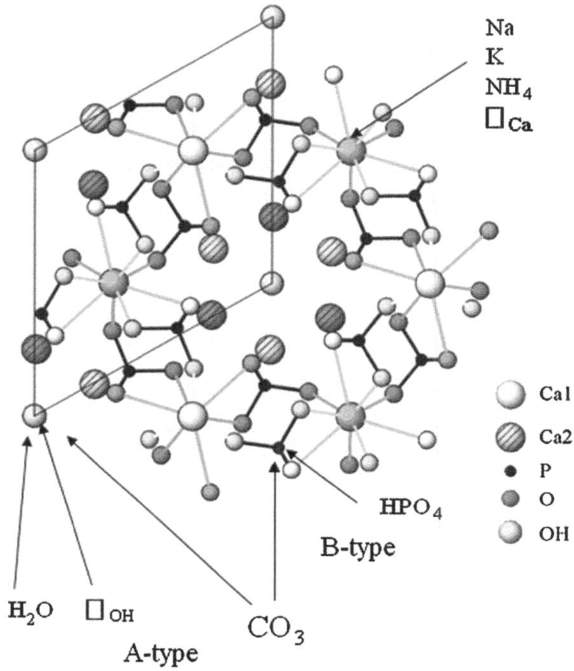


**Fig. 7** Atomic structure of hydroxyapatite (sp.gr. $P6_3/m$). Projection onto (0001) plane. Ion substitutions of biological apatites are shown. The symbols $\square_{Ca}$ and $\square_{OH}$ mean vacancies at Ca and OH sites, respectively (Frank-Kamenetskaya et al. 2011)

$$1/2Ca^{2+} + PO_4^{3-} = 1/2\square_{Ca} + \left[CO_3^{2-}\right]_B \quad (1)$$

$$Ca^{2+} + PO_4^{3-} = R^+ + \left[CO_3^{2-}\right]_B \quad (2)$$

$$1/2Ca^{2+} + OH^- = 1/2\square_{Ca} + (\square_{OH}, H_2O) \quad (3)$$

$$1/2Ca^{2+} + PO_4^{3-} = 1/2\square_{Ca} + HPO_4^{2-} \quad (4)$$

$$Ca^{2+} + PO_4^{3-} = R^+ + HPO_4^{2-} \quad (5)$$

$$Ca^{2+} + OH^- = R^+ + (\square_{OH}, H_2O) \quad (6)$$

Accordingly, there are two type of non-stoichiometry in carbonated apatites of B-type associated with the presence of vacancies in the calcium sites. At first, Ca deficit is realized due to $CO_3^{2-}$ ions of B-type incorporation (Eq. 1). As a result, the lattice constant *a* decreased as the concentration of $CO_3^{2-}$ ions increased. In the second, Ca–deficit is realized due to water molecules and $HPO_4^{2-}$ ions incorporation (Eqs. 3 and 4). As a result, the lattice constant *a* increased as the amount of water molecules at vacant OH sites and $HPO_4^{2-}$ ions increased. The occurrence of

**Table 5** The main elemental composition (wt%) of human body apatites

| Apatite origin | Composition, wt% | | | | | | Reference |
|---|---|---|---|---|---|---|---|
| | Ca | P | $CO_3^{2-}$ | $Na^+$ | $Mg_{2+}$ | $F^+$ | |
| Dental enamel | 33.9–39.3 | 16.1–18.8 | 1.6–7.8 | 0.5–1.0 | 0.2–0.3 | 0.1–0.3 | Elliott (1994), Korago (1992), Frank-Kamenetskaya et al. (2004), Handschin and Stern (1992), Elliott (2002), Gross and Berndt (2002), Meneghini et al. (2003) |
| Dentine | 34.9–40.4 | 16.9–19.5 | 1.0–6.5 | 0.9–1.4 | 0.8–1.0 | 0.07 | |
| Bone | 36.6–37.7 | 17.1–17.7 | 6.5–8.9 | 0.8–1.0 | 0.6–1.1 | 0.03–0.1 | |
| Renal stones | 23.0–33.9 | 15.2–17.5 | ~1 | 0.8–1.6 | 0-1.12 | 0-4.22 | Frank-Kamenetskaya (2008a, b) |
| Dental stones | 29–21 | 16–12 | ~1 | 0.4–1.3 | No data | No data | |
| Salivary stones | 30.3–33.8 | 15.0–17.2 | 1–5 | 0.4–1.0 | 0.5–3.7 | No data | Frank-Kamenetskaya (2008a, b), Pikhur et al. (2008) |
| Calcificates of mitral valves | 23.0–33.5 | 13.4–18.0 | No data | 0.7–1.0 | 0.4–1.8 | No data | Rosseeva et al. (2012) |

$CO_3^{2-}$ ions of the B-type (Eqs. 1 and 2). is accompanied by an increase of the lattice constant *c*. Appearance of vacancies at the calcium sites (Eqs. 1, 3, and 4) is accompanied by decrease of this parameter. So the direction of change lattice constant *c* determines which of the replacements dominates.

All investigated biological apatite formed in human body contain carbonate ion of B-type (determined by infrared spectroscopy) and alkaline cations $R^+$, usually Na (Table 5).

The concentration of $CO_3^{2-}$ ions and alkaline cations in biological apatite does not exceed 9 and 2 wt%, respectively. The unit cell *a* parameter of biological apatites (except the apatite of mitral valves) is usually larger than a parameter of stoichiometric hydroxyapatite (Fig. 2), suggesting a strong effect of water molecules, which occupy nearly all vacant sites of $OH^-$ ions in channels and, to less extent, of $HPO_4^{2-}$ ions. So, the contribution of water molecules and $HPO_4^{2-}$ ions in the formation of vacancies at Ca-sites (Eqs. 3 and 4) is very strong and dominated over the contribution of $CO_3^{2-}$ ions (Eq. 1) The unit cell c parameter is often smaller than that of stoichiometric hydroxyapatite due to the formed thus significant calcium deficits (Eqs. 1, 3, and 4). In apatite of mitral valves directed in different directions changes of *a* parameter under the influence of water molecule's and carbonate ions (Eqs. 1 and 3) cancel each other. The influence of carbonate ions on the c parameter of apatite of mitral valves dominated over the contribution of vacancies at Ca-sites.

The general crystal chemical formula of human body hydroxyapatites can be written as follows: $\left(Me^{2+}_{10-(x+y+z+t)/2}R^+_t\square_{(x+y+z-t)/2}\right)\left[(PO_4)_{6-x-z}(HPO_4)_x\right.$

$(CO_3^{2-})_z]_6\left[(OH)_{2-y}(\square, H_2O)_y\right]_2$, where $Me^{2+}$—Ca, Mg, $R^+$—Na, K, $A^-$—OH, F. In the case of apatite of tooth enamel of permanent teeth of residents of St. Petersburg: $x = 0.51, y = 0.77, z = 0.54$. That is a part of vacancy in the calcium and $OH^-$ ion sites are 8 and 38 %, respectively (Frank-Kamenetskaya et al. 2004).

In general, we can conclude that studied biological apatites are water-containing carbonated hydroxyapatites of B-type with $Ca^{2+}$ and $OH^-$ ions deficiencies. In the most cases (dental hard tissues, renal and salivary stones, and dental calculi) the dominant role in apatite structure belongs to water, the influence of carbonate ions is less. The charge imbalance is mainly compensated by the vacancies at Ca-sites. The contribution of alkali cations is restricted.

Taking into account the range of variation of the unit cell parameters of biological apatites (Fig. 2), it is seen that ion replacements in apatite of pathogenic origin are more profound in comparison with physiogenic dental enamel. In the pathogenic apatite major compositional variations are observed in renal stones. This indicates strong variability of formation conditions.

*Struvite and brushite*

Calcium and magnesium phosphates (struvite and brushite) are also occur in renal stones. Moreover, brushite is the main mineral component of immature dental calculi (Fig. 1c, e).

The determined values of unit cell parameters and results of local elemental analysis of struvite and brushite crystals from renal stones (Tables 3 and 6, respectively) indicate the possibility of ion replacements at different crystallographic sites (Figs. 8 and 9). The value of ratio Mg + Ca/P in struvite (systematically) and Ca/P in brushite (in some cases) less than unity, which indicates the presence of vacancies at corresponding crystallographic sites. At the same time variations of the lattice parameters and consequently the ion replacements in struvite are more pronounced than those in brushite

The study of synthetic analogs of struvite and brushite allowed us to test this conclusion and also to reveal correlations between the elemental compositions and lattice parameters of these minerals and to identify the factors that limit

**Table 6** The main elemental composition (wt%) of renal stone's struvite and brushite (Zhuravlev2009)

| Element | Struvite | | Brushite | |
|---|---|---|---|---|
| | Range | Mean | Range | Mean |
| P | 22.09–24.33 | 23.21 | 19.83–24.97 | 22.65 |
| Mg | 14.44–16.09 | 15.27 | 0.00–1.88 | 0.16 |
| Ca | 0.34–1.89 | 1.12 | 29.87–38.03 | 33.17 |
| S | 0.17–0.22 | 0.20 | 0 | 0 |
| K | 0.28–0.70 | 0.49 | 0.00–1.25 | 0.37 |
| Na | 0.00–0.13 | 0.07 | 0.00–1.27 | 0.48 |
| Cl | 0.00–0.07 | 0.04 | 0.00–0.17 | 0.03 |

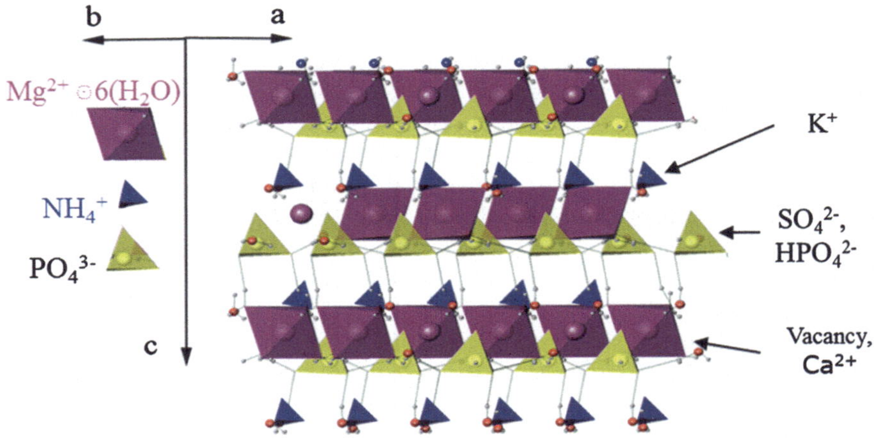


**Fig. 8** Atomic structure of struvite $MgNH_4PO_4$ $6H_2O$ (sp.gr. $Pmn2_1$). The main ion substitutions of human body struvite are shown

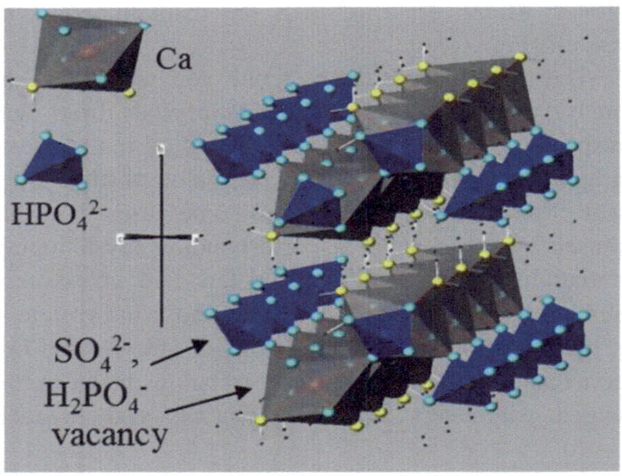


**Fig. 9** Atomic structure of brushite $Ca(HPO_4)2(H_2O)$ (sp.gr.*Cc*). The main ion substitutions of human body brushite are shown

incorporation of impurities in them (Zhuravlev 2009). For Ca-struvites at the level of the trends the direct correlation between the parameters *a*, *b*, *V* and value of ratio Ca + Mg/P (number vacancy in the Mg–site) and reverse correlation between the parameters *a*, *b*, *V* and magnesium content are observed (Table 7), that may be the result of incorparation of water molecules in partially vacant Mg-site struvite structure.

Refinement of the crystal structure of struvite with value of ratio Ca + Mg/P = 0.87 by means of single crystal X-ray diffraction analysis ($Pmn2_1$, $a = 6.946$ (3), $b = 6.135$ (3), $c = 11.213$ (5) Å, $V = 477.8$ (6) $Å^3$) result in the crystal chemical formula:

**Table 7** Coefficients for pair correlation between the elemental composition and lattice parameters of Ca-struvites

| *a* | *b* | *c* | *V* | Mg | Ca | Ca/Mg | (Ca + Mg)/P | |
|---|---|---|---|---|---|---|---|---|
| | **0.64** | 0.43 | 0.89 | **−0.74** | 0.45 | 0.48 | **0.68** | *a* |
| | | 0.53 | 0.87 | **−0.84** | 0.26 | 0.29 | **0.82** | *b* |
| | | | 0.71 | **−0.58** | 0.15 | 0.16 | 0.57 | *c* |
| | | | | **−0.87** | 0.38 | 0.41 | **0.83** | *V* |
| | | | | | −0.27 | −0.31 | | Mg |
| | | | | | | | 0.09 | Ca |
| | | | | | | | 0.13 | Ca/Mg |
| | | | | | | | | (Ca + Mg)/P |

$$NH_4(Mg_{0.86}(H_2O)_{0.09}\square_{0.03}Ca_{0.02})(PO_4)_{0.76}(HPO_4)_{0.24}6H_2O.$$

This is confirmed that in Ca-struvites there is partially vacant Mg-site which is almost completely occupied by water molecules. So, the unit cell parameters of Ca-struvite mostly influenced by $H_2O$ molecules.

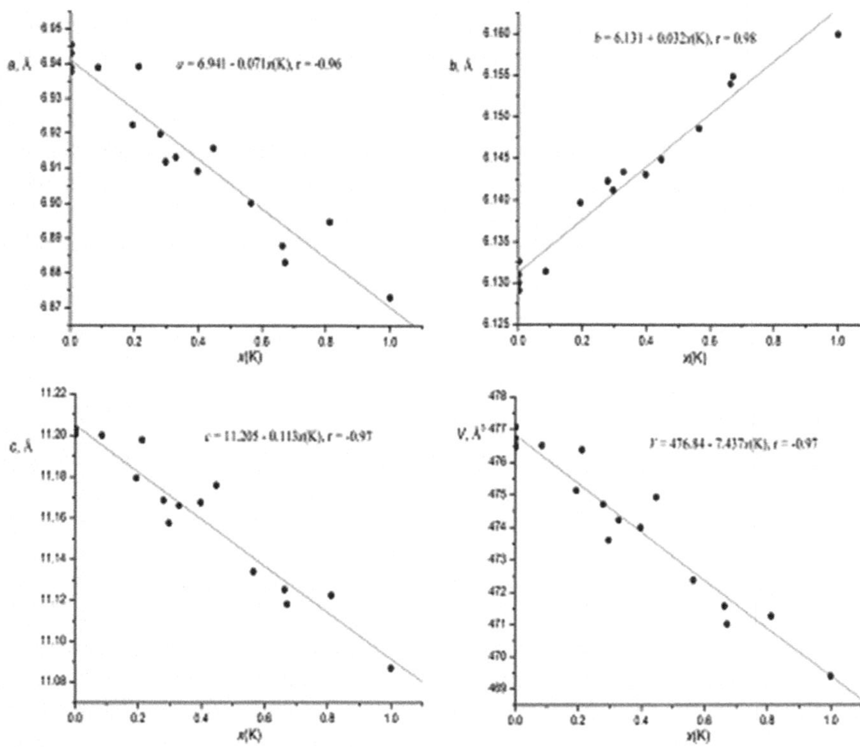


**Fig. 10** Effect of $NH_4^+ \leftarrow K^+$—substitution (*x*(*K*) value) on parameters (*a*, *b*, *c*) and volume (*V*) of struvite unit cell

**Table 8** Maximum concentrations of admixture ions and vacancies in struvite and brushite of renal stones and its synthetic analogs

| Parameter | Struvite | | | | Brushite | | |
|---|---|---|---|---|---|---|---|
| | $(NH_4)^+ \leftarrow K^+$ | $Mg^{2+} \leftarrow Ca^{2+}$ | $Mg^{2+} \leftarrow \square$ | $(PO_4)^{3-} \leftarrow (SO_4)^{2-}$ | $(HPO_4)^{2-} \leftarrow (SO_4)^{2-}$ | $Ca^{2+} \leftarrow K^+$ | $Ca^{2+} \leftarrow \square$ |
| $C_{max}$p.f.u. (renal stone) | 0.02 | 0.06 | 0.08 | 0.01 | 0 | 0.01 | 0.08 |
| $C_{max}$p.f.u. (Synt. cryst.) | 0.84 | 0.02 | 0.14 | 0.01 | 0.05 | 0.02 | 0.10 |
| $(r_{el} - r_{adm})/r_{adm}$% | 4 | 39 | – | 0.5 | | 24.64 | – |

*Note* Designations: $r_{el}$—ionic radius of base element, $r_{adm}$—ionic radius of admixture element
□ means vacancies at crystallographic sites

The substitution $NH_4^+ \leftarrow K^+$ has the strongest influence on struvite unit cell parameters (Fig. 10) Between the $K^+$ ion content and the values of the struvite lattice constants occurred the following correlations: direct with *b* parameter and reverse with *a* and *c* parameters.

Significant reduction of the parameters *a* and *c* can be explained by incorporation of a smaller than $NH_4^+$ ion potassium cation ($rNH_4+$ (1.43 Å) > $r_K^+$ (1.38 Å), Shannon 1976) Increasing of *b* parameter can be attributed to structural deformation, due to the rupture of hydrogen bonds with participation of $NH_4^+$ cation, which are linked polyhedra into chains along this axis.

The comparison of maximum concentrations of impurities in the brushite and struvite of kidney stones with their synthetic analogs (Table 8) demonstrate that concentration of admixture ions in struvite and brushite of human body in most cases is limited by the content of these ions in the physiological solution. In case of Mg replacement by Ca in struvite and Ca by K in brushite ion substitutions are also essentially limited by the size differences of the host and admixture ions.

# 4 Conclusions

Owing to the presented results the Ca-oxalates (whewellite and weddellite) and phosphates (apatite, struvite and brushite) of human body are characterized by variable non-stoichiometry composition that reflects the non- stationarity of formation conditions

Non-stoichiometry of calcium oxalates results from variations in quantities of disorderly distributed water molecules.

Variations in composition of calcium and magnesium phosphates are caused by substitutions at all the crystallographic sites. Ion replacements in apatite and struvite are more prominent in comparison with those in brushite. Non-stoichiometry of apatite and brushite is due to the presence of vacancies at Ca-sites, and that of struvite is due to the vacancies at the Mg-site. Maximum concentration of impurity ions in biological phosphate crystals is mainly limited by the content of these ions in the physiological solution. In case of Mg replacement with Ca in struvite and Ca with K in brushite and apatite, is also essentially limited by the size differences of the host and admixture ions.

In general, the non-stoichiometric chemical composition and different stability of biological minerals is determined by the nonstationary conditions of their formation, in particular the variable composition of saline solutions.

**Acknowledgments** XRD studies had been performed at the X-ray Diffraction Centre of St. Petersburg State University. The work was supported by Saint Petersburg University (project 3.38.243.2015) and RFBR (projects 11-05-90425-Ukr_f _a, 13-05-90432 Ukr_f _a).

The authors are very grateful to V. Yu. El'nikov, E.V. Rosseeva, I.V. Rozhdestvenskaya and S. V. Zhuravlev for active collaboration in this work. They would like also to thank Yu. L. Kretser and M.P. Pavlov for electron probe microanalyses conduction and to S.N. Zimina for gas-volumetric method measurements.

## References

Akselrud LG, Grin YuN, Pecharsky VK, Zavaliy PY et al (1993) Use of the CSD program package for structure determination from powder data. In: Proceedings of 2nd Europe powder diffraction conference Pt 1. Enschede, Trans Tech Pub, The Netherlands

Benramdane L, Bouatia M, Idrissi MOB, Draoui M (2008) Infrared analysis of urinary stones, using a single reflection accessory and a KBr pellet transmission. Spectrosc Lett 41(2):72–80

Conti C, Brambilla L, Colombo C, Dellasega D, Diego GG, Realinia M, Zerbib G (2010) Stability and transformation mechanism of weddellite nanocrystals studied by X-ray diffraction and infrared spectroscopy. Phys Chem Chem Phys 12:14560–14566

Elliott JC (1994) Structure and chemistry of the apatites and other calcium orthophosphates. Elsevier, Amsterdam

Elliott JC (2002) Calcium phosphate biominerals. Rev Min Geochem 48:427–453

Frank-Kamenetskaya OV (2008a) Structure, chemistry and synthesis of carbonate apatites—the main components of dental and bone tissues. In: Krivovichev SV (ed) Minerals as advanced materials I. Springer, Berlin

Frank-Kamenetskaya OV (2008b) Crystal chemistry and synthesis of carbonate apatites—main minerals in living organisms. In: Proceeding of the 9th international congress for applied mineralogy. Brisbane, Australia

Frank-Kamenetskaya OV, Golubtsov VV, Pikhur OL, Zorina ML, Plotkina YuV (2004) Nonstoichiometric apatite of the human dental hard tissues (the age alterations). Proc All-Rus Miner Soc 5:120–130 (in Russian)

Frank-Kamenetskaya O, Kol'tsov A, Kuz'mina M, Zorina M, Poritskaya L (2011) Ion substitutions and non-stoichiometry of carbonated apatite-(CaOH) synthesised by precipitation and hydrothermal methods. J Mol Struct 992: 9–18

Gross KA, Berndt CC (2002) Biomedical applications of apatites. Rev Min Geochem 48:631–672

Handschin RG, Stern WB (1992) Crystallographic lattice refinement of human bone. Calcified Tissue Int 51:111–120

Izatulina AR, Yelnikov VYu (2008) Structure, chemistry and crystallization conditions of calcium oxalates—the main components of kidney stones. In: Krivovichev SV (ed) Minerals as advanced materials I. Springer, Berlin

Izatulina AR, Guzhiy VV, Frank-Kamenetskaya OV (2014) Weddellite from renal stones: Structure refinement and dependence of crystal chemical features on $H_2O$ content. Am Mineral 99:2–7

Kalyuzhniy VP (2003) Electrolytes in health and disease and methods of their study. Terra medica 1:13–18 (in Russian)

Korago AA (1992) Introduction to biomineralogy. Nedra, St. Petersburg (in Russian)

Kuz'mina MA, Zhuravlev SV, El'nikov V.Yu. Frank Kamenetskaya OV (2009) The effect of impurities on crystallization and morphology of struvite (mineral of kidney stones) in the experiment condition. Proc All-Rus Miner Soc 6: 89–95 (in Russian)

Kuz'mina MA, Zhuravlev SV, Frank Kamenetskaya OV (2013) The effect of medium chemistry on the solubility and morphology of brushite crystals. Geol Ore Deposits 8:1–6

Meneghini C, Dalconi MC, Nuzzo S, Mobilio S, Wenk RH (2003) Rietveld refinement on X-ray diffraction patterns of bioapatite in human fetal bones. Biophys J 84:2021–2029

Pikhur OL, Rosseeva EV, Plotkina YuV, Kasbohm J, Golovanova OA, Frank-Kamenetskaya OV, Zorina ML (2008) Features of the morphology and composition of salivary calculi of patients from Saint-Petersburg region. Med Geol Newslett 12:24–29

Rosseeva EV, Frank-Kamenetskaya OV, Golovanova OA, Zorina ML (2009) Simulation of mineral phase formation from human oral-like fluids. Vestnik SPbGU Ser 7(2):12–21

Rosseeva EV, Nikolaev AM, Morozov MV, Frank-Kamenetskaya OV, Lamanova LM (2012) Bioapatites of calcificates of mitral valves. http://www.minsoc.ru/2012-1-111-0 (in Russian)

Rusakov AV, Frank-Kamenetskaya OV, Gurzhiy VV, Zelenskaya MS, Izatulina AR, Sazanova KV (2014) Refinement of the crystal structures of biomimetic weddellites produced by microscopic fungus Aspergillus niger. Crystallogr Rep 59:362–368

Severin ES (ed) (2003) Biochemistry. GEOTAR Med, Moscow (in Russian)

Shannon RD (1976) Revised effective ionic radii and systematic studies of interatomic distances in halides and chaleogenides. Acta Crystallogr A 32:751–767

Sheldrick GM (2004) SADABS. University of Göttingen, Germany

Sheldrick GM (2008) A short history of SHELX. Acta Crystallogr A 64:112–122

Tazzoli V, Domeneghetti C (1980) The crystal structures of whewellite and weddellite: re-examination and comparison. Am Mineral 65:327–334

Zhuravlev SV (2009) Synthesis and crystal chemical study of struvite and brushite (minerals of human kidney stones). Master's thesis, Saint Petersburg State University (in Russian)

# The Role of the Organic Component in the Formation of Organo-Mineral Formations of Human Body

**Olga A. Golovanova, Svetlana A. Gerk and Tatiana S. Mylnikova**

**Abstract** This work examined the protein composition of organic components for different organo-mineral aggregate in the bone tissue of healthy people (without bone disease) and patients with coxarthrosis (arthrosis of the coxofemoral joint), urinary, dental, and salivary calculi. The content of the water-soluble organic substances with a peptide bond in urinary stones depends on the mineral composition of the sample. When monohydrate calcium oxalate crystallizes in the solution in the presence of glutamic acid, the crystal growth is stabilized. When the concentration of amino acid grows, the inhibitory effect grows. Other amino acids (lysine, glycine, aspartic acid) have an inhibitory effect on hydroxyapatite crystal growth in the range of the tested concentrations. By an increase in the electrostatic interaction with non-stoichiometric hydroxyapatite, the tested amino acids can be ranked in order as: lysine > glycine > glutamic acid > aspartic acid. In the course of bone arthrosis, the molecular weight of the organic content was observed to be less than that of collagen. Close packing of the deformed collagen fibers was one of the causes of the increased hardness of arthritic bones. The amino acid composition in the bone tissue of men and women in the age range of 30–79 to 60 years does not depend on the age and gender identity of the samples. After 60 years, the amino acid content varies by age and gender.

**Keywords** Biominerals · Crystallization · Hydroxyapatite · Collagen · Amino acid · Protein compounds · Thermal analysis · Liquid chromatography · Dispersion analysis · Paramagnetic resonance

---

O.A. Golovanova (✉) · S.A. Gerk
Omsk F.M. Dostoevsky State University, Omsk, Russia
e-mail: golovanoa2000@mail.ru

T.S. Mylnikova
National Research Tomsk Polytechnic University, Tomsk, Russia



O.V. Frank-Kamenetskaya et al. (eds.), *Biogenic—Abiogenic Interactions in Natural and Anthropogenic Systems*, Lecture Notes in Earth System Sciences,
DOI 10.1007/978-3-319-24987-2_34

# 1 Introduction

The investigation of minerals formed in the life activity of the human body is an important aspect in current studies. Biominerals and organo-mineral aggregates (OMAs) are either involved in the biological life cycle of a human being or are its product. Minerals of this type are difficult to study because their material and elemental composition is complex, and the minerals (often poorly crystallized) and organic constituents are difficult to separate. Studies have shown that the organic (non-mineral) content in such substances ranges from a few to tens of percent by weight (Dorozhkin 2007; Golovanova 2004, 2007; Luneva 2003; Pyatanova 2004; Zatsepin 2001; Golovanova and Borbat 2005; Kolesnikova and Mikhailova 2004; Gabuda et al. 2005; Pal'chik et al. 2001, 2004; Potapov et al. 2002; Zuzuk et al. 2005). For example, bone and dental tissues differ in mineralization of the extracellular matrix and contain 50 % inorganic compounds by weight, 25 % organic components, and 25 % water.

Currently, there is no unified theory to explain the nature of interactions between mineral and organic components of OMA. Basically three variants of this interaction are considered: direct involvement of the organic matrix in the construction of biominerals by binding the components of mineral phases and initiating their own mineralization (bones, teeth); inhibition and/or promotion of mineral phase crystallization through selective adsorption interaction with these phases (pathogenic OMA); and indirect involvement through interaction in stone-forming solutions under external and internal factors (infections, disorder of calcium and protein metabolism, etc.). For example, the disturbance of bone remodeling has a negative impact on its structure and properties and causes different osteoarticular pathologies. Despite a considerable amount of the research in the bone matrix, diagnosis and correction of such states is a socially significant and not fully solved problem (Luneva 2003; Kovalenko and Bortkewich 2003). To date, there has been no consensus on the changes in the organic composition of the bone tissue that occur with age or as a result of various diseases.

In general, a literature review proves the relevancy of studying the organic component of OMA. Many researchers point out that its specificity, to a considerable degree, controls the process of phase formation in the human body.

To better understand the laws of biomineral genesis in the human body, the structure of biominerals and patterns of their formation need to be studied in more detail. This chapter provides a detailed analysis of the protein composition of the organic components for different OMAs—the bone tissue of healthy people (without bone disease) and patients with coxarthrosis (arthrosis of the coxofemoral joint), as well as urinary, dental, and salivary calculi—and finding the relationship between the amino acid and mineral composition of these formations.

## 2 Materials and Methods

The material for the research was a collection of stones from residents of the Omsk region, Russia: urinary stones (40 samples) taken after open surgery and through extracorporeal lithotripsy, dental stones from patients with generalized periodontitis (85 samples), and salivary stones from the parotid and sublingual salivary glands and their ducts (14 samples), as well as a collection of femoral heads removed from patients with coxarthrosis. The 100 patients were both women and men, aged 30–80 years. The control samples were 4 human femoral heads removed in accordance with regulations (No. 694 dated 21.07.1978, p. 2.24, No. 8-ФЗ dated 12.01.1996, p. 3, and No. 73-ФЗ dated 31.05.2001, pp. 14, 16 Federal law of the State Duma). To identify changes in bone tissue with age and gender, all the material was divided into four age groups (30–49 years, 50–59 years, 60–69 years, and 70–79 years). To assess the dynamics of changes in coxarthrosis, three horizontal sections of separate heads with a thickness of 0.2–0.5 cm were made: upper, middle, and lower sections. After that, they were grinded. The average dried powder samples were analyzed (GOST 17681-82) (Borisov 2001).

The total protein content was determined by the Kjeldahl method (Borisov 2001). The detection limit for protein compounds by this technique was $10^{-3}$ wt%, The relative standard deviation was 0.02. To determine the proportion of the water-soluble components in the protein fraction of urinary calculi, the method of photometric determination of proteins by Benedict was improved (Borisov 2001).

The qualitative and quantitative amino acid composition of the kidney stones was determined by ion exchange column chromatography (T-339 analyzer). The threshold sensitivity of the technique was $10^{-4}$ wt% and the relative standard deviation was 0.04. The amino acid composition of the bone tissue and dental and salivary calculus were studied with amino acid analyzer AAA-39M using high-performance liquid chromatography (HPLC). The detection limit for amino acids was 4.10 wt%.

The thermal analysis of the bone powders was performed with derivatograph SII Diamond-TG-DTA (Perkin Elmer) in a temperature range of 25–1000 °C. The thermogravimetric curve (TG) and its derivative– the differential-gravimetric curve (DTG)—were analyzed to determine the mass loss during annealing. The differential thermal curve (DTA) was analyzed to investigate the thermal transformation energy of the bone components (Egunov 1996; Shestak 1987).

The electron paramagnetic resonance spectra (EPR spectra) of the bone powders were obtained with an ESR70-03 DX/2 spectrometer (Belarus). To activate the paramagnetic centers in powder samples, they were stepwise annealed in the temperature range of 23–600 °C in an electric furnace (SNOL 6.7/1300). Software Origin 7.0 was used to calculate the spectroscopic splitting factor (g-factor), the width of the EPR signal band ($\Delta H$, $G$), and the concentration of the paramagnetic centers ($S_{PMC}$, spin $g^{-1}$). The two latter values were calculated using the reference sample $MgO:Mn^{2+}$ with concentration of the paramagnetic centers equal to

$(1.07 \pm 0.1) \times 10^{15}$ spin/g, and the width of the signal band in the EPR spectra equal to $0.03 \pm 0.64$ $G$.

The dispersion analysis of the model solutions (urine) in the presence of amino acids was carried out with the laser diffraction particle size analyzer Shimadzu SALD-2101. To obtain more reliable measurements, 4–5 fold replication of the analysis was made. The relative standard deviation for the measurement was $S_r = 0.02–0.04$.

StatSoft Statistica 6.0 was used to carry out statistical processing of the data. The results of the cluster analysis were presented as clustering dendrograms by Ward's method and the graphs of distribution of the average values for the characteristics to be defined in each cluster were formed by the $k$-means method (Borovikov 1998; Soshnikova et al. 2004).

## 3 Results and Discussion

### 3.1 Protein Component of Urinary Stones

In mineral composition, urinary stones are distinguished in the following main groups: oxalate (oxalic acid) stones, phosphate (ortho-phosphate) stones, and urate (uric acid and its salts) stones. The most common stones are mixed type: oxalate, phosphate, oxalate, urate, and others (Golovanova 2004, 2007; Pyatanova 2004; Golovanova et al. 2004, 2006; Pal'chik et al. 2001, 2004; Potapov et al. 2002; Zuzuk et al. 2005).

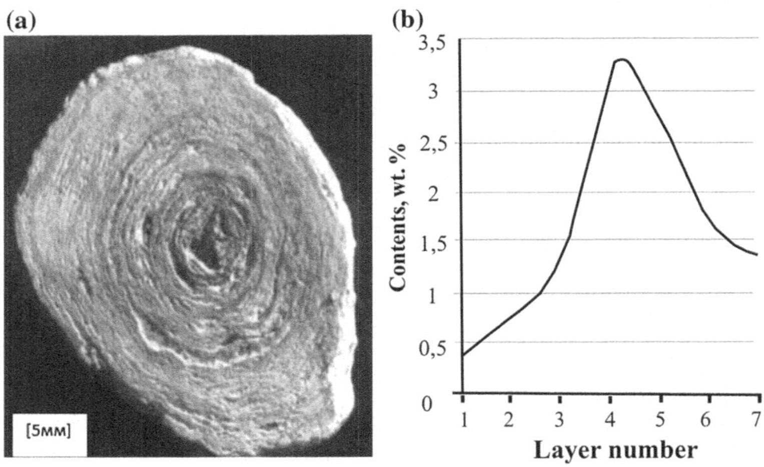


**Fig. 1** Layer-by-layer analysis of the protein compounds (**b**) in a urinary stone (**a**)

The layer-by-layer analysis of the urinary stones of a phosphate-oxalate type with zonal structure (Fig. 1a) shows that the distribution of the water-soluble protein compounds across the stone radius is not uniform (Fig. 1b). This behavior of the concentration of the protein compounds may be due to several causes: resolution caused by the stone growth, treatment, or changes in the conditions of the crystal-forming environment (dissolution stone). We have previously identified three main types of localization of organic protein substances in urinary stones (Golovanova 2007): nuclear (in the center of the aggregate), interlayer (concentric and radial layer), and diffuse (inclusions in mineral crystals). Typically, in kidney stones, all types of organic matter localization are observed simultaneously, with some distinctly prevailing types of localization.

Determination of the total protein content in oxalate (0.9 %) and phosphate-oxalate stones (0.55 %) shows that in oxalate stones, the protein content is almost twice as much than in stones of a phosphate-oxalate type. The statistical analysis shows that these differences are significant ($P = 0.95 = 4.66$ $t_{calc} > t_{tabl} = 2.37$).

The weight content of water-soluble organic substances with a peptide bond in urinary stones is 0.8–3.8 wt%, which depends on the classification group to which the sample belongs. The content of the above organic compounds in samples of a mixed type—phosphate-oxalate in particular—varies over a wider range. This can be caused by different period of change in pH during the formation of stones of a mixed type when organic substances are accumulated (Golovanova 2007; Pyatanova 2004; Zuzuk et al. 2005; Golovanova and Borbat 2005).

The Student *t*-test was used to find out that the difference in the average value of the content of substances with a peptide bond between the phosphate group and groups of oxalate stones, uric acid, urate, and calcium oxalate stones was reliably significant ($P = 0.95$; $t_{exp} = 10.89 > t_{tabl} > 2.37$; $t_{exp} = 8.2 > t_{tabl} = 2.57$; $t_{exp} = 19.6 > t_{tabl} = 2.45$, respectively). The distinction in the mean values of the substance with a peptide bond between the phosphate and phosphate-oxalate groups was not found ($P = 0.95$; $t_{exp} = 1.4 < t_{tabl} = 2.45$).

Thus, the content of the water-soluble organic substances with a peptide bond in urinary stones (Table 1) depends on the mineral composition of the sample. In addition, as we have shown previously, phosphate stones can accumulate trace elements to a greater extent than any other types of stones (Golovanova 2007). Along with the isomorphism of phosphate minerals, it can be due to the fact that

**Table 1** Content of water-soluble organic substances with a peptide bond in urinary calculus

| Stones | Range (wt%) | Mean content of water-soluble organic substances with a peptide bond (wt%) |
|---|---|---|
| Oxalate | 2.2–2.4 | 2.3 |
| Phosphate | 1.4–1.6 | 1.5 |
| Urate | 2.1–2.2 | 2.15 |
| Phosphate-oxalate | 0.88–3.8 | 2.34 |
| Urate-oxalate | 2.0–2.6 | 2.3 |

globular proteins dominate in the composition of the protein fraction in phosphate stones. The primary function of these proteins is to transport low-molecular-weight compounds and metal ions in the body; as a result, their capacity to bind and accumulate minerals is high.

## 3.2 Amino Acid Composition of the Protein Component of Urinary Stones

The study of the collection of urinary stones indicated 14 amino acids. Their quantitative content was determined in the organic component of oxalate, phosphate, urate, urate-oxalate, and phosphate-oxalate stones (Table 2).

The total amino acid content in the samples varies in the range of 0.6101–2.228 wt% and depends on the mineral composition of the stones. The qualitative composition of amino acids (whose number varies from 10 to 14) depends on the phase composition of the stones as well. This fact is proven by the mean values of the amino acid content in the stones from different clusters formed by the *k*-means method (Fig. 2). The data on the amino acid composition of urinary stones enables the identification of three groups of stones: Cluster 1 includes oxalate stones, cluster 2 covers phosphate stones, and cluster 3 contains urate stones.

Analysis of the results indicates that the ratio of the mean concentrations of amino acids in the urinary stones (glutamic acid > lysine > proline > alanine > threonine > valine > glycine > serine > phenylalanine > arginine > leucine > tryptophan > methionine > isoleucine) does not correspond to the ratio of the amino acid content in human urine (Severina 2003; glycine > glutamic acid > lysine > proline, alanine, serine, threonine > arginine, tryptophan > phenylalanine, leucine > valine > isoleucine > methionine). The content of glutamic acid and lysine in urinary stones is much higher than that of other amino acids, and it exceeds their content in the urine. This suggests that the most active role of these amino acids and proteins of high concentration in the formation of urinary stones is due to the presence of the additional amino group in lysine and additional residue carboxyl in glutamic acid. Inhibition of the growth of calcium oxalate monohydrate (basic mineral for urinary stones) by glutamic acid is confirmed by the results obtained in processing the differential curves of particle distribution by size while varying the crystallization time (Table 3).

**Table 2** Total content of amino acids in uroliths

| Stones | Amino acid content (wt%) | |
|---|---|---|
| | Extreme values | Mean value |
| Oxalate | 0.610–0.701 | 0.660 |
| Phosphate | 1.55–2.187 | 1.971 |
| Urate | 2.228 | 2.228 |
| Phosphate-oxalate | 0.918–0.996 | 0.957 |
| Urate-oxalate | 2.024–2.109 | 2.067 |

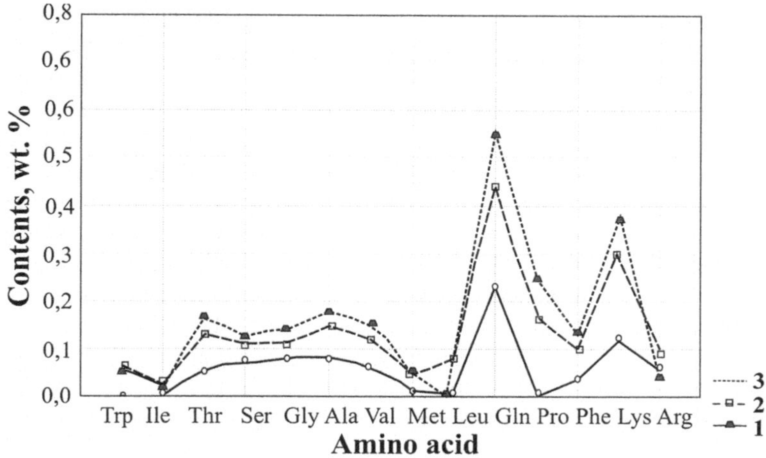


**Fig. 2** Mean values of amino acids in the urinary stones from different clusters: cluster 1 is oxalates, cluster 2 is phosphates, and cluster 3 is urates

**Table 3** Dependence of the average size of calcium oxalate crystals on glutamic acid concentration

| Model solution supersaturation $\gamma = 7$ | The particle diameter (d, μm) |
|---|---|
| Without additives | 30 |
| With glutamic acid under different concentration | |
| $C_{\text{flu}} = 10^{-4}\text{mol/l}$ | 24 |
| $C_{\text{flu}} = 10^{-3}\text{mol/l}$ | 16 |

It can be seen that when monohydrate calcium oxalate crystallizes in the solution in the presence of glutamic acid, the crystal growth is stabilized. When the concentration of amino acid grows, the inhibitory effect grows because the average size of calcium oxalate crystals reduces as the acid concentration rises. Inhibition can be accounted for by the adsorption of glutamic acid in the active centers of the formed crystals; as the acid concentration increases, the proportion of blocked centers grows (Golovanova et al. 2014).

Next, we studied the crystallization of hydroxyapatite (basic mineral for physiogenic and many pathogenic OMA) from the model solution in the presence of the amino acids glycine, lysine, glutamic, and aspartic acid, with concentrations close to their content in biological fluid (urine) 0.004 mol/l. These amino acids were chosen because of their relatively high content, both in the urate stones of a phosphate type and as part of a biological fluid (urine) where the genesis of these aggregates occurs.

To study the effect of amino acids on the crystallization of hydroxylapatite of different stoichiometric compositions, the solid phases were produced. The atomic ratio Ca/P was 1.58 ± 0.01, 1.65 ± 0.01, and 1.67 ± 0.01, respectively (Golovanova 2007). The analysis of variance results of the solid phases are shown in Table 4.

The data analysis shows that as the ratio of $Ca^{2+}/PO_4^{3-}$ in the solid phase and, therefore, in the reference solution grows, the average (from 17.30 μm to 28.35 μm) particle diameter increases. When comparing the results of the experiments on the effect of amino acids on crystallization of hydroxyapatite of different stoichiometric composition, it was found that aspartic and glutamic acids have an inhibitory effect on hydroxyapatite crystal growth in the range of the tested concentrations. It should be noted that the inhibitory effect of glutamic and aspartic acids increases as the ratio of Ca/P grows. This pattern can be attributed to an increase in the electrostatic force between the solid phase and these amino acids due to the change in the surface charge of the hydroxylapatite particles depending on the stoichiometry phase (Golovanova 2007). Thus, the electrostatic interaction between amino acids and the nonstoichiometric hydroxyapatite crystal surface increases as the negative charge of the dominating amino acid in the solution decreases. By the increase in the electrostatic interaction with non-stoichiometric hydroxyapatite, the tested amino acids can be ranked in order as lysine > glycine > glutamic acid > aspartic acid. These patterns are confirmed by the data obtained in the analysis of variance of the synthesized phases (Table 4). The ability of aspartic acid to inhibit the growth of the stoichiometric hydroxyapatite crystal to a greater extent, compared to glutamic acid, can also be attributed to formation of more stable chelate complexes with $Ca^{2+}$ ions (pK $(_{SaAsp})$ 1 = 1.989; pK $(_{CaGlu})$ 1 = 1.474).

The effect of glycine on hydroxyapatite crystallization differs from that of glutamic and aspartic acids, which is obvious from the experimental data on values and character of the synthesized phase particle size distribution. The analysis of the data shows that if the basic ratio in the solution $Ca^{2+}/PO_4^{3-} = 1.90$, the presence of glycine in hydroxyapatite crystallization (concentration = 0.004 mol/L) increases the average particle size compared to the phase without amino acid. However, if the basic values are different, such that $Ca^{2+}/PO_4^{3-} = 1.80$ and 1.67, the average particle size reduces compared with the control samples (Table 4), but to a lesser

**Table 4** Analysis of variance for the synthesized hydroxyapatite phases with amino acids ($t = 37.0 \pm 0.5$ °C; pH = 7.5 ±0.1; $C_{amino\ acide} = 0.004$ mol/l)

| | Amino acids | $Ca/P_{sol}$ = 1.58 ± 0.01 | | $Ca/P_{sol}$ = 1.65 ± 0.01 | | $Ca/P_{sol}$ = 1.67 ± 0.01 | |
|---|---|---|---|---|---|---|---|
| | | D, μm | Sr | D, μm | Sr | D, μm | Sr |
| 1 | – (0) | 17.30 | 0.02–0.03 | 22.54 | 0.02–0,03 | 28.35 | 0.02–0.03 |
| 2 | Glycine (Gly) | 16.94 | | 19.32 | | 32.10 | |
| 3 | Aspartic acid (Asp) | 15.62 | | 17.49 | | 18.20 | |
| 4 | Glutamic acid (Glu) | 14.65 | | 17.81 | | 21.10 | |
| 5 | Lysine (Lys) | 12.35 | | 16.54 | | 21.93 | |

degree than in the presence of aspartic and glutamic acids; however, the adsorption ability of glycine in hydroxyapatite crystals (2.4 nmol/m$^2$) is higher. Therefore, it can be assumed that glycine changes the mechanism of hydroxyapatite crystallization because the anionic form of glycine can replace hydroxyl ions in the hydroxyapatite structure (Rodicheva et al. 2000). Reduction of the phase particle size in the presence of aspartic acid and glutamic acid can be caused by an increase in the particle charge, which decreases aggregation.

## *3.3 Amino Acid Composition of the Protein Component of Dental and Salivary Stones*

The mineral composition of dental and salivary stones is less diverse than that of urinary stones. They are composed primarily of phosphate minerals, and apatite is the predominant mineral. In a study of the collection of dental and salivary stones, 15 amino acids were found, and their quantitative content was determined. It was revealed that the quality characteristics of amino acids were identical in all cases. On average, the total content of amino acids in salivary stones was slightly higher than that in dental stones (Table 5) because the concentration of protein in the environment of their formation was higher, which is consistent with the data in Golovanova (2007). In the oral liquid, the amino acid content (Leontiev and Galiulina 1991) is an order of magnitude less, which indicates their strong concentration in stones in crystallization.

The ratio of amino acids in each of the groups demonstrates some intergroup differences. Discriminant analysis (Fig. 3) was used to show the division into three groups.

The cluster analysis showed that dental and salivary stones are heterogeneous in amino acid composition. As an example, consider a group of salivary stones with three clusters differing in both the absolute content of glutamic acid and serine, as well as in their ratio (Fig. 4).

The amino acid content in cluster 1 averages 10.5 wt% with glutamic acid dominating (3:1). The same ratio of amino acids is characteristic of cluster 2 (samples 2, 3, and 6); however, the average content is much lower and equals 4.79 wt% The average age of the patients is 47 years in cluster 1 and 35 in cluster 2. This suggests that metabolism in the body is less imbalanced at a young age. For cluster 3, the average amino acid content is 6.6 wt% with serine dominating (1:1.3),

**Table 5** Total amino acid content (wt%)

| Object | Amino acid content | Mean amino acid content |
|---|---|---|
| Dental stones | 2.64–13.08 | 6.47 |
| Salivary stones | 4.64–12.31 | 7.54 |
| Oral fluid | 0.28–0.40 | 0.34 |

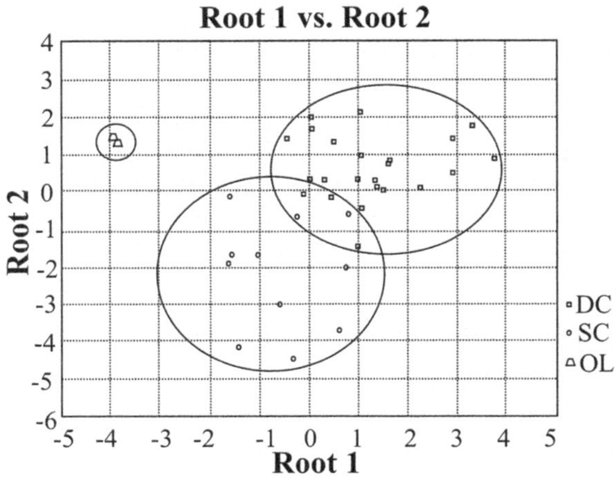


**Fig. 3** Diagram of division into three groups for dental (DC) stones, salivary (SC) stones, and oral liquid (OL) based on the results of the discriminant analysis

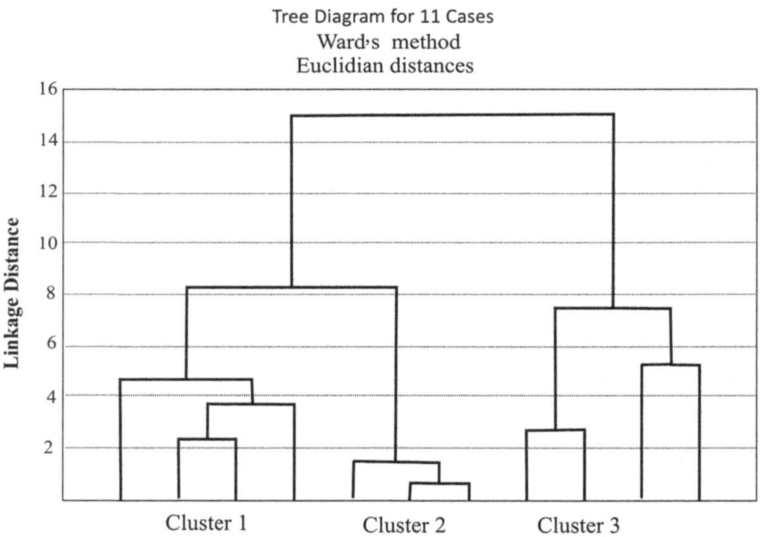


**Fig. 4** Tree diagram for combined samples of salivary stones based on the results of cluster analysis

which may indicate a predominance of this amino acid in the environment for corresponding OMA crystallization.

A comparison of the average amino acid content in dental and salivary stones showed that these series are different from those for oral fluid and different in different types of calculi:

**Oral liquid**: glutamic acid > lysine > leucine > glycine > aspartic acid > serine > alanine > arginine > tyrosine > valine > threonine > isoleucine > histidine > methionine.

**Dental stones**: glutamic acid > serine > alanine > glycine > asparagine > lysine > phenylalanine > valine > leucine > threonine > isoleucine > histidine > arginine > tyrosine > methionine.

**Salivary stones**: glutamic acid > serine > phenylalanine > lysine > arginine > asparagine > tyrosine > leucine > glycine > valine > alanine > histidine > isoleucine > threonine > methionine.

Glutamic acid and serine dominate in both types of stones. A high content of glutamic acid and serine can be due to a strong adsorption interaction with hydroxyapatite, caused by the presence of the additional amino group in glutamine acid and by the presence of the hydroxyl group in serine.

The measurements of the quantitative content of amino acids and elements (Golovanova 2007) (calcium, phosphorus, and others) were used to perform the correlation analysis. The inverse relationship between the total content of amino acids and phosphorus in salivary stones was obtained. This can be accounted for by the competition of the components in the solution for calcium ions in calculus formation and growth. Histidine plays a special role in this process (Table 6) because it is one of the most potent complexing agents among blood and saliva amino acids due to the electron-donor groups in the side chain (LeGeros et al. 1999). The obtained concentration series for different OMAs are compared in Table 6.

**Table 6** Concentration sequence of amino acids OMA from the human body

| OMA | | | Amino acid sequence |
|---|---|---|---|
| Bone section | Under normal conditions | | Glu>Gly>Ile>Leu>Ala>Pro>Lys>Arg>Thr>Val>Phe>Ser>Met |
| | Under pathological conditions | Lower | Glu>Gly>Ile>Leu>Ala>Lys>Pro>Arg>Thr>Val>Phe>Ser>Met |
| | | Intermediate | Glu>Gly>Ile>Leu>Lys>Pro>Ala>Arg>Thr>Val>Phe>Ser>Met |
| | | Upper | Glu>Gly>Ile>Leu>Ala>Lys>Pro>Arg>Thr>Ser>Val>Phe>Met |
| Pathogenic | Dental stones | | Gly>Ser>Ala>Glu>Asp>Lys>Phe>Val>Leu>Thr>Ile>His>Arg>Tyr>Met |
| | Salivary stones | | Gly>Ser>Phe>Lys>Arg>Asp>Tyr>Leu>Glu>Val>Ala>His>Ile>Thr>Met |
| | Renal stones (phosphate type) | | Gly>Lys>Ala>Pro>Thr>Val>Glu>Ser>Phe>Arg>Met>Leu>Ile |

In all OMAs (Table 6), the glutamic acid content is dominant, which indicates its particular role in the formation of these biominerals (Golovanova 2008; Golovanova et al. 2004, 2006; Luneva 2003; LeGeros et al. 1999; Shad et al. 2001). This amino acid is known to be in the body in an anionic form at physiological pH (e.g., blood pH 7.3–7.5), which may result in an interaction with positively charged biomineral regions (LeGeros et al. 1999; Shad et al. 2001).

## 3.4 *Amino Acid Composition of the Protein Component of the Bone Matrix*

The HPLC analysis of human bone tissues showed that the pathologic processes therein do not change the qualitative and quantitative composition of amino acids (Table 6). Therefore, transformation of the organic component in coxarthrosis is degradation, concretion, and, as a result, decrease in the solubility of collagen. A slight decrease in the total amino acid content in the damaged samples ($\Delta = 4$ wt%) can be observed. The amino acids were ranked by their content in powder samples. The obtained concentration sequences were compared with the amino acid composition of pathogenic OMAs from the human body (dental, salivary, and kidney stones, Table 6). The structure-forming role in the formation of both physiogenic (bones and teeth) and pathogenic biominerals belongs to glycine (Golovanova 2008; Fleischmajer et al. 2005; Gabuda et al. 2005; Yushkinan et al. 2007; Zhu et al. 2001). However, the data in Table 6 shows that this amino acid dominates in bone tissues, in contrast to pathogenic OMAs, because it is the main component of the ordered crystalline collagen polypeptide chain (*-Glu-XY-*) and its portions of fibers are the location for bone apatite crystallization.

In physiogenic biominerals, there is a significant quantity of nonpolar amino acids, such as leucine and isoleucine. In pathogenic biominerals, there are polar amino acids such as serine, proline, and lysine, and nonpolar biominerals, such as alanine and phenylalanine.

Statistical analysis of the data was performed on the content of the amino acids close in composition to the control samples of lower damaged sections with the Student *t*-test. The results showed that up to 60 years of age, for both groups of men and women, the amino acid composition of the bone tissue tended to change (Fig. 5a, b).

After 60 years, significant differences in the amino acid content by gender and age were revealed. The bone tissues of men and women of the third and fourth age groups differed significantly in the total content of amino acids ($P = 0.95$ $t_{calc} = 2.14 > t_{tabl} = 2.06$ and $t_{calc} = 2.50 > t_{tabl} = 2\ 06$, respectively). These differences were not found in samples of the bone tissue from the first and second age groups. In bone samples from men, the total amino acid content significantly decreases when transiting from the first and second age groups to the third one (Fig. 5a). A significant difference by the Student *t*-test is found between the first and

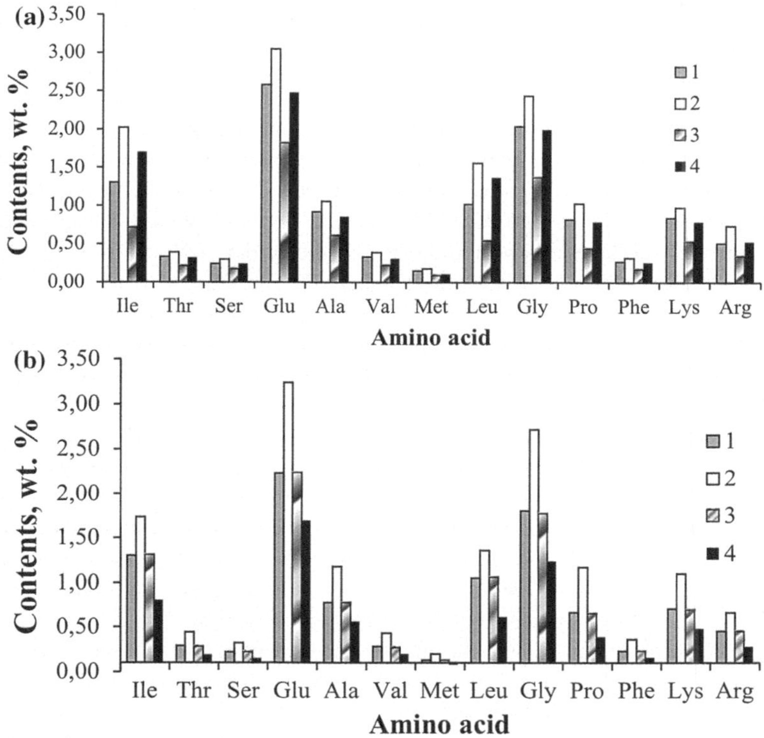


**Fig. 5** The content of amino acids in the bone tissue of men (**a**) and women (**b**) from different age groups: 30–49 years (*1*), 50–59 years (*2*), 60–69 years (*3*), and 70–79 years (*4*)

third age groups ($t_{calc} = 2.38 > t_{tabl} = 2.06$, $P = 0.95$), and the second and third ones ($t_{calc} = 2.85 > t_{tabl} = 2.06$, $P = 0.95$). In women's bone samples, a significant decrease in the concentration of amino acids is observed in the age range of 70–79 (Fig. 5b). The data obtained are confirmed by the cluster analysis.

It should be noted that after 60 years of age, the amount of amino acids (e.g., glutamic acid and glycine, whose biological roles are important in the body's vital processes) is reduced in the bone tissue. The decrease in the average concentrations of amino acids in the bone tissue of the men from the third age group and the women from the fourth age group may be caused by decrease in collagen as a result of the overall processes in the human skeletal system, concomitant diseases (e.g., osteoporosis), etc. (Dedukh et al. 2003).

### 3.5 *Thermal Analysis of the EPR Spectra of the Bone Matrix*

The thermal analysis was used to analyze the content and energy of transformation processes of the bone tissue components in the temperature range of 25–1000 °C. The transformation of the organic component proceeds at temperatures from 200 to 430 °C when low- and high-molecular-weight organic substances are removed (particularly noncollagenous proteins with the molecular weight lower than that of collagen; in association with structural water, II) and from 430 to 600 °C when collagen is disorganized (III). At 25–200 °C, the adsorbed water is lost (I). At temperatures above 700 °C, carbon dioxide releases from the mineral bone components due to the transition of nonstoichiometric carbonate hydroxylapatite to stoichiometric hydroxyapatite.

The content of the organic substances in the upper and middle sections of the damaged femoral heads of the patients from the first and second age groups was found to significantly change compared to the control samples and lower damaged plates. In annealing of the damaged samples, the weight loss of collagen decreases (III), and the content of the organic substances with a lower molecular weight (IV) increases (Fig. 6). The collagen mass loss can be attributed to both degradation of collagen fibers to substances with lower molecular weight (Fig. 7, Curve 2, lower decomposition temperatures III and greater thermal effect in removing III) and to more densely packed collagen fibers due to increased mechanical load on the bone (Fig. 7, Curve 3, higher decomposition temperatures III), which may be one of the causes for high hardness under pathological conditions. In the third and fourth age

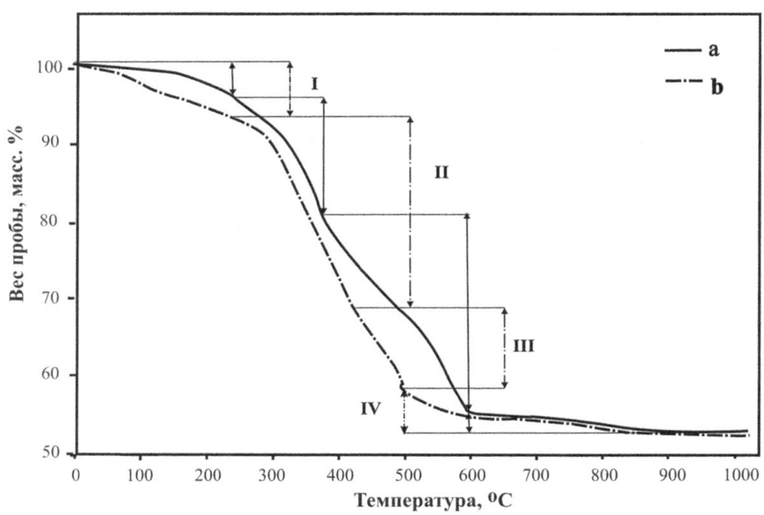


**Fig. 6** Thermogravimetric curves for the bone tissue: control (**a**) and damaged (**b**)

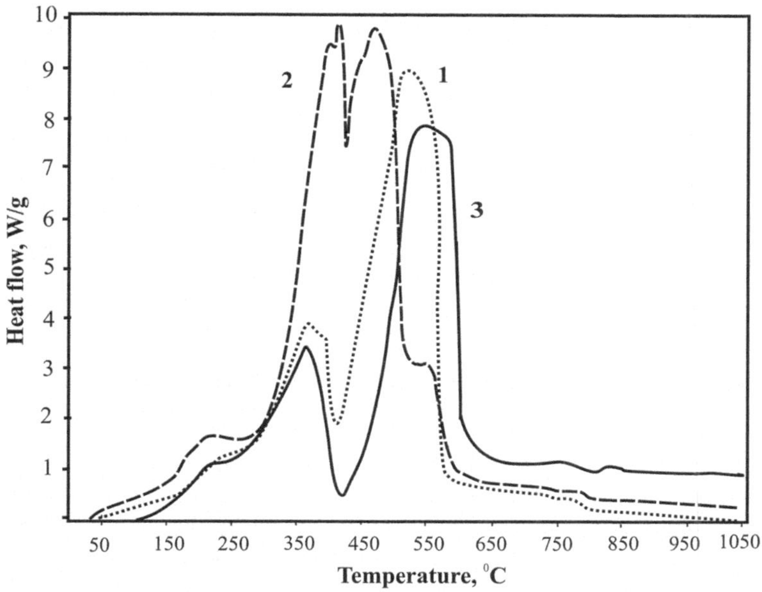


**Fig. 7** The differential thermal curves for the upper sections of the control (*1*) and damaged bone tissues (*2*), (*3*)

groups, wide confidence intervals of organic substance weight loss (II and III) were observed. This may be due to age-related changes in the bone tissue and concomitant diseases (e.g., osteoporosis).

Next, we investigated the nature of the paramagnetic thermochemical radical ions of the organic components. In thermal annealing of powder samples of the bone tissue at temperatures above 200 °C, a single signal "R-centers" of the organic component ($g = 2.0047 \pm 0.0005$, $\Delta H = 5.81 \pm 1.06$ G) can be recorded in the ESR spectra (Gilinskaya et al. 2003). This signal may be caused by unpaired electrons localized on the carbon atoms of the deformed (damaged) and/or broken bonds of the products of thermal transformation of the bone tissue organic component.

The temperature range of the radical ion stability fully coincides with the temperature range for the organic component decomposition (burning in air) of the bone tissue equal to 200–600 °C, where 200–430 °C corresponds to the temperature range of burning high-molecular-weight organic substances with the molecular weight lower than that of collagen (I); and 430 (470)–600 °C is the temperature for thermal transformation of the high-molecular-weight component—that is, collagen (II). In the thermal decomposition of substances whose composition is unordered, compared to collagen (*I*), a large number of paramagnetic radicals is formed (Fig. 8a, b) at 200–430 °C with spin-spin, dipole-dipole, and other interaction types stronger than those for the collagen component (II).

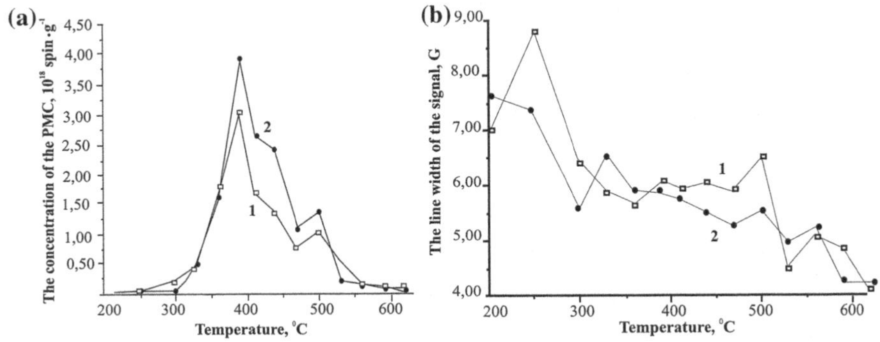


**Fig. 8** Dependence of the concentrations of radical ions in "R-centers" (**a**) and the signal band width in the EPR spectra (**b**) for the bone tissue of persons in the control group (*1*) and coxarthrosis patients (*2*)

In the bone samples of patients with coxarthrosis, organic substances with a low molecular weight dominate (I). This may be caused by degradation of collagen fibers under pathological conditions into amino acids and their degradation products containing atoms with unpaired electrons (e.g., pyruvate, oxaloacetate, acetyl-$K_0A$, amines). Microinteractions between radical ions of the modified collagen fibrils are not distinct enough to be attributed to close packing caused by redistribution of the load on the bone. For instance, the smallest signal band width of the EPR spectra for this type of sample corresponds to collagen decomposition temperatures (390–500 °C, Fig. 8b).

## 4 Conclusion

The obtained results allow us to conclude the following:

1. Significant relationships between the protein and mineral composition of OMAs are revealed.
2. The concentration of amino acids in stones of different types is found to be strong and selective, compared to the body fluids forming them.
3. The protein component actively participates in pathogenic phase crystallization in the human body, which is confirmed by the results of the model experiments on the crystallization of the main phases of urinary stones in the presence of amino acids under conditions close to physiological ones.
4. In the course of bone arthrosis, the molecular weight of the organic content is observed to be less than that of collagen. Close packing of the deformed collagen fibers is one of the causes for the increased hardness of "arthritic" bones.
5. The amino acid composition in the bone tissue of men and women in 60 years does not depend on the age and gender identity of the samples. After 60 years, the amino acid content varies based on age and gender.

**Acknowledgments** The study was partially supported by the Ministry of Education and Science of the Russian Federation within the framework of the state task of the universities in carrying out research work, project number 2953, and a scholarship of President of the Russian Federation for young scientists and graduate students, project number СП-933.2015.4.

## References

Borisov NV (2001) Assessment of meat productivity of cattle. Recommendation: Siberian Branch of RAAS, SibRPTIB, SibRIPGP. Novosibirsk (*in Russian*)

Borovikov V (1998) Program STATISTICA for students and engineers. Moscow, Statistics (*in Russian*)

Dedukh NV, Bengus LM, Basti A (2003) Magnesium and bone. Osteoporosis and Osteopathy 1:18–22 (*in Russian*)

Dorozhkin S (2007) Calcium orthophosphates. J Mater Sci 42(4):1061–1095

Egunov VP (1996) Introduction into thermal analysis. Samara, SamVit (*in Russian*)

Fleischmajer R, Gay S, Wilhelm N, Jerome S, Perlish P (2005) Collagen in the cellular and fibrotic stages of scleroderma. Arthritis Rheum 23(2):190–196

Gabuda SP, Gaydash AA, Drebushchak VA, Kozlova SG (2005) Physical properties and structure of water in fibrous proteins of collagen type according to data of scanning electron calorimetry. Lett JETPH 82(9):693–696 (in Russian)

Gilinskaya LG, Grigorieva TN, Okuneva GN, Vlasov YA (2003) The study of mineral formations in the human heart valves. J Struct Chem 44(4):678–689 (in Russian)

Golovanova OA (2004) Comprehensive study of kidney stones (Review) Proc. of HEI. Series «Chemistry and chemical technology», vol 47, pp 3–12 (Notes 1) (*in Russian*)

Golovanova OA (2007) Pathogenic minerals in the human body. Omsk (*in Russian*)

Golovanova OA (2008) Biomineralogy of urinary, biliary, dental and salivary stones from the human body. Dis. of Dr. Geol.-Min. Sciences, Omsk (*in Russian*)

Golovanova OA, Borbat VF (2005) Kidney stones. Moscow, Medical Book (in Russian)

Golovanova OA, Pyatanova PA, Rosseeva EV (2004) Analysis of the patterns of distribution of the urinary stone protein component. Rep Ac. S 395(5):1–3 (in Russian)

Golovanova OA, Rosseeva EV, Frank-Kamenetskaya OV (2006) The amino acid composition of human urinary stones. Bull SPSU 2(4):123–127 (in Russian)

Golovanova OA, Puning JO, Izatulina AR, Korol'kov VV (2014) Crystallization of monohydrate calcium oxalate in the presence of amino acids, features and patterns, review. J Struct Chem 55:178–193 (in Russian)

Kolesnikova EL, Mikhailova SS (ed) (2004) Human anatomy. Moscow, GEOTAR-Med (*in Russian*)

Kovalenko VN, Bortkewich OP (2003) Osteoarthritis. Moscow (*in Russian*)

LeGeros RZ, Bleiwas CB, Retino M, Rohanizadeh R, LeGeros JP (1999) Zinc effect on the in vitro formation of calcium phosphates: relevance to clinical inhibition of calculus formation. Am J Dent 12(2):65–71

Leontiev VK, Galiulina MV (1991) About micellar state of saliva. Dentistry 5:17–20 (in Russian)

Luneva SN (2003) Biochemical changes in the tissues of the joints in degenerative and dystrophic diseases and biological methods of correction. Dis. … Dr. Biol. Sciences, Tyumen,TSU (*in Russian*)

Pal'chik NA, Stolpovskaya VN, Leonova IV et al (2001) Features of the mineral composition and structure of urinary stones and their prevalence in patients from different regions of Novosibirsk region. Mineralogy of technogenesis. Miass, IMin RAS, vol 1, pp 99–108 (*in Russian*)

Pal'chik TA, Moroz TN, Leonova IV, Miroshnichenko LV (2004) Mineralization in the human body. Biobone interaction: life and stone. In: Proceedings of II international symposium. Saint Petersburg, MS RAN, pp 186–189 (*in Russian*)

Potapov SS, Parshina NV, Chiglintsev AY et al (2002) Statistical study of urolith and urolithiasis of the residents from Chelyabinsk region. Mineralogy of technogenesis—2002. Miass, Nauchmiass, pp 109–123 (*in Russian*)

Pyatanova PA (2004) Physicochemical study of kidney stones, formal genesis: Author's ab. Cand. Chem. Sciences: 02.00.01, Omsk (*in Russian*)

Rodicheva GV, Orlovsky VP, Romanova NM (2000) Synthesis and physicochemical study of glycine-containing calcium hydroxyapatite. J Inorg Chem 45(4):648–651 (in Russian)

Severina ES (ed) (2003) Biochemistry. Moscow, GEOTAR Med (*in Russian*)

Shad M, Ansari T, Afzal U, Kauser S, Rafique M, Khan M (2001) Major Constituents, free Amino acids and metal levels in renal calculi from Multan region. OnLine J Biol Sci 11:1063–1065

Shestak J (1987) Theory of thermal analysis: physical and chemical properties of inorganic solids. Moscow (*in Russian*)

Soshnikova LA, Tamashevich VN, Mahnach LA (2004) Multivariate statistical analysis. Minsk, BSEU (in Russian)

Yushkinan NP, Rakina VI, Kovaleva OV (ed) (2007) The origin of the biosphere and the co-evolution of mineral and biological worlds. Syktyvkar, IG Komi RC UB RUS (*in Russian)*

Zatsepin ST (2001) Bone pathology of adults. Moscow, Medicine (*in Russian*)

Zhu PX, Masuda Y, Koumoto K (2001) Adhesion of hydroxyapatite microparticles on charged surfaces in a supersaturated solution. J Coll Interface Sci 243:31–36

Zuzuk FV, Lvov NI, Franko UL (ed) (2005) Mineralogy of uroliths. Author. dis … Dr. Geol. Science Lutsk (*in Ukraine*)

# Computer Simulation of Defects in Carbonate Fluorapatite and Hydroxyapatites

**Elena A. Kalinichenko, Aleksandr B. Brik, Valentin V. Radchuk, Olga V. Frank-Kamenetskaya and Oleksii Dubok**

**Abstract** Some carbonate apatites were investigated by computer simulation using the General Utility Lattice Program (GULP) with grid techniques. These structures were modeled by a semiempiric method using interatomic potential-based simulation techniques. The supercells of 3 × 3 × 3 elementary cells of carbonate apatite were considered. The arrangement of structural defects (carbonate groups, excess $F_x$ ions, Na ions, Ca vacancies, hydroxyl groups, water molecules $H_2O_{str}$ fixed in structure) in carbonate fluorapatite (CFA) with the chemical composition near to the synthesized A-B type CFA (2.7 wt% $CO_2$, 3.3 wt% F) was established. Structural changes in the A type carbonate hydroxylapatite (CHA) due to formation of the $CO_2^-$-radical from the carbonate ion replaced two adjacent hydroxyls in the hydroxylapatite (HA) channel have been studied. The $CO_2^-$-radical formation is accompanied by the appearance of a Ca2 vacancy and results in the considerable

The original version of this chapter was revised: The spelling of the last author's name was corrected. The erratum to this chapter is available at DOI 10.1007/978-3-319-24987-2_41

E.A. Kalinichenko (✉) · A.B. Brik
Semenenko Institute of Geochemistry, Mineralogy and Ore Formation of NAS of Ukraine, 34 Palladin Ave., Kiev, Ukraine
e-mail: kalinichenko@igmof.gov.ua

A.B. Brik
e-mail: abrik@igmof.gov.ua

V.V. Radchuk
Institute for Telecommunications and Global Information Space of NAS of Ukraine, 13 Chokolovsky Blv., Kiev, Ukraine
e-mail: valentyn.radchuk@gmail.com

O.V. Frank-Kamenetskaya
Institute of Earth Sciences, St-Petersburg State University, 7/9 Universitetskaya Emb., St-Petersburg, Russia
e-mail: ofrank-kam@mail.ru

O. Dubok
Institute for Problems of Material Science of NAS of Ukraine, 3 Kryjanovskogo Str., Kiev, Ukraine
e-mail: dubok@mail.ru



O.V. Frank-Kamenetskaya et al. (eds.), *Biogenic—Abiogenic Interactions in Natural and Anthropogenic Systems*, Lecture Notes in Earth System Sciences, DOI 10.1007/978-3-319-24987-2_35

rearrangement of ions of the structural environment. The anneal effect on the A type CHA structure with the $CO_2^-$-radical ($CHA_r$) and possible accompanying defects has been investigated. The $CHA_r$ structure is most stable at the temperatures of $T = 298$ and 873 K; the CHA structure with the $CO_2^-$-radical and $O^-$-center spaced by one hydroxyl in the same channel ($CHA_{r\text{–}h\text{–}r}$) — at $T = 1073$ and 1173 K. The obtained data complement and are partly in agreement with experimental and theoretical investigations of CHA.

## 1 Introduction

Apatites $[Ca_{10}(PO_4)_6(F, Cl, OH)_2]$ exhibit wide ion variations (Pan and Fleet 2002). Apatites with substitutions by carbonate ions are widespread in nature (Mason et al. 2009; Nokhrin et al. 2006; Pan and Fleet 2002; Yi et al. 2013 and others). Ion substitutions in carbonate apatites characterize the genesis of geological rocks in postmethamorphic processes (Mason et al. 2009; Pan and Fleet 2002). Nonstoichiometric carbonate hydroxyapatite is the main constituent of high-mineralized tissues (bones and teeth) of living organisms (Frank-Kamenetskaya et al. 2011; Pan and Fleet 2002).

The intensive investigations of carbonate substitutions in fluorapatite $Ca_{10}(PO_4)_6F_2$ (FA) and hydroxylapatite $Ca_{10}(PO_4)_6(OH)_2$ (HA) are explained by the necessity of solving mineralogical tasks and developing synthesis technologies of materials for various applications, including biocompatible materials and matrices of sorbents and catalysts on FA and HA basis (Brik et al. 2005, 2013; Fleet and Liu 2008; Frank-Kamenetskaya et al. 2011; Mason et al. 2009 and others).

Numerous studies of carbonate apatites of different origins have shown that carbonate ions can substitute F (OH) ions in the *c*-axis channels (A type replacement) and $PO_4^{3-}$ ions (B type replacement), resulting in the appearance of accompanying structural defects and changes of physical and chemical properties (Fleet and Liu 2008; Frank-Kamenetskaya et al. 2011; Pan and Fleet 2002).

The mechanisms of carbonate ion accommodation in FA and HA differ noticeably but have some similarities (Fleet and Liu 2008; Frank-Kamenetskaya et al. 2011; Mason et al. 2009; Pan and Fleet 2002). The uptake of water molecules fixed in the structure ($H_2O_{str}$) of carbonate apatites is substantially governed by the types of carbonate substitution and channel anions (Frank-Kamenetskaya et al. 2011; Mason et al. 2009; Regnier et al. 1994). The $H_2O_{str}$ molecules are assumed to stabilize the poorly ordered structure of carbonate apatites. These molecules occupy hydroxyl vacancies in CHA channels (Frank-Kamenetskaya et al. 2011) and are incorporated near carbonate ions in CFA: in F vacancies in channels near A carbonates and out of channels near B carbonates (Mason et al. 2009).

The accommodation of fluorine ions out of channels (excess $F_x$ ions) are determined by the chemical composition of CFA and the presence of other structural defects (Fleet and Liu 2008; Mason et al. 2009; Regnier et al. 1994; Yi et al. 2013). The F content in CFA with a high content of B type carbonate can exceed 2 atoms per formula units (apfu) (Nokhrin et al. 2006; Regnier et al. 1994). Vacancies (V) of phosphate groups can appear in such structures (Regnier et al. 1994). Vacancies (V) of hydroxyl groups in channels are typical for the CHA structure (Frank-Kamenetskaya et al. 2011). At the same time, the trends of ion substitutions in carbonate apatites, particularly the arrangement of defects in CFA depending on chemical composition, have been investigated but not enough.

Electron paramagnetic resonance (EPR) is widely used to investigate the crystallochemical features of apatites (Brik et al. 2005; Ikeya 1993; Nokhrin et al. 2006; Oliveira et al. 2000; Schramm et al. 2001; Vorona et al. 2006 and others). The strongest peaks in EPR spectra of biogenic and synthetic carbonate apatites are produced by $CO_2^-$-radicals on crystallite surfaces and in bulk structures (Brik et al. 2005; Ikeya 1993; Oliveira et al. 2000; Schramm et al. 2001 and others). The $CO_2^-$-radicals in apatite channels have been established to form from carbonate ions under irradiation and are responsible for axial and orthorhombic components of EPR signals (Brik et al. 2005; Oliveira et al. 2000; Schramm et al. 2001). Associated formation of $O^-$-centers and about half of $CO_2^-$-radicals in channels of A type CHA have been supposed (Brik et al. 2005). At the same time, forming mechanisms, peculiarities of structural environment, temperature stability of $CO_2^-$-radicals, and $O^-$-centers in apatite channels require further investigation. Such researches are needed to develop methods of analysis of EPR signals of these radicals differed with EPR parameters and temperature stability.

Computer simulation complements the experimental methods of investigation of the structural properties of solids (Gale and Rohl 2003; Peroos et al. 2006; and others). Apatite structures with various substitutions have been simulated by several authors (De Leeuw 2004; Peroos et al. 2006; Schramm et al. 2001; Zahn and Hochrein 2006; and others). The MD simulation has been employed to study the accommodation of the carbonate ion of A and B types in monoclinic HA (Peroos et al. 2006). Schramm et al. (2001) found that the electron density distribution near the carbonate ion and $CO_2^-$ radical in the channel of A type CHA differ significantly, but the effective charges of carbonate oxygen ions are practically equal.

This chapter presents the results of computer modeling of the arrangement of impurity ions and groups and accompanying structural defects in the A-B type CFA and structural changes in the A type CHA caused by the formation of the $CO_2^-$-radical from the A carbonate ion at temperature $T = 298$ K and following annealing at $T = 873$, 1073, and 1173 K.

## 2 Method of Computer Simulation

**Carbonate apatite structures** have been modeled using a semiempirical method considering 3 × 3 × 3 supercells. An ion shell model and interatomic potentials have been used to approximate interactions between ions in apatites with carbonate substitutions (Peroos et al. 2006). The following potentials at parameters listed in Peroos et al. (2006) have been used: the spring potential to take into account polarizability of oxygen ions in the core-shell model; the Coulomb potential for interaction between charge species (the real space range cutoff 10 Å); the Buckingham potential for short-range ionic interaction (the range cutoff 20 Å); the Morse potential for the covalent bonds in phosphate, carbonate, and hydroxide ions; the three-body harmonic angle term for the covalent bond angles in phosphate and carbonate ions; and the four-body term for the torsion energy of carbonate ions.

The supercells of FA ($Ca_{270}$ $[PO_4]_{162}F_{54}$) and HA ($Ca_{270}$ $[PO_4]_{162}OH_{54}$) have been modeled at the used potential parameters for comparison. The atomic coordinates of ions and the lattice parameters of hexagonal FA and HA structures (Hughes et al. 1989) were used as an initial approximation.

The optimization of structure variants with different arrangements of defects have been carried out using the GULP (Gale and Rohl 2003). The Helmholtz free energies $F$ (eV per one elementary cell) of optimized structures have been calculated in order to determine the most stable structures with the min($F$).

**CFA structure**. The simulation model was modified to take into account the polarizability of fluorine ions and interactions of fluorine ions and fixed $H_2O_{str}$ molecules. The additional Lennard-Jones potential for the interaction between oxygen ions of $H_2O_{str}$ molecules was included. The parameters of potentials for fluorine ions and ions of $H_2O$ molecules on the HA surface with $F^- \rightarrow OH^-$ substitutions (De Leeuw 2004) were employed.

The structure of the synthesized A-B type CFA sample with 2.71 wt% $CO_2$, 3.3 wt% F and the lattice parameters of $a = 9.3141$, $c = 6.8991$ Å (Brik et al. 2013) was considered. Nuclear magnetic resonance (NMR) investigations have shown that there were 1.1 wt% $H_2O$ ($H_2O_{str}$ molecules and hydroxyl groups) and some amount of fluorine ions out of channels in this sample. The results of chemical analysis and NMR research (Brik et al. 2013) have allowed us to determine the following crystallochemical formula of the synthesized sample

$$\left[Ca_{9.355}Na_{0.325}(NH_4^+)_{0.012}\right]_{9.692}\left[(PO_4)_{5.196}(CO_3)_{0.554}\right]_{5.75}\left[F_{1.316}OH_{0.61}(CO_3)_{0.037}\right]_2(F_x)_{0.351}\cdot 0.275H_2O_{str}. \quad (1)$$

The relatively large amount of $OH^-$ groups in channels must be noted. Presumably, it is associated with the uptake of some amount of F ions out of channels (by $^{19}F$ MAS NMR data), although the F content in the synthesized sample is below the stoichiometric one (Brik et al. 2013).

The chemical composition of the supercell was based on this crystallochemical formula. Low degrees of substitutions of $F^- \rightarrow O^{2-}$ and $Ca^{2+} \rightarrow NH_4^+$ (Brik et al. 2013) have been neglected. The following supercell of A-B type CFA was simulated at $T = 298$ K:

$$[Ca_{252}Na_9V_9]_{270}\left[(PO_4)_{140}(CO_3)_{15}V_7\right]_{162}\left[F_{36}OH_{16}(CO_3)_1\right](F_x)_9\cdot 7H_2O_{str}. \quad (2)$$

The chemical composition of this CFA structure (wt%: $CO_2$ — 2.73, F — 3.31, CaO — 54.74, $P_2O_5$ — 38.49, $Na_2O$ — 1.08, $NH_3$ — 0) is close to the chemical composition of the synthesized CFA (Brik et al. 2013). The substitution degree of B type carbonate (0.9375) is approximately the same as in the synthesized sample (0.9394).

The following features of the simulated structure have been taken into account: 16 hydroxide and carbonate ions, $7H_2O_{str}$ molecules and vacancies of phosphate groups, 9 excess $F_x$ ions out of the channels, and Na ions and Ca vacancies in the supercell (2). Experimental investigations have shown the approximately even distribution of B carbonate ions and defects of the same type distanced from each other in B type CFA (Fleet and Liu 2008). A similar arrangement of structural defects in the modeling supercell (2) has been considered. The A-B carbonate cluster (closely spaced carbonate ions of A and B types) has been placed approximately in the supercell center. The $H_2O_{str}$ molecules have been supposed to occupy phosphate group vacancies located near B carbonate ions taking into account the uptake of $H_2O_{str}$ molecules out of channels near B carbonates in CFA (Mason et al. 2009) and the equal amount of $H_2O_{str}$ molecules and such vacancies in the supercell (2).

As an initial approximation, the atomic coordinates of ions in FA and hydroxyl ions in HA (Hughes et al. 1989), carbonate group ions and excess $F_x$ ions in CFA (Fleet and Liu 2008), and four variants of excess $F_x$ ion sites, taking into account the possible incorporation of $F_x$ ions in oxygen vacancies near B carbonates were employed. We considered 12 variants of possible orientations of protons of $H_2O_{str}$ molecules in phosphate group vacancies, 9 - hydroxyl protons in channels and 4 - hydroxyl protons near A carbonate ion, 9 (in M1) and 8 (in M2) - Na ions and Ca vacancies in cationic M1 and M2 sites at distributions of Na1 and $V_{Ca2}$ or $V_{Ca1}$ and Na2 for each of 12 variants of distributions of excess $F_x$ ions.

**CHA structure with the $CO_2^-$-radical**. The effective charge of the radical ($q^{eff}(CO_2^-) = -0.955e$) is close to the formula value ($-1e$) at the used potential parameters (Peroos et al. 2006). Therefore, the $CO_2^-$ radical may be modeled at those potential parameters. The atomic coordinates of ions of the A type carbonate group (Fleet and Liu 2005, 2008) with one oxygen vacancy or presumed coordinates of the $CO_2^-$ radical ions (Schramm et al. 2001) in A type CHA have been used to simulate the uptake of the $CO_2^-$ radical in the channel.

The $O^-$-center may be represented by the hydroxide oxygen ion ($O_h$) near the proton vacancy. The following effective charges of the $O^-$-center were considered: $q^{eff}(O^-) = -1.4e$, the effective charge of the $O_h$ ion at the used potential parameters (Peroos et al. 2006); and $q^{eff}(O^-) = -1.0e$, the effective charge obtained as a result of proportional decrease of the core and shell charges of the hydroxide oxygen ion.

The CHA supercell with composition $[Ca_{270-m}V_m]_{270}[PO_4]_{162}[OH_{52-n-p}(CO_t)_1V_nO_p^-]_{54}(0.16\text{wt}\,\%CO_2)$ was considered. The atomic coordinates of ions and the lattice parameters of the hexagonal HA structure (Hughes et al. 1989) were used as an initial approximation.

The simulation of the hexagonal HA structure showed that hydroxyl protons ($H_h$) in the most stable HA structure were oriented in the opposite directions: $O_h$–$H_h$ $H_h$–$O_h$. The same orientation of hydroxyl protons in the channels without a carbonate defect was considered as an initial approach.

The parameters of $n$ and $m = 0$ or 1, $t = 2$ or 3, and $p = 0$ have been considered to study the structural changes caused by forming of the $CO_2^-$-radical from the A carbonate ion substituted for two adjacent hydroxides in the channel at $T = 298$ K. The atomic coordinates of ions of A type carbonate groups in CHA (Fleet and Liu 2005, 2008) were used as an initial approximation.

The parameters of $n$, $m$ and $p = 0$ or 1, $t = 2$ have been used to investigate annealing effects on the CHA structure with the $CO_2^-$-radical in the channel. The lattice parameters of HA at $T = 873$, 1073, and 1173 K (Aminatun et al. 2013) were used to simulate annealing effects at these temperatures.

We considered 12 variants of possible positions of Ca1 and Ca2 vacancies, 4 for the $O^-$-center positions, 4 for hydroxyl vacancy positions, and 5 variants of possible orientations of hydroxyl protons in the channel with the carbonate defect.

## 3 Results and Discussions

### 3.1 Arrangement of Structural Defects in CFA

Grid techniques were employed to accelerate optimization of about of 300 structure variants, from which 35 structures were optimized. The lowest Helmholtz free energies $F$ of the optimized CFA structures are in the range from – 438.7 to – 434.0 eV. These values were approximately the same order of magnitude as the more stable FA structure (with the lowest value of $F = -461.8$ eV) at used potential parameters.

The lowest distortion of $H_2O_{str}$ molecules was found in the CFA structure with $F = -434.0$ eV. The distances between ions of $H_2O_{str}$ molecules in this structure ($|O_w–H_w|^{str} = 1.15–1.55$ Å, $|H_w–H_w|^{str} = 1.96–2.33$ Å) are the closest to those in a water molecule ($O_w–H_w = 0.96$ Å, $H_w–H_w = 1.58$ Å), where the subscript "w" designates ions of a water molecule. The interaction of protons in this CFA structure may cause the doublet line in the $^1H$ NMR spectrum with the split of $\Delta\nu_{H\text{-}H}{}^{gps}\cdot(|H_w–H_w|/|H_w–H_w|^{str})^3 \approx 14.5–24.1$ kHz, where $\Delta\nu_{H\text{-}H}{}^{gps} = 46$ kHz—the doublet split of crystallohydrated $H_2O$ molecules in gypsum structure (Abragam 1961). The experimental investigations have shown a doublet with a split of $\Delta\nu_{H\text{-}H} \approx 34$ kHz caused by $H_2O_{str}$ molecules in the $^1H$ NMR spectrum of the synthesized CFA sample (Brik et al. 2013). Taking into account these results, this CFA structure with the lowest distortion of $H_2O_{str}$ molecules seems to be the most stable.

More low values of $F$ (from −438.7 to −434.6 eV) have been obtained for a few structure variants. However, $H_2O_{str}$ molecules in those structures are substantially distorted: $|O_w–H_w|^{str} = 1.6–3$ Å, $|H_w–H_w|^{str} = 3–5$ Å. The interactions of protons are neglected with such distances between them and cannot be observed in $^1H$ NMR spectra.

The elementary cell of the most stable CFA structure is a bit deformed ($\alpha = 92.2°$, $\beta = 90.9°$, $\gamma = 117.9°$) relative to the hexagonal structure. The lattice parameters of this CFA structure ($a = 9.445$ Å, $b = 9.389$ Å, $c = 7.015$ Å) are a bit higher than the synthesized CFA (Brik et al. 2013). The most stable CFA structure has the most displacements of ions of $H_2O_{str}$ molecules and hydroxyl groups, up to 0.7 Å from initial positions (so that $O_h$–$H_h$ = 1.08 Å, while it is 0.96 Å at initial coordinates); displacements of C ions were up to 0.3 Å; and other ions do not exceed 0.1 Å from initial positions in FA structure.

Figure 1 shows the typical structures of the channels in the most stable CFA structure (ions of Ca1 and phosphate groups are not shown). Excess $F_x$ ions occupy O3 vacant oxygen sites near B type carbonates, including the A-B cluster, so that $F_x$ ions are the most distanced from $H_2O_{str}$ molecules. No dependence was found between the localization of excess $F_x$ ions and $H_2O_{str}$ molecules: four $F_x$ ions are incorporated not far from $H_2O_{str}$ molecules (Fig. 1b), and four those are distanced from $H_2O_{str}$ molecules (Fig. 1c).

The configuration of $[CO_3F_x]_B$ clusters represents a tetrahedron: carbonate oxygen ions ($O_c$) and excess $F_x$ ions occupy vacant sites of phosphate oxygen ions ($O_p$), and carbon ions lie approximately in the centers of faces formed by $O_c$ ions. Conversely, the B type carbonate groups near O3 oxygen vacancies are substantially distorted: carbon ions are substantially displaced from $O_c$ ion planes, although two $O_c$ ions occupy approximately $O_p$ vacant sites.

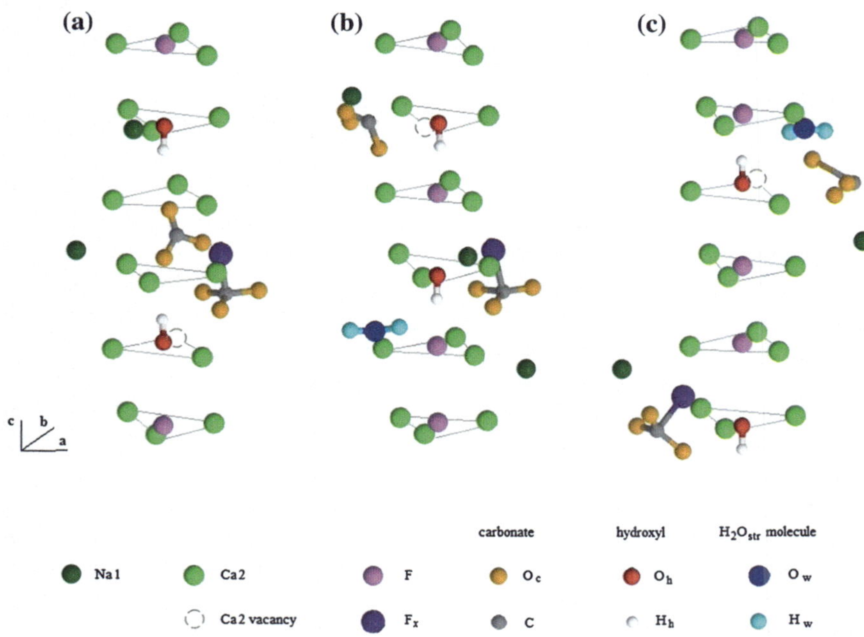


**Fig. 1** The typical channel structures in the most stable CFA structure: **a** the A-B cluster (1 channel) and (**b**, **c**) $[CO_3F_x]_B$ clusters (of 4 similar channels): **b** not far from $H_2O_{str}$ molecules and **c** distanced from $H_2O_{str}$ molecules. The triangles of Ca2 ions are marked. The ions of Ca1 and phosphate groups are not shown

The $H_2O_{str}$ molecules occupy phosphate group vacancies, so that $O_w$ ions lie approximately in vacant phosphorous sites and protons $H_w$ are oriented to vacant oxygen sites of O1 and O2. Hydroxyls in channels are incorporated near B type carbonates, so that $H_h$ protons are oriented toward the carbonate ion planes.

Na1 ions and Ca2 vacancies are the most distanced from each other and located mainly as follows: Na1 ions lie near the $[CO_3F_x]_B$ clusters (except the A-B cluster), close to O1 oxygen ions and $F_x$ ions, Ca2 vacancies form near B type carbonate ions with O3 oxygen vacancies. Two Na1 ions are incorporated near the B type carbonate ions with O3 oxygen vacancies; one Na1 ion and two Ca2 vacancies are relatively distanced from other defects. The obtained data confirm experimental and published data for A-B type CFA.

Computer modeling shows that F ions in the most stable CFA structure occupy two different structural sites—in the channels and out of channels near B carbonate ions (Fig. 1). This confirms the results of $^{19}F$ NMR investigations of the synthesized sample (Brik et al. 2013). The excess $F_x$ ion sites differ depending on the chemical composition of CFA (Fleet and Liu 2008; Mason et al. 2009; Regnier et al. 1994). The similar configuration of $[CO_3F_x]_B$ clusters has been considered by Yi et al. (2013).

The configuration of hydroxyl groups surrounded by F ions in channels (Fig. 1 b, c) corresponds to the results of $^1H$ NMR investigations of the synthesized sample (Brik et al. 2013). The hydrogen-bonded (H-bonded) interactions of the hydroxyl proton with the nearest F ion in the channel ($H_h \cdots F \approx 2.15$ Å) and one of the carbonate oxygen ions ($H_h \cdots O_c \approx 3$ Å) must lead to a substantial decrease of electron density on this proton. It explains the high-frequency shift of the hydroxyl band in the $^1H$ MAS NMR spectrum of the synthesized CFA (Brik et al. 2013). The electron density distributions in the structural environment of hydroxyl groups located near the B carbonate group or $[CO_3F_x]_B$ cluster (Fig. 1 b, c) are distinguished. The presence of the excess $F_x$ atom with the highest electronegativity, relatively not far from the $OH^-$ group ($F_x$–$O_h \sim 3.2$ Å, $F_x$–$H_h \sim 4$ Å), causes a noticeable decrease of electron density on ions of this hydroxyl. This may explain the appearance of the second high-frequency band in the $^1H$ MAS NMR spectrum of the synthesized CFA containing excess $F_x$ ions (Brik et al. 2013).

The water molecule incorporated in the phosphate group vacancy (Fig. 1b, c) with the notably greater volume, presumably fixes in the structure less firmly than in other structural sites. It explains the substantial growth of mobility of $H_2O_{str}$ molecules at $T \geq 400$ K in the synthesized sample by in situ $^1H$ NMR data. The association of Ca2 and O3 vacancies near B carbonate groups (Fig. 1b, c) has been established as a result of investigations of natural CFA with $F > 2$ pfu (Nokhrin et al. 2006). The orientation of hydroxyl protons toward the carbonates planes (Fig. 1) corresponds to modeling results of substitutions by B type carbonates in HA (Peroos et al. 2006).

# 4 The $CO_2^-$-radical in A Type CHA

Grid techniques were employed to accelerate optimization of about 160 structure variants, of which 90 structures have been optimized. The Helmholtz free energies of the most stable CHA structures (with the min ($F$)) with the A carbonate group ($CHA_c$) and $CO_2^-$ radical ($CHA_r$) are approximately the same order of magnitude with higher probability of carbonate substitutions (Table 1) in comparison with the most stable HA structure at used potential parameters ($F_{HA} = -481.68$ eV).

**Table 1** Parameters of the most stable CHA structures, atomic coordinates of the carbonate impurities and some characteristic interatomic distances in these structures

| | Structure | | | |
|---|---|---|---|---|
| | $CHA_c$ | | $CHA_r$ | |
| **Helmholtz free energy (eV per one elementary cell)** | | | | |
| $F$ | –483.43 | | –482.82 | |
| **Lattice constants** (Å) | | | | |
| $a$ | 9.332 | | 9.327 | |
| $b$ | 9.339 | | 9.325 | |
| $c$ | 6.884 | | 6.890 | |
| **Atomic coordinates (x,y,z)** | | | | |
| | **carbonate ion** | | **$CO_2^-$ radical** | |
| C | −0.016, −0.010, 0.499 | | −0.034, −0.033, 0.486 | |
| $O_c$ | −0.096, −0.157, 0.513 | | | |
| $O_c$ | 0.012, 0.070, 0.644 | | −0.002, −0.009, 0.649 | |
| $O_c$ | 0.028, 0.050, 0.342 | | 0.017, 0.016, 0.334 | |
| **Interatomic distances (Å)** | | | | |
| C–$O_c$ | 1.189, 1.192, 1.194 | | 1.149, 1.153 | |
| $O_c$–$O_c$ | 2,044, 2.052, 2.096 | | 2.18 | |
| C–Ca2 | 2.87–3.08 | | 2.71–3.19 | |
| C–P | 3.91–4.10 | | 3.71–4.22 | |
| H–P | 3.68–4.35 | | 3.73–4.26 | |
| **Channel fragment** | **O–H H–O** | **H–O···H–O** | **O–H H–O** | **H–O···H–O** |
| C–$O_h$ | 5.57 | 5.30 | 5.78 | 4.91 |
| $O_c$–$O_h$[a] | 4.61 | 4.25 | 4.72 | 3.78 |
| C–H | 6.45 | 4.19 | 6.85 | 3.79 |
| $O_c$–H[a] | 5.69 | 3.13 | 5.81 | 2.66 |
| $O_h$–$O_h$ | 2.64 | 2.24 | 2.62 | 3.10 |
| $O_h$–H | 1.085, 1.086 | 1.13, 1.12 | 1.08, 1.09 | 1.12 (×2) |
| $O_h$–H[b] | 1.55 (×2) | 1.11 | 1.54, 1.55 | 1.98 |
| H–H[b] | 0.464 | 2.23 | 0.46 | 3.10 |

[a]For axial $O_c$ ions
[b]For ions of different hydroxyls

Figure 2 shows the structures of the channels with the carbonate ion and the $CO_2^-$-radical in the most stable structures (ions of Ca1 and phosphate groups are not shown). The Ca2 vacancy presents near the radical in the $CHA_r$ structure, unlike the $CHA_c$ one (Fig. 2). The $CHA_r$ structure without the Ca2 vacancy is less stable ($F = -481.96$ eV).

Accommodation of the carbonate ion in the channel results in a noticeable decrease in the **a** axis dimension of $CHA_c$ (Table 1), relative to the most stable HA structure at the used potential parameters ($a$ = 9.346, $c$ = 6.884 Å). The lattice parameters of this HA structure conform to the results from the HA structure simulation (Peroos et al. 2006). Radical forming resulted in further contraction of the $a$-dimension and a small expansion of the **c** axis dimension (Table 1).

Hydroxyl ions in the most stable CHA structures are displaced up to 0.7 Å, C ions are displaced up to 0.3 Å, and other ions are displaced no more than 0.1 Å from initial sites in HA structure (Hughes et al. 1989). The ion sites of the A carbonate group in $CHA_c$ (Table 1) are distinguished from those in A type CHA (Fleet and Liu 2005): the oxygen ion was out of channel and the carbon ion ($z \approx 1/2$) was displaced toward to the **c** axis, with axial oxygens - away from this axis. The oxygen ion out of channel is most notably displaced along the hexagonal axis, to the plane $z \geq 1/2$. The sites of the radical ions in $CHA_r$ differ: the carbon ion is displaced on ≈0.33 Å towards the Ca2 vacancy, the radical oxygens - towards the hexagonal axis. One $O_c$ ion is incorporated approximately in the $O_h$ site, whereas another $O_c$ ion is located near the Ca2 vacancy and the other $O_h$ site. The $O_c$–$O_c$ axis is oriented at an angle of 5.4° with the **c** axis.

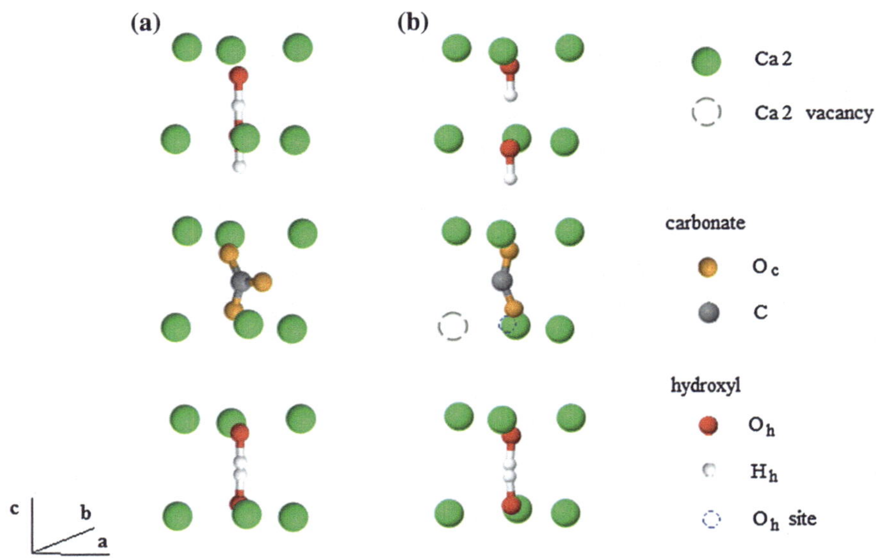


**Fig. 2** The channel structures in the most stable type A CHA structure with: **a** the carbonate group and **b** the $CO_2^-$-radical. Ions of Ca1 and phosphate groups are not shown

The carbonate ion geometry in $CHA_c$ is close to ideal (Fleet and Liu 2005): $O_c$ ions form approximately an equilateral triangle (Table 1) with bond angles of $<O_c–C–O_c = 118–122° \approx 120°$. The radical ions in $CHA_r$ are displaced so that $C–O_c$ bond lengths go down on ≈0.4 Å, $O_c–O_c$ distance increases (Table 1), and the bond angle rises up to 142°.

Accommodation of carbonate impurity leads to noticeable rearrangement of the structure of the nearest environment. The Ca1O9 polyhedrons in $CHA_c$ are the most substantially deformed: interatomic distances of Ca1–O2 and Ca1–O3 are in the range of 2.24–2.98 Å, unlike the HA structure in which these distances are 2.45 and 2.81 Å, accordingly (Hughes et al. 1989). The changes of $P–O_p$ bond lengths in phosphate groups in $CHA_c$ do not exceed 0.01 Å, but the tetrahedron configurations of phosphate groups are skewed: bond angles vary in the range from 102 to 115°.

The $CO_2^-$-radical forming results in displacements of ions that lead to an increase of structure distortion. The most deformed phosphate groups are near the Ca2 vacancy: $P–O_p$ bond lengths are in the range from 1.522 to 1.545 Å. The ranges of distances between the carbon ion and ions of the nearest structural surrounding considerably expand (Table 1). The deformation features of Ca1O9 and $Ca2O6O_h$ polyhedrons change, while the ranges of Ca1–O and Ca2–O distances narrow.

The structures of hydroxyl channels are the most substantially rearranged. One proton is reoriented toward one of the carbonate oxygen ions so that a channel fragment forms with unidirectional O–H bonds (Fig. 2), likely the monoclinic HA structure. This configuration leads to H-bonded interactions between proton and oxygen ions of different hydroxyls: $H\cdots O_h$ distances decrease to 1.11 and 1.98 in $CHA_c$ and $CHA_r$, respectively (Table 1), so that approximately linear $O_h–H\cdots O_h–H$ structures form (Fig. 2). This results in a decrease of $O_h$–H-bond strength: $O_h$–H bond lengths in hydroxyls rise, in relation to those in O–H H–O fragments (Table 1) and in HA structure (1.07 Å—Hughes et al. 1989).

The O–H bonds in unidirectional fragments are oriented at an angle of about 2° with the hexagonal axis. Structures of such fragments in $CHA_c$ and CHAr differ notably (Fig. 2). Hydroxyl ions are displaced towards each other in $CHA_c$ (Table 1) in relation to HA ($O_h–O_h$ = 2.73 Å—Hughes et al. 1989). Radical forming results in the rearrangement of the fragment structure (Table 1): hydroxyls are shifted in opposite directions so that the $O_h–O_h$ distance increases to ≈1/2c, the strength of H-bonding of the hydroxyl proton with the $O_c$ ion rises notably ($O_c\cdots H$ = 2.66 Å).

Accommodation of the carbonate impurity results in noticeable rearranging of the O–H H–O fragment structure. Angles of O–H bonds with the hexagonal axis increase to ≈4° in such fragments in $CHA_c$. The subsequent forming of the radicals leads to an increase of structure distortion: the nearest hydroxyl ion is shifted away from the $O_c$ ion (Table 1), while angles of O–H bonds with the **c** axis increase to ≈7°. Protons are oriented practically along the vertical axis in the most stable HA structure. The distances between ions of hydroxyls in such fragments are practically the same in both CHA structures and are approximately equal to those in the most stable HA structure ($O_h$–H = 1.086 Å, $O_h–O_h$ = 2.64 Å, H–H = 0.46 Å). The proton orientations in other channels of both CHA structures are approximately the same as in the most stable HA structure (O–H H–O). There are observed small deviations of proton orientations from the hexagonal axis in channels which are the nearest to the channel with the carbonate defect. The obtained data complement and are partly in agreement with experimental and theoretical investigations of CHA structures.

The carbonate group configuration in the most stable $CHA_c$ structure is approximately a correct triangle. The noticeable deformation of carbonate geometry in synthesized A type CHA at a high substitution degree was found (Fleet and Liu 2005). The C–H distances in the most stable CHA structures comply with published data (3.65–4.1 Å) (Schramm et al. 2001).

The carbonate substituting in the hexagonal HA channel results in rearrangement of the channel structure so that a fragment of unidirectional O–H-bonds forms. The reorienting of hydroxyl protons resulting in the formation of such fragment channels near structural defects is typical for hexagonal HA (Zahn and Hochrein 2006).

A decrease of the *a*-dimension in the most stable CHA structures contradicts literary data on A type CHA. Modeling of A type substitutions in monoclinic CHA has showed the small increase of lattice parameters in relation to HA (Peroos et al. 2006). Carbonate substitutions lead to expansion of the *a*-dimension in synthesized A type CHA, whereas the *c*-dimension remains approximately constant in relation to HA (Fleet and Liu 2005). Accommodation of carbonate ions in A type CHA may depend substantially on the degree of carbonate substitution. This may explain the contradiction of the *a*-dimension of the most stable $CHA_c$ structure at a low degree of A type substitution (0.16 wt% $CO_2$).

The radical formation and Ca2 vacancy appearance result in a disturbance of the charge balance in the nearest environment. Ion displacements stabilizing $CHA_r$ structure lead to distortions of local structure, configurations of phosphate groups, and the channel with the radical. They also result in the further contraction of the *a*-dimension, a small expansion of the *c*-dimension, and changes of the distances between radical ions (Table 1).

The $CO_2^-$-radical structure in $CHA_r$, with a lower angle of the $O_c$–$O_c$ axis with the **c** axis, conforms to the structure of $CO_2^-$-radicals caused by typical axial signals in the EPR spectra of A type CHA (Schramm et al. 2001). These signals consist, as a rule, of orthorhombic and axial components (Oliveira et al. 2000). Similar two-component signals with distinct parameters were observed in the EPR spectra of B type CHA (Vorona et al. 2006). These signals were assumed to be due to $CO_2^-$-radicals of the same configuration. Differences between components may be caused by structural surroundings of these radicals: axial radical form in more ordered structural regions, whereas orthorhombic radicals form near structural defects (Vorona et al. 2006).

Similarly, the $CO_2^-$-radical localized near the Ca2 vacancy in the most stable A type $CHA_r$ structure (Fig. 2b) may cause the orthorhombic component in EPR spectra of A type CHA. Accordingly, the $CO_2^-$-radical in a more ordered structural region without Ca vacancies in A type CHA can be responsible for the axial component. The computer simulation has shown that orthorhombic $CO_2^-$-radicals in A type CHA are more stable than axial ones ($\Delta F = -0.86$ eV = $-82.9$ kJ mol$^{-1}$), with the above-mentioned assumptions.

## 5 Annealing Effect on the A Type CHA Structure with $CO_2^-$ Radicals

The optimization of approximately 100 structure variants was carried out, from which 25 structures were optimized. Helmholtz free energies of the most stable structures of $CHA_r$ and CHA with the $CO_2^-$ radical and $O^-$ center ($CHA_{r\text{-}h\text{-}r}$) at considered temperatures are listed in Table 2. The higher probability of forming paramagnetic centers in A type CHA at $T$ = 298 K must be noted: optimized structures are more stable than HA structures at the used potential parameters ($F = -481.68$ eV).

The most stable $CHA_r$ structure at $T$ = 298 K (Table 2) was obtained using the HA lattice parameters (Hughes et al. 1989) as an initial approximation. The $CHA_r$ structures optimized at $T$ = 298 K with the other lattice parameters (Aminatun et al. 2013) are less stable. These lattice parameters (Aminatun et al. 2013) have been employed to model $CHA_r$ structures at other considered temperatures.

The most stable $CHA_{r\text{-}h\text{-}r}$ structures at $T$ = 298, 873, and 1173 K (Table 2) were obtained with $q^{eff}(O^-) = -1e$. However, this structure was optimized at $T$ = 1073 K with $q^{eff}(O^-) = -1.4e$ only. It specifies the noticeable change of the electron density distribution in the hydroxyl channel and, especially, in the structural environment of the oxygen ion of the $O^-$-center at $T$ = 1073 K.

Figure 3 shows the structures of the hydroxyl channel with the $CO_2^-$-radical in the most stable structures at considered temperatures. The $CHA_r$ structures are the most stable at $T$ = 298 and 873 K (Fig. 3a, b). There is a Ca2 vacancy near the

**Table 2** Helmholtz free energies $F$ (eV per one elementary cell) of the most stable CHA structures with radicals at considered temperatures

| Hydroxyl channel structure | Temperature (K) | | | |
|---|---|---|---|---|
| | 298 | 873 | 1073 | 1173 |
| | **$CHA_r$** | | | |
| OH HO $CO_2^-$ HO HO $V_{Ca2}$ | $-482.82^a$ | | | |
| OH HO $CO_2^-$ HO HO | | $-492.03$ | | |
| OH HO $CO_2^-$ HO HO $V_{Ca2}$ | | | | $-497.81$ |
| | **$CHA_{r\text{-}h\text{-}r}$** | | | |
| OH OH $CO_2^-$ $O^-$ HO | $-482.39$ | $-491.73$ | | |
| $O^-$ OH $CO_2^-$ HO HO | | | | $-498.05$ |
| OH OH $CO_2^-$ OH $O^{-a}$ | | | $-496.33$ | |

[a] $q^{eff}(O^-) = -1.4e$

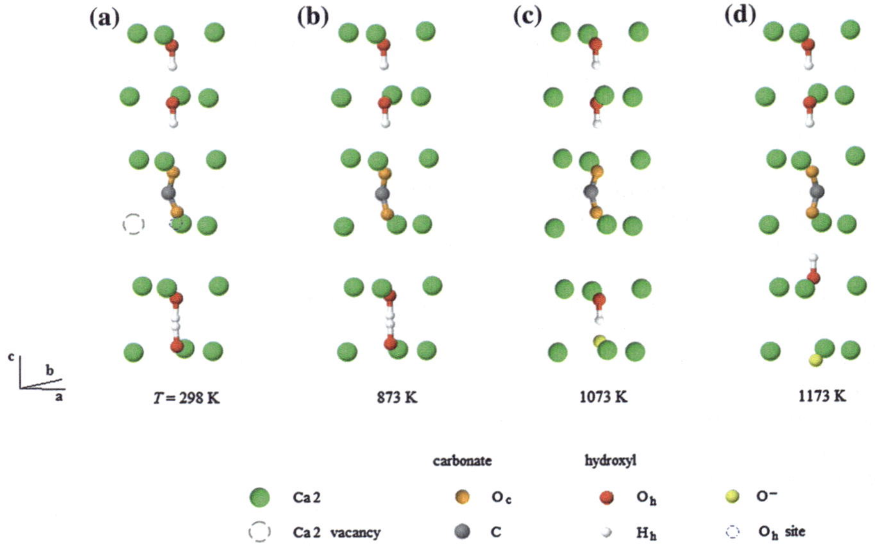


**Fig. 3** The channel structures with radicals in the most stable type A CHA structures at considered temperatures. The ions of Ca1 and phosphate groups are not shown

$CO_2^-$-radical at $T = 298$ K. An increase in temperature to $T = 873$ K leads to small changes of the $CO_2^-$-radical geometry and disappearance of the Ca2 vacancy, but the channel configuration does not rearrange substantially. The $CHA_{r\text{-}h\text{-}r}$ structures at these temperatures are less stable: $\Delta F \approx 0.4$ and 0.3 eV, respectively (Table 2). The paramagnetic centers of $CO_2^-$ and $O^-$ in these structures are incorporated in the nearest sites in the channel (Table 2). The $\Delta F$ differences between the most stable structures of $CHA_r$ and $CHA_{r\text{-}h\text{-}r}$ are the same order as compared with the O–H bond reorientation energy in HA (Zahn and Hochrein 2006).

The $CHA_{r\text{-}h\text{-}r}$ structures are the most stable at $T$ = 1073 and 1173 K. The paramagnetic centers in these structures are more distanced (Fig. 3c, d). This indicates substantial changes of the electron density distribution and, in particular, the $O^-$-center effective charge at $T = 1073$ K. No stable $CHA_r$ structure was found at $T = 1073$ K at the used potential parameters. The $CHA_r$ structure at $T = 1173$ K is less stable: $\Delta F \approx 0.25$ eV (Table 2). The temperature range of existence for the most stable $CHA_{r\text{-}h\text{-}r}$ structure confirms EPR data about the associated growth of the amounts of precursors of $CO_2^-$-radicals and $O^-$-centers in A type CHA annealed at such temperatures (Brik et al. 2005).

Temperature growth leads to substantial changes of the $CO_2^-$ radical configuration and rearrangement of its structural environment in A type CHA. The Ca2 vacancy disappears at $T = 873$ K (Fig. 3b). The $O^-$ center forms in the channel at $T \geq 1073$ K (Fig. 3c, d). These changes conform to temperature ranges of apatite structure rearrangement during dehydration and partial dehydroxylation: the disappearance of vacancies at $T \approx 873$–973 K results in structure ordering, removal of protons and hydroxyl ions at $T \approx 973$–1073 K leads to appearance of oxygen ions (proton vacancies) and hydroxyl vacancies (Aminatun et al. 2013; Brik et al. 2005). The hydroxyl channel structures should be noted to rearrange noticeably but the channel fragment with unidirected O–H bonds is present at all considered temperatures (Fig. 3).

One proton is oriented to the $CO_2^-$radical at $T \leq 1073$ K (Fig. 3a–c) and both protons are oriented at $T = 1173$ K (Fig. 3d). The hydroxyl channel structures near the $O^-$-center change substantially: one proton orients towards the $O^-$-center at $T = 1073$ K, but both protons are directed away from the $O^-$-center at $T = 1173$ K. This may allow the substantial changes in the channel structure rearrangement at these temperatures. A similar structure of hydroxyl channels (with unidirected O–H bonds) near structural defects is typical for apatites of different compositions (Zahn and Hochrein 2006).

## 6 Conclusions

Computer modeling with a semi-empirical method using the GULP program was employed to investigate some carbonated apatite structures. The obtained data complement the results of the experimental and theoretical studies of such structures.

The arrangement of structural defects (carbonate ions, excess $F_x$ ions, Na ions, Ca vacancies, $OH^-$ groups, and fixed $H_2O_{str}$ molecules) in synthesized A-B type CFA (2.7 wt% $CO_2$, 3.3 wt% F) has been studied using grid techniques. The structural defects of a certain type in the most stable CFA structures are located approximately evenly, with most distanced from each other. The $H_2O_{str}$ molecules occupy phosphate group vacancies ($O_w$ ions are in P sites, protons are oriented to O1 and O2 vacant oxygen sites). The hydroxyl ions in channels are incorporated near B carbonate groups. The excess $F_x$ ions occupy O3 oxygen vacancies near B carbonates, including in the A-B cluster. No dependence was found between the localization of excess $F_x$ ions and $H_2O_{str}$ molecules. Na1 ions and Ca2 vacancies are the most distanced from each other and from the A-B cluster. These ions are mainly located in the following way: Na1 ions are near $[CO_3F_x]_B$ clusters, whereas Ca2 vacancies are near B carbonate ions and O3 oxygen vacancies. The obtained data confirm the results of experimental investigations of synthesized CFA (Brik et al. 2013).

The A type CHA structures with the carbonate ion or $CO_2^-$-radical (0.16 wt% $CO_2$) in the channel have been studied using grid techniques. It was shown that $CO_2^-$-radical formation lead to noticeable structure deformations. A Ca2 vacancy was present in the radical structural environment in the most stable $CHA_r$ structure. The hydroxyl channel fragments with unidirected O–H bonds form near the carbonate defect of both structures. The ion sites of carbonate impurity in the most stable $CHA_r$ structure are markedly distinguished from those in the $CHA_c$ structure. The radical configuration is approximately aligned ($O_c$–C–$O_c$): one $O_c$ ion occupies approximately the hydroxyl site, whereas another $O_c$ ion is located near the Ca2 vacancy and the other hydroxyl site. Obtained configuration of the $CO_2^-$-radical confirms the presumed geometry of radicals responsible for axial signals in the EPR spectra of A type carbonate apatites. The structure of the radical near the Ca2 vacancy is assumed to correspond with the orthorhombic $CO_2^-$-radical in A type CHA.

The modeling of A type CHA structures with the $CO_2^-$-radical or, in addition, the $O^-$-center at some temperatures indicates that the $CHA_r$ structure is most stable at $T = 298$ and 873 K; the $CHA_{r\text{-}h\text{-}r}$ structure is most stable at $T = 1073$ and 1173 K. The temperature growth results in substantial changes in the $CO_2^-$ radical configuration and the radical environment in accordance with temperature ranges of apatite structure rearrangement. The temperature range of existence of the most stable $CHA_{r\text{-}h\text{-}r}$ structure confirms EPR data about the associated growth of precursor amounts of $CO_2^-$ radicals and $O^-$ centers in A type CHA annealed at such temperatures (Brik et al. 2005).

The results of our computer simulation of apatite structures are important for the interpretation of results of investigations of natural apatites of different genesis, as well as for recommendations to technologists who synthesize apatite-containing materials with certain properties, such as synthetic analogues of biological apatites. The peculiarities of localization and temperature stability of carbonate radicals in A type $CHA_r$ found by computer simulation are needed for solving the problems of retrospective dosimetry of apatites.

**Acknowledgments** This research was conducted within the framework of the Government Scientific and Technical Program of Introduction and Application of GRID Techniques in 2009–2013, project 38/13, and the Project of Government Found of Fundamental Investigations of Ukraine (project Ф53.6/026) and Saint Petersburg State University (project 3.38.243.2015).

## References

Abragam A (1961) The principles of nuclear magnetism. Clarendon, Oxford

Aminatun S, Penga IYM, Apsari R (2013) The effect of sintering process on the characteristics of hydroxyapatite from cuttlefish bone (Sepia Sp.). RJPBCS 4(4):1431–1442

Brik AB, Shpak AP, Karbovsky VL, Klimenko AP, Dubok O, Kalinichenko AM, Bagmut NN, Bevz VV (2005) The EPR of nanodimensional particles in biogenic and synthetic carbonated apatites. Mineralogie J (Ukraine) 27(1):5–26 (In Russian)

Brik AB, Frank-Kamenetskaya OV, Dubok O, Kalinichenko EA, Kuz'mina MA, Zorina ML, Dudchenko NO, Kalinichenko AM, Bagmut NN (2013) The features of isomorphic substitutions in synthetic carbonatefluorapatites. Mineralogie J (Ukraine) 35(3):3–10 (In Russian)

De Leeuw NH (2004) A computer modeling study of the uptake and segregation of fluoride ions at the hydrated hydroxyapatite (0001) surface: introducing a $Ca_{10}(PO_4)_6(OH)_2$ potential model. PCCP 6:1860–1866

Fleet ME, Liu X (2005) Local structure of channel ions in carbonate apatite. Biomaterials 26 (36):7548–7554

Fleet ME, Liu X (2008) Accommodation of the carbonate ion in fluorapatite synthesized at high pressure. Am Mineral 93:1460–1469

Frank-Kamenetskaya O, Kol'tsov A, Kuz'mina M, Zorina M, Poritskay L (2011) Ion substitutions and non-stoichiometry of carbonated apatite-(CaOH) synthesised by precipitation and hydrothermal methods. J Mol Struct 992:9–18

Gale JD, Rohl AL (2003) The general utility lattice program (GULP). Mol Simul 29:291–341

Hughes JM, Cameron M, Crowley KD (1989) Structural variation in natural F-, OH- and Cl-apatites. Am Mineral 74:870–876

Ikeya M (1993) New applications of electron spin resonance. Dating, dosimetry and microscopy. World Scientific, Singapore

Mason HE, McCubbin FM, Smirnov A, Phillips BL (2009) Solid-state NMR and IR spectroscopic investigation of the role of structural water and F in carbonate-rich fluorapatite. Am Mineral 94:507–516

Nokhrin SM, Pan Y, Nilges MJ (2006) Electron paramagnetic resonance spectroscopic study of carbonate-bearing fluorapatite: new defect centers and constraints on the incorporation of carbonate ions in apatites. Am Mineral 91:1425–1435

Oliveira LM, Rossi AM, Lopes RT (2000) Gamma dose response of synthetic A-type carbonated apatite in comparison with the response of tooth enamel. Appl Radiat Isot 52:1093–1097

Pan Y, Fleet ME (2002) Compositions of the apatite-group minerals: substitution mechanisms and controlling factors. Rev Mineral Geochem 48:13–49

Peroos S, Du Z, de Leeuw NH (2006) A computer modeling study of the uptake, structure and distribution of carbonate defects in hydroxy-apatite. Biomaterials 27:2150–2161

Regnier P, Lasaga AC, Berner RA, Han OH, Zilm KW (1994) Mechanism of $CO_3^{2-}$ substitution in carbonate-fluorapatite: evidence from FTIR-spectroscopy, $^{13}C$ NMR and quantum mechanical calculations. Am Mineral 79:809–818

Schramm DU, Terra J, Rossi AM, Ellis DE (2001) Configuration of $CO_2^-$-radicals in γ-irradiated A-type carbonated apatites: theory and experimental EPR and ENDOR studies. Phys Rev B Condens Matter 63(2):024107

Vorona IP, Ishchenko SS, Baran NP, Petrenko TL, Rudko VV (2006) Evidence of annealing-induced transformation of $CO_2^-$-radicals in irradiated tooth enamel. Radiat Meas 41:577–581

Yi H, Balan E, Gervais C, Segalen L, Fayon F, Roche D, Person A, Morin G, Guillaumet M, Blanchard M, Lazzeri M, Babonneau F (2013) A carbonate-fluoride defect model for carbonate-rich fluorapatite. Am Mineral 98:1066–1069

Zahn D, Hochrein O (2006) The role of substitutional defects in order/disorder phenomena of OH-ions in hydroxyapatite. ZAAC 632(1):79–83

# Regulation of HAP and Iron Oxide Nanoparticle Morphology Using Chelating Agents

**O.M. Osmolowskaya**

**Abstract** The influence of chelating agents (organic acids with different structures) on iron oxides and hydroxyapatite nanoparticle (NP) morphology obtained by two different synthetic procedures (precipitation method and hydrothermal synthesis) were studied. The relationship between the synthetic conditions and the nature of the growth limiters or NP morphology regulation agents was considered. The mechanism of the NP growth regulation and the general peculiarities of this process for materials with different chemical natures are also discussed. It was concluded that the use of chelating agents dramatically changes the NP size and shape and the degree of modification is determined by the binding force between chelating agent and ions on NP surfaces.

**Keywords** Hydrothermal synthesis · Hydroxyapatite · Iron oxides · Precipitation · Nanoparticles · Magnetite · Maghemite

## 1 Introduction

Nanoparticles (NP) for biomedical applications are attracting great interest. The most promising of them are iron oxide NP (for imaging agents, hyperthermia, magnetic separation of biological objects) and hydroxyapatite (HAP) NP (scaffolds for tissue engineering, material for chromatography and sorbents). The main problems of NP synthesis are morphology regulation and prevention against aggregation. The latter is carried out by adsorption of small molecules (e.g., citric acid, oleic acid, amino acids) and also by using macromolecules (e.g., homopolymers, copolymers, natural polymers).

O.M. Osmolowskaya (✉)
Saint Petersburg State University, Saint Petersburg, Russia
e-mail: o_osmolowskaya@mail.ru



O.V. Frank-Kamenetskaya et al. (eds.), *Biogenic—Abiogenic Interactions in Natural and Anthropogenic Systems*, Lecture Notes in Earth System Sciences,
DOI 10.1007/978-3-319-24987-2_36

**Table 1** Surface modification of NPs by using small molecules

| Material | Agent for surface stabilization | Agent for morphology regulation |
|---|---|---|
| $Fe_3O_4$ | Citric acid, oleic acid, amino acids, $SiO_2$ (Laurent et al. 2008) | – |
| $\gamma$-$Fe_2O_3$ | SDS (Wang et al. 2005), DBS, CTAB (Liu et al. 1999) | Succinic acid (Jayarathne et al. 2012) |
| HAP | EDTA (Wei et al. 2014) | Different surfactants (Taheri et al. 2015; Khalid et al. 2013; Wang et al. 2007), amino acid (Gonzalez-McQuire et al. 2004) |

By surface modification of iron oxide NP, protection against interaction with the human immune system and fast elimination of NPs from bloodstream are achieved. For HAP, surface modification minimizes the degree of NP agglomeration. This is very important because phosphates tend to form inorganic polymers, which leads to strong interactions between particles at the suspension.

Table 1 shows the most popular agents for NP surface modification. As shown, morphology regulation is conducted using a limited number of agents. In the case of magnetite, studies are focused mainly on NP surface stabilization. Therefore, the questions remain: What is the mechanism of NP morphology regulation using the various agents? Can the mechanism be the same for different inorganic materials? This chapter attempts to briefly answer these questions.

# 2 Materials and Methods

## *2.1 Materials*

$Ca(NO_3)_2$ (98.5 %), $H_3PO_4$ (98.5 %), $Ca(OH)_2$, NaOH (98.7 %), $FeCl_3$ (98.8 %), $FeCl_2$ (98.8 %), $NH_3$ aqueous solution (98,7 %) (all from Nevareactiv, Russia) were used as received without further purification.

## *2.2 NP Synthesis*

The synthetic conditions for obtained NPs are summarized in Table 2. The synthesis temperature and pH ranges for each case were determined as the interval between NP with good crystallinity formation and the appearance of impurity phases.

In a typical procedure for co-precipitation, the basic initial reagents were placed in a round flask, then the alkali/acid were added dropwise. The pH and temperature values were kept constant by using a Mettler Toledo pH meter and heating the mantle/ice bath correspondingly. In the case of hydrothermal synthesis, the initial reagents were mixed using overhead or magnetic stirrers. The mixture was then

**Table 2** The synthetic conditions for different inorganic materials with water as the solvent

| Material | Method | pH | Temperature range, °C | Initial reagents | Special conditions |
|---|---|---|---|---|---|
| $Fe_3O_4$ | Co-precipitation | 10 | 25–80 | $FeCl_2 + FeCl_3 + NH_4OH$ | Inert atmosphere |
| $\gamma$-$Fe_2O_3$ | Hydrothermal synthesis (Dobrodumov et al. 2014) | 11 | 180 | $Fe(OH)_3$ | – |
| HAP | Co-precipitation (Kozlova et al. 2014) | 8–10 | 15–80 | $Ca(NO_3)_2 + H_3PO_4$ | Preliminary boiled water Ca/P molar ratio = 1.67 |
| | Hydrothermal synthesis (Gordeev et al. 2014) | 10 | 240 | $Ca(OH)_2 + H_3PO_4$ | Preliminary boiled water An excess of $Ca(OH)_2$ |

transferred into a Teflon lined stainless steel autoclave of 180 mL capacity. The sealed tank was heated and held at the needed temperature for a minimum of 1 h.

The resulting black ($Fe_3O_4$), brown ($\gamma$-$Fe_2O_3$), and white (HAP) precipitates were separated by centrifugation (Sigma 2-16P) and washed several times with distilled water by alternating the shaking and collecting steps. After that, the products were freeze dried.

## 2.3 *NP Characterization*

X-ray diffraction (XRD) analysis was performed at Bruker D2 PHASER (Bruker company, Germany) using CuKα and CoKα radiation. The XRD data were collected at room temperature over the 2θ range of 7–100 °. The morphology of the samples was studied by transmission electron microscopy (TEM, JEM 107, JEOLS Ltd). The obtained results were processed by Micran software and the error of mean sizes was calculated. Specific surface area (SSA) estimation was performed by the Brunauer-Emmett-Teller method on a surface area analyzer (Micromeritics ASAP-2020MP). Fourier transform infrared (FTIR) spectra were obtained over the region 400–4000 $cm^{-1}$ on Nicolet 8700 FTIR Spectrophotometer. A Mastersizer 3000 (Malvern, UK) and Nanoparticle Analyzer SZ100 (Horiba Jobin Yvon) were used to determine the size and zeta potential of the NP in a water medium. Before the measurement, the NP were dispersed in distilled water and then ultrasonicated (Bandelin UW 2070) for 1 min.

# 3 Results and Discussion

A large number of methods for NP with different composition and structure synthesis have been developed to date. From the perspective of practical applications, the most promising are hydrothermal (or solvothermal) synthesis and the co-precipitation method. This is due to the fact that the first is easily scaled and the latter does not require special equipment and can be simply produced.

The results of our work on the synthesis of inorganic nanoparticles with different natures (Dobrodumov et al. 2014; Kozlova et al. 2014; Gordeev et al. 2014; Suslonov et al. 2012; Osmolovskii and Osmolovskaya 2011; Petukhova et al. 2014; Osmolovsky and Osmolovskaya 2012) and the data in the literature allow us to offer the following division of agents affecting the NP morphology (Fig. 1).

The main difference between the growth limiters and regulation agents is that the first interacted with the surface of the already formed particles or organized into the micelles acting as the "nanoreactors". In addition, the surface stabilizer may form a chemical bond with the NP surface, especially the organic acid; however, the probability of this is very weak and the binding force is much lower than in the case of chelating agent. Therefore, NP growth is limiting but not regulating. As shown previously (Suslonov et al. 2012; Osmolovskii and Osmolovskaya 2011), the use of growth limiters based on surfactant often leads to the formation of the organic phase on the NP surface. This shell is difficult to remove, so the properties of products may be worse.

The chelating agent may bond with the NP surface and limits its growth. The binding force depends on the agent's nature and the synthetic conditions (pH, temperature, etc.). For example, etidronic acid has the largest binding force at a pH value above 11 because all the acid groups are dissociated.

Based on these occurrences, the choice of chelating agent in each case must be determined by the synthetic parameters. The growth inhibition of chelating agents is based on coordination between the metal ions located on the surface and the acid. Another factor that determined the binding process is the nature of the metal ions. In addition, the role of the chelating agent may consist of the formation of an intermediate with one of the initial reagent, which leads to a decreased reaction rate and the controlled growth of nanoparticles. In the case of iron oxides and

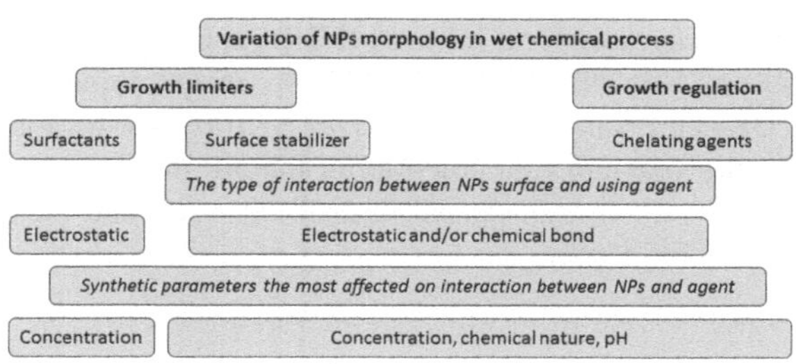


**Fig. 1** The scheme of regulation of the NP morphology using agents of different types

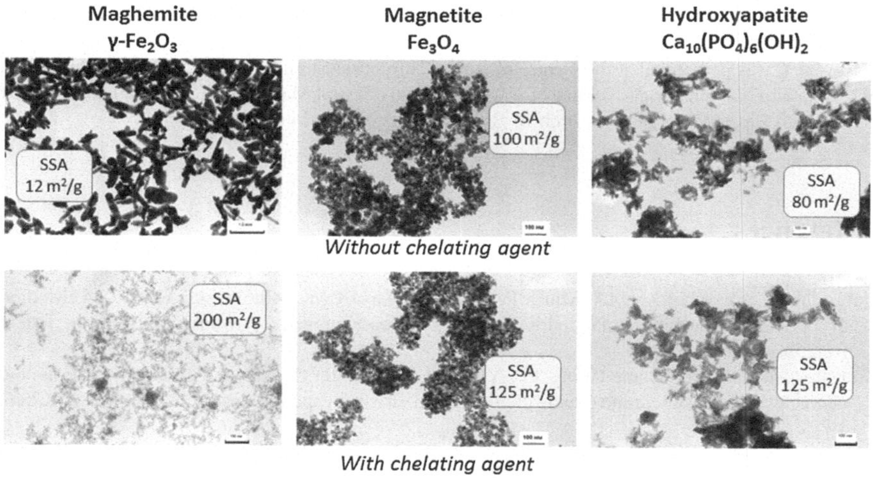


**Fig. 2** The influence of the chelating agent on NP morphology for the example of etidronic acid, as shown in transmission electron microscopy images

hydroxyapatite, the chelating agent works with good efficacy (Fig. 2). Their presence on NP surfaces was confirmed by FTIR analysis.

As shown by transmission electron microscopy images and SSA data, the use of etidronic acid, both in hydrothermal and co-precipitation methods, dramatically changes the NP morphology. This is due to the fact that the etidronic acid formed stable complexes with the iron and calcium ions. This leads to a significant decrease of particle size; in the case of maghemite and HAP, a change in the NP's shape was also found. The same situation was observed when we used glycine as the chelating agent (Kozlova et al. 2014). However, because the complex of the amino acid with calcium is more stable than the same with iron, the HAP morphology changes more intensely—from nanorods to spheres with a size of 10 nm. Similar results were obtained for other acids such as succinic, tartaric, malic, and citric. It was shown that the strongest interaction between the chelating agent and the iron or calcium ion leads to the spherical particle shape. Note also that the use of chelating agents increased the aggregative stability of the NP aqueous suspension and to the decline of zeta-potential.

## 4 Conclusions

Inorganic NP with different morphology were obtained by different synthetic approaches—hydrothermal and co-precipitation methods. It was shown that the use of chelating agents dramatically changes the NP size and shape. The degree of modification is determined by the binding force between the chelating agent and ions on the NP surfaces.

**Acknowledgements** This study was financially supported by RFBR (research project no. 13-03-00943-a).

Research was performed at the Centre of X-ray Diffraction Studies and Innovative Technologies of Composite Nanomaterials, Center of Chemical Analysis and Materials Research, and Centre for Optical and Laser Materials Research of St. Petersburg State University.

## References

Dobrodumov AV, Kozlova MA, Murin IV, Osmolovskaya OM, Osmolovskii MG (2014) Method of obtaining maghemite nanoparticles and superparamagnetic powder composition. Patent RU2533487-C2

Gonzalez-McQuire R, Chane-Ching JY, Vignaud E, Lebugle A, Mann S (2004) Synthesis and characterization of amino acid-functionalized hydroxyapatite nanorods. J Mater Chem 14:2277–2281

Gordeev SV, Nikol'Skii AB, Osmolovskaya OM, Osmolovskii MG (2014) Hydroxyapatite nanoparticles modified by peroxide groups. Russ J Gen Chem 84:789–790

Jayarathne L, Ng WJ, Bandara A, Vitanage M, Dissanayake CB, Weerasooriya R (2012) Fabrication of succinic acid-$\gamma$-$Fe_2O_3$ nano core-shells. Colloids Surf A 403:96–102

Khalid M, Mujahid M, Amin S, Rawat RS, Nusair A, Deen GR (2013) Effect of surfactant and heat treatment on morphology, surface area and crystallinity in hydroxyapatite nanocrystals. Ceram Int 39:39–50

Kozlova MA, Poezzhaev AI, Khramenkova EV, Korzhikov VA, Osmolovskaya OM, Osmolovskii MG (2014) Glycine, biocompatible growth regulator for preparation of inorganic nanomaterials for medicine. Russ J Gen Chem 84:2047–2048

Laurent S, Forge D, Port M, Roch A, Robic C, Vander EL, Muller RN (2008) Magnetic iron oxide nanoparticles: synthesis, stabilization, vectorization, physicochemical characterizations and biological applications. Chem Rev 108:2064–2110

Liu T, Guo L, Tao Y, Wang YB, Wang WD (1999) Synthesis and interfacial structure of nanoparticles $\gamma$-$Fe_2O_3$ coated with surfactant DBS and CTAB. Nanostruct Mater 11:487–492

Osmolovskii MG, Osmolovskaya OM (2011) Generation of nickel nanoparticles in the presence of monobasic saturated acids. Russ J Gen Chem 81:2371

Osmolovsky MG, Osmolovskaya OM (2012) Dimensional effects in the properties of maghemite. Bulletin Russ Academy Sci Phys 76:1146–1148

Petukhova YuV, Osmolowskaya OM, Fedorova AV, Osmolowsky MG (2014) Magnetic behavior of doped $VO_2$ nanoparticles. Bulletin Russ Acad Sci Phys 78:325–327

Suslonov VV, Osmolovskaya OM, Osmolovskii MG (2012) Organic shell influence on physicochemical properties of nickel polyol nanoparticles. Russ J Gen Chem 82:585–1586

Taheri MM, Kadir MRA, Shokuhfar T, Hamlekhan A, Assadian A, Shirdar MR, Mirjalilie A (2015) Surfactant-assisted hydrothermal synthesis of fluoridated hydroxyapatite nanorods. Ceram Int 41:9867–9872

Wang A, Liu D, Yin H, Wu H, Wada Yu, Ren M, Jiang T, Cheng X, Xu Y (2007) Size-controlled synthesis of hydroxyapatite nanorods by chemical precipitation in the presence of organic modifiers. Mater Sci Eng C 27:865–869

Wang D, Cao Ch, Xue Sh, Zhu H (2005) $\gamma$-$Fe_2O_3$ oriented growth by surfactant molecules in microemulsion. J Cryst Growth 277:238–245

Wei W, Cui J, Wei Z (2014) Effects of low molecular weight organic acids on the immobilization of aqueous Pb(II) using phosphate rock and different crystallized hydroxyapatite. Chemosphere 105:14–23

# Calcium Oxalates: Thermodynamic and Kinetic Conditions of Their Formation in the Presence of Organic Components

**Vyacheslav V. Korolkov, Olga A. Golovanova and Marina V. Kuimova**

**Abstract** The paper researches a range of problems related to the peculiarities of calcium oxalate crystallization in the presence of amino acids on the bases of thermodynamic calculations and experimental data. The regularities of phase formation in the system $Ca^{2+}$–$C_2O_4^{2-}$–$H_2O$–amino acid in a widely variable interval of component concentrations and pH are theoretically researched. The influence of pH on the thermodynamic stability of crystalline compounds is considered. The kinetic parameters of crystallization are determined: the induction period, the rate constant, the order of crystals' growth, and the influence of amino acids on the kinetics of crystallization of calcium oxalate is established.

**Keywords** Crystallization · Calcium oxalate · Thermodynamic simulation · Kinetics · Amino acids

## 1 Introduction

The problem of stone formation is becoming more prominent and is a real threat to modern residents of all countries, in particular in metropolitan cities. According to statistics, throughout the twentieth century, various diseases leading to the formation of stones (organo-mineral aggregate—OMA) in human organs has only increased (Golovanova 2007; Larina et al. 2006; Tiktinsky and Alexandrov 2000). Stones are formed at any age, even in very young children from birth. The urinary system stones are the most common that form minerals of pathogenic nature. However, modern medicine has not yet been able to fully understand the causes for the formation of various OMA in the human body.

V.V. Korolkov · O.A. Golovanova (✉)
Omsk F.M. Dostoevsky State University, Omsk, Russia
e-mail: golovanoa2000@mail.ru

M.V. Kuimova
National Research Tomsk Polytechnic University, Tomsk, Russia

 
O.V. Frank-Kamenetskaya et al. (eds.), *Biogenic—Abiogenic Interactions in Natural and Anthropogenic Systems*, Lecture Notes in Earth System Sciences,
DOI 10.1007/978-3-319-24987-2_37

Urinary stone disease (urolithiasis), leading to the formation of stones in the organs of the urogenital system, ranks third in prevalence among all urological diseases. About 5 % of the world's population suffers from urolithiasis, and experts predict that in the future their number will only grow. In Russia, up to 3 % of people suffer from this disease. In 7 % of all cases of urolithiasis, disease occurs in children under the age of 16 years. However, according to statistics, the majority of cases are among adults of working age—from 20 to 49 years; the peak of disease incidence occurs in people aged 35–45 years. This illness is about three times more likely in men than in women (Al Zahrani et al. 2000; Golovanova 2008; Golovanova and Pyatanova 2002; Palchik et al. 2006; Pospekhova et al. 2001; Sevostyanova and Polienko 2004; Zuzuk 2005). Physicians attribute the increased prevalence of urolithiasis in cities with increased consumption of foods rich in fats and animal proteins, as well as with negative environmental conditions in the major metropolitan areas (Assimos and Holmes 2000; Al Zahrani et al. 2000; Arias Funez et al. 2000; Bailly et al. 2000; Bak et al. 2000; Golovanova 2008; Golovanova et al. 2004b; Zuzuk 2005).

Pathogenic organo-mineral aggregates are of complex and heterogeneous composition (Elnikov et al. 2007; Golovanova 2007; Golovanova et al. 2006a; Sevostyanova and Polienko 2004). However, overall, the most common of urinary stones are calcium oxalate stones, composed of minerals $CaC_2O_4{\cdot}H_2O$ whewellite and $CaC_2O_4{\cdot}2H_2O$ weddellite (Golovanova 2007, 2008; Pospekhova et al. 2001; Sevostyanova and Polienko 2004; Tiktinsky and Alexandrov 2000). These minerals may also be a part of dental and salivary stones, gallstones, and others. Moreover, they are found in the composition of mineral deposits in lungs, blood vessels, spleen, prostate and pancreatic glands, muscle, and others (Golovanova 2008). These data highlight the importance of studying the crystallization of calcium oxalate in conditions close to physiological.

Organic materials of proteinic nature, which are components of physiological fluids, have a great influence on the character of the crystallization process of minerals, particularly calcium oxalate (Golovanova et al. 2006b, 2010). Current information about the nature of the crystallization of calcium oxalate complex by composition of physiological solutions is insufficient. New data are needed on the study of crystallization of these minerals from fluids in the presence of amino acids, as they are required from the viewpoint of fundamental understanding of bio-mineralization processes and also medically, to prevent calcium oxalate stone formation in the human body.

The aim of this work is to establish the thermodynamic and kinetic regularities of formation of calcium oxalate in the presence of a number of amino acids.

# 2 Materials and Methods

Model solution, mineral composition (inorganic macro components), temperature, ionic strength, and pH of urine corresponding to that of an average healthy adult person were taken as a thermodynamic model of processes of phase formation (in the system $Ca^{2+}$–$C_2O_4{}^{2-}$–$H_2O$–amino acid, as a prototype of body fluid) (Tiktinsky and Alexandrov 2000). Additives of amino acids were used to research the influence of organic components on the formation of the solid mineral phase. The choice of this set of amino acids is based on their similarities to the physiological fluids of concern, which are the high salinity and organic structures observed in urinary stones (Table 1).

The calculation of amino acids' influence was limited to the description of the process of complexation with calcium ions (Berthon 1995; Kiss et al. 1991; Lurie 1989; Vasiliev 2003; Yamauchi and Odani 1996); while for the calculation of the conditional solubility products, an amendment was introduced, characterized by the fraction of ions $Ca^{2+}$ that are not bound in a complex with amino acids. In calculation, the range of pH was varied over a wide range from 0 to 14. The coefficients of ion activity in these conditions were calculated using the Davis equation (Lurie 1989).

During the calculations to determine the possibilities and conditions of calcium oxalate deposition, the values of the thermodynamic solubility products (Lurie 1989) and the bases of constants of instability of *SC-database* complexes were used. Theoretical definition of the possibilities and conditions of deposition of calcium oxalate monohydrate was made based on the values of the indices of supersaturation of the (SI) system. Stability fields were used for the thermodynamic description of sediment–solution equilibrium in the system, wherein the formation of poorly soluble combination occurs. The principle of design of such diagrams is in determining the functional dependence of the parameter of the minimum concentration of the cation within the sediment; this dependence must be created for the start of the deposition phase at a given pH of the solution and given concentration of the anion: $pM^{q+} = f(pX^{q-}, \text{pH})$.

**Table 1** The amino acids used as impurities, and their characteristics (Dawson et al. 1986)

| Amino acid | Structure | Isoelectric point pI | $pK_a$ |
|---|---|---|---|
| DL-Valine (Val) | 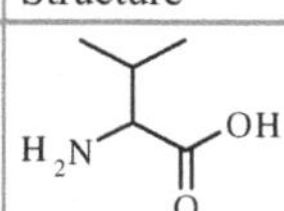 | 5.96 | 2.27<br>9.52 |
| Glycine (Gly) | 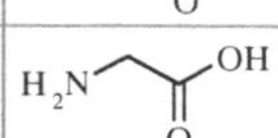 | 5.97 | 2.34<br>9.58 |
| DL-glutamine acid (Glu) | 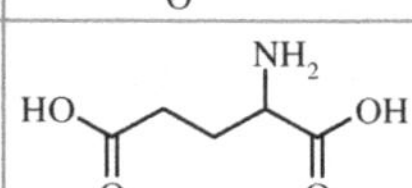 | 3.22 | 2.16<br>4.15<br>9.58 |

The field located above the obtained critical plane characterizes the conditions under which the formation of this phase is thermodynamically impossible. On the basis of the constructed three-dimensional diagrams in the coordinates $pM^{q+}-pX^{q-}-\mathrm{pH}$, conclusions can be drawn about the stability of the system and the prediction of the nature of the change of the balance under different conditions.

The process of crystallization of calcium oxalate monohydrate was studied at 37 °C and the basic values of a supersaturated solution $\gamma = C_0/C_s = 5, 7$; this choice of the basic value of $\gamma$ is justified by the presence of supersaturation in biological media, namely that of the urine of an average healthy adult person (Golovanova 2007; Tiktinsky and Alexandrov 2000).

Supersaturation with calcium oxalate was created in the chemical reaction:

$$Ca^{2+} + C_2O_4^{2-} \rightarrow CaC_2O_4,$$

which is implemented by mixing stock solutions of stoichiometric ratio of soluble compounds—calcium chloride and ammonium oxalate.

The visual method, based on measuring the induction periods ($\tau_{\mathrm{ind}}$), was used to determine the parameters of the nucleation. As it is known, the induction time is inversely proportional to the nucleation rate: $\tau \sim 1/J$. In turn, the dependence of the nucleation rate on supersaturation of the solution is expressed by an exponential function, which includes the surface energy at the interface $\sigma$:

$$J \sim \exp\left[-\frac{16\pi\sigma^3 v^2}{3k_B^3 T^3 (m \ln\gamma)^2}\right] = \exp\left[-\frac{B}{(\ln\gamma)^2}\right], \tag{1}$$

where $v$ is the volume of the molecule; $k_B$ is the Boltzmann constant; $T$ is the temperature; $m = 2$ is the number of ions on which a molecule dissociates in the solutions; $B$ is a constant.

The timing of the induction was carried out in a visual way, as the solution becomes cloudy. The study was conducted in vitro, in a medium close to that of the physiological solution of human urine.

The conductometric method of measurement of solutions' concentration during their crystallization was used to research the crystal growth. Conductometer Anion-4154 was used (Izatulina et al. 2006). According to conductometric analysis, the degree of completeness of the process of crystallization, $\alpha$, as a function of time, was determined from:

$$\alpha = \frac{C_0 - C_\tau}{C_0 - C_s}, \tag{2}$$

where $C_0$ is the initial concentration of calcium oxalate in a supersaturated solution, $C_\tau$ is the concentration of calcium oxalate at time $\tau$, and $C_s$ is the solubility of calcium oxalate.

To determine the kinetic parameters of the growth of crystals of calcium oxalate according to $\alpha = f(\tau)$ dependence, deposition rate was calculated as a function of current absolute supersaturation by the formula:

$$\frac{\mathrm{d}\alpha}{\mathrm{d}\tau} = kA \cdot (C_\tau - C_s)^n, \tag{3}$$

where $A$ is the total surface area of the precipitate, $k$ is the rate constant, and $n$ is the order of the reaction. The estimation of the total surface area for a constant shape of particles is given as

$$A = \beta \cdot N_\tau^{1/3} \cdot V_\tau^{2/3}, \tag{4}$$

where $\beta$ is the form factor, $N_\tau$ is the total number of particles, and $V_\tau$ is the sludge volume at time $\tau$. Considering that $\alpha = V_\tau/V_{\max}$ and $V_{\max}$ is the maximum amount of sediment in full supersaturation, after transformations we obtain the formula for calculation of the kinetic characteristics of the process of crystallization of calcium oxalate (assuming a constant number of particles $N_\tau = N =$ constant):

$$\left(\frac{\mathrm{d}\alpha}{\mathrm{d}\tau}\right) \cdot \alpha^{-2/3} = k'(C_\tau - C_s)^n, \tag{5}$$

where $k'$ includes all constants ($V_{\max}$, $\beta$, $N$, and $k$) and is constant for the given initial conditions. Applying the logarithm we obtain

$$\lg\left(\frac{\mathrm{d}\alpha}{\mathrm{d}\tau}\right) - \frac{2}{3}\lg\alpha = \lg k' + n\lg(C_\tau - C_s). \tag{6}$$

The structure of dependence in the coordinates $\lg(d\alpha/d\tau) - 2/3\lg\alpha = f(\lg(C_\tau - C_s))$ should give a straight line. The jog, chipped off by the straight line on the $Y$-axis, gives the value lg $k'$, and the slope ratio corresponds to the order of the growth rate $n$.

The kinetics of crystallization was researched both in solutions of calcium oxalate without organic components and with additives of the amino acids in the concentration of 0.004 mol/l, corresponding to their location in physical solution.

The phase composition of precipitate deposits obtained during synthesis was studied by X-ray diffraction (XRD), the diffraction patterns were obtained by powder diffraction technique on X-ray DRON-3, and phase identification was carried out using a database ICDD PDF for powder diffraction. IR spectra were obtained on the spectrophotometer "FT-801", the samples were prepared by pressing tablets with KBr. The spectrum of test samples was recorded in the range of 4000–470 $\text{cm}^{-1}$. In order to obtain information about the interaction of amino acids with calcium oxalate crystals, experiments were carried out on adsorption of amino acids on the synthesized samples (Golovanova and Vysotsky 2011).

# 3 Results and Discussion

## 3.1 Thermodynamic Simulation

Based on the available data about the thermodynamic values of solubility products at 310 K, functional dependences were determined $p\text{Ca}^{2+} = f\left(p\text{C}_2\text{O}_4^{2-}, \text{pH}\right)$, and three-dimensional diagrams ("stability field") were constructed for calcium oxalate monohydrate $CaC_2O_4 \cdot H_2O$ (Fig. 1).

To evaluate the influence of acidic medium and the concentration of amino acid additives on the possible formation of sparingly soluble compounds in solution, we constructed graphical dependences (Fig. 2) of the index of supersaturation of these factors: SI $= f(\text{pH})$ and SI = $f(pC$ amino acid) It is assumed that if SI > 0, precipitation of this phase from the solution is thermodynamically more likely.

Medium acidity has the greatest influence on the thermodynamic stability, the state of the system, and processes of crystallization of the substances. Thus, with increasing pH values, the driving force of crystallization of $CaC_2O_4 \cdot H_2O$ is greatly increased. The described regularity is explained by the fact that when pH increases, relative concentration of $C_2O_4^{2-}$—in the system increases, so a positive correlation between the supersaturation of the medium and pH is observed.

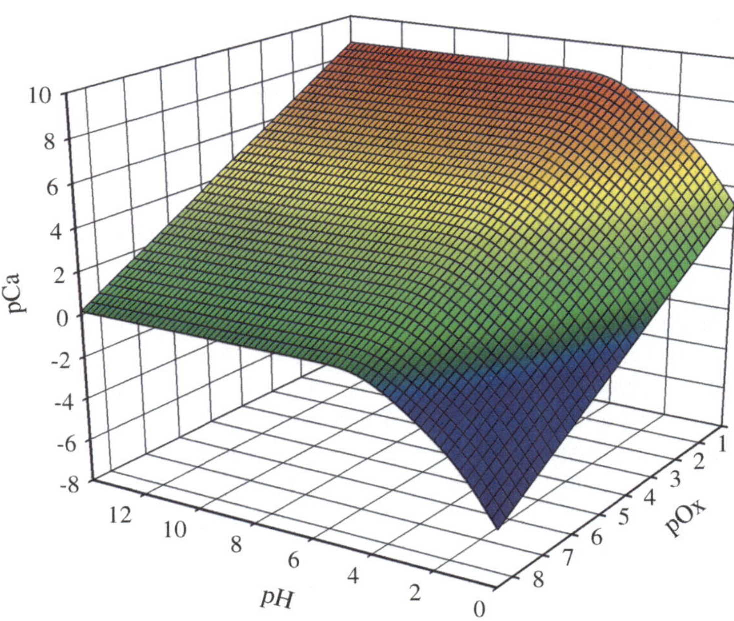


**Fig. 1** Field of stability in the formation of the poorly soluble compound system $Ca^{2+}$–$C_2O_4^{2-}$–$H_2O$

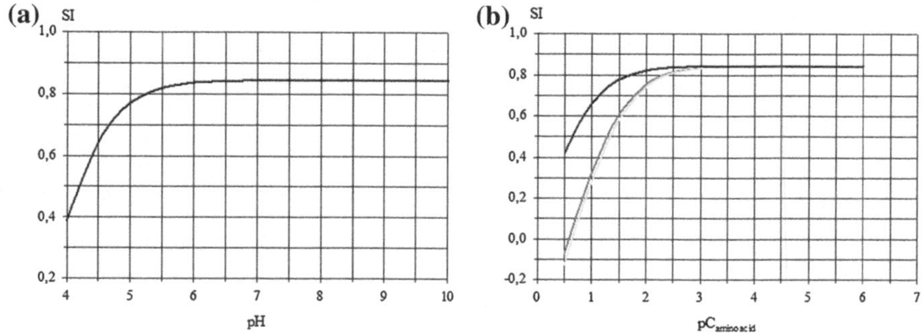


**Fig. 2** Plots of the SI index of supersaturation of calcium oxalate: **a** from pH solution; **b** from amino acids concentration: DL-glutamic acid; glycine; L-lysine

It was established that the introduction of amino acids in concentrations corresponding to physiological solution (0.004 mol/l) in the system, has little effect on the formation of mineral solids due to low values of stability constants of complexes with calcium ions for all considered amino acids.

It should be noted that the constructed thermodynamic model reflects the possibility of phase formation, only on the basis of data on their thermodynamic stability in the standard state, and does not take into account, in particular, kinetic factors influencing the formation of the solid phase in the real world. Therefore, conducting a model experiment in vitro is a necessary condition for determining the possibility of the formation of sparingly soluble compounds in solution, ionic composition (inorganic macro components), temperature, and pH close to the parameters typical to the biological fluid.

The study, with the help of X-ray analysis of the solid phases formed during crystallization from the model solution, showed that the obtained precipitate in all cases is calcium oxalate monohydrate (Rashkovich and Petrova 2006); whereas the presence of other phases, as impurities, was not detected (Fig. 3). Additionally, several studies have reported the formation of di- and tri-hydrate calcium oxalate in similar experiments.

The influence of pH (4.5; 6.0; 7.5) and temperature (20, 37, 58 °C) on the crystallization of calcium oxalate was studied. It is shown that the obtained precipitates are calcium oxalate monohydrate in all cases. The synthesis with additives of glycine, glutamic acid, and urea (their concentrations were chosen corresponding with their content in the biological fluid, human urine) indicate the formation of calcium oxalate monohydrate.

Thus, it can be concluded that the composition of deposition precipitate $CaC_2O_4{\cdot}H_2O$ does not change under the chosen conditions and in the presence of these additives. The obtained results are consistent with the known data (Golovanova 2007; Palchik et al. 2006; Sevostyanova and Polienko 2004; Zuzuk 2005) that oxalate urinary stones, more than any other (phosphate stones and uric acid), are conservative with respect to the accumulation of micro-impurities.

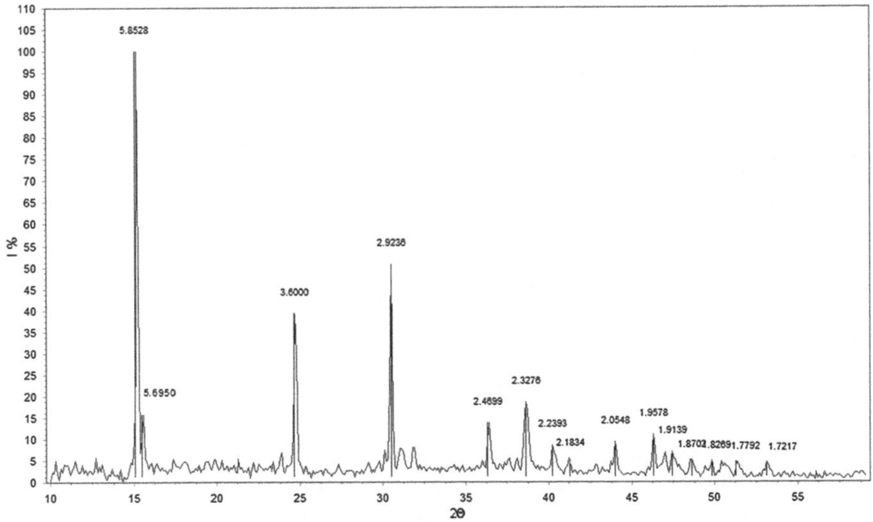


**Fig. 3** XRD pattern of calcium oxalate monohydrate

## 3.2 *Kinetics of Nucleation*

In the first stage the influence of supersaturation on the values of the induction periods of calcium oxalate monohydrate in the absence of extraneous additives was studied (Golovanova et al. 2004a). A linear relationship was obtained $\ln \tau = f\left((\ln \gamma)^{-2}\right)$. This nonstandard dependence, as it was shown, is a composition of two exponential dependences with differing values of the exponent. The determination of surface energy $\sigma$, a member of the constant $B$ for two plots of the kinetic curve, gives the values of 15.3 and 36.0 mJ/m$^2$ corresponding to heterogeneous and homogeneous nucleation, respectively.

These values are close to those obtained earlier (Golovanova et al. 2013), but significantly different from those found here, which can be explained by the absence of impurities in the system. The influence of organic additives of amino acids on the induction period of calcium oxalate monohydrate was very diverse and depended on the type of acid. The obtained data are shown in Fig. 4, illustrating the values of induction period of calcium oxalate monohydrate in the presence of amino acids.

Figure 4 shows that amino acids can both inhibit the crystallization of calcium oxalate monohydrate (DL-glutamic acid, glycine, L-lysine) and promote it (DL-valine). It is assumed that the inhibitory effect of amino acid is due to their adsorption on the active centers of the surface of the formative crystals. Adsorption is carried out due to the interaction between the positively charged surface of the crystals of calcium oxalate and the amino acid which is in the most probable conformation

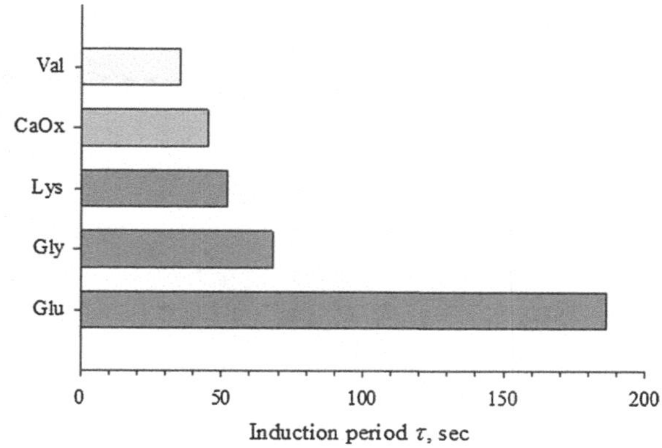


**Fig. 4** The influence of amino acids on the kinetics of nucleation of calcium oxalate: *Val* DL-valine; *Lys* L-lysine; *Gly* glycine; *Glu* DL-glutamic acid; *CaOx* calcium oxalate

under these coitiondns. In this case, we can expect that with the growth of inhibitory action, the content of amino acids in the kidney stones will increase. Indeed, at least for the basic amino acids, this is true (Golovanova et al. 2013b).

Comparative analysis of the effects of the studied amino acids on the induction period shows that this effect is probably connected with the main characteristics of the amino acids defining their adsorption on the surface of the crystals of calcium oxalate; i.e., with the structure (in particular, with the number of carboxyl groups) and with protolytic properties which determine the state and forms of locations of the amino acids in the solution at various values of pH. However, it is not yet possible to identify any general correlation between the characteristics of the amino acid and its effect on the nucleation of calcium oxalate monohydrate. Thus, the opposite effect of amino acids (similar in structure and properties) on the nucleation of calcium oxalate can be explained by the fact that both the inhibition and promotion of nucleation are realized by a single mechanism—tight binding of the amino acid with calcium ions on the surface of the nucleus (inhibition) or in the solution (promotion).

## 3.3 Kinetics of Crystallization

According to the obtained kinetic curves (Fig. 5), the dependences were constructed according to the form $\lg(d\alpha/d\tau) - 2/3 \lg \alpha = f(\lg(C_\tau - C_s))$ in which one can select several linear sections with different angles of inclination (Fig. 6). Plot A corresponds to the increase in the total number of particles due to the formation of nuclei; plot B corresponds to the growth of formed particles without increase in their total number; section C corresponds to secondary processes: reduction of the overall number of generating particles owing to the dissolution of small and growth of larger crystals, as well as aggregation of particles (Golovanova et al. 2013a).

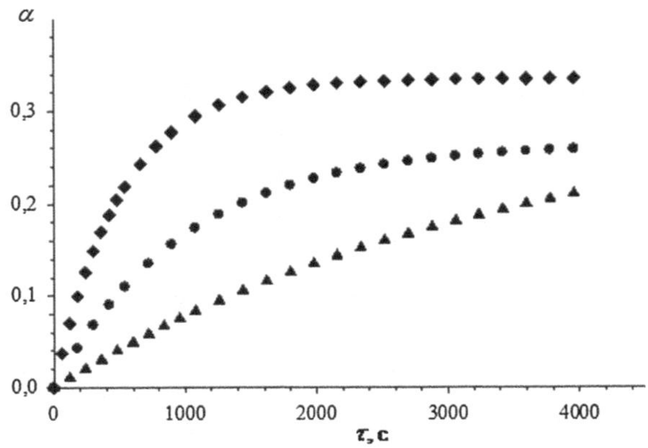


**Fig. 5** Kinetic curves of crystallization of calcium oxalate ($\gamma = 7$) in the presence of amino acids: *filled circle* without additives; *filled triangle* L-lysine; *filled diamond* DL-valine

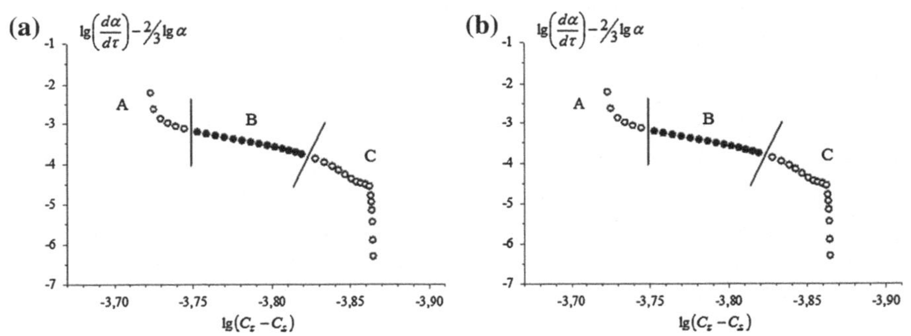


**Fig. 6** Kinetic parameters of crystallization calcium oxalate ($\gamma = 7$) in the presence of amino acids: **a** without additives; **b** in the presence of glycine

Plot B has the greatest interest for kinetics, which is why it is used for the calculation of basic kinetic parameters of calcium oxalate crystallization. The intersection of this segment with the ordinate axis gives the rate constant of deposition reaction, and the obliquity of the segment determines the order of the reaction. The constants lg $k'$ and $n$, defined during the processing of experimental data, are shown in Table 2.

First of all, one should note the great importance of the order of $n$-reaction in the crystallization of calcium oxalate. Apparently, this is a consequence of the approximation of power law dependence on the exponential law, describing kinetics of crystal growth by a mechanism of two-dimensional nucleation. For high supersaturation this mechanism of growth is quite real. It can be seen that with the increase of the initial supersaturation, the rate of crystallization increases. This can be explained by both the increase of the total number of nucleation centers and by the increase of the average rate of crystal growth.

**Table 2** Kinetic parameters of the process of calcium oxalate crystallization in the presence of amino acids

| Additive | Time of induction $\tau$, c | $\gamma = 5$ | | $\gamma = 7$ | | $\gamma = 10$ | |
|---|---|---|---|---|---|---|---|
| | | $n$ | lg $k'$ | $n$ | lg $k'$ | $n$ | lg $k'$ |
| CaOx | 45 | 7.9 | 26.5 | 10.1 | 33.1 | 12.0 | 38.6 |
| Glu | 186 | 5.4 | 16.3 | 7.5 | 23.9 | 9.1 | 28.6 |
| Gly | 68 | 7.1 | 23.5 | 10.0 | 32.8 | 9.9 | 31.0 |
| Lys | 52 | 5.3 | 16.9 | 7.5 | 23.6 | 9.3 | 28.4 |
| Val | 35 | 8.0 | 27.0 | 10.2 | 34.1 | 11.6 | 36.9 |

The presence of amino acids in the solution has, as in the case of nucleation, a different effect on the process of crystallization of calcium oxalate. Glutamic acid, lysine, and glycine have an inhibitory effect; valine has an expressed promoting effect.

The inhibitory effect of amino acids is more obviously explained by their adsorption on the growing crystals of calcium oxalate monohydrate. Analyzing the structure of the amino acids and their state in the solution (Table 1), it can be concluded that the inhibition of the growth of the crystals of whewellite increases with the increase of the length of the hydrocarbon radical, growth of the number of carboxyl groups, and the amino acid presence in the solution at physiological values of pH in the form of charged ions.

On the other hand, the amino acids can serve as new centers of whewellite nucleation, which increase the number of crystals, as they are able to fix calcium ions. In addition, being adsorbed on the surface of crystals of calcium oxalate, the amino acids may stimulate two-dimensional nucleation, thereby increasing the growth rate of crystals. These effects explain the promotional effect of amino acids on the crystallization of whewellite.

These data indicate that the same amino acid can affect the crystallization both as an inhibitor and as a promoter. The predominance of a particular impact determines the subdivision of amino acids on promoters and inhibitors. This dual effect explains the contradictions observed in the literature in the views of different authors on the role of amino acids in the process of crystallization of calcium oxalate.

With the increase of initial supersaturation, inhibition of the growth of amino acids is enhanced. This behavior is unexpected, as usually an increase in supersaturation reduces the inhibition of crystallization with impurities. Thus, the dual effect of amino acids is due to the competition of effects of inhibition and promotion of crystallization. If the proposed mechanism for the acceleration of crystal growth is valid, then for a given amino acid, with increase of supersaturation, the promoting effect should be attenuated and the rate of two-dimensional nucleation increased rapidly). Consequently, the inhibition of crystallization of whewellite by amino acid is enhanced

It is assumed that the inhibitory effect of amino acids is due to their adsorption on active centers of the surface of the forming crystals, owing to the interaction between the positively charged surface of calcium oxalate and the amino acid, which is in the most probable conformation. Hereby, in the initial stage, the interaction of uncharged amino acid with positively charged surface of crystals of calcium oxalate is possible. This can lead to the binding of a carboxyl group with the calcium ion, which causes deprotonation of the second carboxyl group and its interactions with another calcium ion. Additional stabilization of forming surface compounds may be due to the ability of amino acids with two or more carboxyl groups to form a multidentate chelate complex with calcium ions. Thus, the internal complexation, protonation of amino acid, and stereochemical factors play a significant role in the process of adsorption of amino acids on the surface of crystals of calcium oxalate.

To confirm the possibility of the adsorption influence of amino acids, the adsorption of glutamic acid on synthesized calcium oxalate monohydrate has been studied. The obtained experimental data show that adsorption can be described within the Langmuir model ($R^2$ = 0.994), and the Freundlich model ($R^2$ = 0.985).

According to the Langmuir equation $1/G = 1/G_\infty + 1/G_\infty KC$, $G_\infty$ value was calculated corresponding to the complete monomolecular coating of surface (Table 3).

It can be considered that in experiments at the maximum concentration of glutamic acid in the solution equal to 0.020 M, almost saturated monolayer is formed on the surface of the crystals, as the value of $G_{max}$ is very close to the calculated $G_\infty$ ($G_{max}$ = 0.010 mol/kg, $G_\infty$ = 0.012 mol/kg). The constant of adsorption, obtained from the Langmuir equation is very high ($K$ = 227.7 1/M), which indicates a strong adsorption of glutamic acid on the crystals of calcium oxalate.

The Freundlich equation $\ln G = \ln k + (1/n) \ln Cp$ enables indirect characterization of the value of the specific surface area of the adsorbent through the constant $k$ (Table 4).

The low value of the coefficient $k$ in the Freundlich equation indicates good crystallinity of the adsorbent. The adsorbent is synthesized from calcium oxalate monohydrate.

The obtained results of adsorption explain the strong inhibitory effect of amino acids on the processes of crystallization of calcium oxalate. Based on this, the

**Table 3** The calculation of the coefficients in the Langmuir equation

| Equation | $1/G_\infty$, kg/mol | $G_\infty$, kg/mol | $1/K \cdot G_\infty$, kg/mol | K, 1/M |
|---|---|---|---|---|
| $Y = 83.5X + 0.366$ | 83.5 | 0.012 | 0.366 | 227.7 |

**Table 4** The calculation of coefficients in the Freundlich equation

| Equation | $1/n$ | $n$ | $\ln k$ | $k$ |
|---|---|---|---|---|
| $Y = 0.38X + 3.01$ | 0.38 | 2.63 | −3.01 | 0.5 |

concentrations of amino acids in saline are sufficient to inhibit the processes of nucleation and for the growth of crystals of whewellite. In the process of stone growth, the concentration of lithogenic components and pH of the physiological solution change. Temporary decrease in supersaturation of urine in calcium oxalate to level $\gamma = 7$ will result in the normal amino acid composition of urine, in a complete inhibition of the stone growth owing to the adsorption of the impurity and deposition of organic matter on the surfaces of the stone. The increase in supersaturation (according to the obtained data to $\gamma \geq 10$) will cause a zone of regrowth with the capture of organic matter. As a result, there is a thin oscillating zone of organic matter which is characteristic of calcium oxalate stones (Izarulina et al. 2014).

Strong affinity of amino acids with dicarboxylic group to oxalate monohydrate indicates that the proteins, rich in these amino acids, play a functional role in the pathogenic formation of kidney stones.

With IR spectroscopy it was shown that these amino acids are actually adsorbed on powders of calcium oxalate monohydrate: on the spectra of the adsorbent there are characteristic amino acid bands at 1200–1000 cm$^{-1}$, 3300–3200 cm$^{-1}$ and 1400–1300 cm$^{-1}$ (Fig. 7).

The study of the conditions of calcium oxalate formation at reduced concentration of glutamic acid, up to 11 times its physiological pH, enables establishing the fact of its formation, within these conditions, of the mixture of monohydrate and dihydrate calcium oxalate (Fig. 8). Such combination is often found in the composition of urinary stones (Golovanova 2007, 2008; Palchik et al. 2006; Sevostyanova and Polienko 2004; Zuzuk 2005).

In the samples at the concentration of glutamic acid less than the physiological pH of 6–8, crystals of calcium oxalate dihydrate of various sizes appear. And at the concentration of 10–11, the microparticles have approximately the same sizes, although their number increases.

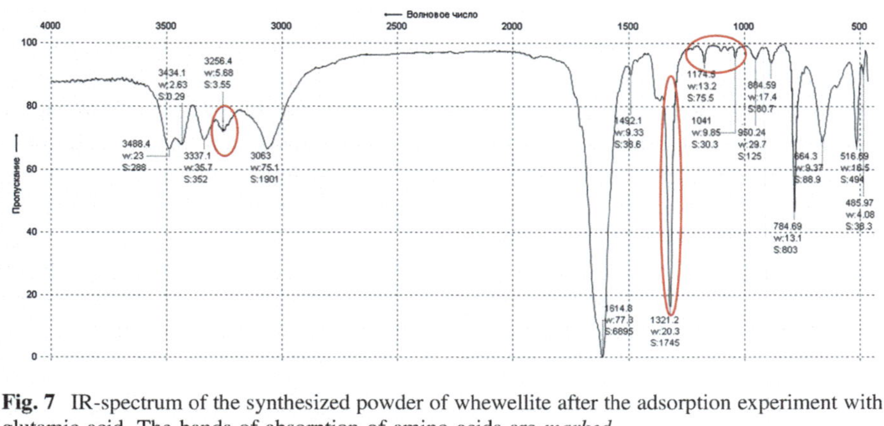


**Fig. 7** IR-spectrum of the synthesized powder of whewellite after the adsorption experiment with glutamic acid. The bands of absorption of amino acids are *marked*

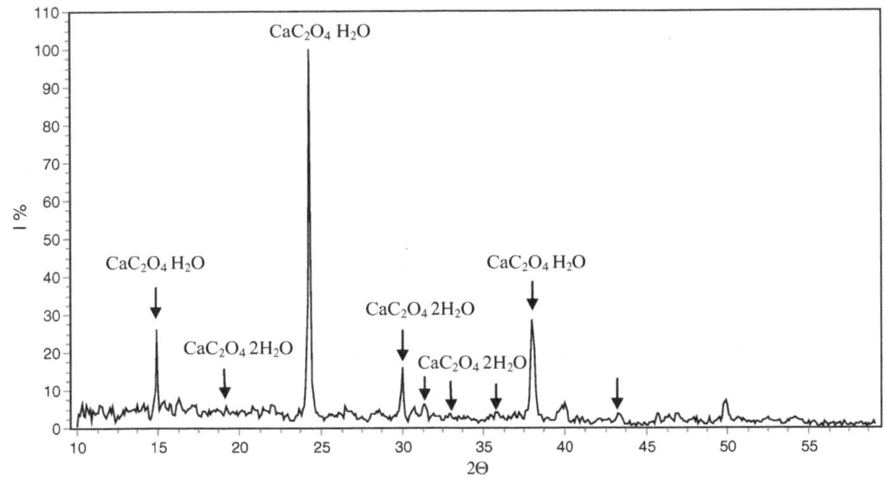


**Fig. 8** X-ray powder diagram of calcium oxalate in the presence of the glutamic acid in the solution

It is known that in urinary stones, rhythmic alternation of monohydrate and dihydrate zones of calcium oxalate is often observed. The structure of the stone may be of bi-zonal texture with whewellite nucleus and weddellite outer area that sometimes takes up to 3/4 diameter of the stone. Weddellite is a metastable phase in the presence of stone formation, and usually its crystals are replaced by fine-grained crystals whewellite (Izarulina et al. 2014).

# 4 Conclusion

1. The possibility of phase formation in the $Ca^{2+}$–$C_2O_4{}^{2-}$–$H_2O$–amino acid, the nature of crystallizing compounds, their thermodynamic stability, as well as the depth of the deposition reaction is determined mainly by the value of medium acidity. It was established that the role of amino acids as a ligand in the processes of complexation in the system has a slight impact on the formation of mineral rigid phase due to low values of the constants of stability of complexes with calcium ions.
2. In the study of the nucleation processes of calcium oxalate monohydrate in model solutions without impurities and with additives of amino acids in concentrations close to physiological, the following results were obtained:

- in solutions without impurities, the transition from heterogeneous to homogeneous nucleation at the increase of the supersaturation in excess of $\gamma = 12$ is observed; the effective surface energy increases more than twofold;
- various amino acids have both inhibiting and promoting action on the nucleation of calcium oxalate monohydrate;
- the increase of inhibitory properties of amino acids correlates with the increase of their content in calcium oxalate kidney stones.

3. Studies of the kinetics of crystallization of calcium oxalate monohydrate in the presence of amino acids additives in concentrations close to the physiological, gave the following results:

   - the growth of crystal of whewellite occurs by the mechanism of two-dimensional nucleation;
   - different amino acids have both inhibiting or promoting action on the crystallization of whewellite;
   - the influence of amino acids on the growth of crystals of whewellite coincides with their effect on the induction time of nucleation of whewellite;
   - the inhibiting properties of amino acids increase and promoting ones fall with the increase of supersaturation owing to the competition of these effects;
   - both effects can be explained in the adsorption of amino acids on the crystals of whewellite: inhibition by preventing the points of growth, promotion by the creation of centers of two-dimensional nucleation on the surface of crystals.

## References

Al Zahrani H, Norman RW, Thompson C, Weerasinghe S (2000) The dietary habits of idiopathic calcium stone-formers and normal control subjects. Brit J Urol Int 2000 85(6):616–620

Arias Funez F, Garcia Cuerpo E, Lovaco Castellanos F, Escudero Barrilero A, Avila Padilla S, Villar Palasi J (2000) Epidemiologia de la litiasis urinaria en nuestra Unidad. Evolucion en el tiempo y factores predictivos. [Epidemiology of urinary lithiasis in our Unit. Clinical course in time and predictive factors]. Arch Esp Urol 53(4):343–347

Assimos DG, Holmes RP (2000) Role of diet in the therapy of urolithiasis. Urol Clin North Am 27 (2):255–268

Bailly GG, Norman RW, Thompson C (2000) Effects of dietary fat on the urinary risk factors of calcium stone disease. Urology 56(1):40–44 (in Russian)

Bak M, Thomsen JK, Jakobsen HJ, Petersen SE, Petersen TE, Nielsen NC (2000) Solid-state $^{13}C$ and $^{31}P$ NMR analysis of urinary stone. J Urol 164:856–863 (in Russian)

Berthon G (1995) The stability constants of metal complexes of amino acids with polar side chains. Pure Appl Chem 67(7):1117–1240

Dawson R, Elliot D, Elliot W, Jones K (1986) Data for biochemical research. Clarendon Press, Oxford, p 544

Elnikov VY, Rosseeva EV, Golovanova OA, Frank-Kamenetskaya OV (2007) Thermodynamic and experimental modeling of the formation of major mineral phases of uroliths. Russ J Inorg Chem 2007 52(2):150–157

Golovanova OA (2007) Pathogenic minerals in the human body. Omsk, 395 (in Russian)

Golovanova OA (2008) Biomineralogy urinary, biliary, dental and salivary stones from the human body. The dissertation of Dr. geol.-mineral. Sciences, St. Petersburg State University, St. Petersburg (in Russian)
Golovanova OA, Pyatanova PA (2002) The influence of the chemical composition of drinking water on the trace element composition of kidney stones. In: Proceedings of the universities. A series of "Chemistry and chemical technology", vol 45, 2nd edn, pp 136–139 (in Russian)
Golovanova OA, Vysotsky AS (2011) Effect of organic and inorganic additives to the crystallization of calcium oxalate monohydrate. Butlerov Commun 26(12):45–54 (in Russian)
Golovanova OA, Achkasova EY, Kogut VV, Zhelyaev EV (2004a) Development of software for the simulation of nucleation in multiphase systems. Math Struct Model 13:144–149 (in Russian)
Golovanova OA, Pyatanova PA, Rosseeva EV (2004b) Analysis of the patterns of distribution of the protein component of urinary stones. Reports Acad Sci 395(5):1–3 (in Russian)
Golovanova OA, Achkasova EY, Punin YO, Zhelya EV (2006a) Basic laws of crystallization of calcium oxalate in the presence of amino acids. Crystallography 51(2):376–382 (in Russian)
Golovanova OA, Rosseeva EV, Frank-Kamenetskaya OV (2006b) The amino acid composition of human urinary stones. Bulletin of St. Petersburg State University 2(4):123–127 (in Russian)
Golovanova OA, Puning YO, Bielsko LV, Frank-Kamenetskaya OV (2010) Comparative characteristics of rocks of different genesis of the human body at the protein component. Butlerov Commun 22(11):53–63 (in Russian)
Golovanova OA, Korolkov VV, Puning YuO, Vysotsky AS (2013a) Effect of amino acids on the kinetics of crystallization of calcium oxalate monohydrate. Chem Sustain Dev 21(4):401–409 (in Russian)
Golovanova OA, Korolkov VV, Puning YO, Vysotsky AS (2013b) Features crystallization of calcium oxalate in the presence of valine and lysine. Bullet Omsk Univ (2):34–41 (in Russian)
Izatulina AR, Golovanova OA, Puning YO, Voytenko NN, Drozdov VA (2006) A study of factors influencing the crystallization of calcium oxalate monohydrate. Bulletin Omsk Univ (3):45–47 (in Russian)
Izatulina AR, Punin YO, Golovanova OA (2014) To the formation of aggregate structures of kidney stones. J Struct Chem 55(7):1225–1231
Kiss T, Sovago I, Gergely A (1991) Critical survey of stability constants of complexes of glycine. Pure Appl Chem 63(4):597–638
Larina TA, Kuznetsova TA, Koroleva LY (2006) Risk factors for urolithiasis in children Orel Region. In: Scientific notes Oryol State University. Scientific papers scientific—Research Center of Pedagogy and Psychology, vol 7 Oryol, pp 135–138 (in Russian)
Lurie Y (1989) Handbook of analytical chemistry. M.: Chemistry, 448 p (in Russian)
Palchik NA, Moroz TN, Maximova NV, Darin AV (2006) Mineral and microelement compositions of urinary stones. Russ J Inorg Chem 51(7):1098–1105
Pospekhova GP, Shailieva LO, Fedoseyev GB, Petrova MA (2001) Chronic obstructive pulmonary disease with respiratory oksalozom. General Practitioner (1):174–190 pp (in Russian)
Rashkovich LN, Petrova EV (2006) Crystallization of calcium oxalate. J Chem Life (1):158–168 (in Russian)
Sevostyanova OA, Polienko AK (2004) The mineral composition of uroliths. Bulletin Tomsk Polytech Univ 307(2):62–64 (in Russian)
Tiktinsky OL, Alexandrov VP (2000) Urolithiasis. St. Petersburg: Piter, 384 p (in Russian)
Vasiliev VP (2003) Analytical chemistry. In: Drofa M (ed) Titrimetric and gravimetric methods of analysis, vol. 1, 368 p (in Russian)
Yamauchi O, Odani A (1996) Stability constants of metal complexes of amino acids with charged side chains—part 1: positively charged side chains. Pure Appl Chem 68(2):469–496
Zuzuk FW (2005) Mineralogiya urolitiv. Author dissertation of Dr. Geol. Sciences Lviv, National. University I. Franka

# Kinetic Characteristics of Crystallization in Prototypes of Biological Fluids

**Ekaterina S. Chikanova, Olga A. Golovanova and Marina V. Kuimova**

**Abstract** This chapter presents the results of research on the kinetics of nucleation and crystal growth in solutions simulating the composition of human saliva and plasma. The kinetic characteristics of the processes (order, constant, activation energy, specific surface energy, the size of the critical nucleus) were obtained. The influence of some inorganic (magnesium ions) and organic (protein, alanine and glucose) additives on the kinetics of crystallization is shown.

**Keywords** Kinetic · Crystallization · Biological fluids · Saliva · Plasma

## 1 Introduction

According to statistics, in recent years, the percentage of pathogenic mineral formation in the human body has increased (Tas 2000; Larsen and Pearce 2003; Frank-Kamenetskaya et al. 2011; Pearce et al. 2002; Rosseeva et al. 2009; Titov et al. 2001, 2002). In most cases, the crystallization of organo-mineral aggregates (OMAs) occurs in complex compositions of biological fluids, both under physiogenic and in pathogenic conditions (Danil'chenko 2007; Rosseeva et al. 2009; Titov et al. 2002). This phenomenon is due to a number of factors of exogenetic and endogenic nature (Golovanova 2011). At present, little information is available on the patterns of OMA crystallization in multicomponent physiological solutions.

Crystallization from simulated solutions of biological fluids is difficult to investigate because these solutions include a large number of organic and inorganic compounds, and many factors affect the releasers of crystallization. In addition, OMA formation most commonly occurs in nonequilibrium conditions, and kinetic

---

E.S. Chikanova · O.A. Golovanova (✉)
Omsk F.M. Dostoevsky State University, Omsk, Russia
e-mail: golovanoa2000@mail.ru

M.V. Kuimova
National Research Tomsk Polytechnic University, Tomsk, Russia

 
O.V. Frank-Kamenetskaya et al. (eds.), *Biogenic—Abiogenic Interactions in Natural and Anthropogenic Systems*, Lecture Notes in Earth System Sciences,
DOI 10.1007/978-3-319-24987-2_38

factors account for their appearance. The establishment of crystallization patterns from bioliquid prototypes enables one to reveal the mechanisms of formation and growth of crystal phases, which promotes understanding of the nature of these processes and implementation of measures to prevent a range of socially significant diseases (Golovanova 2007, 2008; Golovanova et al. 2006; Safronova et al. 2009).

For example, the main inorganic phase of pathogenic calcification of collagen and muscle tissue, as well as of bone and dental tissue, is calcium phosphate. With a certain extent of approximation and idealization, calcium phosphate is referred to carbonate-containing hydroxyapatite, which is typically slightly crystallized and non-stoichiometric due to the presence of significant quantities of foreign ions. Some of these ions are a part of the apatite crystal lattice, and others are adsorbed on the apatite surface. This apatite is a typical biogenic mineral, which spatially, genetic-structurally, and morphologically is closely associated with proteins, lipids, and polysaccharides of the body tissue.

In turn, pathogenic apatite does not have such a close association with metabolic processes in the body, as a physiogenic one. The degree of its imperfection is always high, and to a greater degree it depends on local processes rather than on the body state as a whole (Lamanova 2010; Gilinskaya et al. 2003a, b, c; Gilinskaya et al. 2007; Rosseeva et al. 2012; Titov et al. 2001; Wiechert et al. 1970).

This chapter examines the kinetics of crystallization from simulated solutions of human saliva and plasma under near physiological conditions.

## 2 Materials and Methods

The solutions containing ions, whose joint presence does not yield slightly soluble compounds, were used to study the crystallization of a solution simulating the composition of human blood plasma and saliva (Severin and Solov'eva 1979).

The choice of the initial reagents and their ratios in the solution were determined so that the concentration of ions and the ionic strength of the solution were as close as possible to the given parameters of the simulation system. Supersaturation was achieved by means of a chemical reaction (1), which was realized by mixing the solutions of slightly soluble compounds of calcium chloride and hydrophosphates of ammonium and potassium in a crystallizer.

$$CaX_2 + M_2HPO_4 + 2H_2O \rightarrow CaHPO_4\,2H_2O \downarrow + 2MX \tag{1}$$

Brand pure salts for analysis and chemically pure and distilled water were used as initial reagents. For each series of the experiments, the solutions containing cations and anions, whose joint presence does not form slightly soluble compounds under these conditions, were prepared. In each series of the experiments, the pH values were corrected to physiological pH (6.93–7.00 ± 0.01) by adding a 20 % solution of NaOH (KOH) or HCl (concentrated). After mixing equivalent volumes

**Table 1** Concentrations of variation of the studied additives

| Component | Concentration | |
|---|---|---|
| | Saliva | Blood plasma |
| Magnesium ions (mol/l) | 3.25–9.75 | 0.0019–0.0057 |
| Glucose (g/l) | 0.054–0.162 | 0.93–93.0 |
| Protein (g/l) | 1.0–5.0 | – |
| Alanine (mol/l) | – | 0.005–0.049 |

of solution, the following parameters were determined: induction periods, optical density, electrical conduction, and calcium concentration in solution.

To study the influence of inorganic and organic compounds, the following components were added to the simulation system: magnesium ions, glucose, alanine, and protein (casein) in physiological concentrations and exceeding their pH (Table 1). The visual method to measure the induction periods was used to determine the parameters of nucleation. The regulation of the nucleation process was carried out by observing the uniformity of the conditions of mixing. The time of the turbidity of solutions was determined with a stopwatch, and 4–5 parallel measurements were performed for each concentration. The obtained data were processed using Eq. (2).

$$\lg t = \lg k - n.\lg S \tag{2}$$

where $t$ is induction time, $S$ is supersaturation, and $n$ is the order of the reaction.

The dependence of the induction period on supersaturation enables one to experimentally determine the free surface energy of the forming nucleuses. The value $\sigma$ was found by Eq. (3):

$$\sigma = \left[\frac{3(RT)^3(\ln\tau_1 - \ln\tau_2)\ln^2 S_1 \ln^2 S_2}{16\pi\sigma^3 V_M^2\left(\ln^2 S_1 - \ln^2 S_2\right)}\right]^{1/3} \tag{3}$$

The crystallization rate was assessed by different methods (turbidimetric, potentiometric). The measurements were carried out with KFK-2, and distilled water was used as a blank reagent. The measurements of the optical density of solutions were made at a wavelength $\lambda$ = 670 pm in Baly tubes with a light path $l$ = 2.007 cm. Three parallel measurements were made for each experiment. The error did not exceed 5 %.

The potentiometric method (ionomer "I-150-MI") was used to study partial dependencies of the kinetics of crystallization by deposit-forming ions (calcium ions). The calcium ion concentration in the crystallization process was determined by the direct potentiometry method using an ion-selective electrode. The crystallization was conducted in a temperature-controlled solution at 25 ± 0.5 °C. The potential in the sample solution was measured after certain time periods, and the concentration of $Ca^{2+}$ ions was determined using the calibration curve. This experiment was repeated three times. The measurement error was 5–10 %.

At the end of the experiment, the hard phases were filtered, dried, and analyzed by X-ray diffraction analysis (XRD) using qualitative and quantitative phase analysis. XRD was conducted with the Bruker D8 Advance X-ray diffractometer (Germany) in monochromate Cu-$k_\alpha$ radiation. The qualitative analysis of the phase composition of the sample was carried out by comparing the experimental values of the interplanar spaces and relative intensities of the diffraction peaks with a set of corresponding values for each of the proposed phases in the international database obtained by powder diffractometry PDF-2. The detection limit was 0.5–5 wt%. The analysis of variance was conducted with a laser diffraction particle size analyzer (Shimadzu SALD-2101).

The results of the analysis were used to obtain a curve of particle size distribution and average particle size in micrometers using special software (WING-2; WING-3). For more reliable results, a four to five-fold repetition of the analysis was carried out. The relative standard deviation for the measurement data was $S_r = 0.02 - 0.04$. Statistical data processing was carried out using software StatSoft Statistica 6.0.

# 3 Results and Discussion

## 3.1 Nucleation Parameters

As the result of the research on the dependence of the induction time on supersaturation (Fig. 1), it was established that the curves are not linear and consist of three parts: for higher supersaturations, the induction times are very small; for

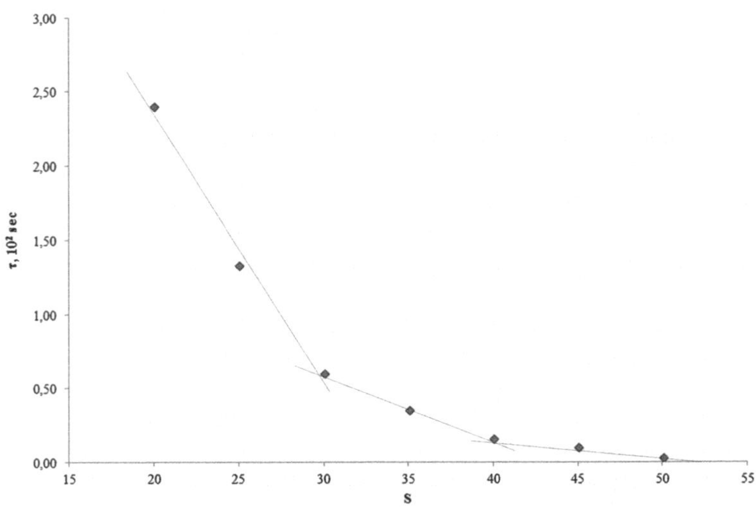


**Fig. 1** Dependence of the induction period on supersaturation in simulated solutions of saliva

**Table 2** Parameters of nucleation for simulated systems

| Parameter of nucleation | Saliva | Blood plasma |
|---|---|---|
| $n_N$ | 4.5 ± 0.1 | 1.6 ± 0.1 |
| k ($s^{-1}l^{n}mol^{-n}$) | 2.78 × $10^8$ | 0.88 × $10^8$ |

average supersaturations, the induction times range from several seconds to several minutes; and for small supersaturations, the induction times increase rapidly, tending to infinity. This regularity can be explained by the fact that, with an increase of supersaturation, the concentration of deposit-forming ions increases. The obtained values of the induction periods were used to calculate the overall order of the reaction by a graphical method (Hamsky 1969; Todes et al. 1984). The value of the order indicates the number of particles that make up the composition of the nucleus, and the constant value e characterizes the total number of the formed particles (Hamsky 1969; Todes et al. 1984).

It has been found that the order of nucleation is 4.5 for the simulation system of saliva and 1.6 for the simulated system of blood plasma. The nucleation constant for the saliva system is three-fold more than the corresponding constant for the solution of blood plasma (Table 2).

The specific surface energy was calculated according to the Gibbs-Volmer theory based on the data of dependence of the nucleation on supersaturation (Frolov 1982). As a result, it was discovered that specific surface energy is low at low supersaturation and increases at high supersaturation (Tables 3 and 4). This points to the fact that, with an increase in supersaturation, the transition from heterogeneous nucleation of crystallites to a homogeneous one is observed. Herewith, a smaller value of the specific surface energy is "effective," reflecting the adhesion of the nucleus in the active center. The comparison of the systems shows that the specific surface energy of saliva is magnitudes greater than that for the simulated solution of blood plasma.

On the basis of the thermodynamic model of nucleation and the calculated value of the specific surface energy, the size of the critical nucleus can be calculated according to Eq. (4) (Frolov 1982). Nuclei that are larger than the critical one are capable of further growth because their free energy decreases.

**Table 3** Dependence of the specific surface energy on supersaturation of saliva

| $S$ | Time of induction (s) | $\sigma$ (mJ/$m^2$) |
|---|---|---|
| 15 | Stable system | 46.0 |
| 20 | 240.0 ± 2.3 | |
| 25 | 133.0 ± 3.8 | |
| 30 | 60.0 ± 2.3 | |
| 35 | 35.0 ± 0.6 | |
| 40 | 16.0 ± 0.6 | |
| 45 | 10.0 ± 1.1 | |
| 50 | 3.0 ± 0.6 | |

**Table 4** Dependence of parameters of nucleation on supersaturation of blood plasma

| $S$ | Time of induction (s) | $\sigma$ (mJ/m$^2$) |
|---|---|---|
| 3 | Stable system | 22.0 |
| 4 | 10.56 | |
| 5 | 8.48 | |
| 7 | 5.02 | |
| 10 | 3.54 | |
| 12 | 0.92 | 24.0 |
| 15 | 0.80 | |
| 20 | 0.70 | |
| 25 | 0.52 | |
| 30 | 0.10 | 25.0 |
| 40–90 | 0.06 | |

$$r^* = \frac{2\sigma}{RT \ln S} \tag{4}$$

The obtained results show that the size of the critical nucleus for two systems is in the range of 3–20 μm. With the increase of supersaturation in the solution, the size of the critical nucleus reduces, which is in agreement with the data reported elsewhere (Moshkin1984; Larichev et al. 2006).

It is known that inorganic and organic additives influence the processes of crystallization in biological matrixes (Golovanova and Vysotsky 2011; Golovanova et al. 2011). In the study of the influence of magnesium ions on the stage of nucleation for two simulation solutions, it was found that the inhibitory effect of this additive increases as the ion concentration increases. It should be noted that for the saliva, the influence of magnesium ions does not depend on supersaturation. The inhibitory effect of magnesium ions can be due to the formation of slightly soluble compounds of magnesium, $Mg_3(PO_4)_2$ ($_pK_s = 13$).

Glucose, alanine, and protein (casein) were chosen from the organic components contained in the composition of bioliquids. For the simulated solution of saliva, it was found that both glucose and protein affect the process of crystallization of slightly soluble compounds at the nucleation stage. In the process, glucose reduces the induction period (i.e., the nucleation rate increases) and the protein increases the induction time; consequently, the rate of nucleation reduces.

The following patterns were revealed while studying the influence of glucose on the nucleation stage: under high supersaturations in the simulated systems, as the additive ratio increases, the induction period exponentially decreases. For low values of supersaturation, the dependence of the induction period on the concentration of glucose has a linear correlation. In our opinion, the formation of calcium gluconate in the solution contributes to heterogeneous nucleation by means of inoculation required for crystallization of calcium phosphates.

In the case of the protein additives (casein), we found that the induction time linearly increases as the concentration of casein grows. Thus, the introduction of protein in physiological concentrations reduces the nucleation rate by 1.5–1.7 times. As the concentration of protein additives (casein) increases by 5 times, the rate decreases by 6.6–6.9 times. The presence of casein in the medium can be assumed to stabilize the colloidal calcium phosphate particles of the simulated saliva due to forces of adsorption interaction, thereby preventing coagulation and formation of aggregates, which can be the centers of crystallization. In addition, the general viscosity of the system increases, which prevents the diffusion of the particles from the volume and the formation of crystallites.

In the simulated solution of blood plasma, both glucose and alanine inhibit nucleation at all concentrations, and the effect increases as the additive ratio in the mother liquor increases. A series to reduce the impact of additives on the nucleation process is obtained: alanine > glucose > magnesium ions.

Thus, it was experimentally found that all of the selected additives affect the nucleation stage.

## 3.2 *Kinetics of Crystallization*

To study the kinetics of crystal growth, a range of supersaturation was selected and the kinetic curves were taken (Fig. 2). In all the observed dependences for both considered systems, two distinct parts can be observed: the first corresponds to the increase in the number of particles in the volume, whereas the second corresponds to the growth of the formed particles.

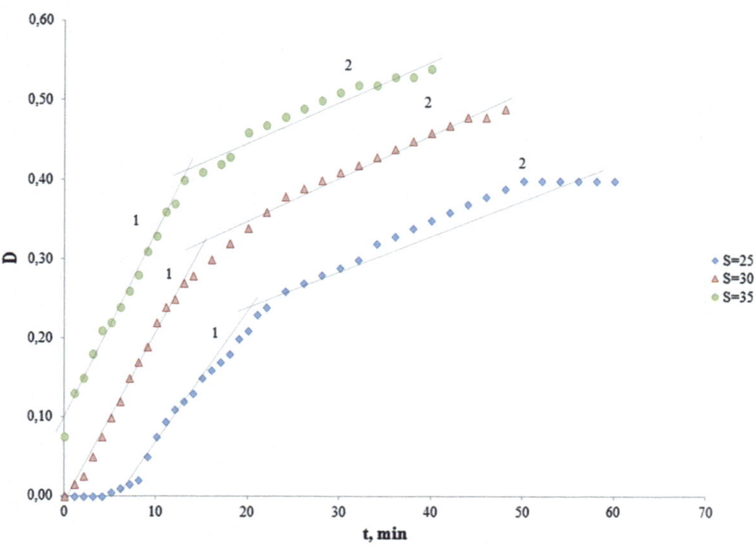


**Fig. 2** Dependence of the optical density on the crystallization time for the simulated solution of saliva

**Table 5** Kinetic parameters of saliva crystallization

| Section | Supersaturation | Order | $R^2_{tabl}$ | $R^2_{cal}$ | $k \times 10^3$ ($s^{-1}$) |
|---|---|---|---|---|---|
| 1 | $S = 25$ | $n = 0$ | 0.514 | 0.987 | 13.0 ± 4.2 |
| | $S = 30$ | | | 0.967 | 18.0 ± 4.0 |
| | $S = 35$ | | | 0.966 | 18.5 ± 2.9 |
| 2 | $S = 25$ | $n = 0$ | 0.514 | 0.980 | 4.6 ± 0.6 |
| | $S = 30$ | | | 0.988 | 5.5 ± 0.5 |
| | $S = 35$ | | | 0.970 | 5.1 ± 0.8 |

**Table 6** Kinetic parameters of plasma crystallization

| Section | Supersaturation | Order | $R^2_{tabl}$ | $R^2_{cal}$ | $k \times 10^3$ ($s^{-1}$) |
|---|---|---|---|---|---|
| 1 | 5 | $n = 0$ | 0.514 | 0.978 | 18.36 ± 0.76 |
| | 10 | | | 0.939 | 6.50 ± 0.39 |
| | 20 | | | 0.977 | 6.60 ± 0.43 |
| | 30 | | | 0.904 | 6.90 ± 0.19 |
| | 40 | | | 0.946 | 6.85 ± 0.27 |
| 2 | 5 | $n = 0$ | 0.514 | 0.870 | 4.45 ± 0.13 |
| | 10 | | | 0.891 | 0.20 ± 0.05 |
| | 20 | | | 0.900 | 0.25 ± 0.07 |
| | 30 | | | 0.853 | 0.15 ± 0.02 |
| | 40 | | | 0.891 | 0.15 ± 0.01 |

The order was determined by the graphical method for each of the sections (Tables 5 and 6). Both for the first and for the second sections, the order is equal to zero, which is characteristic of ion-exchange heterogenous reactions that occur in the solution.

The rate constants were found. The calculations showed that, for the first section, the rate constant is by an order of magnitude greater than for the second section for all considered supersaturations. The observed dependences and statistical processing of data indicate that an increase in supersaturation over 10 does not influence significantly the magnitude of the rate constant. In our opinion, this is due to the high content of deposit-forming components: their concentration is so high that the expenditure of concentration on the formation and growth of crystals does not affect their contents.

An important characteristic of the crystallization kinetics is the value of the activation barrier, which is the activation energy. Its value indicates the rate-limiting step of the process that enables evaluation of the control methods. To calculate the value of activation energy, the dependence of the rate constant of crystallization on the temperature was used. The calculation was done by the Arrhenius equation (Simiohin 2001) (Table 7). The obtained values of the activation energy indicate that the first stage occurs in the region of mixed-order kinetics, and the second stage is limited by the rate of ion diffusion to the surface of the substance hard phase.

**Table 7** Activation energy for the simulated solution of saliva

| Section | $k \times 10^3$ at $t = 25$ °C ($s^{-1}$) | $k \times 10^3$ at $t = 50$ °C ($s^{-1}$) | $E_{act}$ (kJ/mol) |
|---|---|---|---|
| N1 | 16.5 | 114.4 | 62.0 |
| N2 | 5.1 | 7.9 | 14.1 |

**Table 8** Kinetic characteristics of the crystallization process in saliva

| Supersaturation | Order | $R^2_{tabl}$ | $R^2_{cal}$ | lg $K'$ |
|---|---|---|---|---|
| $S = 25$ | 0.45 ± 0.10 | 0.514 | 0.991 | 0.20 ± 0.09 |
| $S = 30$ | 0.64 ± 0.11 | | 0.988 | 0.72 ± 0.04 |
| $S = 35$ | 0.87 ± 0.19 | | 0.922 | 1.45 ± 0.18 |

**Table 9** Kinetic characteristics of the crystallization process in blood plasma

| Supersaturation | Order | $R^2_{tabl}$ | $R^2_{cal}$ | lg $K'$ |
|---|---|---|---|---|
| $S = 5$ | 0.59 ± 0.12 | 0.514 | 0.6585 | 2.56 ± 0.20 |
| $S = 10$ | 0.58 ± 0.14 | | 0.7641 | 2.01 ± 0.35 |
| $S = 20$ | 0.46 ± 0.11 | | 0.8035 | 2.00 ± 0.53 |
| $S = 30$ | 0.41 ± 0.17 | | 0.6367 | 2.04 ± 0.58 |
| $S = 40$ | 0.32 ± 0.09 | | 0.8452 | 2.00 ± 0.46 |

To compare the crystallization processes that occur at different supersaturations, the value of the efficiency of the crystallization behavior (conversion $\alpha$) was used (Frolov 1982). The second section is of greatest interest in the study of kinetic curves (Fig. 2). It corresponds to the crystal growth, so it was used to calculate the basic kinetic characteristics. The order of the rate constant (for the second stage) obtained in the experimental data processing is shown in Tables 8 and 9.

The analysis showed that the particular order by calcium ions for simulated systems does not exceed a unit; this value is typical of the systems where heterogeneous nucleation and further growth prevail. Furthermore, it is important to note that with the growth of supersaturation, the particular order and the constant increase for the simulated solution of saliva and for the blood plasma system, they slightly reduce. This can be explained by the fact that as the calcium content in the solution increases, the number of the crystallization nuclei grows; therefore, the crystal growth rate slows down. The values of the constants are no longer statistically different at supersaturation above 20.

Analysis of variance was performed to obtain the differential curves of particle distribution by sizes at different intervals of crystallization, characterizing the formation of particles of hard phase (Fig. 3). The change in the average particle size was defined after mathematical processing of the differential curves. The analysis showed that during crystallization the particle size increases, and the confidence intervals for different time intervals do not overlap. With the increase of supersaturation of deposit-forming ions, the size of crystal particles decreases, which is in

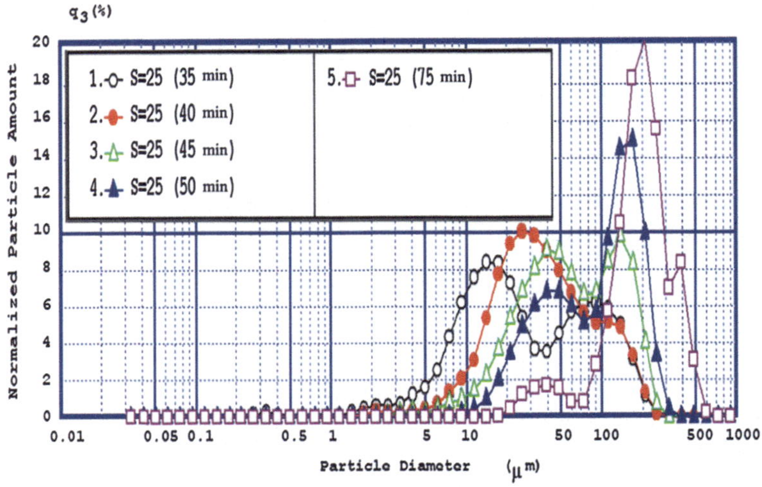


**Fig. 3** Diagrams of the particle size distribution at different time intervals (saliva $S = 25$)

agreement with the data reported elsewhere (Frolov 1982; Moshkin 1984; Hamsky 1969; Golovanova et al. 2011; Larichev et al. 2006; Todes et al. 1984). The obtained dependencies have a linear correlation in the coordinates $d^3 = f(t)$ (Fig. 4). Consequently, in terms of the theory of Lifshitz–Slyozov–Wagner (Frolov 1982), it can be assumed that the growth of the particles in these systems occurs owing to isothermal distillation (Ostwald ripening)—that is, the matter is transferred from small particles to large particles because the chemical potential of the latter is smaller (Kelvin effect). As a result, small particles gradually dissolve and large particles grow. Thus, the aggregation of particles is due to the diffusion transfer of the matter.

It is established that the dependence of particle size on supersaturation in the simulated solutions is nonlinear; it has a maximum point of extremum, which indicates that two-dimensional nucleation is characteristic of the simulated liquids (Frolov 1982), wherein the calculated value is $n = 2.00 \pm 0.06$. The solutions with supersaturation were used to study the influence of various additives on the stages of crystal growth: saliva $S = 30$ and blood plasma $S = 10$. The data processing showed that the order for the systems in the presence of additives, as well as for the system in the absence of additives, is less than a unit. The value of the rate constant is less than that for the system in the absence of additives.

Magnesium ions, at concentrations of five or ten times greater than the value of biological pH, reduce the rate of crystallization proportionally to the increase in the content of this component, but herewith do not influence the corresponding constants and the order. If the concentration increases 15-fold, not only the rate of crystallization but also the constants in each of the plots decrease. Glucose enhances the rate of crystallization in the simulated solution of saliva, and its effect appears

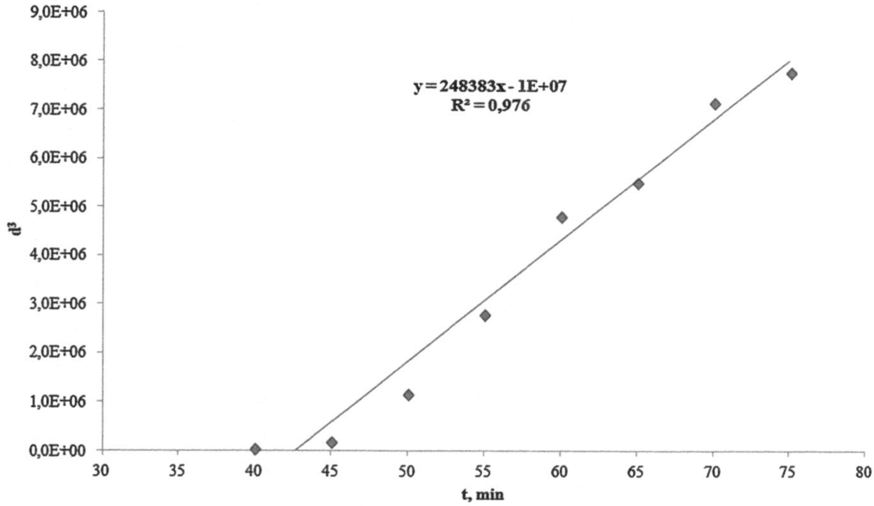


**Fig. 4** Dependence of the cube of particle size on crystallization time (saliva $S = 25$)

for both studied stages. This may occur because of the formation of complex compounds of calcium ions with glucose (Gergel 2005; Gilmiyarov 2002).

For the simulated solution of blood plasma, it was found that the greater the concentration of glucose and alanine in the solution, the lower the trough concentration of calcium ions. The series to lessen the influence of additives on private kinetic characteristics was obtained: magnesium ions > alanine > glucose. Comparing the influence of additives on the nucleation phases and on the crystal growth, it should be noted that the effect is more distinct at the stage of nucleation.

The pH of the solution plays an important role in the formation of a mineral component from the biolifluid prototypes. It is known (Solonenko and Golovanova 2011) that in a three-component system $Ca(OH)_2$–$H_3PO_4$–$H_2O$, 11 known unsubstituted calcium orthophosphates can be found with an atomic ratio of Ca/P in the range 0.5 and 2.0. The most important parameters are the value of the atomic ratio of Ca/P, basicity/acidity, and solubility. These parameters strongly depend on the pH of the solution. The lesser the value of Ca/P, the more acidic and water-soluble calcium orthophosphate is. Due to the three-fold ionic equilibrium in the solution containing phosphate ions, the pH change leads to a change in relative concentrations of the four polymorphic forms of phosphoric acid and, hence, the chemical composition and the amount of calcium orthophosphate, which are formed as a result of direct deposition.

In the study of kinetic regularities depending on the pH, we found that as pH grows, the order and the constant of crystallization remain the same but the rate of the deposit formation changes. Thus, at pH = 6.80 ± 0.01, the rate of crystallization in both regions for two systems is reduced on average by 1.3 times; however, as pH increases to 7.10 ± 0.01, the rate increases in both stages by 1.3–1.6 times. This can

**Table 10** Composition of the deposit obtained after 24 h of crystallization at pH = 6.50 ± 0.01

| Phase | Content % wt. |
|---|---|
| $CaHPO_4 \cdot 2H_2O$ | 71.8 |
| $Ca_4H(PO_4)_3 \cdot 2.5H_2O$ | 17.9 |
| $Ca_{10}(PO_4)_6(OH)_2$ | 10.3 |

**Table 11** Composition of the deposit obtained after 24 h of crystallization at pH = 8.50 ± 0.01

| Phase | Content % wt. |
|---|---|
| $CaHPO_4 \cdot 2H_2O$ | Insignificantly |
| $Ca_4H(PO_4)_3 \cdot 2.5H_2O$ | 25.1 |
| $Ca_{10}(PO_4)_6(OH)_2$ | 74.9 |

be explained by the predominance of one of the forms of phosphate ions in the solution.

It was found that in the simulated solution of blood plasma, the pH variation does not influence the induction periods. In our opinion, this is due to the presence of powerful buffer systems in the blood plasma: carbonate and phosphate. Additionally incorporated $H^+$ and $OH^-$ ions interact with the components of the buffer systems, and at that, no significant changes in the pH value and, therefore, in the properties of the system can be observed. The values of the constants are independent of pH both in the first and in the second regions. Thus, the change of pH in the range of 7.0–8.0 is found to have no influence on the parameters of the process of crystallization of the simulated human blood plasma system.

The analysis of the solid phases, obtained in simulation at average concentrations and examined physiological pH (6.50–8.50) of saliva a day after crystallization, showed that the mixture of calcium phosphate is formed with a predominance of a mineral component—brushite (kinetically controlled phase) (Table 10). The presence of impure, slightly soluble phosphates can be observed: $Ca_4H(PO_4)_3 \cdot 2.5H_2O$ and $Ca_{10}(PO_4)_6(OH)_2$ (the content of other phosphates is insignificant). As pH grows, their content increases (Table 11).

X-ray diffraction analysis of the synthesized solid phases obtained from the simulated solution of the blood plasma showed that carbonated hydroxyl-apatite ($Ca_{10-x/2}(PO_4)_{6-x}(CO_3)_x(OH)_2$) and whitlockite ($Ca_9Mg(PO_4)_6(PO_3OH)$) are found in the deposit. Low intensity and half-width of the diffraction reflections indicate low crystallinity of the obtained powders.

The diffraction patterns of solid phases, synthesized in the presence of the investigated additives, were obtained. The peaks of the diffraction patterns of the samples in the presence of additives became fuzzier as the additive ratio increased. The size of the crystallites in the presence of additives and without them was calculated. It was found that the size of the crystallites in the presence of additives decreases by an order of magnitude. This may be due to the adsorption of molecules on the surface of the faces, crystal face poisoning, and deceleration in growth of the rate of crystallites, which is in agreement with the previously obtained kinetic data.

# 4 Conclusion

As the result of this research, the crystallization process from solutions simulating the composition of saliva and human plasma under near physiological conditions is studied. The kinetic characteristics of the processes are obtained (order, constant, activation energy, the specific surface energy, the size of the critical nucleus). The influence of a number of additives in biological fluids on the processes of crystallization in the prototype of blood plasma and human saliva was found.

# References

## Journal Article

Danil'chenko SN (2007) The structure and properties of calcium apatite with the viewpoint biomineralogy (review). The Vistn SSU. Ser Phys Math Mech 2:33–59 (In Russian)

Frank-Kamenetskaya O, Kol'tsov A, Kuz'mina M, Zorina M and Poritskaya L (2011) Ion substitutions and non-stoichiometry of carbonated apatite-(CaOH) synthesised by precipitation and hydrothermal methods. J Mol Struct 992(1–3):9–18

Gilinskaya LG, Grigoryeva TN, Okuneva NG, Vlasov YuA (2003a) The study of pathogenic mineral formations in the human heart valves. I. The chemical and phase composition. J Struct Chem 44(4):678–689 (In Russian)

Gilinskaya LG, Okuneva NG, Vlasov YuA (2003b) The study of pathogenic mineral formations in the human heart valves. II. EPR spectroscopy. J Struct Chem 44(5):882–889 (In Russian)

Gilinskaya LG, Rudina NA, Okuneva NG, Vlasov YuA (2003c) The study of pathogenic mineral formations in the human heart valves. III. Electron microscopy. J Struct Chem 44(6):1122–1129 (In Russian)

Gilinskaya LG, Zanan YN, Rudina NA (2007) Bacterial genesis of phosphate in the human body and nature. Lithol Miner Resour 1:63–75 (In Russian)

Golovanova OA (2008) Features pathogenic mineral formation. Tomsk State Univ J 313:215–224 (In Russian)

Golovanova OA (2011) Biomineral composites of the human body: theory, practice and prospects. Butlerov Comm 24(3):113–122 (In Russian)

Golovanova OA, Vysotsky AS (2011) Effect of organic and inorganic impurities on crystallization of calcium oxalate monohydrate. Butlerov Comm 26(12):45–54 (In Russian)

Golovanova OA, Achkasova EYu, Punin YuO, Zhelyaev EV (2006) Main regularities of crystallization of calcium oxalate in the presence of amino acids. Crystallogr Rep 51(2): 348–354

Golovanova OA, Punin YO, Vysotsky AS, Khannanov VR (2011) Effect of organic and inorganic contaminants on the nucleation of calcium oxalate monohydrate. Chem Sustainable Dev 5:463–470

Lamanova LM (2010) Tissue calcification in the cardiovascular system. Tomsk State Univ J 337:194–197 (In Russian)

Larsen MJ, Pearce EIF (2003) Saturation of human saliva with respect to calcium salts. Arch Oral Biol 48(4):317–322

Moshkin SV (1984) Features of the mechanism of crystal growth of sparingly soluble compounds at high supersaturation. Crystallogr Crystal Chem Coll Art 3:100–119 (In Russian)

Pearce EIF, Dong YM, Yue L, Gao XJ (2002) Plaque minerals in the prediction of caries activity. Community Dent Oral Epidemiol 30:61–69

Rosseeva EV, Frank-Kamenetskaya OV, Golovanova OA, Zorina ML (2009) The formation of mineral phases in the prototype human oral fluid. Vestn St Petersburg Univ Geol Geogr 2:12–20 (In Russian)

Safronova TV, Putlyaev VI, Sergeeva AI, Kunenkov ĖV, Tretyakov YuD (2009) Synthesis of nanocrystalline hydroxyapatite, calcium saccharate, calcium and ammonium hydrophosphate. Dokl Chem 426(4):491–496 (In Russian)

Solonenko AP, Golovanova OA (2011) Thermodynamic modeling of forming calcium orthophosphate. Butlerov Comm 24(2):106–112 (In Russian)

Tas AC (2000) Synthesis of biomimetic Ca-hydroxyapatite powders at 37 °C in synthetic body fluids. Biomaterials 21:1429–1438

Titov AT, Larionov PM, Shchukin VS, Zaikovsky VI (2001) The mechanism of mineralization of the heart valves. J Surf Invest X-ray Synchrotron Neutron Tech 3:74–79 (In Russian)

Titov AT, Larionov PM, Zaikovsky VI, Ivanova AS (2002) Hydroxyapatite in human blood. J Surf Invest X-ray Synchrotron Neutron Tech 7:72–76 (In Russian)

## Book

Golovanova OA (2007) Pathogenic minerals in the human body. OmSU, Omsk

Wiechert AM, Sedov KR, Sokolova RI (1970) Calcification of the arteries. Medicine, Moscow (In Russian)

## Paper Presented at a Conference

Roseeva EV, Nikolaev AM, Morozov MV, Frank-Kamenetskaya OV, Lamanova LM (2012) Bioapatites of calcificates of mitral valves. Paper presented at the International scientific conference «Annual Meeting of the Russian Mineralogical Society combined with the Fedorov Session 2012», RMS, Saint Petersburg, 9–11 October 2012 (In Russian)

## Dissertation

Gergel NI (2005) Visualization of pathochemical disorders in the body in the physico-chemical and metabolic indicators of oral fluid. Dissertation, Samara State Medical University (In Russian)

Gilmiyarov EM (2002) Dental and somatic status of the organism in terms of oral fluid. Dissertation, Samara State Medical University (In Russian)

## Institutional Author (Book)

Frolov YG (1982) Course of colloidal chemistry. Textbook for high schools, Chemistry, Moscow (In Russian)

Hamsky EV (1969) Crystallization from a solution. Science, Leningrad (In Russian)

Larichev TA, Sotnikova LV, Sechkarev BA, Breslav YA, Utekhin AN (2006) Mass crystallization in inorganic systems. Textbook, Kuzbassvuzizdat, Kemerovo (In Russian)

Severin SE, Solov'eva GA (1979) Workshop on biochemistry. MSU, Moscow (In Russian)
Simiohin IA (2001) Physical chemistry. Textbook for high schools. MSU, Moscow (In Russian)
Todes OM, Seballo VA, Goltsiker AD (1984) Mass crystallization from solutions. Chemistry, Leningrad (In Russian)

# Biomimetic Synthesis of Strontium-Containing Apatite and Its Peculiar Properties

**Rinat R. Izmailov, Olga A. Golovanova and Marina V. Kuimova**

**Abstract** This paper examines the influence of strontium chloride on the process of crystallization of strontium-substituted carbonated hydroxyapatite from prototype human synovial fluid. Products of synthesis were studied by Fourier transform infrared spectroscopy, X-ray diffraction, and differential thermal analysis. The amount of strontium in the sample using atomic emission analysis was determined. For the synthesized phases, Ca/P-ratio, the parameters of crystallinity, the crystallite size using the formula of Selyakov-Scherrer, and crystal lattice parameters were calculated. It was established that increasing the concentration of strontium ions in the analytical model increases the proportion of brushite in the solid phase. It was found that with increasing formation of a precipitate beneath the mother liquor, the kinetically stable-phase of brushite transforms to the thermodynamically more stable phase (Sr-containing carbonated hydroxyapatite).

**Keywords** Hydroxyapatite · Carbonated apatite · Brushite · Bone apatite · Sr-containing materials · Ca/P ratio · Synovial fluid · Crystallite size

## 1 Introduction

There is much interest in hydroxyapatite $Ca_{10}(PO_4)_6(OH)_2(OH)$ due to its high value to biological research and practical medicine; it is the major inorganic component of human bone tissue and is highly biocompatible (Danilchenko 2007; Mazurenko 2002; Medvedev 1993; Shchepetkin 1995; Sidelnikov 2000; Tas 2008; Fazan and Marquis 2000; Izmaylov and Golovanova 2012, 2013, 2014; Orlovskii and Barinov 2001; Putlyaev and Safronova 2006; Rodrrigues-Lorenzo and Vallet-Regi 2000;

R.R. Izmailov (✉) · O.A. Golovanova
Omsk F.M. Dostoevsky State University, Omsk, Russia
e-mail: r.r.izmailov@gmail.com

M.V. Kuimova
National Research Tomsk Polytechnic University, Tomsk, Russia



O.V. Frank-Kamenetskaya et al. (eds.), *Biogenic—Abiogenic Interactions in Natural and Anthropogenic Systems*, Lecture Notes in Earth System Sciences,
DOI 10.1007/978-3-319-24987-2_39

Balamurugan et al. 2009; Gnedenkov et al. 2011; Guicheux et al. 1995). One of the main structural characteristics of OH is its suitability for use in a variety of isomorphic substitutions, while keeping the hexagonal space group $P6_3/m$ (Guicheux et al. 1995). Such substitutions may change its thermal stability, solubility, and textural properties, as well as the reactivity of its surface (Izmailov et al. 2014; Landi et al. 2008; Li et al. 2010a, b; Liu et al. 2001, 2013; Oliveira et al. 2007; Safronova et al. 2007; Shi et al. 2005; Sopyan et al. 2008; Titov et al. 2001; Veresov et al. 2000). Inorganic components play an important role in the biological composition of hydroxyapatite, particularly cations of sodium, strontium, zinc, iron, and magnesium and anions of fluorine, carbonate, chloride, and others (Guicheux et al. 1995).

The literature describes the biological effects of inorganic ions and their influence on the processes associated with bone formation, which are important for the treatment of diseases and the development of models of mineralization and demineralization of bone (Vodianova et al. 2011). It is known that one of the most common diseases, occupying a leading place in the disease distribution of the musculoskeletal system, is osteoporosis. Osteoporosis is a disease in which there is a disturbance of bone remodeling, weight loss, and destruction (Vodianova et al. 2011; Zhang et al. 2011). Experimental and clinical studies have provided evidence that strontium compounds and medicinal products based on them stimulate the process of bone formation and inhibit the process of its destruction (Sidelnikov 2000; Tas 2008; Fazan and Marquis 2000; Izmaylov and Golovanova 2012, 2013). However, methods for the introduction of strontium into the crystal structure of bone apatite of humans have been insufficiently studied. Currently, there are two points of view on the mechanism of incorporation of strontium into the structure of the inorganic matrix of human bone tissue (Oliveira et al. (2007):

1. A rapid surface exchange mechanism between the blood and bone hydroxyapatite
2. A slow heteroionic replacement mechanism leading to the incorporation of strontium into the crystal lattice of apatite bone with the substitution of calcium ions therein

Thus, the aim of this work is to study the processes of bone mineralization for a prototype of human synovial fluid in the presence of strontium compounds and to establish the influence of strontium ions on the crystalline substance.

## 2 Materials and Methods

The synthesis of Sr-substituted carbonated hydroxyapatite was carried out on the model solution approximated by ion-electrolyte composition, pH, and ionic strength in human synovial fluid. Deposition of the solutions was performed at pH = 7.4, which corresponds to the physiological value of acidity of synovium in a norm (Izmaylov and Golovanova 2014). To synthesize Sr-containing materials based on carbonated hydroxyl-apatite, varying contents of strontium ions and calcium were

**Table 1** Mean concentration of inorganic ions of human synovial fluid, mmol/l Izmaylov and Golovanova (2014)

| pH | 7.40 ± 0.05 |
|---|---|
| Component_ion | |
| Calcium | 12.1–0 |
| Phosphates | 20.9 |
| Sodium | 140.00 |
| Magnesium | 1.10 |
| Potassium | 4.60 |
| Chlorides | 103.00 |
| Carbonates | 27.00 |
| Sulfates | 11.40 |
| Strontium | 0–12.1 |

introduced; their concentration ranged from 0 to 100 wt% of the maximum possible concentration of calcium ions. As sources of $Sr^{2+}$ ions, inorganic salt was used: $SrCl_2{*}6H_2O$. Concentrations of the other mineral components conform to those of human synovial fluid (Table 1). Crystallization of the solid phase was carried out for 7, 14, 21, and 28 days.

After aging of the precipitate in the model solution for the selected time period, the solid phase was separated from the solution by filtration, dried at a temperature of about 100 °C to constant weight for the complete removal of chemically free water, weighed, and examined with the use of complex physical and chemical methods.

X-ray diffraction (XRD; D8 Advance, Bruker; program TOPAS 3.0 (Bruker), quantitative phase analysis of multicomponent samples, calculation of crystallite size [coherent scattering regions (CSR)], and phase identification were carried out using a database (ICDD PDF) for powder diffraction. All precipitates were analysed with Fourier-transform infrared spectroscopy (FTIR spectrometer IRPrestige-21, Shimadzu; tablets with KBr). The software package PeakFit_v 4.11 was used for mathematical analysis of the spectra, which consists of the expansion of the spectral range of 400–650 $cm^{-1}$ into three elementary absorption bands (Lorentz distribution, $P = 0.99$). According to the results of decomposition, the parameter of infrared splitting antisymmetric deformation vibration $V_4$ communication O–P–O was calculated as the ratio of the intensities of the two peaks to the intensity of the "trough" between them (1):

$$k = \frac{I(564\,\mathrm{cm}^{-1}) + I(604\,\mathrm{cm}^{-1})}{I(584\,\mathrm{cm}^{-1})} \tag{1}$$

According to its value, the degree of crystallinity of the investigated mineral was estimated. Analysis of the specific surface area was performed (BET SORBTOMETR, Katakon). Differential thermal analysis (DTA) (STA-449C, NETZSCH) included the steps of starting thermal decomposition at ambient air conditions, at a current of 20 % $O_2$ in Ar flowing at a rate of 70 ml/min, heating rate of 10 °C/min, over the temperature interval 100–900 °C. The elemental composition of the synthesis products was determined by atomic emission analysis (Varian 710-ES, Agilent Technologies). In the supernatant, the residual calcium content [RD 52.24.403-94] and phosphate [RD 52.24.433-2005] were determined by the method of added-found; the ratio of Ca/P was evaluated in the sample.

## 3 Results and Discussion

The production of a modified hydroxyapatite was based on the following chemical reaction:

$$10M^1X_2 + 6M_2HPO_4 + 8MOH \rightarrow Ca_{10}(PO_4)_6(OH)_2\downarrow + 20MX + 6H_2O,$$

where $M = Na^+, K^+$; $M^1 = Ca^{2+}, Mg^{2+}, Sr^{2+}$; $X = Cl^-$.

According to the results of XRD, the obtained precipitate contained hydroxyapatite and brushite (Fig. 1). When interpreting the diffraction patterns (Fig. 1), it was revealed that increasing amounts of strontium ions in the initial solution decreased the content of hydroxyapatite in the sediment and the transition to a monophasic brushite precipitation, indicating the inhibition of ions in the strontium hydroxyapatite crystallization processes. We observed a diffuse X-ray diffraction pattern, which indicates a high dispersion of hydroxyapatite particles.

A rough estimate of the size of the formed crystallites was made using the Selyakov-Scherrer formula (Table 2). For the axis [002] corresponding to the crystallographic direction perpendicular to the basal plane of the hexagonal unit cell of apatite, the sizes of crystallite were calculated (Table 2). It was found that with an increasing concentration of strontium ions in the initial solution, there is a growth of crystallite size, which may indicate an isomorphic substitution of calcium by the strontium ions in the structure of the obtained hydroxyapatite Danilchenko (2007). It is also found that the axis corresponding to perpendicular directions (e.g., [300] and [310]) broadened to a greater extent, indicating a smaller width/thickness of the crystallites. Thus, the form of hydroxyapatite crystallites is elongated along the hexagonal axis, which is characteristic of biogenic apatite.

Based on the data, the contents of calcium and phosphorus in the solid phase and the mother liquor after filtration were determined by the ratio of Ca/P (Fig. 2). It was found that increasing the content of strontium ions in the initial solution produced a regular decrease in the Ca/P ratio. The stoichiometric ratio of such a change is explained by isomorphic substitutions in the cation sublattice of calcium ions for strontium ions (Table 3). Furthermore, it is noted that at low concentrations of calcium ions and an increase of strontium ions, in the initial solution, the mass of the solid phase decreased.

In solid phases, atomic emission analysis was conducted to determine the content of strontium ions. The results of this analysis show that the composition of the

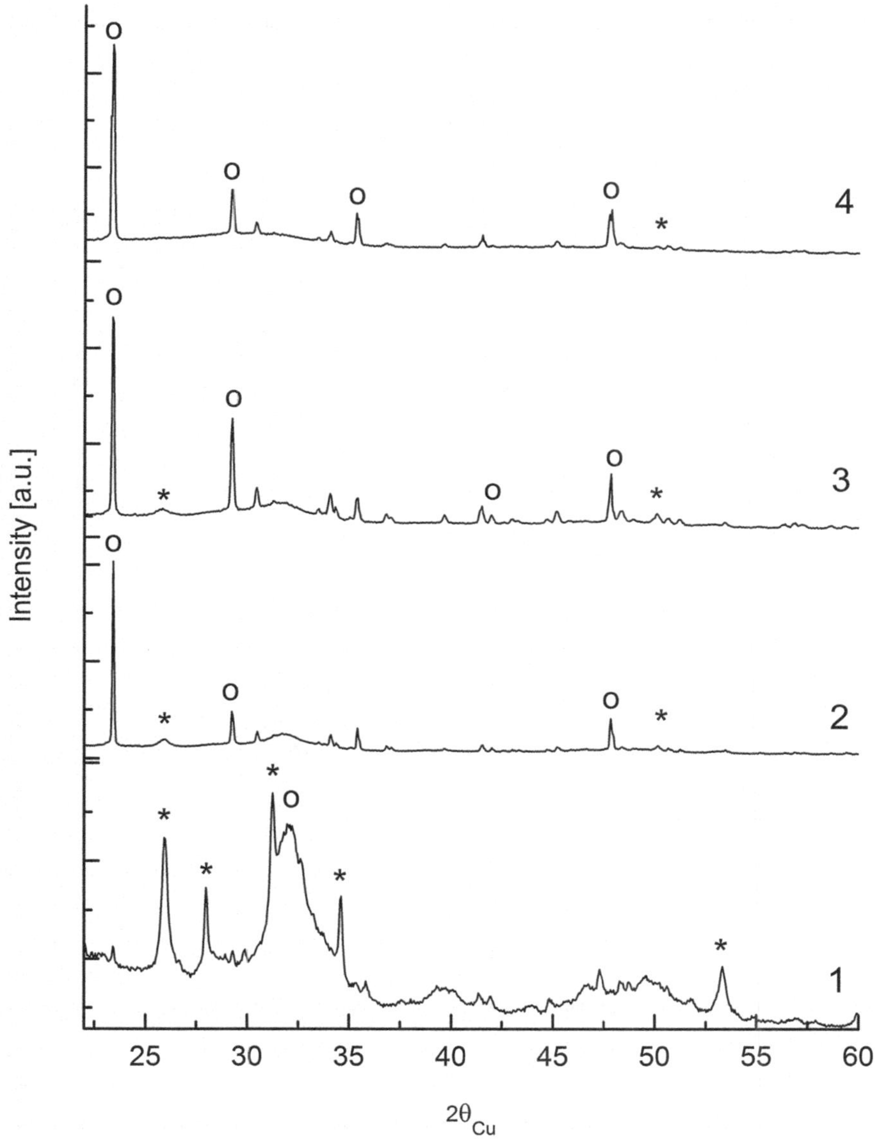


**Fig. 1** XRD pattern of calcium phosphates obtained after 7 days of experiment from model solutions with different content of $Sr^{2+}$ ions (wt%): 1–10, 2–20, 3–40, 4–50. *Asterisk* hydroxyapatite; *open circle* brushite

obtained samples includes strontium (Table 4). The content of strontium ions in precipitate is in direct proportion to the concentration of strontium-containing agents in the initial model solution. This is in agreement with the XRD results and

**Table 2** The dimensions of the crystallites of calcium phosphate with varying concentrations of $Sr^{2+}$

| Ratio of the ions in of the stock solution | | D, Å (hydroxyapatite) | D, Å (brushite) |
|---|---|---|---|
| $Ca^{2+}$ | $Sr^{2+}$ | | |
| 1 | 0 | 14.0 | – |
| 0.8 | 0.2 | 13.7 | 28 |
| 0.6 | 0.4 | 21.2 | 28 |
| 0.5 | 0.5 | – | 28 |

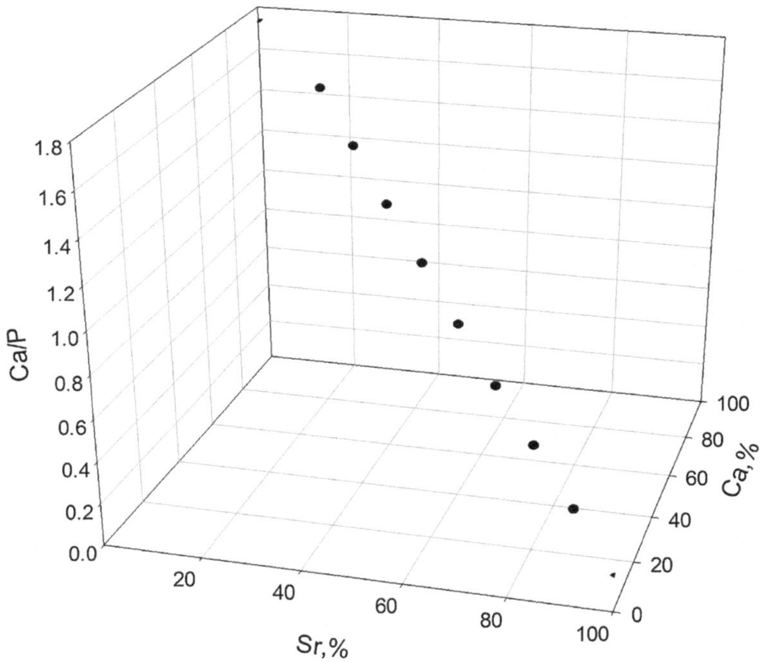


**Fig. 2** Dependence of the Ca/P ratio on the solid phase based on the composition of the starting solution

**Table 3** Dependence of the Ca/P and mass of the precipitate (m) on the concentration of calcium and strontium ions

| Ratio of the ions in of the stock solution | $Ca^{2+}$, wt% | 100 | 90 | 80 | 70 | 60 | 50 | 40 | 30 | 20 | 10 | 0 |
|---|---|---|---|---|---|---|---|---|---|---|---|---|
| | $Sr^{2+}$, wt% | 0 | 10 | 20 | 30 | 40 | 50 | 60 | 70 | 80 | 90 | 100 |
| m, gram | | 2.21 | 1.18 | 1.99 | 1.60 | 1.59 | 1.50 | 1.22 | 1.54 | 1.44 | 1.22 | 1.36 |
| Ca/P | | 1.7 | 1.6 | 1.4 | 1.2 | 1.0 | 0.8 | 0.7 | 0.5 | 0.3 | 0.2 | – |

**Table 4** Contents of Sr in precipitates

| Ratio of the ions in of the stock solution | | Content Sr in the precipitation, wt% |
|---|---|---|
| $Ca^{2+}$ | $Sr^{2+}$ | |
| 1 | 0 | 0 |
| 0.8 | 0.2 | 5.4 ± 0.1 |
| 0.6 | 0.4 | 9.5 ± 0.2 |
| 0.5 | 0.5 | 15.0 ± 0.4 |

indicates that the developed synthesis allows one to obtain Sr-substituted hydroxyapatite with a strontium content of 5–15 wt%.

The quality of the synthesized material was determined by infrared spectroscopy (Fig. 3). All IR spectra of obtained powders are characterized by bands of valence vibrations of OH groups in the wavelength range of 3570–3730 $cm^{-1}$ and deformation vibrations at 3000–3600 $cm^{-1}$ for O–H groups involved in hydrogen bonding formation. It can also be noted that the bands of deformation vibrations υ4 O–P–O in $PO_4^3$ were 640–550 $cm^{-1}$, the asymmetric stretching vibrations υ3 PO in $PO_4^{3-}$ were 1060–1030 $cm^{-1}$, and the deformation vibrations of the O–C–O in $CO_3^{2-}$ were 870–879 $cm^{-1}$ and 1400–1500 $cm^{-1}$. It was found that carbonate ions are incorporated in the structure of hydroxyapatite in the position of hydroxyl groups ($OH^-$), which indicates that they resulted from synthesizing of the A-type carbonated hydroxyapatite.

Processing the IR spectra revealed that the maximum crystallinity index (IRSF) was observed in the synthesized sample at a concentration of strontium ions in solution of 50 wt%, which correlates with the XRD data, because when the concentration of strontium ions is 50 wt%, there is a monophase deposition of brushite at experimental conditions (Table 5).

For the obtained samples, specific surface areas were determined by the specific surface-point nitrogen adsorption methods of calculation conducted by the conventional Brunauer-Emmett-Teller (N2-BET) equation (Table 3). In the concentration range from 0 to 50.0 wt% strontium ions, we found that increasing the content of the precipitate reduced the specific surface area of the particles. This is due to an increase in the proportion of brushite in the solid phase, which is in agreement with the results of XRD (Table 6).

The synthesized powders were analyzed by thermal analysis in the thermogravimetric–differential gravimeter–differential thermal analysis (TG-DTG-DTA). Analysis of the obtained curves established that samples underwent a crystal phase transformation, as follows: At $t$ = 100–280 °C, loosely adsorbed water was removed from the solid phase. At $t$ = 280–550 °C, water was removed during crystallization in the maximum temperature range of endothermic effect, which corresponds to the process of structural formation of monetite $CaHPO_4 * 2H_2O = CaHPO_4 + 2H_2O + Q$. At $t$ = 500–900 °C, removal of carbonate ions from the structure of hydroxyapatite formed the mixed-phase strontium-containing β- tricalcium phosphate. According to the results of the DTA, during the heat treatment, there is a mass loss of precipitation

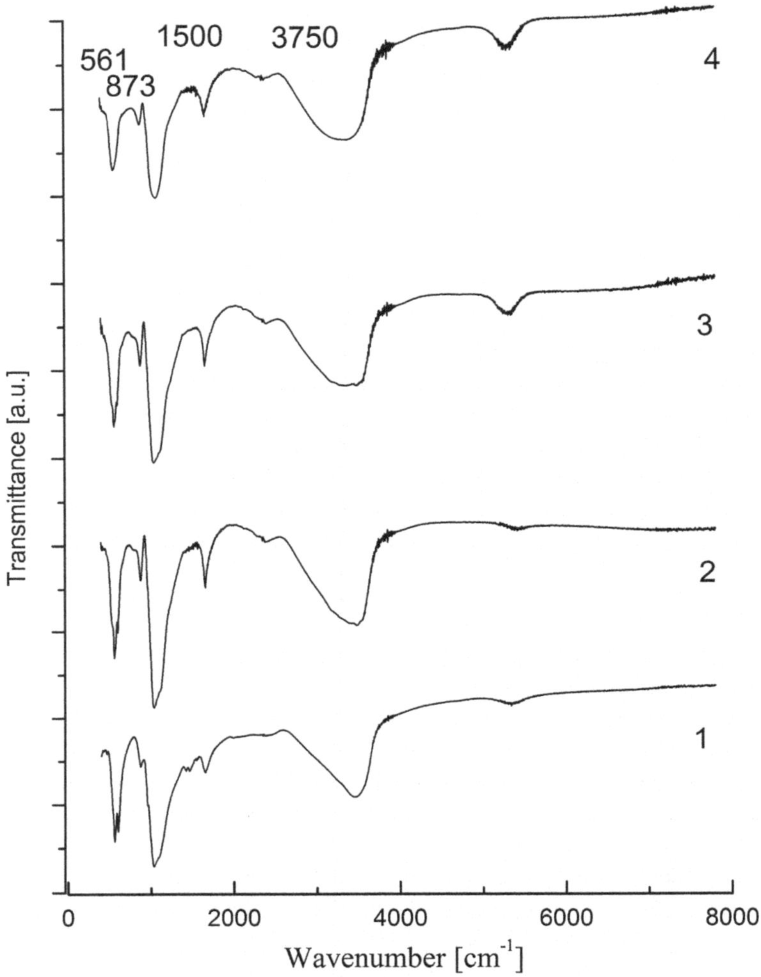


**Fig. 3** FTIR spectra of precipitations obtained after 7 days of experiment from model solutions with different content of strontium ions (wt%): 1–10, 2–20, 3–40, 4–50

**Table 5** Dependence of the degree of crystallinity on strontium ion content (% of calcium phosphates)

| Ratio of the ions in of the stock solution | | IRSF |
|---|---|---|
| $Ca^{2+}$ | $Sr^{2+}$ | |
| 1 | 0 | 1.59 |
| 0.8 | 0.2 | 1.82 |
| 0.6 | 0.4 | 1.94 |
| 0.5 | 0.5 | 2.04 |

**Table 6** Dependence of the specific surface area of calcium phosphates of the strontium concentration in the original model solution

| Ratio of the ions in of the stock solution | | $S_{SA}$ ($м^2/g$) |
|---|---|---|
| $Ca^{2+}$ | $Sr^{2+}$ | |
| 1 | 0 | 80 ± 5 |
| 0.8 | 0.2 | 70 ± 5 |
| 0.6 | 0.4 | 50 ± 3 |
| 0.5 | 0.5 | 30 ± 2 |

**Table 7** Mass loss (Δm) at different temperatures and concentrations of Sr ions (% wt of calcium phosphates)

| Ratio of the ions in of the stock solution | | Δm (100…280 °C) % | Δm (280…550 °C) % | Δm (550…900$^0$) % | ∑Δm % |
|---|---|---|---|---|---|
| $Ca^{2+}$ | $Sr^{2+}$ | | | | |
| 1 | 0 | 5.89 | 1.85 | 10.18 | 17.92 |
| 0.8 | 0.2 | 7.10 | 7.85 | 15.94 | 30.89 |
| 0.6 | 0.4 | 13.43 | 1.77 | 17.20 | 32.40 |
| 0.5 | 0.5 | 16.48 | 3.65 | 22.30 | 42.43 |

(Table 7). It has been established that the maximum weight loss was observed in the temperature range of 500–900 °C, corresponding to the removal of water of crystallization.

Analysis of the data (Table 7) showed that the observed increase in the total mass loss with increasing strontium content in the samples is due to the change of the proportion of phases present and the associated phase transformations.

## 3.1 *Time and Formation Sr-Containing Apatite*

To study the effect of time on the formation of phases in the solution of synovial fluid model, an experiment was conducted with a fixed content of Sr-containing salt (50 wt% strontium ion and 50 wt% calcium) but different holding times of the solid phase under the mother liquor. According to XRD of samples after 2 weeks and 3 weeks, we found peaks of brushite $CaHPO_4 (H_2O)_2$ with the monoclinic lattice [subgroup Ia (9)] and broad reflects of poorly crystallized phases with a structure similar to hexagonal structure of hydroxyapatite (Fig. 4).

Hydroxyapatite is known to apply to the space group $P6_3/m$ (176) with lattice parameters $a$ = 9.424 Å, $c$ = 6.879 Å (01-074-0565). Because the initial

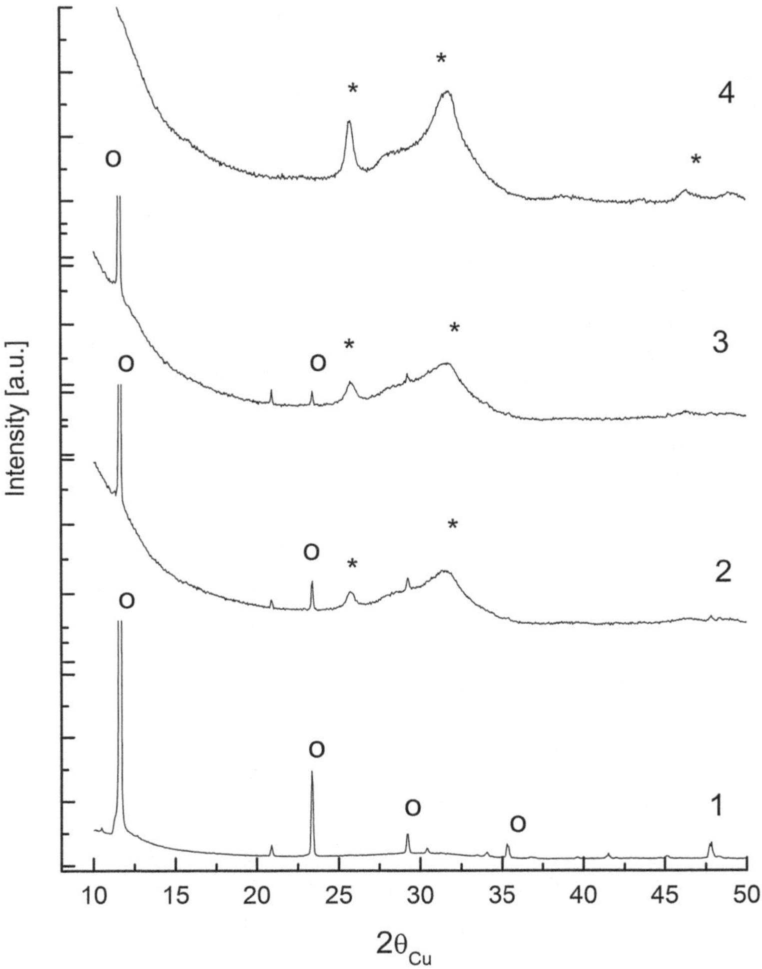


**Fig. 4** XRD pattern of calcium phosphates synthesized from model solutions with different content of $Sr^{2+}$ ions and $Ca^{2+}$ ions in the medium in a ratio (wt%) 50 and 50 (days): 1–7; 2–14; 3–21; 4–28. *Asterisk* hydroxyapatite; *open circle* brushite

composition of the synovial fluid includes carbonate ions, it can be assumed that the resulting compound may be a Sr-Ca-carbonated-phosphate-hydrate with possible formula $Ca_{8-n}Sr_nH_2(PO)_4 \cdot 6H_2O \cdot NaHCO_3 \cdot H_2O$. With respect to the brushite phase, besides the presence of texturing in the direction [010], degradation of peaks is observed in samples with increasing diffraction angles above 30°. For the sample synthesized after 4 weeks present only peaks phase structure of hydroxyapatite with

CSR size = 41 ± 7, and parameters of the crystal lattice, and = 9.506 ± 0.005 Å, c = 6.919 ± 0.006 Å.

The obtained X-ray results indicate that the substitution of Ca by Sr increases the unit cell parameters without changing the structure and displacement of reflections [002] and [300] towards narrow angles 2θ. It was established that with increasing incubation time of precipitate under the mother liquor, kinetically stable phase brushite transforms into the thermodynamically more stable Sr-containing carbonated hydroxyl-apatite. The phase relationships were as follows: at 2 weeks, 88 % of brushite and 12 % Sr-carbonated hydroxyl-apatite; at 3 weeks, 71 % brushite and 29 % Sr- carbonated hydroxyl-apatite; at 4 weeks, 100 % carbonated hydroxyl-apatite. The results of this study can be used in traumatology and orthopaedics in the treatment of patients with osteoporosis and investigations into Sr-containing drugs, as well as to develop new forms of drugs to facilitate targeted drug delivery.

## 4 Conclusions

XRD analysis showed that increasing the concentration of strontium ions in the model system increases the proportion of brushite in the solid phase. Infrared spectroscopy was used to measure solid phase crystallinity and confirmed its marked increase with increasing concentration of strontium ions in the initial solution; using atomic emission analysis demonstrated that increasing the concentration of strontium ions in the model solution corresponds to an increase of its content in the sediment. It was found that the ratio of Ca/P in the synthetic samples increased from 0.84 to 1.74 with decreasing concentrations of strontium ions in the model solutions. DTA was used to show that in the heat-treated samples, mass loss increases with increasing strontium concentration of ions in the initial solution. It was also found that with increasing incubation time of precipitate under the mother liquor, kinetically stable phase of brushite transforms into thermodynamically more stable Sr-containing carbonated hydroxyapatite.

**Acknowledgments** The study was supported by Russian Foundation for Basic Research (RFBR) within the scientific project №14_03_31506 mol_a.

## References

Balamurugan A, Balossier G, Torres P, Michel J (2009) Sol–gel synthesis and spectrometric structural evaluation of strontium substituted hydroxyapatite. Mater Sci Eng 29:1006–1009

Danilchenko SN (2007) Structure and properties of apatite from the point of view of biomineralogy and biomaterials. Bull SumDU 2:33–59 (in Russian)

Fazan F, Marquis PM (2000) Dissolution behavior of plasma-sprayed hydroxyapatite coating. J Mater Sci Mater Med 11:787–793

Gnedenkov SV, Scharkeev YuP, Sinebryukhov SL, Khrisanfova OA, Legostaeva EV, Zavidnaya AG, Puz AV, Khlusov IA (2011) Formation and properties of bioactive surface layers on titanium. Inorg Mater Appl Res 2(5): 474–481

Guicheux J, Grimandi G, Trecant M, Faivre A, Takahashi S, Daculsi G (1995) Bi-phasic calcium phosphate ceramics as a bone drag delivery system for human growth hormone. In: 12 European conference on biomaterials, Porto, Portugal

Izmailov RR, Golovanova OA (2014) Adhesive and morphological characteristics of Carbonate Hydroxyapatite prepared from a model human synovial fluid on Titanium alloys. Inorg Mater 50(6):592–598

Izmailov RR, Golovanova OA, Panova TV (2014) Carbonate hydroxylapatite synthesis and crystallization on metallic implant. J Struct Chem 55(5):946–953 (in Russian)

Izmaylov RR, Golovanova OA (2012) Solubility carbonate hydroxyapatite and hydroxyapatite obtained from a model solution synovial fluid. Bull Omsk Univ 4:109–113 (in Russian)

Izmaylov RR, Golovanova OA (2013) Crystallization of Calcium phosphates from a model solution of synovial liquid in the presence of Titanium compounds. Glass Phys Chem 39 (4):458–461

Landi E, Sprio S, Sandri M, Celotti G, Tampieri A (2008) Development of Sr and $CO_3$ cosubstituted hydroxyapatites for biomedical applications. Acta Biomater 4(3):656–663

Li C, Paris O, Siegel S et al (2010a) Strontium is incorporated into mineral crystals only in newly formed bone during strontium ranelate treatment. J Bone Miner Res: 968—975

Li Z, Lu WW, Deng L, Chiu PKY, Fang D, Lam RWM, Leong JCY, Luk KDK (2010b) The morphology and lattice structure of bone crystal after strontium treatment in goats. J Bone Miner Metab 28:25–34

Liu C, Huang Y, Shen W, Cui O (2001) Kinetics of hydroxyapatite precipitation at pH 10 to 11. Biomaterials 22:301–306

Liu Q, Huang S, Matinlinna JP, Chen Zh, Pan H (2013) Insight into biological apatite: physiochemical properties and preparation approaches. BioMed Research International

Mazurenko SO (2002) Osteoporosis in patients with renal insufficiency. Bulletin of St. Petersburg State Medical Academy. II Mechnikov 3: 70–72 (in Russian)

Medvedev EF (1993) Ceramic and glass-ceramic materials for bone implants. Glass Ceram 2:18–20 (in Russian)

Oliveira A, Reis R, Li P (2007) Strontium-substituted apatite coating grown on Ti6Al4 V substrate through biomimetic synthesis. J Biomed Mater Res B 83(1):258–265

Orlovskii VP, Barinov SM (2001) H and hydroxyapatite ydroxyapatite-matrix ceramics: a survey. Rus J Inor Chem 46(2):129–149

Putlyaev VI, Safronova TV (2006) A new generation of calcium phosphate biomaterials: the role of phase and chemical compositions. Glass Ceram 63(3–4):99–102

Rodrrigues-Lorenzo LM, Vallet-Regi M (2000) Controlled crystallization of phosphate apatites. Chem Mater 12:2460–2465

Safronova TV, Shekhirev MA, Putlyaev VI, Tretyakov YuD (2007) Hydroxyapatite-based ceramic materials prepared using solutions of different concentrations. Inorg Mater 43(8):901–909 (in Russian)

Shchepetkin IA (1995) Calcium phosphate materials in biological fluids. Biol Bull Rev 115(1):60–73 (in Russian)

Shi J, Klocke A, Zhang M, Bismayer U (2005) Thermally-induced structural modification of dental enamel apatite: decomposition and transformation of carbonate groups. Eur J Min 17:769–775

Sidelnikov AI (2000) Comparative characteristics of the group of titanium, used in the production of modern dental implants. Info-Dent 5:10–12 (in Russian)

Sopyan I, Singh R, Hamdi M (2008) Ind J Chem 47:1626–1631

Tas AC (2008) Monodisperse calcium carbonate microtablets forming at 70 & #xB0;C in prerefrigerated $CaCl_2$-gelatin-urea solutions. Int J Appl Ceram Technol 6(1):53–59

Titov AT, Larionov PM, Shchukin VS, Zaikovskii VI (2001) The mechanism of mineralization of the heart valves. J Surf Investig: x-ray Synchrotron Neutron Tech 3:74–79 (in Russian)

Veresov AG, Putlyaev VI, Tretyakov YD (2000) Advances in the field ceramic materials. Russ J Gen Chem 94(6):32–46 (in Russian)

Vodianova OV, Shepelkevich AP, Vasiliev NA (2011) Experience in the use of strontium ranelate in the complex therapy of postmenopausal osteoporosis. Med News 7:49–53 (in Russian)

Zhang W, Shen Y, Pan H et al (2011) Effects of strontium in modified biomaterials. Acta Biomater 7(2):800–808

# Part VI
# History of Science

# Some Aspects of Geological Microbiology in the Scientific Heritage of V.O. Tauson (1894–1946)

**Natalia N. Kolotilova**

**Abstract** The name of V.O. Tauson occupies an important place in the history of Russian school of microbiology. Tauson finished the Moscow University in 1924. His diploma work was linked with the destruction of the paraffin by the mould *Aspergillus flavus*. This topic was proposed by Prof. E.E. Uspensky (1889–1938), the founder of the Department of microbiology at Moscow University. The pioneer works of Tauson were linked with the microbial oxidation of phenanthrene, naphthalene, saturated, and nonsaturated hydrocarbons, wax, oils. He paid much attention to the role of microorganisms in geological processes, such as weathering of rocks, transformation, and destruction of several caustobioliths (peat, coal, petroleum). Tauson is also the author of some interesting hypotheses on the bioenergy of microbial metabolism, linked with the thermodynamics of anabolic processes.

**Keywords** History of microbiology · Geological microbiology · Hydrocarbons biodegradation

Vladimir Ottovitch Tauson (in other transcription W.O. Tausson) occupies an important place among the Russian microbiologists of the first half of the twentieth century whose investigations created the base for the development of geological microbiology, microbial ecology, and other branches of science linked with the biogenic–abiogenic interactions on our planet.

V. Tauson (Fig. 1) was born in Ryazan in the family of an engineer. In 1912, he finished the technical college (the so-called real school, Realschule) in Moscow and entered the Moscow University and at the same time working in the Polytechnic museum and in a cadet school as a chemist. The studies were interrupted by the World War I (1914). Tauson was mobilized and sent to West Siberia in the chemistry army service (1916). During the Civil War (1918), he worked as a

N.N. Kolotilova (✉)
Moscow University, Moscow, Russia
e-mail: kolotilovan@mail.ru

 
O.V. Frank-Kamenetskaya et al. (eds.), *Biogenic—Abiogenic Interactions in Natural and Anthropogenic Systems*, Lecture Notes in Earth System Sciences,
DOI 10.1007/978-3-319-24987-2_40

**Fig. 1** Vladimir O. Tauson (1894–1946)

chemist in the Kolchak's army in Siberia, Ural, Altay, and later—in the chemical laboratories of the different soviet institutions.

In 1922, Tauson came back to Moscow and returned to the Moscow University. The theme of his diploma work, "The decomposition of paraffin by the mould *Aspergillus flavus*" (1924) was inspired by his scientific tutor, the founder of the department of microbiology in the Moscow University (1924), Prof. E.E. Uspenski.

The history of this issue is interesting. Being a student, E. Uspenski worked on the problem of the role of Si in the nutrition of plants. In his experiments, he utilized special vessels of a paraffin as he could not utilize any flasks of a glass because it contains silicates. Once Uspenski observed on the inner surface of the paraffin vessel a mould and isolated a micromycet, *A. flavus*. This strain was able to grow on a paraffin and became the object of Tauson's investigations.

His investigation became one of the first works on the utilization of hydrocarbons by microorganisms. It must be mentioned that the first investigation of the ability of microorganisms to oxidize a hydrocarbon as methane, was made in the laboratory of M. Bejerinck (Delft) by N. Zöhngen (1904). The same year Rahn demonstrated the ability of *Penicillum glaucum* to grow on alkanes, later this possibility was also shown for some bacteria. But in general, there was very little knowledge of this subject.

After graduating the Moscow University Tauson continued his studies of the microbial degradation of such stable compounds as a paraffin, naphthalene, phenanthrene, aromatic hydrocarbons, wax, oil, etc. (Tausson 1925, 1927, 1928a, b, 1929). In 1925 he defended the thesis called "On the oxidation of phenantrene by bacteria".

Tauson realized many expeditions to Caucasia on the oil fields (1925–1928), investigated the coal weathering in Moscow region (1930). In 1930, he became the head of the laboratory of bioenergetics and geological activity of microorganisms in the Microbiological institute associated with the Moscow University ("Narcompros"). He observed the participation of microbes in the formation and transformation of different caustobioliths (peat, coal, oil); the role of microbes in the geological processes from the evolutionary viewpoint had been discussing. His investigations demonstrated the important role of microbes in the processes of mineral and rock leaching, indicated the possibility to utilize it in the purification of waste-waters, etc. His works also elucidated some important biochemical steps in the processes of degradation some aliphatic and aromatic hydrocarbons, such as the so-called β-oxidation of the chain, the broke of the aromatic hydrocarbons etc. (Tauson 1934).

Some of Tauson's works were also dedicated the problem of a possible microbe role in the genesis and chemical structure of oils. He proposed the original hypothesis of a microbe evolution and microbial processes on Earth, which was in fact the theory of co-evolution of biosphere and geosphere (Tauson 1936a).

During the Tauson expedition with his students to Taman peninsula (1932) the close interaction between the processes of mud volcanism and the activity of anaerobic bacteria was demonstrated (Tauson and Aleshina 1932).

The study of the conditions necessary for the plant debris humification and the genesis of caustobioliths demonstrated the dependence of these processes on the redox potential. The influence of the physicochemical factors like a redox potential and pH on the microbial processes was largely discussed by E.E. Uspenski, and he was the first who introduced these parameters into the microbiology.

In 1934, Tauson and his students participated in the expedition on Pamir to investigate the microorganisms living in the high mountains. Tauson hypothesized about specific conditions of the existence in these at the big altitudes such as extremely high insolation, low temperatures, absence of water and organic substances and the natural isolation of these habitats from the other regions may contribute to the election of specific, peculiar microorganisms, mostly chemolithotrophs, which could be even probably the beings at the early stages of evolution. Some of his hypotheses were affirmed. On rocks and in primitive soils of Pamir Tauson found a rather high level of nitrification, the occurrence of nitrogen fixation (presence of *Azotobacter*) was also demonstrated. Extremely low growth of many bacteria groups was noted (Tauson 1936b). The issue of the evolution bacteria stage in high mountains remained unresolved.

Most of the mentioned investigations became topics (served the base for) of the lectures on the geological activity of microorganisms that Tauson had been reading for students of the department of microbiology in the Moscow University. In 1935, he received a degree of the doctor of biological sciences and the status of professor. Basing on his lectures Tauson wrote two popular books, "The heritage of microbes" (Tauson 1947) and "The great affairs of small beings" (Tauson 1948) which were published only in 1948, after his death.

In the middle of 1930s, several scientific institutions in Moscow were reorganized, and the laboratory guided by Tauson in the Microbiological Institute was transported to the Institute of Plant Physiology, Academy of Sciences, USSR. The latest works of his life were mainly linked with the problems of bioenergetics. On the basis of thermodynamics he proposed the theory of so-called exothermic synthesis in microorganisms (Tauson 1950). It explained the possibility of several unusual metabolic reactions in microorganisms. This interesting and veritable original theory occupied a special page in the history of science.

The most important and the well-known investigations of Tauson were linked with the microbial degradation and utilization of hydrocarbons and other chemically stable compounds. They had a great fundamental and practical importance. It elucidated some biochemical ways of the degradation of hydrocarbons. Their results were used in several processes linked with the activity of man such as purification of industrial waste-waters; utilization of microorganisms using hydrocarbons as indicators in screening of oil fields; production of microbial protein on the base of hydrocarbons of oil. The latter process became the base of a well-known program of a large-scale microbial industry in the USSR. Tauson's investigations enlarged our knowledge about the role of microbes in the biogenic–abiogenic interactions in nature.

## Reference

Tauson VO (1934) On the degradation of hydrocarbons by microorganisms. Priroda 6:43–54 (in Russian)

Tauson VO (1936a) On the evolution of microorganisms during geological epochs. Arch Biol Nauk 43:267–286 (in Russian)

Tauson VO (1936) Materials on microbiology of Pamir. Trudy Sredneasiatsk. Gos. Universiteta 32:3–17 (in Russian)

Tauson VO (1947) The heritage of microbes. Izd. Acad. Nauk SSSR, Moscow-Leningrad (in Russian)

Tauson VO (1948) The Great affairs of small beings. Izd. Acad. Nauk SSSR, Moscow-Leningrad (in Russian)

Tauson VO (1950) The basic propositions of plant bioenergetics. Izd. Acad. Nauk SSSR, Moscow-Leningrad (in Russian)

Tauson VO, Aleshina VI (1932) On the bacterial reduction of sulfates in presence of hydrocarbons. Microbiologia 1:229–261 (in Russian)

Tausson WO (1925) Zur Frage über die Assimilation des Paraffins durch Mikroorgamismen. Biochem Ztschr 155:356–358

Tausson WO (1927) Naphthalin als Kohlenstoffquelle für Bakterien. Planta 4:214–256
Tausson WO (1928a) Die Oxidation des Phenantrens durch Bakterien. Planta 5:239–273
Tausson WO (1928b) Über die Oxydation der Wachse durch Mikroorganismen. Biochem Zeutschr 193:85–93
Tausson WO (1929) Über die Oxydation der Benzolkohlenwasserstoffe durch Bakterien. Planta 7:735–758

# Erratum to: Computer Simulation of Defects in Carbonate Fluorapatite and Hydroxyapatites

**Elena A. Kalinichenko, Aleksandr B. Brik, Valentin V. Radchuk, Olga V. Frank-Kamenetskaya and Oleksii Dubok**

## Erratum to:
## O.V. Frank-Kamenetskaya et al. (eds.), *Biogenic—Abiogenic Interactions in Natural and Anthropogenic Systems*, Lecture Notes in Earth System Sciences, DOI 10.1007/978-3-319-24987-2_35 Erratum DOI 10.1007/978-3-319-24987-2_41

The spelling of the Aleksej V. Dubok name was incorrect. The name should read as "Oleksii Dubok".

---

The updated original online version for this chapter can be found at DOI 10.1007/978-3-319-24987-2_35

---

E.A. Kalinichenko (✉) · A.B. Brik
Semenenko Institute of Geochemistry, Mineralogy and Ore Formation of NAS of Ukraine, 34 Palladin Ave., Kiev, Ukraine
e-mail: kalinichenko@igmof.gov.ua

A.B. Brik
e-mail: abrik@igmof.gov.ua

V.V. Radchuk
Institute for Telecommunications and Global Information Space of NAS of Ukraine, 13 Chokolovsky Blv., Kiev, Ukraine
e-mail: valentyn.radchuk@gmail.com

O.V. Frank-Kamenetskaya
Institute of Earth Sciences, St-Petersburg State University, 7/9 Universitetskaya Emb., St-Petersburg, Russia
e-mail: ofrank-kam@mail.ru

O. Dubok
Institute for Problems of Material Science of NAS of Ukraine, 3 Kryjanovskogo Str., Kiev, Ukraine
e-mail: dubok@mail.ru

 
O.V. Frank-Kamenetskaya et al. (eds.), *Biogenic—Abiogenic Interactions in Natural and Anthropogenic Systems*, Lecture Notes in Earth System Sciences, DOI 10.1007/978-3-319-24987-2_41

# Index





O.V. Frank-Kamenetskaya et al. (eds.), *Biogenic—Abiogenic Interactions in Natural and Anthropogenic Systems*, Lecture Notes in Earth System Sciences, DOI 10.1007/978-3-319-24987-2

The manufacturer's authorised representative in the EU is Springer Nature Customer Service Centre GmbH, Europaplatz 3, 69115 Heidelberg, Germany. If you have any concerns regarding our products, please contact ProductSafety@springernature.com

Printed and bound by CPI Group (UK) Ltd, Croydon, CR0 4YY
25/09/2026
02222852-0005